COOPERATIVE PHENOMENA NEAR PHASE TRANSITIONS

The MIT Press
Cambridge, Massachusetts, and London, England

COOPERATIVE PHENOMENA NEAR PHASE TRANSITIONS

A Bibliography with Selected Readings

edited by
H. Eugene Stanley

Printed and bound by Semline, Inc.
in the United States of America.

Library of Congress Cataloging in Publication Data

Stanley, Harry Eugene.
Cooperative phenomena near phase transitions.

1. Phase transformations (Statistical physics) --
Bibliography. I. Title.
Z7144.S76S7 016.536'401 72-13838
ISBN 0-262-19114-8
ISBN 0-262-69039-X (pbk)

CONTENTS

III. ABSTRACTS OF THEORETICAL ARTICLES 273

Griffiths, R. B.
"Thermodynamics near the Two-Fluid Critical Mixing Point in He^3-He^4," Phys. Rev. Lett. 24, 715-717 (1970) 287

Griffiths, R. B.
"Dependence of Critical Indices on a Parameter," Phys. Rev. Lett. 24, 1479-1482 (1970) 287

Griffiths, R. B. and Wheeler, J. C.
"Critical Points in Multicomponent Systems," Phys. Rev. A2, 1047-1064 (1970) 287

Guggenheim, E. A.
"The Principle of Corresponding States," J. Chem. Phys. 13, 253-261 (1945) 288

Gunton, J. D. and Buckingham, M. J.
"Condensation of the Ideal Bose Gas as a Cooperative Transition," Phys. Rev. 166, 152-158 (1968) 288

Halperin, B. I. and Hohenberg, P. C.
"Scaling Laws for Dynamic Critical Phenomena," Phys. Rev. 177, 952-970 (1969) 288

(a) Hankey, A. and Stanley, H. E.
"Systematic Application of Generalized Homogeneous Functions to Static Scaling, Dynamic Scaling, and Universality," Phys. Rev. B6, 3515-3542 (1972) 289

(b) Hankey, A., Stanley, H. E., and Chang, T. S.
"Geometric Predictions of Scaling at Tricritical Points," Phys. Rev. Lett. 29, 278-281 (1972) 289

Hohenberg, P. C.
"Existence of Long-Range Order in One and Two Dimensions," Phys. Rev. 158, 383-386 (1967) 289

Hohenberg, P. C. and Barmatz, M.
"Gravity Effects near the Gas-Liquid Critical Point," Phys. Rev. A6, 289-313 (1972) 290

Ising, E.
"Beitrag zur Theorie des Ferromagnetismus," Zeit. Physik. 31, 253-258 (1925) 290

Jasnow, D. and Wortis, M.
"High-Temperature Critical Indices for the Classical Anisotropic Heisenberg Model," Phys. Rev. 176, 739-748 (1968) 290

(a) Josephson, B. D.
"Equation of State near the Critical Point," J. Phys. C2, 1113-1115 (1969) 291

(b) Schofield, P.
"Parametric Representation of the Equation of State near a Critical Point," Phys. Rev. Lett. 22, 606-608 (1969) 291

PUBLISHER'S NOTE

The aim of this format is to close the time gap between the preparation of certain works and their publication in book form. A large number of significant though specialized manuscripts make the transition to formal publication either after a considerable delay or not at all. The time and expense of detailed text editing and composition in print may act to prevent publication or so to delay it that currency of content is affected.

The text of this book has been photographed directly from the author's typescript. It is edited to a satisfactory level of completeness and comprehensibility though not necessarily to the standard of consistency of minor editorial detail present in typeset books issued under our imprint.
The MIT Press

PREFACE

This volume is intended to serve as a companion to my Introduction to Phase Transitions and Critical Phenomena (Oxford University Press, London and New York, 1971) and to provide the student using that book with additional source material too extensive for inclusion there. Moreover, I feel that there is no better source for acquiring understanding in this field than a study of the original articles themselves. Accordingly, this volume begins with a bibliography of about 2500 articles, many of them quite fundamental to the field. Since many articles would fall under more than one heading, I have arranged them alphabetically. I have used the Oxford "name and date" system, with full titles and inclusive page numbers provided.

Obviously no collection such as this can be all inclusive, and the selections made inevitably reflect the special interests of the editor. Although Introduction to Phase Transitions and Critical Phenomena was largely limited to a description of magnetic and liquid-gas phase transitions, I felt it would also be useful to include works concerned with a wider variety of cooperative phenomena near phase transitions. To this end, there are some articles on cooperative phenomena in binary alloys, helium, structural phase transitions, liquid crystals, and biological systems. Quick referencing by categories is provided by a coded letter appearing in the left margin of many entries.

It is hoped that the present bibliography is current to June 1972 and includes publications from a wide variety of journals regularly publishing articles on critical phenomena. Since many students--even those whose work is directly concerned with this field--may easily overlook many of these, the bibliography opens with a list of 61 journals. Thereafter, it lists 50 relatively recent conference proceedings dealing with aspects of critical phenomena; in many cases, individual articles from these proceedings have not been included in the bibliography.

The field of critical phenomena has developed via an interplay of theory and experiment. In order more fully to cover the experimental aspects, 35 short experimental articles have been reproduced. Considerations of space require that the 65 remaining theoretical articles be represented by abstract only. All 100 reprinted articles are arranged alphabetically by author for the convenience of the reader, although certain very closely related or sequential articles are grouped and listed under the name of the author of the earliest article cited.

It should be added that in those cases where authors have prepared both a short and a long publication on the same subject (for example, a Phys. Rev. Letter followed some time later by a Phys. Rev. article) both items should be found in the bibliography. In the case of experimental papers, the shorter form is reproduced, whereas in the case of a theoretical work, the abstract of the longer publication is chosen.

Any bibliographer must be aware that he may have omitted significant material, but I feel that the lists presented here will be of value to those working in the field of cooperative phenomena near phase transitions.

I wish to express my sincere thanks to all who have assisted me in the preparation of this volume--especially to Ms. Idahlia Stanley for her considerable help in assembling much of the material included, to H. Birecki, M. Collins, R. Daga, F. Harbus, C.S. Hui, D. Karo, R. Krasnow, D. Lambeth, M. H. Lee, L. L. Liu, S. Milošević, G. Tuthill and K. G. Wilson for their assistance in updating the bibliography and in proofreading, to R. J. Birgeneau, M. E. Fisher, G. Hawkins, and P. C. Hohenberg for their advice in the selection of the 100 works singled out for special attention, and to Mrs. Janet Nadeau and Mrs. Carole Solomon for their assistance and their patience in preparing the typescript.

H. Eugene Stanley

I
BIBLIOGRAPHY

LIST OF JOURNALS REGULARLY PUBLISHING ARTICLES ON CRITICAL PHENOMENA, TOGETHER WITH ABBREVIATION USED IN BIBLIOGRAPHY

Acta Physica Polonica	Acta Phys. Polon
Advances in Chemical Physics	Adv. Chem. Phys.
Advances in Physics	Adv. Phys.
American Journal of Physics	Am. J. Phys.
Annals of Physics	Ann. Phys.
Annual Review of Physical Chemistry	Ann. Rev. Phys. Chem.
Bulletin of the American Physical Society	Bull. Am. Phys. Soc.
Canadian Journal of Physics	Can. J. Phys.
Chemical Physics Letters	Chem. Phys. Lett.
Communications in Mathematical Physics	Commun. Math. Phys.
Contemporary Physics	Contemp. Phys.
Czechoslovak Journal of Physics: Sec. B	Czech. J. Phys. B
Ferroelectrics	Ferroelectrics
Helvetica Physica Acta	Helv. Phys. Acta
Indian Journal of Physics	Indian J. Phys.
International Journal of Magnetism	Int. J. Magnetism
International Journal of Quantum Chemistry	Int. J. Quantum Chem.
JETP Letters (English transl. of ZhETF Pis'ma v Redaktsiyu	JETP Lett.
Journal de Physique	J. de Physique
Journal of the American Chemical Society	J. Am. Chem. Soc.
Journal of Applied Physics	J. Appl. Phys.
Journal of Chemical Physics, The	J. Chem. Phys.
Journal of Low Temperature Physics	J. Low Temp. Phys.
Journal of Mathematical Physics	J. Math. Phys.
Journal of Physical Chemistry	J. Phys. Chem.
Journal of the Physical Society of Japan	J. Phys. Soc. Japan
Journal of Physics (formerly Proc. Phys. Soc.)	J. Phys.
Journal of the Physics and Chemistry of Solids	J. Phys. Chem. Solids
Journal of Research of the National Bureau of Standards	J. Res. Natl. Bur. Std. (U.S.)
Journal of Statistical Physics	J. Stat. Phys.
Molecular Physics	Mol. Phys.
Nuovo Cimento	Nuovo Cimento
Optics Communications	Optics Commun.
Philosophical Magazine	Phil. Mag.
Physica	Physica
Physica Norvegica	Phys. Norveg.
Physica Status Solidi	Phys. Stat. Sol.
Physical Review, The	Phys. Rev.
Physical Review Letters	Phys. Rev. Lett.
Physics	Physics
Physics and Chemistry of Liquids	Phys. Chem. Liquids
Physics of Condensed Matter	Phys. Cond. Mat.
Physics Letters	Phys. Lett.
Physics Today	Phys. Today
Proceedings of the National Academy of Sciences	Proc. Natl. Acad. Sci. U.S.
Proceedings of the Physical Society (London)	Proc. Phys. Soc.
Proceedings of the Royal Society (London)	Proc. Roy. Soc.
Progress of Theoretical Physics (Kyoto)	Progr. Theoret. Phys.
Reports on Progress in Physics	Rept. Progr. Phys.
Review of Modern Physics	Rev. Mod. Phys.
Revista Mexicana de Fisica	Rev. Mex. Fis.
Revue Roumaine de Physique	Rev. Roum. Phys.
Science	Science
Science Reports of the Research Institutes Tohoku University	Sci. Rept. Res. Inst. Tohoku
Solid State Communications	Sol. State Commun.
Soviet Physics-JETP (English transl. of Zhurnal Eksperimental'noi i Teoreticheskoi Fiziki)	Sov. Phys.-JETP
Soviet Physics-Solid State (English transl. of Fizika Tverdogo Tela)	Sov. Phys.-Solid State
Sov. Physics-Uspekhi (English transl. of Uspekhi Fizicheskikh Nauk)	Sov. Phys.-Usp.
Transactions of the Faraday Society	Trans. Faraday Soc.
Zeitschrift für Physik	Z. Physik
Zeitschrift für Physikalische Chemie (Frankfurt)	Z. Physik. Chem. (Frankfurt)

SELECTED CONFERENCES SINCE 1964 IN WHOSE PROCEEDINGS THE READER WILL FIND ARTICLES ON COOPERATIVE PHENOMENA AND PHASE TRANSITIONS

An asterisk denotes the fact that conference was focused mainly on critical phenomena.

1964:

a. Univ. Colorado Institute on Theoretical Physics, August, 1964. Proceedings: Lectures in Theoretical Physics, Univ. Colorado Press, Boulder, 1965.

b. International Conference on Magnetism, Nottingham, England, September, 1964.

c. Tenth Annual Conference on Mangetism and Magnetic Materials, Minneapolis, Minn., USA, November, 1964. Proceedings: J. Appl. Phys. 36, March 1965.

1965:

* a. Conference on Phenomena in the Neighborhood of Critical Points, Washington, D.C., USA, April, 1965. Proceedings: Critical Phenomena, M. S. Green and J. V. Sengers, Eds., N.B.S. Misc. Publication No. 273 (available from Superintendent of Documents, U.S. Government Printing Office, Washington, D.C. 20402, for $2.50).

b. Eleventh Annual Conference on Magnetism and Magnetic Materials, San Francisco, California, USA, November, 1965. Proceedings: J. Appl. Phys. 37, March, 1966.

1966:

a. Enrico Fermi Summer School on Magnetism, Varenna, Italy, June, 1966. Proceedings: Theory of Magnetism in Transition Metals, W. Marshall, Ed., Academic Press, London, 1967.

b. Brandeis Univ. Summer Institute in Theoretical Physics, Waltham, Mass., USA, July, 1966. Proceedings: Statistical Physics, Phase Transitions, and Superfluidity, M. Chrétien, E. P. Gross, and S. Deser, Eds., Gordon and Breach Science, Publ., 1968.

* c. Gordon Conference on Cooperative Phenomena in Liquids and Solids, August, 1966. No proceedings published.

d. Twelfth Annual Conference on Magnetism and Magnetic Materials, Washington, D. C., USA, November, 1966. Proceedings: J. Appl. Phys. 38, March, 1967.

1967:

a. Scottish Universities Summer School of Physics, July, 1967. Proceedings: Mathematical Methods in Solid State and Superfluid Theory, R. C. Clark and G. H. Derrick, Eds., Plenum Press, N.Y., 1968.

b. Latin American School of Physics, Santiago, Chile, July, 1967. Proceedings: Solid State Physics, Nuclear Physics, and Particle Physics, I. Saavedra, Ed., W. A. Benjamin, Inc., N.Y., 1968.

c. International Conference on Magnetism and Thirteenth Annual Conference on Magnetism and Magnetic Materials (joint meeting), Boston, Massachusetts, September, 1967. Proceedings: J. Appl. Phys. 39, March, 1968.

1968:

a. Conference on Fluctuations in Superconductors, Asilomar, Pacific Grove, California, 13-15 March, 1968. Proceedings: Fluctuations in Superconductors, W. S. Goree and F. Chilton, Eds., Stanford Research Institute, Menlo Park, California, 1968. Available on request.

b. Trieste Symposium on Contemporary Physics, Trieste, Italy, July, 1968. Proceedings: Contemporary Physics, International Atomic Energy Agency, Vienna, 1969.

* c. Banff Summer School on Critical Phenomena, Banff, Alberta, Canada, August, 1968. Prof. D. D. Betts, Director. No proceedings.

d. Eleventh International Conference on Low-Temperature Physics, St. Andrews, Scotland, August, 1968. Proceedings: Proc. LT-11, J. F. Allen, D. M. Finlayson, and D. M. McCall, Eds., St. Andrews Univ. Press, Scotland, 1969.

e. International Conference on Statistical Mechanics, Kyoto, Japan, September, 1968. Proceedings: January, 1969, supplement to J. Phys. Soc. Japan, Volume 26.

f. International Conference on Light Scattering Spectra of Solids, New York Univ., N.Y., September, 1968. Proceedings: Light Scattering Spectra of Solids, G. B. Wright, Ed., Springer, N.Y., 1969.

g. Fourteenth Annual Conference on Magnetism and Magnetic Materials, N.Y., N.Y., November, 1968. Proceedings: J. Appl. Phys. 40, March, 1969.

1969:

* a. Fourteenth Conference on Chemistry at the University of Brussels (Solvay Conference), Université Libre de Bruxelles, Brussels, Belgium, May, 1969. Proceedings: Phase Transitions, Interscience Publishers, N.Y., 1971.

b. International Conference on Superconductivity, Stanford University, Stanford, California, USA, August, 1969. Proceedings: Physica 55, October, 1971.

c. Fifteenth Annual Conference on Magnetism and Magnetic Materials, Philadelphia, Pa., USA, November, 1969. Proceedings: J. Appl. Phys. 41, March 1970.

d. Symposium on Statistical Mechanics at the Turn of the Decade (in Honor of Professor George Uhlenbeck), Northwestern University, Evanston, Illinois, USA, October, 1969. Proceedings: Physics at the Turn of the Decade, E. G. D. Cohen, Ed., Marcel Dekker, Publ., N.Y., 1971.

1970:

* a. International Conference on Dynamical Aspects of Critical Phenomena, Fordham University, New York, N.Y., USA, June, 1970. Proceedings: Dynamical Aspects of Critical Phenomena, J. I. Budnick and M. P. Kawatra, Eds., Gordon and Breach Publishers, N.Y., 1972.

* b. MIT Summer Course in Phase Transitions and Critical Phenomena, Cambridge, Massachusetts, July, 1970. No proceedings.

c. Les Houches Summer School on Mathematical Physics, Les Houches, France, July, 1970. Proceedings to be published by Gordon and Breach, N.Y., 1972.

* d. Enrico Fermi Summer School on Critical Phenomena, Varenna, Italy, July, 1970. Proceedings: Critical Phenomena, M. S. Green, Ed., Academic Press, London (in press).

e. NATO Summer Institute in Mathematical Physics, Istanbul, Turkey, August, 1970. Proceedings to be published by Gordon and Breach Publishers under the editorship of A. O. Barut.

f. NATO Summer Institute on Magnetism, Nice, France, August, 1970. Proceedings to be published by Gordon and Breach Publishers under the editorship of S. Foner.

g. International Conference on Magnetism, Grenoble, France, September, 1970. Proceedings: J. Phys. Paris 32S, 1971.

h. Twelfth International Conference on Low-Temperature Physics, Kyoto, Japan, September, 1970. Proceedings: Proc. LT-12, E. Kanda, Ed., Academic Press, 1971.

* i. Battelle Conference on Critical Phenomena in Alloys, Magnets, and Superconductors, Gstaad, Switzerland, September, 1970. Proceedings: Critical Phenomena in Alloys, Magnets, and Superconductors, R. E. Mills, E. Ascher, and R. I. Jafee, Eds., McGraw-Hill Book Co., N.Y., 1971.

j. Sixteenth Annual Conference on Magnetism and Magnetic Materials, Miami, Florida, USA, November, 1970. Proceedings: J. Appl. Phys. 42, March, 1971.

1971:

a. Sixth IUPAP International Conference on Statistical Mechanics, University of Chicago, Chicago, Ill., USA, March, 1971. Proceedings: Statistical Mechanics: New Concepts, New Problems, New Applications, S. A. Rice, K. F. Freed, and J. C. Light, Eds., Univ. Chicago Press, Chicago, 1972.

* b. NATO Winter Institute on Structural Phase Transitions and Soft Modes, Geilo, Norway, April, 1971. Proceedings: Structural Phase Transitions and Soft Modes, E. J. Samuelsen, E. Andersen, and J. Feder, Eds., Universitetsforlaget, Oslo, 1971.

* c. MIT Summer Course on Phase Transitions and Critical Phenomena, Cambridge, Mass., USA, June, 1971. No proceedings.

d. MIT Summer Course on Biomedical Physics and Biomaterials Science, Cambridge, Mass., USA, July, 1971. Proceedings: Biomedical Physics and Biomaterials Science, H. E. Stanley, Ed., M.I.T. Press, Cambridge, 1972.

e. International Conference on Light Scattering, Paris, France, July, 1971. Proceedings: Light Scattering in Solids, M. Balkanski, Ed., Flammarion Sciences, Paris, 1971.

f. International Conference on Statistical Mechanics and Field Theory, Technion-Israel Institute of Technology, Haifa, Israel, July, 1971. Proceedings: Statistical Mechanics and Field Theory, R. Sen, Ed., Keter Publ. Co., Jerusalem, 1972.

g. Battelle Rencontres in Mathematics and Physics, Seattle, Washington, August, 1971. Proceedings to be published by Springer Verlag.

h. Seventeenth Annual Conference on Magnetism and Magnetic Materials, Chicago, Illinois, USA, November, 1971. Proceedings: A.I.P. Conference Proceedings Volume No. 5, 1972.

1972:

a. Mayo Clinic Conference on Me branes, Viruses, and Immune Mechanisms in Clinical and Experimental Diseases, Univ. Minnesota, Minneapolis, Minn., June, 1972. Proceedings: Membranes, Viruses, and Immune Mechanisms, S. B. Day and R. Good, Eds., Academic Press, N.Y., 1972.

* b. MIT Summer Course on Phase Transitions and Critical Phenomena, Cambridge, Mass., June, 1972. No proceedings published.

c. International Conference on Padé Approximants, Canterbury, England, July, 1972. Proceedings: Proc. International Conf. on Padé Approximants, P. R. Graves-Morris, Ed., Academic Press, London, 1973.

d. Thirteenth International Conference on Low-Temperature Physics, Boulder, Colorado, USA, August, 1972. Proceedings to be published by Colo. Assoc. Universities Press, Boulder, under the editorship of E. Hammel, W. O'Sullivan, and K. Timmerhaus.

e. Fourth International Congress on Biophysics, Moscow, USSR, August, 1972. Proceedings to be published.

f. Eighteenth Annual Conference on Magnetism and Magnetic Materials, Denver, Colo., USA, November, 1972. Proceedings: A.I.P. Conference Proceedings, Volume No. 10, 1973.

1973 (partial list):

* a. Enrico Fermi Summer School on Local Aspects of Phase Transitions, K. A. Müller, Director, Varenna, Italy, July, 1973. Proceedings to be published by Academic Press.

b. International Congress on Magnetism, Moscow, USSR, August, 1973.

c. Nineteenth Annual Conference on Magnetism and Magnetic Materials, Boston, Mass., USA, November, 1972. Proceedings to be published.

THE FOLLOWING ABBREVIATIONS ARE USED IN THE LEFT MARGIN OF THE BIBLIOGRAPHY TO DENOTE EASILY CATEGORIZED SUBJECTS:

a = alloys (binary, ternary, ...).

b = biologically-relevant cooperative phenomena (macromolecules, membranes, ...).

d = systems of lattice dimensionality lower than 3, excluding Ising model calculations (see the notation "1" below).

f = ferroelectrics, structural and displacive phase transitions, including models growing out of the exact solution of the KDP model of a ferroelectric for d = 2.

g = renormalization group, including expansions in the variables (4-d) and 1/D.

h = helium, helium mixtures. Both critical point and λ-line phenomena.

L = liquid crystals.

m = multicritical (tricritical, tetracritical, ...)points. Critical spaces of order greater than 2. Binary mixtures are given this notation.

r = articles including general material of a review nature. It should be noted that books are given this notation, as well as some review articles that include considerable original material.

s = superconductors.

u = articles relating to the universality hypothesis concerning what parameters critical properties depend upon and which parameters are 'irrelevant.'

1 = theoretical work on systems with effective spin dimensionality D = 1 (Ising-like systems).

2 = theoretical work on systems with effective spin dimensionality D = 2 (planar-like systems). Both the quantum-mechanical S = 1/2 XY model and the classical plane rotator model are given this notation.

3 = theoretical work on systems with effective spin dimensionality D = 3 (Heisenberg-like systems).

∞ = theoretical work on systems with effective spin dimensionality D = ∞ (spherical model, Bose gas).

l ABE, R. (1965) "Statistical mechanics of a system of Ising spins near the transition point," Prog. Theoret. Phys. 33, 600-613.

l --- (1967) "Logarithmic singularity of specific heat near the transition point in the Ising model," Ibid. 37, 1070-1079 (1967).

l --- (1967) "Note on the critical behavior of the Ising ferromagnets," Ibid. 38, 72-80.

--- (1967) "Singularity of specific heat in the second order phase transition," Ibid. 38, 322-331.

l --- (1967) "Critical behavior of pair correlation function in Ising ferromagnets," Ibid. 38, 568-575.

l --- (1968) "Dynamics of the Ising model near the transition point," Ibid. 39, 947-956.

u --- (1970) "Some remarks on perturbation theory and phase transition with an application to anisotropic Ising model," Ibid. 44, 339-347.

l --- (1972) "Critical exponent of the Ising model in the high density limit," Ibid. 47, 62-68.

l --- (1972) "Critical exponent of the Ising model in the high density limit. II," Ibid. 47, 1200-1203.

g --- (1973) "Expansion of a critical exponent in inverse powers of spin dimensionality," Ibid. 49.

l --- and HATANO, A. (1969) "Dynamics of the Ising model near the transition point. II," Ibid. 41, 941-948.

l ABRAHAM, D. B. (1972) "High temperature susceptibility bounds for the two-dimensional Ising model," Phys. Lett. 39A, 357-359 (1972).

f --- and LIEB, E. H. (1971) "Anomalous specific heat of sodium trihydrogen selenite - an associated combinatorial problem," J. Chem. Phys. 54, 1446-1450.

g ABRIKOSOV, A. A. and MIGDAL, A. A. (1970) "On the theory of the Kondo effect," J. Low Temp. Phys. 3, 519.

h ACHTER, E. (1968) "Ion mobility in ^{3}He in the critical region," Phys. Lett. 27A, 687-688.

h --- and MEYER, L. (1969) "X-ray scattering from liquid helium," Phys. Rev. 188, 291-300.

r ACZÈL, J. (1966) Lectures on functional equations and their applications, Academic Press, N. Y.

r --- (1969) On applications and theory of functional equations, Academic Press, N. Y.

ADAM, M., SEARBY, G., and BERGÉ, P. (1972) "Rayleigh scattering from entropy fluctuations in monocrystalline succinonitrile," Phys. Rev. Lett. 28, 228-230

h AHLERS, G. (1968) "Effect of the gravitational field on the superfluid transition in He^4," Phys. Rev. 171, 275-282.

h --- (1968) "Thermal conductivity of He I near the superfluid transition," Phys. Rev. Lett. 21, 1159-1162.

h --- (1968) "Second sound velocity in He II, thermal conductivity in He I, and scaling laws near the superfluid transition," Phys. Lett. 28A, 507-508.

h --- (1969) "Thermodynamics of the isentropic sound velocity near the superfluid transition in He^4," Phys. Rev. 182, 352-362.

h --- (1969) "Heat capacity at constant pressure near the superfluid transition in He^4," Phys. Rev. Lett. 23, 464-468.

h --- (1970) "Thermal conductivity of a He^3-He^4 mixture near the superfluid transition," Ibid. 24, 1333-1336.

h --- (1971) "Experimental tests of scaling laws near the superfluid transition in He^4 under pressure," Proc. LT12, Kyoto (ed. by E. Kanda, Keigaku Publ. Co., Tokyo, 1971), p. 21-27.

h --- (1971) "Heat capacity near the superfluid transition in He^4 at saturated vapor pressure," Phys. Rev. A3, 696-716.

h --- (1971) "On the viscosity of ^{4}He near the superfluid transition," Phys. Lett. 37A, 151-153.

h --- EVENSON, A., and KORNBLIT, A. (1971) "On the depression of the superfluid transition temperature in He^4 by a heat current," Phys. Rev. A4, 804-806.

AKHIEZER, I. A. and GINZBURG, A. E. (1971) "Critical fluctuations near transition points between ferromagnetic and antiferromagnetic phases," Phys. Lett. 37A, 63-64.

ALBEN, R. (1970) "Liquid-vapor-like critical Points in anisotropic ferrimagnets," Phys. Rev. Lett. 24 68-71.

ALDER, B. J. and WAINWRIGHT, T. E. (1962) "Phase transition in elastic disks," Phys. Rev. 127, 359-361.

ALEKHIN, A. D. and KRUPSKII, N. P. (1971) "Experimental verification of scale law in the critical region of cyclopentane," JETP Lett. 14, 403-406.

ALEXANIAN, M. and WORTMAN, D. E. (1966) "New series expansion in statistical mechanics," Phys. Rev. 143, 96-103.

l ALLAN, G. A. T. (1970) "Critical temperatures of Ising lattice films," Ibid. B1, 352-356.

--- and BETTS, D. D. (1967) "Spin one exchange interaction model of ferromagnetism," Proc. Phys. Soc. 91, 341-352.

l ---, --- (1968) "The frequency-dependent initial perpendicular susceptibility of the Ising Model," Can. J. Phys. 46, 15-24.

l ---, --- (1968) "The temperature and frequency dependence of the inelastic neutron scattering from an Ising magnet," Ibid. 46, 799-802.

t ALLEN, J. W. (1971) "Optical observation of stress-induced spin flop in Cr_2O_3," Phys. Rev. Lett. 27, 1526.

ALLEN, S. J. (1968) "Spin-lattice interaction in UO_2. II. Theory of the first-order phase transition," Phys. Rev. 167, 492.

m ALPERT, S. S., YEH, Y., and LIPWORTH, E. (1965) "Observation of time-dependent concentration fluctuations in a binary mixture near the critical temperature using a He-Ne laser," Phys. Rev. Lett. 14, 486-488.

a ALS-NIELSEN, J. (1969) "Investigation of scaling laws by critical neutron scattering from beta-brass," Phys. Rev. 185, 664-666.

--- (1970) "Analysis of critical neutron-scattering data from iron and dynamical scaling theory," Phys. Rev. Lett. 25, 730-734.

a --- and DIETRICH, O. W. (1967) "Pair-correlation function in disordered β-brass as studied by neutron diffraction," Phys. Rev. 153, 706-711.

a ---, --- (1967) "Long-range order and critical scattering of neutrons below the transition temperature in β-brass," Ibid. 153, 717-721.

---, --- (1969) "Spin waves and the order-disorder transition in chromium," Phys. Rev. Lett. 22, 290-292.

---, ---, KUNNMANN, W., and PASSELL, L. (1971) "Critical behavior of the Heisenberg ferromagnets EuO and EuS," Ibid. 27, 741-744.

---, ---, MARSHALL, W., and LINDGARD, P. A. (1967) "Inelastic critical scattering of neutrons from terbium," Sol. State Commun. 5, 607-611.

m ALVESALO, T., BERGLUND, P., ISLANDER, S., PICKETT, G. R., and ZIMMERMAN, W., Jr. (1969) "Specific heat of liquid He^3/He^4 mixtures near the junction of the lambda and phase-separation curves," Phys. Rev. Lett. 22, 1281-1283.

m ---, ---, ---, ---, --- (1971) "Specific heat of liquid He^3/He^4 mixtures near the junction of the λ and phase-separation curves. I.," Phys. Rev. A4, 2354-2368.

AMBLER, E., EISENSTEIN, J. C., and SCHOOLEY, J. F. (1962) "Traces of products of angular momentum natrices," J. Math. Phys. 3, 118-130.

L AMER, N. B., SHEN, Y. R., and ROSEN, H. (1970) "Raman study of para-azoxydianisole at the phase transitions," Phys. Rev. Lett. 24, 718-720.

d AMIT, D. J. (1968) "Phase transition in HeII films," Phys. Lett. 26A, 448-449.

m ANANTARAMAN, A. V., WALTERS, A. B. WALTERS, EDMONDS, P. D., and PINGS, C. J. (1966) "Absorption of sound near the critical point of the nitrobenzene-iso-octane system," J. Chem. Phys. 44, 2651-2658.

ANDERSON, E. E. (1966) "Power laws near the Curie point of yttrium iron garnet," Phys. Rev. Lett. 17, 375-376.

a ---, ARAJS, S., STELMACH, A. A., TEHAN, B. L., and YAO, Y. D. (1971) "Critical exponent β for nickel and nickel-copper alloys," Phys. Lett. 36A, 173-175.

ANDERSON, N. S. and DELSASSO, L. P. (1951) "The propagation of sound in carbon dioxide near the critical point," J. Acoust. Soc. Am. 23, 423-429.

l ANDERSON, P. W. and YUVAL, G. (1971) "Some numerical results on the Kondo problem and the inverse square one-dimensional Ising model," J. Phys. C. 4, 607-608.

---, ---, and HAMANN, D. R. (1970) "Exact results in the Kondo problem. II. Scaling theroy, qualitatively correct solution, and some new results on one-dimensional classical statistical models," Phys. Rev. B1, 4464-4473.

ANDREWS, T. (1869) "The Bakerian lecture - On the continuity of the gaseous and liquid states of matter," Phil. Trans. 159, 575-591.

ANGELESCU, N. COSTACHE, G., and NENCIU, G. (1972) "Molecular field theory and phase transitions in partially finite spin systems," Phys. Stat. Sol. 51, 205-214.

ANISIMOV, M.A., VORONEL', A. V., and GORODETSKIĬ, E. E. (1971) "Isomorphism of critical phenomena," Sov. Phys.-JETP 33, 605-612.

m ---, ---, and OVODOVA, T. M. (1972) "Experimental investigation of the singularity of specific heat at the critical stratification point of a binary mixture," Ibid. 34, 583-587.

ANTONINI, M. (1967) "Diffusion magnétique des neutrons par MnF_2 près du Point de Néel," J. Phys. Chem. Solids 28, 11-16.

b APPLEQUIST, J. (1966) "True phase transitions in macromolecules of the DNA type," J. Chem. Phys. 45, 3459-3461.

b --- (1969) "Higher-order phase transitions in two-stranded macromolecules," Ibid. 50, 600-609.

b --- (1969) "First-order phase transitions in multistranded macromolecules," Ibid. 50, 609-612.

ARAJS, S. (1965) "Paramagnetic behavior of nickel just above the ferromagnetic Curie temperature," J. Appl. Phys. 36, 1136-1137.

--- and COLVIN, R. V. (1964) "Ferromagnetic-paramagnetic transition in iron," Ibid. 35, 2424-2426.

---, TEHAN, B. L., ANDERSON, E. E., and STELMACH, A. A. (1970) "Power laws for the magnetization of iron," Int. J. Magnetism 1, 41-44.

---, ---, ---, --- (1970) "Critical magnetic behavior of nickel near the Curie point," Phys. Stat. Sol. 41, 639.

ARAKI, H. and LIEB, E. H. (1970) "Entropy inequalities," Commun. Math. Phys. 18, 160-170.

h ARCHIBALD, M., MOCHEL, J. M., and WEAVER, L. (1968) "Size dependence of the thermal conductivity of helium I near the λ point," Phys. Rev. Lett. 21, 1156-1158.

m ARCOVITO, G., FALOCI, C., and ROBERTI, M. (1969) "Shear viscosity of the binary system aniline-cyclohexane near the critical point," Ibid. 22, 1040-1042.

m ARROTT, A. (1968) "Existence of a critical line in ferromagnetic to paramagnetic transitions," Ibid. 20, 1029-1032.

--- (1971) "Problem of using kink-point locus to determine the critical exponent β," J. Appl. Phys. 42, 1282-1283.

--- (1971) "Parametric equation of state applied to the magnetization of nickel near its Curie temperature," Ibid. 42, 1288-1289.

--- and NOAKES, J. E. (1967) "Approximate equation of state for nickel near its critical temperature," Phys. Rev. Lett. 19, 786-789.

ARTYUKHOVSKAYA, L. M., SHIMANSKAYA, E. T., and SHIMANSKIĬ, YU. I. (1971) "Investigation of the thermodynamic properties of pentane near the liquid-vapor critical point," Sov. Phys.-JETP 32, 375-379.

ASANO, T. (1968) "Generalization of the Lee-Yang theorem," Prog. Theoret. Phys. 40, 1328-1336.

l ASHKIN, J. and LAMB, W. E., Jr. (1943) "The propagation of order in crystal lattices," Phys. Rev. 64, 159-178.

ASHKIN, J. and TELLER, E. (1943) "Statistics of two-dimensional lattices with four components," Ibid. 64, 178-184.

a ASHMAN, J. and HANDLER, P. (1969) "Specific heat of the order-disorder transition in β brass," Phys. Rev. Lett. 23, 642-644.

d ASO, K. and MIYAHARA, S. (1966) "Magnetic susceptibility of $SrLaFeO_4$ with K_2NiF_4 structure," J. Phys. Soc. Japan 21, 1833.

m ATACK, D. and RICE, O. K. (1954) "Critical phenomena in the cyclohexane-aniline system," J. Chem. Phys. 22, 382-385.

AUSLOOS, M. and KAWASAKI, K. (1971) "Comment on Marcelja's 'resistivity of antiferromagnets above the Néel temperature'," J. Phys. C. 4, L132-L133.

AVENARIUS, M. (1874) "Ueber innere latente Wärme," Ann. Physik 151, 303-316.

F AXE, J. D., DORNER, B., and SHIRANE, G. (1971) "Mechanism of the ferroelectric phase transformation in rare-earth molybdates," Phys. Rev. Lett. 26, 519-523.

BAER, S. and LEBOWITZ, J. L. (1964) "Convergence of fugacity expansion and bounds on molecular distributions for mixtures," J. Chem. Phys. 40, 3474-3478.

BAGATSKIĬ, M. I., VORONEL', A. V., and GUSAK, V. G. (1963) "Measurement of the specific heat C_v of argon in the immediate vicinity of the critical point," Sov. Phys.-JETP 16, 517-518.

m BAK, C. S. and GOLDBURG, W. I. (1969) "Light scattering in an impure binary liquid mixture near the critical point," Phys. Rev. Lett. 23, 1218-1220.

m ---, ---, and PUSEY, P. N. (1970) "Light-scattering study of the critical behavior of a three-component liquid mixture," Ibid. 25, 1420-1422.

1 BAKER, G. A., Jr. (1961) "One-dimensional order-disorder model which approaches a second-order phase transition," Phys. Rev. 122, 1477-1484.

1 --- (1961) "Application of the Padé approximant method to the investigation of some magnetic properties of the Ising model," Ibid. 124, 768-774.

--- (1962) "Certain general order-disorder models in the limit of long-range interactions," Ibid. 126, 2071-2078.

--- (1963) "Further applications of the Padé approximant method to the Ising and Heisenberg models," Ibid. 129, 99-102.

3 --- (1964) "Padé approximant bounds for the magnetic susceptibility in the three-dimensional, spin-1/2 Heisenberg model," Ibid. 136, A1376-A1380.

r --- (1965) "The theory and application of the Padé approximant method," Adv. Theor. Phys. 1, 1.

1 --- (1967) "Convergent, bounding approximation procedures with applications to the ferromagnetic Ising model," Phys. Rev. 161, 434-445.

--- (1968) "Some rigorous inequalities satisfied by the ferromagnetic Ising model in a magnetic field," Phys. Rev. Lett. 20, 990-992.

--- (1969) "Best error bounds for Padé approximants to convergent series of Stieltjes," J. Math. Phys. 10, 814-820.

r --- (1971) "Singularity structure of the perturbation series for the ground-state energy of a many fermion system," Rev. Mod. Phys. 43, 479-531.

1 --- (1972) "Ising model with a scaling interaction," Phys. Rev. B5, 2622-2633.

--- and ESSAM, J. W. (1970) "Effects of lattice compressibility on critical behavior," Phys. Rev. Lett. 24, 447-449.

a ---, --- (1971) "Statistical mechanics of a compressible Ising model with application to β brass," J. Chem. Phys. 55, 861-880.

3 ---, EVE, J., and RUSHBROOKE, G. S. (1970) "Magnetic phase-boundary of the spin-1/2 Heisenberg ferromagnetic model," Phys. Rev. B2, 706-721.

--- and GAMMEL, J. L. (1961) "The Padé approximant," J. Math. Analysis and Applic. 2, 21-30.

---, ---, and HILL, G. J. (1963) "Study of the perturbation series for the ground-state energy of a many-fermion system," Phys. Rev. 132, 1373-1387.

---, ---, and WILLS, J. G. (1961) "An investigation of the applicability of the Padé approximant method," J. Math. Analysis and Applic. 2, 405-418.

1 --- and GAUNT, D. S. (1967) "Ising-model critical indices below the critical temperature," Phys. Rev. 155, 545-552.

3 ---, GILBERT, H. E., EVE, J., and RUSHBROOKE, G. S. (1966) "On the Heisenberg spin-1/2 ferromagnetic models," Phys. Lett. 20, 146-147.

3 ---, ---, ---, --- (1966) "On the field-dependent susceptibility of the Heisenberg spin-1/2 ferromagnet," Ibid. 22, 269-271.

3 ---, ---, ---, --- (1967) "A data compendium of linear graphs with application to the Heisenberg model," Brookhaven National Laboratory Report BNL 50053, 1-251.

3 ---, ---, ---, --- (1967) "High-temperature expansions for the spin-1/2 Heisenberg model," Phys. Rev. 164, 800-817.

d ---, ---, ---, --- (1967) "On the two-dimensional, spin-1/2 Heisenberg ferromagnetic models," Phys. Lett. 25A, 207-209.

---, RUSHBROOKE, G. S., and GILBERT, H. E. (1964) "High-temperature series expansions for the spin-1/2 Heisenberg model by the method of irreducible representations of the symmetric group," Phys. Rev. 135, A1272-A1277.

BAKRI, M. M. (1971) "Cluster expansions for physical adsorption," Physica 52, 16-20.

BALE, H. D., DOBBS, B. C., LIN, J. S., and SCHMIDT, P. W. (1970) "X-ray scattering studies of critical opalescence in argon at constant density," Phys. Rev. Lett. 25, 1556-1559.

BALLY, D., GRABCEV, B., LUNGU, A. M., POPOVICI, M., and TOTIA, M. (1967) "Small-angle critical magnetic scattering of neutrons in iron," J. Phys. Chem. Solids 28, 1947-1955.

---, POPOVICI, M., TOTIA, M., GRABCEV, B., and LUNGU, A. M. (1967) "Critical magnetic scattering of neutrons in cobalt," Phys. Lett. 25A, 595-596.

---, ---, ---, ---, --- (1968) "Evidence for Fisher's correlation function in iron from critical neutron scattering," Ibid. 26A, 396-397.

BALTZER, P. K, WOJTOWICZ, P. L., ROBBINS, M., AND LOPATIN, E. (1966) "Exchange interactions in ferromagnetic chromium chalcogenide spinels," Phys. Rev. 151, 367-377.

b BARGERON, C. B., MC CALLY, R. L., TATHAM, P. E. R., CANNON, S. M., and HART, R. W. (1972) "Light-beating spectrum of erythrocyte ghosts," Phys. Rev. Lett. 28, 1105-1107.

r BARKER, J. A. and HENDERSON, D. (1968) "The equation of state of simple liquids," J. Chem. Educ. 45, 1-6.

h BARMATZ, M. (1970) "Ultrasonic measurements near the critical point of He^4," Phys. Rev. Lett. 24, 651-654.

--- and HOHENBERG, P. C. (1970) "Test of a parametric equation of state and calculation of gravity effects at the gas-liquid critical point," Ibid. 24, 1225-1229.

h --- and RUDNICK, I. (1968) "Velocity and attenuation of first sound near the λ point of helium," Phys. Rev. 170, 224-238.

h --- and RUDNICK, I. (1968) "Measurement of the density maximum of He^4 at saturated vapor pressure," Ibid. 173, 275-277.

l BAROUCH, E. (1971) "On the two-dimensional Ising model with random impurities," J. Math. Phys. 12, 1577-1578.

f --- (1971) "Long-range order of Wu's modified F-model," Phys. Lett. 34A, 347-349.

2 --- and DRESDEN, M. (1969) "Exact time-dependent analysis for the one-dimensional XY model," Phys. Rev. Lett. 23, 114-117.

2 --- and MC COY, B. M. (1971) "Statistical mechanics of the XY model. II. Spin-correlation functions," Phys. Rev. A3, 786-804.

2 ---, ---, and DRESDEN, M. (1970) "Statistical mechanics of the XY model. I," Ibid. A2, 1075-1092.

BARRY, J. H. (1966) "Magnetic relaxation near a second-order phase-transition point," J. Chem. Phys. 45, 4172-7144.

--- (1968) "Statistical-mechanical theory of resonance susceptibility in Heisenberg antiferromagnetism," Phys. Rev. 174, 531-539.

BARTIS, F. J. (1970) "Thermodynamic equations for dysprosium-aluminum garnet near its Néel point," Phys. Stat. Sol. 39, 665.

BARTIS, F. J. (1971) "The analogue of Fisher's relation in the coexistence region," Ibid. 48, K163.

BARTKOWSKI, R, SAGE, J. P., and LE CRAW, R. C. (1968) "Spin-wave relaxation in ferromagnetic $CdCr_2Se_4$," J. Appl. Phys. 39, 1071-1072.

BATA, L. and KROÓ, N. (1968) "Investigation of the dynamical properties of liquid-gas systems by coherent cold neutron scattering near the critical point," Neutron Inelastic Scattering I, 615-622.

b BAUR, M. E. and NOSANOW, L. H. (1963) "Simple model for the helix-random coil transition in polypeptides," J. Chem. Phys. 38, 578-582.

f BAXTER, R. J. (1970) "Exact isotherm for the F model in direct and staggered electric fields," Phys. Rev. B1, 2199-2202.

f --- (1971) "Eight-vertex model in lattice statistics," Phys. Rev. Lett. 26, 832-836.

f --- (1972) "Partition function of the eight-vertex lattice model," Ann. Phys. 70, 193-228.

--- (1972) "One-dimensional anisotropic Heisenberg chain," Ibid. 70, 323-337.

BEAN, C. P. and RODBELL, D. S. (1962) "Magnetic disorder as a first-order phase transformation," Phys. Rev. 126, 104-115.

BEDEAUX, D., MILOŠEVIĆ, S., and PAUL, G. (1971) "Linear response theory for systems obeying the master equation," J. Stat. Phys. 3, 39-51.

BELL, G. M. (1972) "Statistical mechanics of water: lattice model with directed bonding," J. Phys. C. 5, 889.

BELLEMANS, A. and NIGAM, R. K. (1966) "Phase transitions in the hard-square lattice gas," Phys. Rev. Lett. 16, 1038-1039.

BELOV, K. P., NIKITIN, S. A., MURAV'EVA, V. M., SOLNTSEVA, L. I., and CHUPRIKOV, G. E. (1972) "Causes of the anomalous temperature dependence of the critical field in terbium," Sov. Phys.-JETP 34, 787-789.

f BELOZOROV, D. P. (1970) "Coupled electromagnetic and spin waves in ferrodielectrics near the critical point," Ibid. 31, 365.

h BENDINER, W., ELWELL, D., and MEYER, H. (1968) "Properties of ^{3}He near the critical point," Phys. Lett. 26A, 421-422.

r BENEDEK, G. B. (1968) "Thermal fluctuations and the scattering of light," In Statistical Physics, Phase Transitions, and Superfluidity, M. Chrétien, E. P. Gross, and S. Deser, Eds. (Gordon and Breach, N. Y.) Vol. 2.

r --- (1969) "Optical mixing spectroscopy, with applications to problems in physics, chemistry, biology, and engineering," In Polarization, Matter, and Radiation (Presses Universitaire de France, Paris), p. 49-84.

r --- (1969) "Phase transitions," Proc. Fourteenth Conf. on Chemistry at the University of Brussels.

b --- (1971) "Theory of transparency of the eye," Appl. Opt. 10, 459-473.

--- and GREYTAK, T. (1965) "Brillouin scattering in liuqids," Proc. IEEE 53, 1623-1629.

3 BENNETT, H. S. (1966) "Behavior of the Heisenberg ferromagnet for temperatures near and above the transition point," Ann. Phys. 39, 127-175.

3 --- (1968) "Diffuse and propagating modes in the Heisenberg paramagnet," Phys. Rev. 174, 629-639.

3 --- (1968) "Phenomenology of neutron scattering in Heisenberg systems," Ibid. 176, 659-654.

--- (1969) "Magnetic scattering of neutrons from Heisenberg antiferromagnets," J. Appl. Phys. 40, 1552-1553.

3 --- (1969) "Ultrasonic attenuation in Heisenberg magnets," Phys. Rev. 181, 978.

3 --- and MARTIN, P. C. (1965) "Spin diffusion in the Heisenberg paramagnet," Ibid. 138, A608-A617.

--- and PYTTE, E. (1967) "Ultrasonic attentuation in the Heisenberg paramagnet," Ibid. 155, 553-562.

m BERESTOV, A. T., VORONEL', A. V., and GITERMAN, M. SH. (1972) "Singularities of the diagram of state of an He^3-He^4 mxiture due to the hydrostatic effect," JETP Lett. 15, 190-193.

d BEREZINSKIǏ, V. L. (1971) "Destruction of long-range order in one-dimensional and two-dimensional systems having a continuous symmetry group I. Classical systems," Sov. Phys.-JETP 32, 493-500.

d --- (1972) "Destruction of long-range order in one-dimensional and two-dimensional systems possessing a continuous group. II. Quantum systems," Ibid. 34, 610-616.

m BERGÉ, M., CALMETTES, P., DUBOIS, M., and LAJ, C. (1970) "Experimental observation of the complete Rayleigh central component of the light scattered by a two-component fluid," Phys. Rev. Lett. 24, 89-90.

m ---, ---. LAJ, C., TOURNARIE, M., and VOLOCHINE, B. (1970) "Dynamics of concentration fluctuations in a binary mixture in the hydrodynamical and nonhydrodynamical regimes," Ibid. 24, 1223-1225.

m ---, ---, ---, and VOLOCHINE, B. (1969) "Experimental evidence of the 'critical region' of a binary mixture by means of inelastic scattering of light," Ibid. 23, 693-694.

m ---, ---, VOLOCHINE, B., and LAJ, C. (1969) "A study of dynamics of concentration fluctuations in a binary mixture, by means of inelastic scattering of light, in the immediate vicinity of its critical temperature," Phys. Lett. 30A, 7-8.

--- and DUBOIS, M. (1971) "Experimental confirmation of the Kawasaki-Einstein-Stokes formula; measurement of small correlation lengths," Phys. Rev. Lett. 27, 1125-1127.

∞ BERLIN, T. and KAC, M. (1952) "The spherical model of a ferromagnet," Phys. Rev. 86, 821-835.

--- and MONTROLL, E. (1952) "On the free energy of a mixture of ions: An extension of Kramer's theory," J. Chem. Phys. 20, 75-84.

--- and THOMSEN, J. (1952) "Dipole-dipole interaction in simple lattices," Ibid. 20, 1368-1374.

---, WITTEN, L., and GERSCH, H. A. (1953) "The imperfect gas," Phys. Rev. 92, 189-201.

BERLINSKY, A. J. and HARRIS, A. B. (1970) "High-temperature expansion for the orientational specific heat of solid H_2 and D_2," Ibid. A1, 878-887.

a BERNASCONI, J. and RYS, F. (1971) "Critical behavior of a magnetic alloy," Ibid. B4, 3045-3048.

r BERNE, B. J. and FORSTER, D. (1971) "Topics in time-dependent statistical mechanics," Ann. Rev. Phys. Chem. 22, 563.

f BERRE, B., FOSSHEIM, K., and MÜLLER, K. A. (1969) "Critical attenuation of sound by soft modes in $SrTiO_3$," Phys. Rev. Lett. 23, 589-591.

BERSOHN, R., PAO, Y., and FRISCH, H. L. (1966) "Double-quantum light scattering by molecules," J. Chem. Phys. 45, 3184-3198.

BESTGEN, W., GROSSMANN, S., and ROSENHAUER, W. (1968) "Phase transitions and zeros in several physical variables," Phys. Lett. 28A, 117-118.

BETHE, H. A. (1935) "Statistical theory of superlattices," Proc. Roy. Soc. A150, 552-575.

∞ BETTONEY, G. V. and MAZO, R. M. (1970) "Equivalence of some generalizations on the spherical model," J. Math. Phys. 11, 1147-1149.

BETTS, D. D. (1964) "The exact solution of some lattice statistics models with four states per site," Can. J. Phys. 42, 1564-1572.

1 --- and ALLAN, G. A. T. (1969) "Some dynamical properties of the Ising model," J. Phys. Soc. Japan 26, 163-164.

1 --- and DITZIAN, R. V. (1968) "High-temperature properties of the Ising model on the cristobalite lattice," Can. J. Phys. 46, 971-975.

2 ---, ---, ELLIOTT, C. J., and LEE, M. H. (1971) "Critical behavior of the XY model of a ferromagnet," J. de Physique 32, C1-356-358.

d ---, ELLIOTT, C. J., and DITZIAN, R. V. (1971) "Phase transitions in two-dimensional spin systems," Can. J. Phys. 49, 1327-1334.

---, ---, and LEE, M. H. (1969) "High-temperature series expansion for the partition function and specific heat of the XY model on a face-centered cubic lattice," Phys. Lett. 29A, 150-152.

2 ---, ---, --- (1970) "Exact high temperature series expansions for the XY model," Can. J. Phys. 48, 1566-1577.

u ---, GUTTMANN, A. J., and JOYCE, G. S. (1971) "Lattice-lattice scaling and the generalized law of corresponding states," J. Phys. C4, 1994-2008.

--- and HUNTER, D. L. (1965) "A lattice statics model derived from the two-layer adsorption problem," Can. J. Phys. 43, 980.

2 --- and LEE, M. H. (1968) "Critical properties of the XY model," Phys. Rev. Lett. 20, 1507-1510.

d BHAGAVAN, M. R. (1969) "Superconductivity in one-dimensional systems," J. Phys. C. 2, 1092-1096.

1 BIALYNICKI-BIRULA, I. and PIASECKI, J. (1967) "Dynamics of the Ising chain. I" Bull. de L'Academie Pol. des Sci. 15, 211-216.

m BIDAUX, R., CARRARA, P., and VIVET, B. (1967) "Antiferromagnetisme dans un champ magnetique I. Traitement de champ moleculaire," J. Phys. Chem. Solids 28, 2453-2469.

m BIENENSTOCK, A. (1966) "Variation of the critical temperatures of Ising antiferromagnets with applied magnetic field," J. Appl. Phys. 37, 1459-1461.

a --- and LEWIS, J. (1967) "Order-disorder of nonstoichoimetric binary alloys and Ising antiferromagnets in magnetic fields," Phys. Rev. 160, 393-403.

3 BINDER, K. (1969) "A Monte-Carlo method for the calculation of the magnetization of the classical Heisenberg model," Phys. Lett. 30A, 273-274.

--- (1969) "High temperature expansions of spin correlation functions for Ising and Heisenberg ferromagnets," Phys. Stat. Sol. 32, 891-903.

--- and RAUCH, H. (1970) "Spin-correlation effects above the Curie point and their action on neutrons," Z. Angew. Phys. 28, 325-331.

---, ---, and WILDPANER, V. (1970) "Monte Carlo calculation of the magnetization of superparamagnetic particles," J. Phys. Chem. Solids 31, 391-397.

BIRGENEAU, R. J., ALS-NIELSEN, J., and BUCHER, E. (1971) "Magnetic excitons in singlet-ground-state ferromagnets," Phys. Rev. Lett. 27, 1530-1533.

d ---, DINGLE, R., HUTCHINGS, M., SHIRANE, G., and HOLT, S. L. (1971) "Spin correlations in a one-dimensional Heisenberg antiferromagnet," Ibid. 26, 718-721.

d ---, GUGGENHEIM, H. J., and SHIRANE, G. (1969) "Neutron scattering from K_2NiF_4: A two-dimensional Heisenberg antiferromagnet," Ibid. 22, 720-723.

d ---, ---, --- (1970) "Neutron scattering investigation of phase transitions and magnetic correlations in the two-dimensional antiferromagnets K_2NiF_4, Rb_2MnF_4, and Rb_2FeF_4," Phys. Rev. B1, 2211-2230.

d ---, SKALYO, J., and SHIRANE, G. (1970) "Phase transitions and magnetic correlations in two-dimensional antiferromagnets," J. Appl. Phys. 41, 1303.

d ---, ---, --- (1971) "Critical magnetic scattering in K_2NiF_4," Phys. Rev. B3, 1736-1749.

m ---, YELON, W. B., COHEN, E., and MAKOVSKY, J. (1972) "Magnetic properties of $FeCl_2$ in zero field. I. Excitations," Ibid. B5, 2607-2615.

BLATT, J. M. and MATSUBARA, T. (1958) "Quasi-chemical equilibrium theory, part II," Progr. Theoret. Phys. 20, 553-575.

m BLAZEY, K. W., MÜLLER, K. A., ONDRIS, M., and ROHRER, H. (1970) "Antiferromagnetic resonance truncated by the spin-flop transition," Phys. Rev. Lett. 24, 105-107

BLECH, I. A. and AVERBACH, B. L. (1964) "Spin correlations in MnO," Physics 1, 31-44.

---, --- (1966) "Long-range magnetic order in MnO," Phys. Rev. 142, 287-290.

L BLINC, R., HOGENBOOM, D. L., O'REILLY, D. E., and PETERSON, E. M. (1969) "Spin relaxation and self-diffusion in liquid crystals," Phys. Rev. Lett. 23, 969-972.

f --- and SVETINA, S. (1966) "Cluster approximations for order-disorder-type hydrogen-bonded ferroelectrics. I. Small clusters," Phys. Rev. 147, 423-429.

f ---, --- (1966) "Cluster aprroximations for order-disorder-type hydrogen-bonded ferroelectrics. II. Application to KH_2PO_4," Ibid. 147, 430-438.

f --- and ŽEKŠ, B. (1968) "On the pressure dependence of the ferroelectric Curie temperatures of KH_2PO_4 and KD_2PO_4," Phys. Lett. 26A, 468-469.

BLINOWSKI, K. and CISZEWSKI, R. (1968) "The temperature dependence of the 'Curie point' for iron observed in the critical scattering of neutrons," Ibid. 28A, 389-390.

l BLOCH, C. and LANGER, J. S. (1965) "Diagram renormalization, variational principles, and the infinite-dimensional Ising model," J. Math. Phys. 6, 554-572.

m BLOCH, D., CHAISSE, F., and PAUTHENET, R. (1966) "Effects of hydrostatic pressure on the magnetic ordering temperatures and the magnetization of some ionic compounds," J. Appl. Phys. 37, 1401-1402.

BLOCH, M. (1962) "Magnon renormalization in ferromagnets near the Curie point," Phys. Rev. Lett. 9, 286-287.

d BLOEMBERGEN, P., TAN, K. G., LEFEVRE, F. H. J., and BLEYENDAAL, A. H. M. (1971) "Specific heat of the two-dimensional spin 1/2 Heisenberg ferromagnet," J. de Physique 32, 878-879.

l BLOOD, F. A., Jr. (1969) "Approximate calculations for the two-dimensional Ising model," J. Stat. Phys. 2, 301-304.

l BLOTE, H. W. J. and HUISKAMP, W. J. (1969) "Heat capacity measurements on Rb_3CoCl_5 compared with the Ising model," Phys. Lett. 29A, 304-305.

m BLUME, M. (1966) "Theory of the first-order magnetic phase change in UO_2," Phys. Rev. 141, 517-524.

m ---, EMERY, V. J., and GRIFFITHS, R. B. (1971) "Ising model for the λ transition and phase separation in He^3-He^4 mixtures," Ibid. A4, 1071-1077.

--- and HUBBARD, J. (1970) "Spin correlation functions at high temperatures," Ibid. B1, 3815-3830.

--- and WATSON, R. E. (1967) "First-order phase transition in a rigorously solvable magnetic system," J. Appl. Phys. 38, 991-992.

g BOCCARA, N. (1969) "On the violation of Abelian Groups in second order phase transitions," Phys. Lett. 28A, 474-475.

--- (1969) "Exact thermodynamic relations and rigorous exponent equalities in the vicinity of a second-order transition line," Sol. State Commun. 7, 331-333.

f --- (1971) "Self-consistent theory of displacive phase transitions in magnetic materials," Phys. Stat. Sol. 43, K11.

BOERSTOEL, B. M., ZWART, J. J., and VAN BAARLE, C. (1970) "Observation of a sharp transition in the specific heat of dilute Pd-Mn alloys and the influence of large magnetic fields, Phys. Lett. 31A, 378-379.

d BOGOLJUBOW, N. N. (1962) "Quasimittelwerte in Problemen der statistischen Mechanik," Phys. Abh. aus der Sow. Union 6, 1-24; 113-138; 229-252.

BONNER, J. C. and FISHER, M. E. (1962) "The entropy of an antiferromagnet in a magnetic field," Proc. Phys. Soc. 80, 508-515.

d ---, --- (1964) "Linear magnetic chains with anisotropic coupling," Phys. Rev. 135, 640-658.

m --- and NAGLE, J. F. (1971) "Phase behavior of models with competing interactions," J. Appl. Phys. 42, 1280-1282.

BOSE, S. M. and FOO, E-NI. (1972) "Density of spin wave states in disordered Heisenberg ferromagnetic chains," J. Phys. C. 5, 1082.

m BOTCH, W. and FIXMAN, M. (1962) "Viscosity of critical mixtures: Dependence on viscosity gradient," J. Chem. Phys. 36, 3100-3101.

---, --- (1965) "Heat capacity of gases in the critical region," Ibid. 42, 196-198.

2 BOWERS, R. G. and JOYCE, G. S. (1967) "Lattice model for the λ transition in a Bose fluid," Phys. Rev. Lett. 19, 630-632.

a --- and MC KERRELL, A. (1970) "Magnetic binary alloys and lattice gases: Exact results in one dimension for Heisenberg interactions," Phys. Lett. 32A, 167-168.

3 --- and WOOLF, M. E. (1969) "Some critical properties of the Heisenberg model," Phys. Rev. 177, 917-932.

a BRAGG, W. L. and WILLIAMS, E. J. (1934) "The effect of thermal agitation on atomic arrangement in alloys," Proc. Roy. Soc. A145, 699-730.

BRAUN, P., HAMMER, D., TSCHARNUTER, W., and WEINZIERL, P. (1970) "Rayleigh scattering by SF_6 in the critical region," Phys. Lett. 32A, 390-391.

d BREED, D. J. (1967) "Experimental investigation of two two-dimensional antiferromagnets with small anisotropy," Physica 37, 35-46.

g BREZIN, E., WALLACE, D. J., and WILSON, K. G. (1972) "Feynman-Graph expansion for the equation of state near the critical point (Ising-like case)," Phys. Rev. Lett. 29, 591-594.

BRILLOUIN, L. (1914) "Diffusion of light through a homogeneous transparent body," Cr. Acad. Sci. Paris 158, 1331-1336.

--- (1922) "Diffusion de la lumiere et des rayons X par un corps transparent homogene influence de l'agitation thermique," Ann. Phys. (Paris) 17, 88-122.

BRODKORB, W. (1966) "Zur Theroie des Ferromagnetismus magnetisch anisotroper Schichten mit kubischer Kristallstruktur. I. Berechnung der Magnetisierung und der Curietemperatur mit der Methode der Greenschen Funktionen," Phys. Stat. Sol. 16, 225-236.

f BRODY, E. M. and CUMMINS, H. Z. (1968) "Brillouin-scattering study of the ferroelectric transition in KH_2PO_4," Phys. Rev. Lett. 21, 1263-1266.

f BROPHY, J. J. and WEBB, S. L. (1962) "Critical fluctuations in triglycene sulfate," Phys. Rev. 128, 584-588.

BROUT, R. (1959) "Statistical mechanical theory of a random ferromagnetic system," Ibid. 115, 824-835.

--- (1960) "Statistical mechanical theory of ferromagnetism. High density behavior," Ibid. 118, 1009-1019.

∞ --- (1961) "Statistical mechanics of ferromagnetism; spherical model as high-density limit," Ibid. 122, 469-474.

r --- (1965) Phase transitions (W. A. Benjamin, Inc., New York).

∞ --- (1967) "Molecular field theory, the Onsager reaction field and the spherical model," Physics 3, 317-329.

--- (1971) "Scaling in the critical region: An additional relation among critical parameters (δ=5)," Phys. Lett. 34A, 115-117.

--- (1971) "Static scaling in the critical region," Proc. Int. Conf. Magnetism, Chicago.

--- and ENGLERT, F. (1960) "Linked-clusters expansion in quantum statistics: Petit canonical ensemble," Phys. Rev. 120, 1519-1527.

f ---, MÜLLER, K. A., and THOMAS, H. (1966) "Tunnelling and collective excitations in a microscopic model of ferroelectricity," Sol. State Commun. 4, 507-510.

3 BROWN, H. A. (1956) "Ferromagnetic Curie temperatures by the Kramers-Opechowski method," Phys. Rev. 104, 624-625.

3 --- (1968) "The critical and high-temperature behavior of a Heisenberg ferromagnet," Phys. Stat. Sol. 25, K11-K14.

--- (1971) "Heisenberg ferromagnet with biquadratic exchange," Phys. Rev. B4, 115-121.

3 --- and LUTTINGER, J. M. (1955) "Ferromagnetic and antiferromagnetic Curie temperatures," Ibid. 100, 685-692.

BRUCH, L. W. (1970) "Susceptibility and fluctuation. II. Determination of the frequency moments," Ibid. B2, 721-723.

r BRUSH, S. G. (1967) "History of the Lenz-Ising model," Rev. Mod. Phys. 39, 883-893.

d BUCCI, C,. GUIDI, G., and VIGNALI, C. (1972) "Two-dimensional antiferromagnets: Rb^{87} NMR in Rb_2CoF_4," Sol. State Commun. 10, 803-810.

f BUCHER, E., BIRGENEAU, R. J., MAITA, J. P., FELCHER, G. P., and BRUN, T. O. (1972) "Magnetic and structural phase transition in DySb," Phys. Rev. Lett. 28, 746-749.

BUCK, H. and ALEFELD, G. (1972) "Hydrogen in Palladium - Silver in the neighbourhood of the critical point," Phys. Stat. Sol. 49, 317 (1972).

h BUCKINGHAM, M. J. and FAIRBANK, W. M. (1961) "The nature of the λ-transition in liquid helium," In Progress in low-temperature physics, C. J. Gorter, Ed., North-Holland, Amsterdam, Vol. 3, p. 80.

l --- and GUNTON, J. D. (1969) "Correlations at the critical point of the Ising model," Phys. Rev. 178, 848-853.

BUFF, F. P., LOVETT, R. A., and STILLINGER, F. H., Jr. (1965) "Interfacial density profile for fluids in the critical region," Phys. Rev. Lett. 15, 621-623.

BURCH, T., CRAIG, P. P., HEDRICK, C., KITCHENS, T. A., BUDNICK, J. I., CANNON, J. A., LIPSICAS, M., and MATTIS, D. (1969) "Swtiching in magnetite: A thermally driven magnetic phase transition," Ibid. 23, 1444-1447.

l BURGOYNE, P. N. (1963) "Remarks on the combinatorial approach to the Ising problem," J. Math. Phys. 4, 1320-1326.

BURKE, T. and LEBOWITZ, J. L. (1968) "Note on statistical mechanics of random systems," Ibid. 9, 1526-1534.

---, ---, and LIEB, E. (1966) "Phase transition in a model quantum system: Quantum corrections to the location of the critical point," Phys. Rev. 149, 118-122.

l BURLEY, D. M. (1960) "Some magnetic properties of the Ising model," Phil. Mag. 5, 909-919.

b BUSH, R. and THOMPSON, C. J. (1971) "Time-dependent model for hemoglobin and allosteric enzymes. I. General formulation," Biopolymers 10, 961-972.

b ---, --- (1971) "Time dependent model for hemoglobin and allosteric enzymes. II.," Ibid. 10, 1331-1349.

BYKOV, A. M., VORONEL', A. V., SMIRNOV, V. A., and SHCHEKOCHIKHINA, V. V. (1971) "Experimental verification of similarity theory at the critical point of Ar," JETP Lett. 21-24.

L CABANE, B. and CLARK, W. G. (1970) "Effects of order and fluctuations on the N^{14} NMR in a liquid crystal," Phys. Rev. Lett. 25, 91-93.

m CADIEU, F. J. and DOUGLASS, D. H., Jr. (1968) "Effects of impurities on higher order phase transitions," Ibid. 21, 680-682.

a CAHN, J. W. (1968) "Spinodal decomposition," Trans. Met. Soc. AIME 242, 166-180.

CALLEN, H. (1965) "Thermodynamic fluctuations," Am. J. Phys. 33, 919-922.

CALMETTES, P. and LAJ, C. (1971) "Study of heat diffusion in transparent media by means of a thermal-lens effect," Phys. Rev. Lett. 27, 239-242.

CAMP, W. J. and FISHER, M. E. (1971) "Behavior of two-point correlation functions at high temperatures," Ibid. 26, 73-77.

CANNELL, D. and BENEDEK, G. (1970) "Brillouin spectrum of xenon near its critical point," Ibid. 25, 1157-1161.

CANNON, J. A., BUDNICK, J. I., KAWATRA, M. P., MYDOSH, J. A., and SKALSKI, S. (1971) "Critical behavior of the magnetic susceptibility of $CdNi_2$," Phys. Lett. 35A, 247-248.

m CAPEL, H. W. (1966) "On the possibility of first-order phase transitions in Ising systems of triplet ions with zero-field splitting," Physica 32, 966-988.

m --- (1966) "Phase transitions in spin-one Ising systems," Phys. Lett. 23, 327-328.

m --- (1967) "On the possibility of first-order transitions in Ising systems of triplet ions with zero-field splitting II," Physica 33, 295-331.

m --- (1967) "On the possibility of first-order transitions in Ising systems of triplet ions with zero-field splitting III," Physica 37, 423-441.

r CAPOCACCIA, D. and VINCENTINI-MISSONI, M. (1971) "Recent developments on the equilibrium thermodynamics of critical points," Riv. Nuovo Cimento 1, 519-545.

CARMI, G. (1968) "Mechanism of first order phase transition in quantum Coulomb plasmas," Phys. Lett. 27A, 8-9.

L CARR, E. F. (1970) "Anomalous alignment in the smectic phase of a liquid crystal owing to an electric field," Phys. Rev. Lett. 24, 807-811.

CARUSO, A., DE ANGELIS, A., GATTI, G., GRATTON, R., and MARTELLUCI, S. (1971) "Change in the scaling laws for laser-produced plasmas through the critical regime," Phys. Lett. 35A, 279-281.

CASEY, L. M. and RUNNELS, L. K. (1969) "Model for correlated molecular rotation," J. Chem. Phys. 51, 5070-5089.

CASHER, A. and LEBOWITZ, J. L. (1971) "Heat flow in regular and disordered harmonic chains," J. Math. Phys. 12, 1701-1711.

--- and REVZEN, M. (1967) "Bose-Einstein condensation of an ideal gas," Am. J. Phys. 35, 1154-1158.

d CHADDHA, G. and TRIPATHI, R. S. (1971) "The effect of crystal field anisotropy on the Néel temperature T_N of layered Heisenberg antiferromagnets," Phys. Lett. 37A, 277-279.

CHANDLER, D. and OPPENHEIM, I. (1968) "Fluctuation theory and critical phenomena," J. Chem. Phys. 49, 2122-2127.

l CHANG, C. H. (1952) "The spontaneous magnetization of a two-dimensional rectangular Ising model," Phys. Rev. 88, 1422.

f CHANG, K. S., WANG, S-Y., and WU, F. Y. (1971) "Circle theorem for the ice-rule ferroelectric models," Ibid. A4, 2324-2327.

b CHANG, R. F., KEYES, P. H., SENGERS, J. V., and ALLEY, C. O. (1971) "Dynamics of concentration fluctuations near the critical mixing point of a binary fluid," Phys. Rev. Lett. 27, 1706-1709.

b ---, ---, ---, --- (1972) "Non-local effects in the diffusion coefficient of a binary fluid near the critical mixing point," Berichte der Bunsen-Gesellschaft fur Physikalische Chemie 76, 260-267.

m CHANG, T. S., HANKEY, A., and STANLEY, H. E. (1972) "Scaling in multicomponent systems," Bull. Am. Phys. Soc. 17, 277.

m ---, ---, --- (1973) "Scaling laws for fluid systems using generalized homogeneous functions of strong and weak directions," Phys. Rev. B7.

m ---, ---, --- (1973) "Double-power scaling functions near tricritical points," Proc. LT13, Boulder, Colo.

m ---, ---, --- (1973) "Generalized scaling hypothesis in multicomponent systems, with application to tricritical points and critical points of higher order," Phys. Rev. B7.

b CHANGEUX, J. and PODLESKI, T. R. (1968) "On the excitability and cooperativity of the electroplax membrane," Proc. Natl. Acad. Sci. U.S. 59, 944-950.

b ---, THIERY, J., TUNG, Y., and KITTEL, C. (1967) "On the cooperativity of biological membranes," Ibid. 57, 335-341.

CHARLES, H. K., Jr., and JOSEPH, R. I. (1972) "Critical properties of antiferromagnets," Phys. Rev. Lett. 28, 823-825.

h CHASE, C. E., WILLIAMSON, R. C., and TISZA, L. (1964) "Ultrasonic propagation near the critical point in helium," Ibid. 13, 467-469.

h --- and ZIMMERMAN, G. O. (1965) "Dielectric constant of ^{3}He in the critical region," Ibid. 15, 483-485.

CHEN, H. H. and JOSEPH, R. I. (1970) "Possibility of multipolar ordering in the exchange interaction model of ferromagnetism," Phys. Rev. B2, 2706.

---, --- (1970) "Critical properties of the exchange interaction model of ferromagnetism," Sol. State Commun. 8, 459.

---, --- (1972) "Exchange interaction model of ferromagnetism," J. Math. Phys. 13, 725.

--- and LEVY, P. M. "Quadrupole phase transitions in magnetic solids," Phys. Rev. Lett. 27, 1383-1385.

m CHEN, S. H. and POLONSKY, N. (1968) "Observation of anomalous damping and dispersion of hypersounds in a binary liquid mixture near the solution critical point," Ibid. 20, 909-911.

m ---, --- (1969) "Elastic and inelastic scattering of light from a binary liquid mixture near the critical point," J. Phys. Soc. Japan 26, 179-182.

m --- and POLONSKY-OSTROWSKY, N. (1969) "Intensity correlation measurement of light scattered from a two-component fluid near the critical mixing point," Optics Commun. 1, 64-66.

l CHENG, H. and WU, T. T. (1967) "Theory of Toeplitz determinants and the spin correlations of the two-dimensional Ising model. III," Phys. Rev. 164, 719-735.

CHESTER, G. V. (1970) "Speculations on Bose-Einstein condensation and quantum crystals," Ibid. A2, 256-258.

d ---, FISHER, M. F., and MERMIN, N. D. (1969) "Absence of anomalous averages in systems of finite nonzero thickness or cross section," Ibid. 185, 760-762.

d CHINN, S. R., DAVIES, R. W., and ZEIGER, H. J. (1971) "Spin-wave theory of two-magnon Raman scattering in a two-dimensional antiferromagnet," Ibid. B4, 4017-4023.

a CHIPMAN, D. R. and WALKER, C. B. (1971) "Long-range order in β-brass," Phys. Rev. Lett. 26, 233-236.

CHISHOLM, J. S. R. (1966) "Approximation by sequences of Padé approximants in regions of meromorphy," J. Math. Phys. 7, 39-44.

CHISHOLM, R. C. and STOUT, J. W. (1962) "Heat capacity and entropy of $CoCl_2$ and $MnCl_2$ from 11° to 300°K. Thermal anomaly associated with antiferromagnetic ordering in $CoCl_2$," J. Chem. Phys. 36, 972-978.

L CHISTYAKOV, I. G. (1967) "Liquid crystals," Sov. Phys.-Usp. 9, 551-573.

f CHOCK, D. P., RÉSIBOIS, P., DEWEL, G., and DOGONNIER, R. (1971) "Proton dynamics in H-bonded ferroelectrics. I. Kinetic equations for correlation functions above T_c," Physica 53, 364-392.

CHOY, T. F. (1967) "Asymptotic behavior for the particle distribution functions of simple fluids near the critical point. II," J. Chem. Phys. 47, 4296-4319.

--- (1971) "Counting theorem for the generating function of self-avoiding walks," Ibid. 54, 1852-1865.

--- and MAYER, J. E. (1967) "Asymptotic behavior for the particle distribution function of simple fluids near the critical point," Ibid. 46, 110-122.

--- and REE, F. H. (1968) "Inequalities for critical indices near gas-liquid critical point," Ibid. 49, 3977-3987.

m CHU, B. (1963) "Critical opalescence of methanol-cyclohexane, transmission measurements," Ibid. 67, 1969-1972.

m --- (1964) "Critical opalescence of a fluorocarbon-hydrocarbon liquid mixture. Normal perfluoroheptane-normal heptane," J. Am. Chem. Soc. 86, 3557-3561.

m --- (1964) "Critical opalescence of a binary liquid mixture, n-decane-β,β'-dichloroethyl ether. I. Light scattering," J. Chem. Phys. 41, 226-234.

m --- (1965) "X-ray study of critical opalescence of polystyrene in cyclohexane," Ibid. 42, 426-429.

m --- (1965) "Critical opalescence of a binary liquid mixture, n-decane-β, β'-dichloroethyl ether. II. Small-angle X-ray scattering," Ibid. 42, 2293-2294.

m --- (1965) "Critical opalescence of binary liquid mixtures. The Debye molecular interaction range," Ibid. 69, 2329-2332.

r --- (1967) Molecular forces, Wiley Interscience, New York.

m --- (1967) "Correlation effect in the time-dependent concentration fluctuations in the isobutyric acid-water system near its critical mixing point," J. Chem. Phys. 47, 3816-3820.

m --- (1967) "Observation of time-dependent concentration fluctuations in critical mixtures," Phys. Rev. Lett. 18, 200-202.

L ---, BAK, C. S., and LIN, F. L. (1972) "Coherence length in the isotropic phase of a room-temperature nematic liquid crystal," Ibid. 28, 1111-1114.

--- and BEARMAN, R. J. (1968) "Effects of sample configuration on compressible fluids in the earth's field," J. Chem. Phys. 48, 2377-2378.

--- and DUISMAN, J. A. (1967) "Effects of isothermal compressibility and gravity on critical opalescence studies of one-component systems," Ibid. 46, 3267-3268.

m --- and KAO, W. P. (1965) "Critical opalescence of a binary liquid mixture, n-dodecane-β, β'-dichloroethyl ether. III. Very near the critical mixing point," Ibid. 42, 2608-2610.

m ---, --- (1965) "Light scattering of a critical binary liquid mixture: triethylamine-water," Can. J. Chem. 43, 1803-1811.

m ---, PALLESEN, M., and KAO, W. P. (1965) "Critical opalescence of a fluorocarbon-hydrocarbon binary liquid mixture: normal perfluoroheptan-Iso-octane," J. Chem. Phys. 43, 2950-2953.

m --- and SCHOENES, F. J. (1968) "Diffusion coefficient of the isobutyric-acid-water system in the critical region," Phys. Rev. Lett. 21, 6-10.

m ---, ---, and FISHER, M. E. (1969) "Light scattering and pseudospinodal curves: the isobutyric-acid-water system in the critical region," Ibid. 185, 219-226.

m ---, ---, and KAO, W. P. (1968) "Spatial and time-dependent concentration fluctuations of the isobutyric acid-water system in the neighborhood of its critical mixing point," J. Am. Chem. Soc. 90, 3042-3048.

CHYNOWETH, A. G. and SCHNEIDER, W. G. (1952) "Ultrasonic propagation in xenon in the region of its critical temperature," J. Chem. Phys. 20, 1777-1783.

CISZEWSKI, R. and BLINOWSKI, K. (1971) "Spin-pair correlation in iron by neutron diffraction technique," Acta Phys. Polon. A39, 465.

CLARK, R. K. and NEECE, G. A. (1968) "A decorated lattice model for ternary systems," J. Chem. Phys. 48, 2575.

l CLAYTON, D. B. and WOODBURY, G. W., Jr. (1971) "Interface structure of the two-dimensional, square Ising model," J. Chem. Phys. 55, 3895-3901.

h CLOW, J. R. and REPPY, J. D. (1966) "Temperature dependence of the superfluid density in HeII near T_λ," Phys. Rev. Lett. 16, 887-889.

h ---, --- (1967) "Temperature dependence of superfluid critical velocities near T_λ," Ibid. 19, 291-293.

COBB, C. H., JACCARINO, V., REMEIKA, J. P., SILBERGLITT, R., and YASUOKA, H. (1971) "Field dependence of the magnetization and spin-wave correlations in ferromagnetic $CrBr_3$," Phys. Rev. B3, 1677-1688.

f COCHRAN, W. (1960) "Crystal stability and the theory of ferroelectricity," Adv. Phys. 9, 387-423.

f COCHRAN, W. (1969) "Dynamical, scattering and dielectric properties of ferroelectric crystals," Ibid. 18, 157-192.

r COHEN, M. H. (1970) "Review of the theory of amorphous semiconductors," J. Non-Crystalline Solids 4, 391-409.

3 COLLINS, M. F. (1970) "High-temperature wavelength-dependent properties of a Heisenberg paramagnet," Phys. Rev. B2, 4552-4558.

3 --- (1971) "Series expansions for high-temperature dynamics of Heisenberg paramagnets," Ibid. B4, 1588-1593.

---, MINKIEWICZ, V. J., NATHANS, R., PASSELL, L., and SHIRANE, G. (1969) "Critical and spin-wave scattering of neutrons from iron," Ibid. 179, 417-430.

---, NATHANS, R., PASSELL, L., and SHIRANE, G. (1968) "Critical behavior of iron above its Curie temperature," Phys. Rev. Lett. 21, 99-102.

COLVIN, R. V. and ARAJS, S. (1965) "Magnetic susceptibility of face-centered cubic cobalt just above the ferromagnetic Curie temperature," J. Phys. Chem. Solids 26, 435-437.

COMES, R., DENOYER, F., DESCHAMPS, L., and LAMBERT, M. (1971) "Critical anisotropic fluctuations at the 184°K transition of $KMnF_3$ single crystals," Phys. Lett. 34A, 65-67.

u CONIGLIO, A. (1972) "Universality for three dimensional lattice models with long range interactions," Ibid. 38A, 105-107.

u --- (1972) "Scaling parameter and universality," Physica 58, 489-510.

b CONWAY, A. and KOSHLAND, D. E., Jr. (1968) "Negative cooperativity in enzyme action. The binding of diphosphopyridine nucleotide to glyceraldehyde 3-phosphate dehydrogenase," Biochemistry 7, 4011-4022.

b COOK, R. A. and KOSHLAND, D. E., Jr. (1970) "Positive and negative cooperativity in yeast glyceraldehyde 3-phosphate dehydrogenase," Biochemistry 9, 3337-3342.

s COOPER, L. N. and STÖLAN, B. (1971) "Origin of the pairing interaction in the theory of superconductivity," Phys. Rev. B4, 863-864

COOPER, M. J. (1968) "Comments on the scaling-law equation of state," Ibid. 168, 183-184

--- (1971) "Expanded formulation of thermodynamic scaling in the critical region," J. Res. Natl. Bur. Std. (U.S.) 75A, 103-107.

--- and GREEN, M. S. (1968) "Generalized scaling and the critical Eigenvector in ideal Bose condensation," Phys. Rev. 176, 302-309.

---, VICENTINI-MISSONI, M., and JOSEPH, R. I. (1969) "Linear parameterization of the equation of state near the critical point," Phys. Rev. Lett. 23, 70-73.

COOPERSMITH, M. H. (1967) "Crunodes, Acnodes and phase transitions," Phys. Lett. 24A, 700-701

--- (1967) "On the behavior of thermodynamic functions near the critical point of a ferromagnet," Ibid. 25A, 66-68.

--- (1968) "Holomorphy and phase transitions: a scaling law for magnetic systems," Phys. Rev. 167, 478-479.

--- (1968) "Analytic free energy: a basis for scaling laws," Ibid. 172, 230-240.

--- (1968) "Apparent violation of Griffiths' inequality for He^4," Phys. Rev. Lett. 20, 432-434.

--- (1971) "Multiple-valuedness of the free energy in the neighborhood of a critical point," Phys. Lett. 36A, 453-455.

a --- and BROUT, R. (1961) "Threshold concentration for the existence of ferromagnetism in dilute alloys," J. Phys. Chem. Solids 17, 254-258.

---, --- (1963) "Statistical mechanical theory of condensation," Phys. Rev. 130, 2539-2559.

r CORCIOVEI, A., COSTACHE, G., and VAMANU, D. (1972) "Ferromagnetic thin films," Solid State Physics 27, 237-350.

CORLISS, L. M., DELAPALME, A., HASTINGS, J. M., LAU, H. Y., and NATHANS, R. (1969) "Critical magnetic scattering from $RbMnF_3$," J. Appl. Phys. 40, 1278.

f CORROCHANO, F., JIMÉNEZ-DÍAZ, B., and MAURER, E. (1970) "Effect of several parameters on the ferroelectric-paraelectric transition in triglycine sulfate," Electronica Fis. Aplic. 13, 83-87.

b COSTA, S. M., FROINES, J. R., HARRIS, J. M., LEBLANC, R. M., ORGER, B. H., and PORTER, G. (1972) "Model systems for photosynthesis," Proc. Roy. Soc. 326, 503-519.

COTTAM, M. G. (1971) "Theory of dipole-dipole interactions in ferromagnets: I. The thermodynamic properties," J. Phys. C. 4, 2658-2672.

--- (1971) "Theory of dipole-dipole interactions in ferromagnets: II. The correlation functions," Ibid. 4, 2673-2683.

--- and STINCHCOMBE, R. B. (1970) "Correlation functions of a Heisenberg antiferromagnet," Ibid. 3 2305.

d COTTERILL, R. M. J. and PEDERSEN, L. B. (1972) "A molecular dynamics study of the melting of a two-dimensional crystal," Sol. State Commun. 10, 439-442.

COX, D. E., SHIRANE, G., BIRGENEAU, R. J., and MAC CHESNEY, J. B. (1969) "Neutron-diffraction study of magnetic ordering in Ca_2MnO_4," Phys. Rev. 188, 930-932.

CRACKNELL, A. P. (1971) "The application of Landau's theory of continuous phase transitions to magnetic phase transitions," J. Phys. C. 4, 2488-2500.

CRACKNELL, M. F. and SEMMENS, A. G. (1971) "On the temperature dependence of the sound attenuation maximum as a function near the Néel point in MnF_2," Ibid. 4, 1513-1518.

f CRAIG, P. P. (1966) "Critical phenomena in ferroelectrics," Phys. Lett. 20, 140-142.

r --- and GOLDBURG, W. I. (1969) "Transport properties near magnetic critical points," J. Appl. Phys. 40, 964-971.

---, ---, KITCHENS, T. A., and BUDNICK, J. I. (1967) "Transport properties at critical points: the resistivity of nickel," Phys. Rev. Lett. 19, 1334-1337.

CROTHERS, D. M. (1968) "Calculation of melting curves for DNA," Biopolymers 6, 1391-1404.

b --- and KALLENBACH, N. R. (1966) "On the helix-coil transition in heterogeneous polymers," J. Chem. Phys. 45, 917-927.

b ---, ---, and ZIMM, B. H. (1965) "The melting transition of low-molecular-weight DNA: theory and experiment," J. Mol. Biol. 11, 802-820.

d CROW, J. E., THOMPSON, R. S., KLENIN, M. A., and BHATNAGAR, A. K. (1970) "Divergent fluctuations in superconducting films," Phys. Rev. Lett. 24, 371-375.

CULLEN, J. R. and CALLEN, E. (1970) "Collective electron theory of the metal-semiconductor transition in magnetite," J. Appl. Phys. 41, 879-880.

---, --- (1971) "Band theory of multiple ordering and the metal-semiconductor transition in magnetite," Phys. Rev. Lett. 26, 236-238.

---, --- (1971) "Multiple ordering and first order transitions: magnetite," Sol. State Commun. 9, 1041-1043.

---, CHAUDHURY, B., and CALLEN, E. (1972) "A calculation of the conductivity of magnetite above the Verwey temperature," Phys. Lett. 38A, 113-114.

CUMMINS, H. Z. and GAMMON, R. W. (1966) "Rayleigh and Brillouin scattering in liquids: the Landau-Placzek ratio," J. Chem. Phys. 44, 2785-2796.

--- and SWINNEY, H. L. (1966) "Critical opalescence: the Rayleigh linewidth," Ibid. 45, 4438-4444.

---, --- (1970) "Dispersion of the velocity of sound in xenon in the critical region," Phys. Rev. Lett. 25, 1165-1169.

r ---, --- (1970) "Light beating spectroscopy," Progress in Optics 8, 135-203.

r CUSACK, N. E. (1967) "The physics of liquid metals," Contemp. Phys. 8, 583-606.

CYROT, P. (1970) "Phase transition in Hubbard model," Phys. Rev. Lett. 25, 871-874.

CZAJA, W. (1970) "An apparent Landau-type second order transition in the spinel $CdIn_2S_4$," Kondens. Mat. 10, 299-316.

DABOUL, J. (1971) "Scaling symmetry in thermodynamics or generalization of the law of corresponding states," Phys. Rev. A4, 1583.

D'ABRAMA, G., RICCI, F. P., and MENZINGER, F. (1972) "Shape of the coexistence curve of the Ga-Hg system near T_c," Phys. Rev. Lett. 28, 22-24.

DAHL, D. and MOLDOVER, M. R. (1971) "Metastable thermodynamic states near the critical point of He^3," Ibid. 27, 1421-1424.

DALTON, N. W. (1966) "Second interactions in the Ising and Heisenberg models of ferromagnetism," Proc. Phys. Soc. 88, 659-678.

--- (1966) "Distant-neighbour interactions in ferromagnetic systems (BPW approximation)," Ibid. 89, 845-857.

--- and DOMB, C. (1966) "Crystal statistics with long-range forces II. Asymptotic behaviour of the equivalent neighbour model," Ibid. 89, 873-891.

---, ---, and SYKES, M. F. (1964) "Dependence of critical concentration of a dilute ferromagnet on the range of interaction," Ibid. 83, 496-498.

--- and RIMMER, D. E. (1969) "High temperature series expansion for the general spin anisotropic Heisenberg model," Phys. Lett. 29A, 611-612.

--- and WOOD, D. W. (1965) "Critical properties of the Heisenberg ferromagnet with higher neighbor interactions (S=1/2)," Phys. Rev. A138, 779-792.

---, --- (1967) "Critical behaviour of the simple anisotropic Heisenberg model," Proc. Phys. Soc. 90, 459-474.

---, --- (1968) "Padé approximant results for the spherical and classical Heisenberg models," Phys. Lett. 28A, 417-418.

---, --- (1969) "Critical point behavior of the Ising model with higher-neighbor interactions present," J. Math. Phys. 7, 1271-1302.

DANIELIAN, A. and STEVENS, K. W. H. (1957) "High temperature susceptibilities," Proc. Phys. Soc. B70, 326-328.

m D'ARRIGO, G., MISTURA, L., and TARTAGLIA, P. (1971) "Sound absorption in critical mixtures," Phys. Rev. A3, 1718-1721.

--- and SETTE, D. (1968) "Ultrasonic absorption and velocity near the critical region of nitrobenzene-η-hexane mixtures," J. Chem. Phys. 48, 691-698.

DASH, J. G., DUNLAP, B. D., and HOWARD, D. G. (1966) "Internal field of Fe^{57} in nickel from 77°K to the Curie point," Phys. Rev. 141, 376-378.

DAVIS, B. W. and RICE, O. K. (1967) "Thermodynamics of the critical point: liquid-vapor systems," J. Chem . Phys. 47, 5043-5051.

DAVIS, H. L. and NARATH, A. (1964) "Spin-wave renormalization applied to ferromagnetic $CrBr_3$," Phys. Rev. 134, 433-441.

L DAVISON, L. (1969) "Linear theory of heat conduction and dissipation in liquid crystals of the nematic type," Ibid. 180, 232-237.

DE BLOIS, R. W. and RODBELL, D. S. (1963) "Magnetic first-order phase transition in single-crystal MnAs," Phys. Rev. 130, 1347-1360.

DE BOER, J. (1948) "Quantum theory of condensed permanent gases. I. The law of corresponding states," Physica 14, 139-148.

--- (1949) "Molecular distribution and equation of state of gases," Rept. Progr. Phys. 12, 305-374.

--- and BLAISSE, B. S. (1948) "Quantum theory of condensed permanent gases. II. The solid state and the melting line," Physica 14, 149-164.

b DE GENNES, P. G. (1968) "Critical opalescence of macromolecular solutions," Phys. Lett. A26, 313-314.

b --- (1969) "Nucleic acid denaturation," Comments on Solid State Physics 2, 149-154.

L --- (1969) "Liquid crystals," Ibid. 2, 213-217.

b --- (1969) "Some conformation problems for long macromolecules," Rept. Progr. Phys. 32, 187-205.

L --- (1971) "Motion of walls in a nematic under rotating fields," J. de Physique 32, 789-792.

g --- (1972) "Exponents for the excluded volume problem as derived by the Wilson method," Phys. Lett. 38A, 339-340.

DE GIORGIO, V. and SCULLY, M. O. (1970) "Analogy between the laser threshold region and a second-order phase transition," Phys. Rev. A2, 1170-1177.

d DE JONGH, L. J. (1972) "Critical behaviour of the perpendicular susceptibility of antiferromagnetic $(C_2H_5NH_3)_2\,CuCl_4$," Sol. State Commun. 10, 537-541.

d ---, BLOEMBERGEN, P., and COLPA, J. H. P. (1972) "Transition temperatures of weakly anisotropic layer-type magnets. Dependence on anisotropy," Physica 58, 305-314.

d ---, MIEDEMA, A. R., and WIELINGA, R. F. (1970) "Measurement of the magnetic susceptibility of $Cu(NH_4)_2Br_4 \cdot 2H_2O$ near its Curie temperature," Ibid. 46, 44-58.

d --- and VAN AMSTEL, W. D. (1971) "Two-dimensional Heisenberg ferromagnet: susceptibility," J. de Physique 32, 880-881.

d ---, ---, and MIEDEMA, A. R. (1972) "Magnetic measurements on $(C_2H_5NH_3)_2\ CuCl_4$ ferromagnetic layers coupled by a very weak antiferromagnetic interaction," Physica 58, 277-304.

DE SOBRINO, L. (1967) "On the kinetic theory of a van der Waals gas," Can. J. Phys. 45, 363-385.

--- (1968) "On the transport properties of a van der Waals gas near the critical point," Ibid. 46, 2821-2841.

d DE VRIES, G., BREED, D. J., MAARSCHALL, E. P., and MIEDEMA, A. R. (1968) "Experiments on two-dimensional Heisenberg systems," J. Appl. Phys. 39, 1207-1208.

d DE WIJN, H. W., WALSTEDT, R. E., WALKER, L. R., and GUGGENHEIM, H. J. (1970) "Sublattice magnetization of quadratic layer antiferromagnets," Phys. Rev. Lett. 24, 832-835.

b DEAL, W., ERLANGER, B., and NACHMANSOHN, D. (1969) "Photoregulation of biological activity by photochromic reagents III. Photoregulation of bioelectricity by acetylcholine receptor inhibitors," Proc. Natl. Acad. Sci. U.S. 64, 1230-1234.

m DEBYE, P. (1959) "Angular dissymmetry of the critical opalescence in liquid mixtures," J. Chem. Phys. 31, 680-687.

--- and KLEBOTH, K. (1965) "Electrical field effect on the critical opalescence," Ibid. 42, 3155-3162.

l DEKEYSER, R. and HALZEN, F. (1969) "Influence of the pair correlations on the phase transition in an Ising lattice," Phys. Rev. 181, 949-953.

l --- and ROGIERS, J. (1972) "Higher-order correlations in the two-dimensional Ising model," Physica 59 23-28.

3 DE LEENER, M. and RÉSIBOIS, P. (1969) "Irreversibility in Heisenberg spin systems. IV. Time-dependent fluctuations at the Curie point," Phys. Rev. 178, 819-828.

m DEUTCH, J. M. and ZWANZIG, R. (1967) "Anomalous specific heat and viscosity of binary van der Waals mixtures," J. Chem. Phys. 46, 1612-1620.

DEVELEY, G. (1965) "Variation de la susceptibilité magnétique du fer, du gadolinium et du nickel au voisinage du point de Curie," Compt. Rend. 260, 4951-4953.

g DI CASTRO, C. (1971) "Relevance and limitations of scaling laws in the theory of phase transitions: field-theoretic approach," Riv. Nuovo Cimento 1, 199-226.

g --- and JONA-LASINIO, G. (1969) "On the microscopic foundation of scaling laws," Phys. Lett. 29A, 322-323.

---, FERRO-LUZZI, F., and TYSON, J. A. (1969) "Dynamical scaling laws and time dependent Landau-Ginzburg equation," Ibid. 29A, 458-459.

a DIETRICH, O. W. and ALS-NIELSEN, J. (1967) "Temperature dependence of short-range order in β-brass," Phys. Rev. 153, 711-717.

d DIETZ, R. E., MERRITT, F. R., DINGLE, R., HONE, D., SILBERNAGEL, B. G., and RICHARDS, P. M. (1971) "Exchange narrowing in one-dimensional systems," Phys. Rev. Lett. 26, 1186-1188.

DILLON, J. F., Jr., KAMIMURA, H., and REMEIKA, J. P. (1962) "Magnetic rotation of visible light by ferromagnetic $CrBr_3$," Ibid. 9, 161-163.

---, ---, --- (1963) "Magneto-optical studies of chromium tribromide," J. Appl. Phys. 34, 1240-1245.

2 DITZIAN, R. V. (1972) "High-temperature series for the susceptibility of the Ashkin-Teller model," J. Phys. C. 5, L250-L251.

f --- (1972) "Estimates for the critical index of the susceptibility for Baxter's model as a function of a parameter," Phys. Lett. 38A, 451-453.

--- (1972) "Universality and a 3d Ising model whose critical exponent γ varies continuously with a parameter," Phys. Lett. 42A, 67-78.

2 --- and BETTS, D. D. (1970) "Estimates for the critical gap index for the three dimensional spin 1/2 XY model," Phys. Lett. 32A, 152-153.

2 ---, --- (1972) "Dynamical properties of the three-dimensional XY model," Can. J. Phys. 50, 129-138.

L DOANE, J. W. and VISINTAINER, J. J. (1969) "Proton spin-lattice relaxation in liquid crystals," Phys. Rev. Lett. 23, 1421-1423.

DOBBS, B. C. and SCHMIDT, P. W. (1972) "Gravitational effects in X-ray studies of critical opalescence," J. Chem. Phys. 56, 2421-2426.

1 DOBSON, J. F. (1969) "Many-neighbored Ising chain," J. Math. Phys. 10, 40-45.

1 DOMB, C. (1949) "Statistical mechanics of some cooperative phenomena," Nature 163, 775-780.

1 --- (1949) "Order-disorder statistics I.," Proc. Roy. Soc. 196, 36.

1 --- (1949) "Order-disorder statistics II. A two-dimensional model," Ibid. 199, 199-221.

--- (1951) "The melting curve at high pressures," Phil. Mag. 42, 1316.

1 --- (1952) "L'influence de la structure du réseau sur l'anomalie de la chaleur spécifique de modèle d'Ising," Changement des Phases 193-196.

--- (1954) "On multiple returns in the random-walk problem," Proc. Camb. Phil. Soc. 50, 586-591.

--- (1955) "Statistical physics and its problems," Science Progress 43, 402-417.

--- (1957) "Specific heat of compressible lattices and the theory of melting," J. Chem. Phys. 25, 783.

3 --- (1956) "On high temperature expansions in the Heisenberg theory of ferromagnetism," Proc. Phys. Soc. B. 49, 486.

--- (1957) "Melting curves of helium and hydrogen isotopes," Ibid. 70, 1950.

--- (1958) "Some theoretical aspects of melting," Nuovo Cimento Suppl. 9, 9.

--- (1959) "Fluctuation phenomena and stochastic processes," Nature 184, 509.

r --- (1960) "On the theory of cooperative phenomena in crystals," Adv. Phys. 9, 149-361.

--- (1963) "Excluded-volume effect for two- and three-dimensional lattice models," J. Chem. Phys. 38, 2957-2963.

--- (1964) "Some statistical problems connected with crystal lattices," J. Royal Statistical Soc. B. 26, 367-397.

r --- (1965) "Statistical mechanics of critical behavior in magnetic systems," In Magnetism, G. Rado and H. Suhl, Eds. (Academic Press, N.Y.), Vol. IIA, p. 1.

d --- (1965) "On co-operative effects in finite assemblies," Proc. Phys. Soc. 86, 933-938.

r --- (1966) "Some recent developments in the theory of magnetic transitions," Ann. Acad. Sci. Fenn. AVI, [Proc. 1966 Low Temperature Calorimetry Conf.] 210, 167-179.

--- (1966) "First-order and λ-point transitions in ordering assemblies," Proc. Phys. Soc. 88, 260-262.

--- (1967) "Specific heat of fluids in the critical region," Faraday Soc. Discussions 43, 85.

--- (1968) "Configurational studies of the Ising and classical Heisenberg models," J. Appl. Phys. 39, 614-615.

--- (1968) "Equation of state of a ferromagnet in the critical region," Ibid. 39, 620-622.

--- (1968) "Multiple-correlation functions and scaling laws," Phys. Rev. Lett. 20, 1425-1428.

r --- (1969) "Self-avoiding walks on lattices," Adv. Chem. Phys. 15, 229.

--- (1969) "Pair correlation fucntion in the critical region," J. Phys. C. 2, 2434-2435.

r --- (1970) "Series expansions for ferromagnetic models," Adv. Phys. 19, 339-370.

--- (1970) "Self-avoiding walks and the Ising and Heisenberg models," J. Phys. C. 3, 256-284.

--- (1970) "Spin-pair correlation function and susceptibility of the planar Ising model," Ibid. 3, L85-L87.

--- (1971) "Cluster expansion for the dilute Ising ferromagnet," Ibid. 4, L325-L328.

--- (1971) "The Curie point," In Statistical Mechanics at the Turn of the Decade, E. G. D. Cohen, Ed. (Marcel Dekker, N.Y.), p. 81-128.

--- (1972) "A note on the series expansion method for clustering problems," Biometrika 59, 209.

--- (1972) "Term structure of series expansions for the Ising and classical vector models and dilute magnetism," J. Phys. C. 5, 1399-1417.

--- (1972) "Weighting of graphs for the Ising and classical vector models," Ibid. 5, 1417-1428.

--- (1972) "Definition of the critical region--correction terms in the scaling form of the equation of state," In Proc. Varenna Summer School, M.S. Green, Ed. (Academic Press, N.Y.).

3 --- and BOWERS, R. G. (1969) "Specific heat of europium sulphide: A comparison between theory and experiment," J. Phys. C. 2, 755-758.

--- and DALTON, N. W. (1966) "Crystal statistics with long-range forces I. The equivalent neighbour model," Proc. Phys. Soc. 89, 859-871.

--- and FISHER, M. E. (1958) "On random walks with restricted reversals," Proc. Camb. Phil. Soc. 54, 48-59.

b ---, GILLIS, J., and WILMERS, G. (1965) "On the shape and configuration of polymer molecules," Proc. Phys. Soc. 85, 625-645.

r --- and GREEN, M. S., Eds. (1972) Phase Transitions and Critical Points, Academic Press, London.

l --- and GUTTMANN, A. (1970) "Low-temperature series for the Ising model," J. Phys. C. 3, 1652-1660.

--- and HEAP, B. R. (1967) "The classification and enumeration of multiply connected graphs," Proc. Phys. Soc. 90, 985-1001.

--- and HILEY, B. J. (1962) "On the method of Yvon in crystal statistics," Ibid. A268, 506-526.

--- and HIOE, F. T. (1969) "Mean-square intrachain distances in a self-avoiding walk," J. Chem. Phys. 51, 1915-1919.

---, --- (1969) "Correlations in a self-avoiding walk," Ibid. 51, 1920-1928.

---, --- (1970) "The transition matrix approach to self-avoiding walks," J. Phys. C. 3, 2223-2232.

--- and HUNTER, D. L. (1965) "On the critical behaviour of ferromagnets," Proc. Phys. Soc. 86, 1147-1151.

b --- and JOYCE, G. S. (1972) "Cluster expansion for a polymer chain," J. Phys. C. 5, 956-977.

r --- and MIEDEMA, A. R. (1964) "Magnetic transitions," Prog. Low Temp. 4, 296-343.

--- and POTTS, R. B. (1951) "Order-disorder statistics. IV. A two-dimensional model with first and second interactions," Proc. Roy. Soc. A210, 125-141.

3 --- and SYKES, M. F. (1956) "On high temperature expansions in the Heisenberg theory of ferromagnetism," Pooc. Phys. Soc. B69, 486.

---, --- (1956) "On metastable approximations in co-operative assemblies," Proc. Roy. Soc. A235, 247-259.

---, --- (1957) "The calculation of lattice constants in crystal statistics," Phil. Mag. 2, 733-749.

---, --- (1957) "Specific heat of a ferromagnetic substance above the Curie point," Phys. Rev. 108, 1415-1416.

---, --- (1957) "On the susceptibility of a ferromagnetic above the Curie point," Proc. Roy. Soc. A240, 214-228.

---, --- (1957) "High temperature susceptibility expansions for the close packed hexagonal lattice," Proc. Phys. Soc. B. 70, 896-897.

l ---, --- (1961) "Use of series expansions for the Ising model susceptibility and excluded volume problem," J. Math. Phys. 2, 63-67.

---, --- (1961) "Cluster size in random mixtures and percolation processes," Phys. Rev. 122, 77-78.

---, --- (1962) "Effect of change of spin on the critical properties of the Ising and Heisenberg models," Ibid. 128, 168-173.

3 --- and WOOD, D. W. (1964) "New method of deriving high temperature expansions for the Heisenberg model," Phys. Lett. 8, 20-21.

3 ---, --- (1965) "On high temperature expansions for the Heisenberg model," Proc. Phys. Soc. 86, 1-16.

--- and WYLES, J. A. (1969) "Anomalous specific heat near the Curie temperature," J. Phys. C. 2, 2436-2437.

f DRAEGERT, D. A. and SINGH, S. (1971) "Dielectric susceptibility and the order of ferroelectric phase transitions," Sol. State Commun. 9, 595-597.

d DRUMHELLER, J. E., DICKEY, D. H., REKLIS, R. P., ZASPEL, C. E., and GLASS, S. J. (1972) "Exchange-energy constants in some S=1/2 two-dimensional Heisenberg ferromagnets," Phys. Rev. B5, 4631-4636.

b DUBIN, S. G., BENEDEK, G. B., BANCROFT, F. C., and FREIFELDER, D. (1970) "Molecular weights of coliphages and coliphage DNA II. Measurement of diffusion coefficients using optical mixing spectroscopy, and measurement of sedimentation coefficients," J. Mol. Biol. 54, 547-556.

b ---, CLARK, N. A., and BENEDEK, G. B. (1971) "Measurement of the rotational diffusion coefficient of lysozyme by depolarized light scattering: configuration of lysozyme in solution," J. Chem. Phys. 54, 5158-5164.

m DUBOIS, M. and BERGE, P. (1971) "Measurement of correlation lengths in the perfluoroheptane iso-octane critical mixture," Phys. Lett. 37A, 155-157.

m ---, --- (1971) "Experimental study of Rayleigh scattering related to concentration fluctuations in binary solutions; evidence of a departure from ideality," Phys. Rev. Lett. 26, 121-124.

---, ---, and LAJ, C. (1970) "Measurement of the diffusion coefficient of small molecules by means of quasi-elastic scattering of light," Chem. Phys. Lett. 6, 227-230.

DUFFY, W., JR., and BARR, K. P. (1968) "Theory of alternating antiferromagnetic Heisenberg linear chains," Phys. Rev. 165, 647-654.

m DURCZEWSKI, K. (1971) "Field-dependent phase transitions of uniaxial ferromagnets," Acta Phys. Polon. A40, 505.

DYKHNE, A. M. and KRIVOGLAZ, M. A. (1970) "Fluctuations near critical points and near second-order phase transitions," Sov. Phys.-Solid State 12, 1349.

DYSON, F. J. (1969) "Existence of a phase-transition in a one-dimensional Ising ferromagnet," Commun. Math. Phys. 12, 91-107.

--- (1969) "Non-existence of spontaneous magnetization in a one-dimensional ferromagnet," Ibid. 12, 212-215.

--- (1970) "Correlations between eigenvalues of a random matrix," Ibid. 19, 235.

--- (1971) "An Ising ferromagnet with discontinuous long-range order," Ibid. 21, 269-283.

--- (1971) "Phase transitions in ferromagnets," In Statistical Mechanics at the Turn of the Decade, E. G. D. Cohen, Ed. (Marcel Dekker, N.Y.), p. 129-144.

r --- (1972) "Existence and nature of phase transitions in one-dimensional Ising ferromagnets," A.M.S. Symposium talk (to be published in A.M.S. symposium volume).

f --- and LENARD, A. (1967) "Stability of matter. I," J. Math. Phys. 8, 423-434.

b EDELSTEIN, A. S. (1964) "Linear Ising models and the antiferromagnetic behavior of certain crystalline organic free radicals," J. Chem. Phys. 40, 488-495.

l --- (1965) "Superparamagnetism in the low-temperature limit of Ising models," Ibid. 42, 2879-2884.

--- (1966) "Ordering in linear antiferromagnetic chains with anisotropic coupling," Phys. Rev. 142, 259-263.

EDEN, D., GARLAND, C. W., and THOEN, J. (1972) "Sound absorption and dispersion along the critical isochore in xenon," Phys. Rev. Lett. 28, 726-729.

EDWARDS, C., LIPA, J. A., and BUCKINGHAM, M. J. (1968) "Specific heat of xenon near the critical point," Ibid. 20, 496-499.

h EDWARDS, M. H. (1956) "The index of refraction of liquid helium," Can. J. Phys. 34, 898-900.

h --- (1957) "Refractive index of He^4: saturated vapor," Phys. Rev. 108, 1243-1245.

h --- (1965) "Nonanalytic form of the coexistence curve of helium at the critical point," Phys. Rev. Lett. 15, 348-351.

h --- (1967) "The coexistence curve of ^4He above 0.98 T_c," Proc. LT20 H93, 522-526.

h --- and WOODBURY, W. C. (1963) "Saturated He^4 near its critical temperature," Phys. Rev. 129, 1911-1918.

EDWARDS, S. F. (1959) "The statistical thermodynamics of a gas with long and short-range forces," Phil. Mag. 4, 1171-1182.

b --- (1965) "The statistical mechanics of polymers with excluded volume," Proc. Phys. Soc. 85, 613-624.

b --- (1967) "The statistical mechanics of polymerized material," Ibid. 92, 9-16.

r EGELSTAFF, P. A., (1969) Introduction to the liquid state, Academic Press, London.

r --- and RING, J. W. (1968) "Experimental data in the critical region," In Phys. of Simple Liquids, p. 253-297.

m --- and WIDOM, B. (1970) "Liquid surface tension near the triple point," J. Chem. Phys. 53, 2667-2669.

d EHRENFREUND, E., RYBACZEWSKI, E. F., GARITO, A. F., and HEEGER, A. J. (1972) "Enhanced nuclear relaxation in a one-dimensional metal near the Mott transition," Phys. Rev. Lett. 28, 873-877.

EIBSCHÜTZ, M., SHTRIKMAN, S., and TREVES, D. (1966) "Internal field in orthoferrities and the one third power law," Sol. State Commun. 4, 141-145.

EINSTEIN, A. (1910) "Theorie der Opaleszenz von homogenen Flüssigkeiten und Flüssigkeitsgemischen in der Nähe des kritischen Zustandes," Ann. Physik. 33, 1275-1298.

EISENMAN, G., SANDBLOM, J. P., and WALKER, J. L., Jr. (1967) "Membrane structure and ion permeation," Science 155, 965-974.

EISENSTEIN, A. and GINGRICH, N. S. (1942) "The diffraction of X-rays by argon in the liquid, vapor, and critical regions," Phys. Rev. 62, 261-270.

ELLIOTT, R. J., HEAP, B. R., MORGAN, D. J., and RUSHBROOKE, G. S. (1960) "Equivalence of the critical concentrations in the Ising and Heisenberg models of ferromagnetism," Phys. Rev. Lett. 5, 366-367.

r --- and MARSHALL, W. (1958) "Theory of critical scattering," Rev. Mod. Phys. 30, 75-89.

l ---, PFEUTY, P., and WOOD, C. (1970) "Ising model with a transverse field," Phys. Rev. Lett. 25, 443-446.

l --- and WOOD, C. (1971) "The Ising model with a transverse field - I. High temperature expansion," J. Phys. C. 4, 2359-2369.

b ELLIS, E. L. and DELBRUCK, M. (1939) "The growth of bacteriophage," J. Gen. Phys. 22, 365-384.

ELVEY, J. S. and PENROSE, O. (1968) "The Yang-Lee distribution of zeros for a classical one-dimensional system," Phys. Rev. 126, A456-457.

ENGLERT, F. (1963) "Linked cluster expansions in the statistical theory of ferromagnetism," Ibid. 129, 567-577.

--- and BROUT, R. (1960) "Dielectric formulation of quantum statistics of interacting particles," Ibid. 120, 1085-1092.

u ENTING, I. G. and OITMAA, J. (1972) "On the validity of scaling theory for the anisotropic Ising model," Phys. Lett. 38A, 107-108.

2 EPSTEIN, A., GUREWITZ, E., MAKOVSKY, J., and SHAKED, H. (1970) "Magnetic structure and two-dimensional behavior of Rb_2MnCl_4 and Cs_2MnCl_4," Phys. Rev. B2, 3703-3706.

h ERBEN, K. D. and POBELL, F. (1968) "λ-point shift in turbulently flowing ^4He," Phys. Lett. 26A, 368-369.

ERBER, T., GURALNICK, S. A., and LATAL, H. G. (1972) "A general phenomenology of hysteresis," Ann. Phys. 69, 161-192.

ERDÖS, P. and RHODES, E. (1968) "Magnetically coupled impurities in a linear chain Heisenberg ferromagnet," Helv. Phys. Acta 41, 785-792.

l ESIPOV, V. S. (1969) "The Ising model containing an impurity," Sov. Phys.-Solid State 11, 1256-1261.

--- and MIKULINSKII, M. A. (1970) "Influence of sample inhomogeneity on the behavior of the susceptibility of a system near the critical point," Sov. Phys.-JETP 31, 1175-1178.

ESSAM, J. W. (1970) "Linked-cluster expansions for the correlation functions of lattice systems," Proc. Camb. Phil. Soc. 67, 523-534.

l --- and FISHER, M. E. (1963) "Padé approximant studies of the lattice gas and Ising ferromagnet below the critical point," J. Chem. Phys. 38, 802-812.

---, --- (1970) "Some basic definitions in graph theory," Rev. Mod. Phys. 42, 271-288.

--- and GARELICK, H. (1967) "Critical behaviour of a soluble model of dilute ferromagnetism," Proc. Phys. Soc. 92, 136-149.

l ---, --- (1969) "Time-dependent susceptibility and neutron scattering cross section of the Ising model with general spin," J. Phys. C. 2, 317-328.

--- and GWILYM, K. M. (1971) "The scaling laws for percolation processes," J. Phys. C. 4, L228-L232.

l --- and HUNTER, D. L. (1968) "Critical behaviour of the Ising model above and below the critical temperature," J. Phys. C. 1, 392-407.

--- and SYKES, M. F. (1963) "The crystal statistics of the diamond lattice," Physica 29, 378-388.

---, --- (1966) "Percolation processes. I. Low-density expansion for the mean number of clusters in a random mixture," J. Math. Phys. 7, 1573-1581.

b ESSIG, A., KEDEM, O., and HILL, T. L. (1966) "Net flow and tracer flow in lattice and carrier models," J. Theoret. Biol. 13, 72-89.

EZAWA, H. and LUBAN, M. (1967) "Criterion for Bose-Einstein condensation and representation of canonical commutation relations," J. Math. Phys. 8, 1285-1311.

FALICOV, L. M., GONCALVES DA SILVA, C. E. T., and HUBERMAN, B. A. (1972) "Metal-insulator and magnetic phase transitions: a thermodynamic model," Sol. State Commun. 10, 455-458.

FALK, H. (1964) "Variational treatment of the Heisenberg antiferromagnet," Phys. Rev. A133, 1382-1389.

3 --- (1964) "Inequalities relating the nearest-neighbor spin correlation and the magnetization for the Heisenberg Hamiltonian," J. Math. Phys. 5, 1478-1480.

--- (1966) "Upper and lower bounds for canonical ensemble averages," Ibid. 7, 977-979.

1 --- (1966) "Ising chain with a spin impurity," Phys. Rev. 151, 304-311.

--- (1968) "Lower bound for the isothermal magnetic susceptibility," Ibid. 165, 602-605.

--- (1970) "Inequalities of J. W. Gibbs," Am. J. Phys. 38, 858-869.

--- (1972) "Entropy decomposition and transfer-matrix problems," J. Math. Phys. 13, 608.

--- and BRUCH, L. W. (1969) "Susceptibility and fluctuation," Phys. Rev. 180, 442-444.

--- and FUIJGROK, TH. W. (1965) "Variational approaches to the antiferromagnetic linear chain," Ibid. A139, 1203-1208.

--- and SUZUKI, M. (1970) "Spin correlation and entropy," Ibid. B1, 3051-3057.

FAN, C. (1972) "On critical properties of the Ashkin-Teller model," Phys. Lett. 39A, 136-137.

L ---, KRAMER, L., and STEPHEN, M. J. (1970) "Fluctuations and light scattering in cholesteric liquid crystals," Phys. Rev. A2, 2482-2489.

1 --- and MC COY, B. M. (1969) "One-dimensional Ising model with random exchange energy," Ibid. 182, 614-623.

1 --- and WU, F. Y. (1969) "Ising model with second-neighbor interaction. I. Some exact results and an approximate solution," Ibid. 179, 560-570.

---, --- (1970) "General lattice model of phase transitions," Ibid. B2, 723-733.

FARRELL, R. A. and MEIJER, P. H. E. (1965) "First order transitions in simple magnetic systems," Physica 31, 725-748.

1 ---, --- (1969) "High temperature specific heat of the Ising model," Phys. Rev. 180, 579-581.

1 ---, MORITA, T., and MEIJER, P. H. E. (1966) "Cluster expansion for the Ising model," J. Chem. Phys. 45, 349-363.

FAY, D. and LAYZER, A. (1968) "Superfluidity of low-density fermion systems," Phys. Rev. Lett. 20, 187-190.

f FEDER, J. (1971) "On the critical behaviour at second order structural phase transitions," Sol. State Commun. 9, 2021-2024.

f --- and PYTTE, E. (1970) "Theory of a structural phase transition in Perovskite-type crystals. II. Interaction with elastic strain," Phys. Rev. B1, 4803-4810.

1 FEDJANIN, V. K. (1969) "Application of the Bogoliubov inequality in the Ising model," Phys. Lett. 29A, 40-41.

1 --- (1971) "An inequality for the thermal heat capacity at constant magnetization in the Ising model," Ibid. 34A, 323-325.

h FEENBERG, E. (1970) "Microscopic quantum theory of the helium liquids," Am. J. Phys. 38, 684-704.

FEKE, G. T., FRITSCH, K., and CAROME, E. F. (1969) "Low frequency sound velocity in CO_2 near the critical point," Phys. Rev. Lett. 23, 1282-1286.

---, HAWKINS, G. A., LASTOVKA, J. B., and BENEDEK, G. B. (1971) "Spectrum and intensity of light scattered from sulfur hexafluoride along the ciritcal isochore," Ibid. 27, 1780-1783.

FELDERHOF, B. U. (1966) "Onsager relations and the spectrum of critical opalescence," J. Chem. Phys. 44, 602-609.

--- (1970) "Phase transitions in one-dimensional cluster-interaction fludis III. Correlation functions," Ann. Phys. 58, 281-300.

--- (1970) "Surface effects in one-dimensional classical fluids with nearest-neighbor interactions," Phys. Rev. A1, 1185-1195.

m --- (1972) "Time-dependent statistics of binary linear lattices," J. Stat. Phys. 6, 21.

--- and FISHER, M. E. (1970) "Phase transitions in one-dimensional cluster-interaction fluids II. Simple logarithmic model," Ann. Phys. 58, 268-280.

FERDINAND, A. E. (1967) "Statistical mechanics of dimers on a quadratic lattice," J. Math. Phys. 8, 2332-2339.

1 --- and FISHER, M. E. (1969) "Bounded and inhomogeneous Ising models. I. Specific-heat anomaly of a finite lattice," Phys. Rev. 185. 832-846.

3 FERER, M. (1971) "Spin-spin correlations of the classical Heisenberg ferromagnet on the fcc lattice for temperatures above T_c," Ibid. B4, 3964-3970.

1 ---, MOORE, M. A., and WORTIS, M. (1969) "Scaling form of the spin-spin correlation function of the three-dimensional Ising ferromagnet above the Curie temperature," Phys. Rev. Lett. 22, 1382-1385.

1 ---, ---, --- (1971) "Universality of critical correlations in the three-dimensional Ising ferromagnet," Phys. Rev. B3, 3911-3914.

3 ---, ---, --- (1971) "Some critical properties of the nearest-neighbor, classical Heisenberg model for the fcc lattice in finite field for temperatures greater than T_c," Ibid. B4, 3954-3963.

d FERNANDEZ, J. F. (1969) "Lack of crystalline order in two dimensions," Phys. Lett. 30A, 40-41.

d --- (1971) "Bose-Einstein condensation in two dimensions," Phys. Rev. A3, 1104-1108.

3 --- and GERSCH, H. A. (1968) "Divergences in the space-time correlation functions for the Heisenberg magnet in one dimension," Ibid. 172, 341-344.

FERRELL, R. A. (1968) "Spectrum of light doubly scattered by an opalescent fluid," Ibid. 169, 199-200.

--- (1969) "Fluctuations and the superconducting phase transition: critical specific heat and paraconductivity," J. Low Temp. Phys. 1, 241-271.

--- (1969) "Fluctuations and the superconducting phase transition: II. Onset of Josephson tunneling and paraconductivity of a junction," Ibid. 1, 423-442.

m --- (1970) "Decoupled-mode dynamical scaling theory of the binary-liquid phase transition," Phys. Rev. Lett. 24, 1169-1172.

h ---, MENYHÁRD, N., SCHMIDT, H., SCHWABL, F., and SZÉPFALUSY, P., (1967) "Dispersion in second sound and anomalous heat conduction at the lambda point of liquid helium," Ibid. 18, 891-894.

h ---, ---, ---, ---, --- (1968) "Fluctuations and lambda phase transition in liquid helium," Ann. Phys. 47, 565-613.

FIELD, J. J. (1972) "Ferromagnetism in a narrow-band solid," J. Phys. C. 5, 664.

FIERZ, M. (1951) "Zur Theorie der Kondensation," Helv. Phys. Acta 24, 357-366.

--- (1956) "Uber die statistischen Schwankungen in einem kondensierenden System," Ibid. 29, 47-54.

b FISHER, H. F., GATES, R. E., and CROSS, D. G. (1970) "A ligand exclusion theory of allosteric effects," Nature, 228, 247-249.

1 FISHER, M. E. (1959) "The susceptibility of the plane Ising model," Physica 25, 521-524.

1 --- (1959) "Transformations of Ising models," Phys. Rev. 113, 969-981.

--- (1960) "Perpendicular susceptibility of an anisotropic antiferromagnet," Physica 26, 618-622.

--- (1960) "Lattice statistics in a magnetic field. I. A two-dimensional super-exchange antiferromagnet," Proc. Roy. Soc. A254, 66-85.

--- (1960) "Lattice statistics in a magnetic field II. Order and correlations of a two-dimensional super-exchange antiferromagnet," Ibid. A256, 502-513.

--- (1961) "Critical probabilities for cluster size and percolation problems," J. Math. Phys. 2, 620-627.

--- (1961) "Statistical mechanics of dimers on a plane lattice," Phys. Rev. 124, 1664-1672.

--- (1961) "Relation between the specific heat and susceptibility of an antiferromagnet," Phil. Mag. 7, 1731-1743.

--- (1962) "On the theory of critical point density fluctuations," Physica 28, 172-180.

1 --- (1963) "Perpendicular susceptibility of the Ising model," J. Math. Phys. 4, 124-135.

--- (1963) "Lattice statistics - A review and an exact isotherm for a plane lattice gas," Ibid. 4, 278-286.

3 --- (1964) "Magnetism in one-dimensional systems - The Heisenberg model for infinite spin," Am. J. Phys. 32, 343-346.

--- (1964) "The free energy of a macroscopic system," Archive for Rational Mechanics and Analysis 17, 377-410.

--- (1964) "Deviations from van der Waals behavior on the critical isobar," J. Chem. Phys. 41, 1877-1888.

r --- (1964) "Correlation functions and the critical region of simple fluids," J. Math. Phys. 5, 944-962.

--- (1964) "Specific heat of a gas near the critical point," Phys. Rev. A136, 1599-1604.

--- (1965) "The shape of a self-avoiding walk or polymer chain," J. Chem. Phys. 44, 616-622.

--- (1965) "Correlation functions and the coexistence of phases," J. Math. Phys. 6, 1643-1653.

r --- (1965) "The nature of critical points," Lectures in Theoretical Physics Vol. VIIC, (Univ. Colo. Press, Boulder, Colo.).

--- (1965) "Bounds for the derivatives of the free energy and the pressure of a hard-core system near close packing," J. Chem. Phys. 42, 3852-3856.

b --- (1966) "Effect of excluded volume on phase transitions in biopolymers," Ibid. 44, 1469-1473.

l --- (1966) "On the dimer solution of planar Ising models," J. Math. Phys. 7, 1776-1781.

--- (1966) "Quantum corrections to critical-point behaviour," Phys. Rev. Lett. 16, 11-14.

r --- (1966) "Critical point singularities: notes, definitions and formulas for lattice gases and Ising magnets," Proc. Conf. on Phenomena in the Neighborhood of Critical Points, NBS Misc. Publ. 273, 21-26.

--- (1966) "Theory of critical fluctuations and singularities," Ibid. 273, 108-115.

--- (1966) "Cluster size and percolation theory," Proc. IBM Scientific Computing Symp. on Combinatorial Problems, 3, Chapter 11.

--- (1967) "Magnetic critical point exponents - Their interrelations and meaning," J. Appl. Phys. 38, 981-990.

--- (1967) "Critical temperatures of anisotropic Ising lattices. II. General upper bounds," Phys. Rev. 162, 480-485.

--- (1967) "The theory of condensation and the critical point," Physics 3, 255-283.

r --- (1967) "The theory of equilibrium critical phenomena," Rept. Progr. Phys. 30, 615-730.

h --- (1968) "The decay of superflow in helium," In Fluctuations in Superconductors, W. S. Goree and F. H. Chilton, Eds., Stanford Res. Inst. Menlo Park.

u --- (1968) "Renormalization of critical exponents by hidden variables," Phys. Rev. 176, 257-272.

--- (1969) "Phase transitions in one-dimensional classical fluids with many-body interactions," Centre Natl. Recherche Sci. 181, 87-103.

r --- (1969) "Aspects of equilibrium critical phenomena," J. Phys. Sco. Japan Suppl. 26, 87-93.

--- (1969) "Rigorous inequalities for critical-point correlation exponents," Phys. Rev. 180, 594-600.

r --- (1969) "Phase transitions and critical phenomena," Proc. Internat. Symp. on Contemp. Phys., Trieste, Vol. I, 19-46.

--- (1972) "On discontinuity of the pressure," Commun. Math. Phys. 26, 4-16.

d --- (1972) "Scaling theory for finite-size effects in the critical region," Phys. Rev. Lett. 28, 1516-1519.

l --- and BURFORD, R. J. (1967) "Theory of critical-point scattering and correlations. I. The Ising model," Phys. Rev. 156, 583-622.

--- and CAMP., W. J. (1971) "Behavior of two-point correlation functions near and on a phase boundary," Phys. Rev. Lett. 26, 565-568.

---, --- (1972) "Estimation of spectra from moments - Application to the Hubbard model," Phys. Rev. B5, 3730-3737.

--- and ESSAM, J. W. (1961) "Some cluster size and percolation problems," J. Math. Phys. 2, 609-619.

--- and FELDERHOF, B. U. (1970) "Phase transitions in one-dimensional cluster-interaction fluids IA. Thermodynamics," Ann. Phys. 58, 176-216.

---, --- (1970) "Phase transitions in one-dimensional cluster-interaction fluids IB. Critical behavior," Ibid. 58, 217-267.

d --- and FERDINAND, A. E. (1967) "Interfacial, boundary, and size effects at critical points," Phys. Rev. Lett. 19, 169-172.

--- and GAUNT, D. S. (1964) "Ising model and self-avoiding walks on hypercubical lattices and 'high-density' expansions," Phys. Rev. 133, 224-239.

b --- and HILEY, B. J. (1961) "Configuration and free energy of a polymer molecule with solvent interactions," J. Chem. Phys. 32, 1257-1267.

d --- and JASNOW, D. (1971) "Decay of order in isotropic systems of restricted dimensionality. II. Spin systems," Phys. Rev. B3, 907-924.

--- and LANGER, J. S. (1968) "Resistive anomalies at magnetic critical point," Phys. Rev. Lett. 20, 665-668.

--- and LEBOWITZ, J. L. (1970) "Asymptotic free energy of a system with periodic boundary conditions," Commun. Math. Phys. 19, 251-272.

g --- and PFEUTY, P. (1972) "Critical behavior of the anisotropic n-vector model," Phys. Rev. B6, 1889-1891

--- and RUELLE, D. (1966) "The stability of many-particle systems," J. Math. Phys. 7, 260-270.

--- and SCESNEY, P. E. (1969) "Mobile-electron Ising ferromannets," J. Appl. Phys. 40, 1554.

---, --- (1970) "Visibility of critical-exponent renormalization," Phys. Rev. 2, 825-835.

--- and STEPHENSON, J. (1963) "Statistical mechanics of dimers on a plane lattice. II. Dimer correlations and monomers," Ibid. 132, 1411-1431.

--- and SYKES, M. F. (1959) "Excluded-volume problem and the Ising model of ferromagnetism," Ibid. 114, 45-58.

---, --- (1962) "Antiferromagnetic susceptibilities of the simple cubic and body-centered cubic Ising lattice," Physica 28, 939-956.

--- and WIDOM, B. (1969) "Decay of correlations in linear systems," J. Chem. Phys. 50, 3756-3772.

FISK, S. and WIDOM, B. (1969) "Structure and free energy of the interface between fluid phases in equilibrium near the critical point," Ibid. 50, 3219.

FIXMAN, M. (1955) "Molecular theory of light scattering," Ibid. 23, 2074-2079.

--- (1960) "Density correlations, critical opalescence, and the free energy of nonuniform fluids," Ibid. 33, 1357-1362.

--- (1960) "Ultrasonic attenuation in the critical region," Ibid. 33, 1363-1370.

m --- (1962) "Viscosity of critical mixtures," Ibid. 36, 310-318.

m --- (1962) "Heat capacity of critical mixtures," Ibid. 36, 1957-1960.

m --- (1962) "Absorption and dispersion of sound in critical mixtures," Ibid. 36, 1961-1964.

--- (1962) "Correlations at the critical point," Ibid. 36, 1965-1968.

r --- (1964) "The critical region," in Adv. Chem. Phys. I. Prigogine, Ed. (Interscience Publishers, Inc., New York), Vol. VI, pp. 175-228.

b --- (1965) "Dynamics of polymer chains," J. Chem. Phys. 42, 3831-3837.

b --- (1966) "Polymer dynamics: Boson representation and excluded-volume forces," Ibid. 45, 785-792.

b --- (1966) "Polymer dynamics: Non-Newtonian intrinsic viscosity," Ibid. 45, 793-803.

--- (1967) "Transport coefficients in the gas critical region," Ibid. 47, 2808-2818.

FLAX, L. and RAICH, J. C. (1969) "Calculation of the generalized Watson sums with an application to the generalized Heisenberg ferromagnet," Phys. Rev. 185, 797.

d FLEURY, P. A. and GUGGENHEIM, H. J. (1970) "Magnon-pair modes in two dimensions," Phys. Rev. Lett. 24, 1346-1349.

FLICKER, M. and LEFF, H. S. (1968) "On the Bethe hypothesis for the anisotropic Heisenberg chain," Phys. Rev. 168, 578-587.

--- and LIEB, E. H. (1967) "Delta-function Fermi gas with two-spin devices," Ibid. 161, 179-188.

FLORY, P. J. and MILLER, W. G. (1966) "A general treatment of helix-coil equilibria in macromolecular systems," J. Mol. Biol. 15, 284-297.

FONER, S. (1966) "First- and higher-order magnetic phase transitions in dysprosium aluminum garnet," J. Appl. Phys. 37, 1120-1121.

FORD, N. C., Jr., and BENEDEK, G. B. (1965) "Observation of the spectrum of light scattered from a pure fluid near its critical point," Phys. Rev. Lett. 15, 651-653.

---, LANGLEY, K. H., and PUGLIELLI, V. G. (1968) "Brillouin linewidths in CO_2 near the critical point," Ibid. 21, 9-12.

L FORSTER, D., LUBENSKY, T. C., MARTIN, P. C., SWIFT, J., and PERSHAN, P. S. (1971) "Hydrodynamics of liquid crystals," Ibid. 26, 1016-1019.

--- and MARTIN, P. C. (1970) "Kinetic theory of a weakly coupled fluid," Phys. Rev. A2, 1575-1590.

FORTUIN, C. M. (1972) "On the random-cluster model. II. The percolating model," Physica 58, 393-418.

---, GINIBRE, J., and KASTELEYN, P. W. (1971) "Correlation inequalities on some partially ordered sets," Commun. Math. Phys. 22, 89-103.

h FRANKEL, N. E. (1968) "Order parameter for superfluid helium," Phys. Lett. 28A, 7-8.

1 --- (1968) "Order parameter for the Ising ferromagnet," Ibid. 28A, 440-441.

--- (1969) "Critical point behavior of a phenomenological model of cooperativity," Ibid. 29A, 446-447.

FREDERICKS, G. E. (1971) "Order-disorder transition in NH_4Cl. II. Thermal expansion," Phys. Rev. B4, 911-919.

3 FREEMAN, S. and WOJTOWICZ, P. J. (1968) "Critical-point behavior of the susceptibility of classical Heisenberg ferrimagnets," J. Appl. Phys. 39, 622-623.

---, --- (1968) "Critical index of the susceptibility for the B-site spinel ferromagnet," Phys. Lett. 26A 231-232.

---, --- (1969) "Critical-point behavior of classical Heisenberg ferrimagnets," Phys. Rev. 177, 882-888.

---, --- (1972) "High-temperature susceptibility and critical-point behavior of Heisenberg ferrimagnets," Ibid. B6, 304-307.

r FRENCH, M. J., ANGUS, J. C., and WALTON, A. G. (1969) "Laser beat frequency spectroscopy," Science 163, 345-351.

FRENKEL, J. (1939) "Statistical theory of condensation phenomena," J. Chem. Phys. 7, 200-201.

--- (1939) "A general theory of heterophase fluctuations and pretransition phenomena," Ibid. 7, 538-547.

FREUD, P. J. and HED, A. Z. (1969) "Dynamics of the electric-field-induced conductivity transition in magnetite," Phys. Rev. Lett. 23, 1440-1443.

FREUND, I. (1967) "Critical harmonic scattering in NH_4Cl," Ibid. 19, 1288-1291.

FRIEDMAN, H. L. and RAMANTHAN, P. S. (1970) "Theory of mixed electrolyte solutions and application to a model for aqueous lithium chloride-cesium chloride," J. Phys. Chem. 74, 3756-3765.

FRISCH, H. L. (1966) "High frequency linear response of classical fluids," Physics 2, 209-215.

--- (1967) "Study of turbulence by spectral fine structure of scattered light," Phys. Rev. Lett. 19, 1278-1279.

--- and BRADY, G. W. (1962) "On small-angle critical scattering," J. Chem. Phys. 37, 1514-1521.

--- and HELFAND, E. (1960) "Conditions imposed by gross properties on the intermolecular potential," Ibid. 32, 269-270.

--- and MC KENNA, J. (1965) "Double scattering of electromagnetic radiation by a fluid," Phys. Rev. 139, A68-A77.

FUCHIKAMI, N. and OGUCHI, T. (1970) "Residual entropy of $NaH_3(SeO_3)_2$," Progr. Theoret. Phys. 44, 1500-1508.

FUKUYAMA, Y. (1967) "Sound attenuation in the critical gases," Ibid. 39, 26-36.

FULIŃSKI, A. (1970) "Virial expansion, equation of state and phase transition," Acta Phys. Polon A37, 177.

--- (1971) "One-chain approximation for the Gaussian model of a gas," Ibid. A39, 181.

--- and JURKIEWICZ, M. (1972) "Phase transitions in binary mixtures," Ibid. A41, 193.

m ---, --- (1972) "Statistical-mechanical theory of mixtures in the one-chain approximation. II. multicomponent systems," Ibid. A41, 205.

GALGANI, L. and SCOTTI, A. (1968) "Remarks on convexity of thermodynamic functions," Physica 40, 150.

---, --- (1969) "Further remarks on convexity of thermodynamic functions," Ibid. 42, 242.

GALLAVOTTI, G. (1969) "Divergences and the approach to equilibrium in the Lorentz and the Wind-Tree models," Phys. Rev. 185, 308-322.

--- (1970) "High-temperature properties of random spin systems," J. Math. Phys. 11, 141-146.

--- (1971) "Boundary conditions and correlation functions in the ν-dimensional Ising model at low temperature," Commun. Math. Phys. 23, 275-284.

---, LANFORD, O. E., III, and LEBOWITZ, J. L. (1970) "Thermodynamic limit of time-dependent correlation functions for one-dimensional systems," J. Math. Phys. 11, 2898-2905.

1 --- and MARTIN-LÖF, A. (1972) "Surface tension in the Ising model," Commun. Math. Phys. 25, 87-126.

--- and MIRACLE-SOLE, S. (1968) "Correlation functions of a lattice system," Ibid. 7, 274-288.

---, ---, and RUELLE, D. (1968) "Absence of phase transitions in one-dimensional systems with hard cores," Phys. Lett. 26A, 350-351.

3 GAMMEL, J., MARSHALL, W., and MORGAN, L. (1963) "An application of Padé approximants to Heisenberg ferromagnetism and antiferromagnetism," Proc. Roy. Soc. 275, 257-270.

GAMMON, R. W., SWINNEY, H. L., and CUMMINS, H. Z. (1967) "Brillouin scattering in carbon dioxide in the critical region," Phys. Rev. Lett. 19, 1467-1469.

GANDEL'MAN, G. M., DMITRIEV, N. A., and SARRY, M. F. (1971) "Quantitative criterion for ferromagnetism," Sov. Phys.-JETP 32, 1093.

m GANGULY, B. N. and GRIFFIN, A. (1968) "Scattering of light from entropy fluctuations in ^{3}He-^{4}He mixtures," Can. J. Phys. 46, 1895-1903.

GARCIA-COLIN, L. and LEY-KOO, M. (1969) "Behaviour of the pair correlation function in the critical region," Phys. Lett. 29A, 688-689.

l GARELICK, H. and ESSAM, J. W. (1968) "Critical behaviour of the three-dimensional Ising model specific heat below T_c," J. Phys. C. 1, 1588-1595.

GARLAND, C. W. (1964) "Generalized Pippard equations," J. Chem. Phys. 41, 1005-1008.

a --- (1964) "Comment on the elastic properties of β brass," Phys. Rev. A135, 1696-1697.

r --- (1970) "Ultrasonic investigation of phase transitions and critical points," Physical Acoustics 7, 51-148.

---, EDEN, D., and MISTURA, L. (1970) "Critical sound absorption in xenon," Phys. Rev. Lett. 25, 1161-1165.

--- and JONES, J. S. (1964) "Effect of ordering on the lattice energy and compressibility of ammonium chloride," J. Chem. Phys. 41, 1165-1166.

---, --- (1965) "Ultrasonic attentuation in ammonium chloride," Ibid. 42, 4194-4199.

--- and NOVOTNY, D. B. (1969) "Ultrasonic velocity and attentuation in KH_2PO_4," Phys. Rev. 177, 971-978

l --- and RENARD, R. (1966) "Order-disorder phenomena. I. Instability and hysteresis in an Ising model near its critical point," J. Chem. Phys. 44, 1120-1124.

---, --- (1966) "Order-disorder phenomena. III. Effect of temperature and pressure on the elastic constants of ammonium chloride," Ibid. 44, 1130-1139.

--- and SCHUMAKER, N. E. (1967) "Effect of ordering on the infra-red spectrum of ammonium chloride," J. Phys. Chem. Solids 28, 799-803.

--- and SNYDER, D. D. (1970) "Ultrasonic attenuation near the lambda transition in NH_4Cl at high pressures," Ibid. 31, 1759-1764.

--- and WEINER, B. B. (1971) "Changes in the thermodynamic character of the NH_4Cl order-disorder transition at high pressures," Phys. Rev. B3, 1634-1637.

--- and YARNELL, C. F. (1966) "Temperature and pressure dependence of the elastic constants of ammonium bromide," J. Chem. Phys. 44, 1112-1120.

---, --- (1966) "Order-disorder phenomena. IV. Ultrasonic attenuation near the lambda point in ammonium chloride," Ibid. 44, 3678-3681.

--- and YOUNG, R. A. (1968) "Order-disorder phenomena. VI. Anomalous changes in the volume of ammonium choloride," Ibid. 48, 146-148.

f GARNIER, P. R. (1971) "Specific heat of $SrTiO_3$ near the structural transition," Phys. Lett. 35A, 413-414.

--- and SALAMON, M. B. (1971) "First-order transition in chromium at the Néel temperature," Phys. Rev. Lett. 27, 1523-1526.

GARSIDE, D. H., MOLGAARD, H. V., and SMITH, B. L. (1968) "Refractive index and Lorentz-Lorenz function of xenon liquid and vapour," J. Phys. B1, 449-457.

GARTENHAUS, S. and STRANAHAN. G. (1968) "Critical-point behavior of a model describing a system of interacting fermions," Phys. Rev. 173, 260-270.

m GASPARINI, F. and MOLDOVER, M. R. (1969) "Specific heat of He^3-He^4 mixtures very near the λ line," Phys. Rev. Lett. 23, 749-752.

GATES, D. J. and PENROSE, O. (1969) "The van der Waals limit for classical systems. I. A variational principle," Commun. Math. Phys. 15, 255-276.

---, --- (1970) "The van der Waals limit for classical systems. II. Existence and continuity of the canonical pressure," Ibid. 16, 231-237.

---, --- (1970) "The van der Waals limit for classical systems. III. Deviation from the van der Waals-Maxwell theory," Ibid. 17, 194-209.

GAUDIN, M. (1971) "Boundary energy of a Bose gas in one dimension," Phys. Rev. A4, 386-394.

--- (1971) "Thermodynamics of the Heisenberg-Ising ring for $\Delta \geq 1$," Phys. Rev. Lett. 26, 1301-1304.

l GAUNT, D. S. (1967) "The critical isotherm and critical exponents of the three-dimensional Ising ferromagnet," Proc. Phys. Soc. 92, 150-158.

--- (1969) "Exact series-expansion study of the monomer-dimer problem," Phys. Rev. 179, 174-186.

l --- and BAKER, G. A. (1970) "Low-temperature critical exponents from high-temperature series: The Ising model," Ibid. B1, 1184-1210.

l --- and DOMB, C. (1968) "The specific heat of the three-dimensional Ising model below T_c," J. Phys. C1, 1038-1045.

l ---, --- (1970) "Equation of state of the Ising model near the critical point," Ibid. C3, 1442-1461.

---, FISHER, M. E., SYKES, M. F., and ESSAM, J. W. (1964) "Critical isotherm of a ferromagnet and of a fluid," Phys. Rev. Lett. 13, 713-715.

---, --- (1965) "Bounds for the derivatives of the free energy and the pressure of a hard core system near close packing," J. Chem. Phys. 42, 3852-3856.

∞ GERSCH, H. A. (1963) "One-dimensional phase transition in the spherical model of a gas," Phys. Fluids 6, 599-608.

∞ --- and BERLIN, T. H. (1962) "Spherical lattice gas," Phys. Rev. 127, 2276-2283.

GEY, W. and DOHNLEIN, D. (1969) "Effect of pressure on T_c and band structure in transition metal alloys," Phys. Lett. 29A, 450-451.

d GHOSH, D. K. (1971) "Nonexistence of magnetic ordering in the one- and two-dimensional Hubbard model," Phys. Rev. Lett. 27, 1584-1587.

GIBBERD, R. W. (1968) "A reformulation of the dimer problem," Can. J. Phys. 46, 1681-1684.

f --- (1968) "Combinatorial approach to the ferroelectric problem," Phys. Rev. 171, 563-566.

--- (1969) "Application of many-fermion techniques to lattice statistical problems," Can. J. Phys. 47, 809-821.

--- (1969) "Some exact results for the anisotropic triangular lattice," Ibid. 47, 2445-2448.

l --- (1969) "Next-nearest-neighbor Ising model," J. Math. Phys. 10, 1026-1029.

l --- (1970) "High-temperature specific heat series of the Ising model on the crystobalite lattice," Can. J. Phys. 48, 307-312.

b GIBBS, J. H. and DI MARZIO, E. A. (1959) "Statistical mechanics of helix-coil transitions in biological macromolecules," J. Chem. Phys. 30, 271-282.

GIGLIO, M. and BENEDEK, G. B. (1969) "Angular distribution of the intensity of light scattered from xenon near its critical point," Phys. Rev .Lett. 23, 1145-1149.

GIJSMAN, H. M., POULIS, N. J., and VAN DEN HANDEL, J. (1959) "Magnetic susceptibilities and phase transitions of two antiferromagnetic manganese salts," Physica 25, 954-968.

m GILMER, G. H., GILMORE, W., HUANG, J., and WEBB, W. W. (1965) "Diffuse interface in a critical fluid mixture," Phys. Rev. Lett. 14, 491-494.

GILLIS, J. (1955) "Correlated random walk," Proc. Camb. Phil. Soc. 51, 639-651.

GINIBRE, J. (1969) "Existence of phase transitions for quantum lattice systems," Commun. Math. Phys. 14, 205-234.

--- (1969) "Simple proof and generalization of Griffiths' second inequality," Phys. Rev. Lett. 23, 828-830.

--- (1970) "General formulation of Griffiths' inequalities," Commun. Math. Phys. 16, 310-328.

---, GROSSMANN, A., and RUELLE, D. (1966) "Condensation of lattice gases," Ibid. 3, 187-193.

f GINZBURG, V. L. (1960) "Some remarks on phase transitions of the second kind and the microscopic theory of ferroelectric materials," Sov. Phys.-Solid State 2, 1824-1834.

GIRARDEAU, M. D. (1969) "Fermi versus Bose condensation in superconductors and liquid ^{4}He," Phys. Lett. 29A, 64-65.

GITERMAN, M. SH. and KONTOROVICH, V. M. (1965) "The effect of spatial dispersion on the propagation and scattering of waves near the critical point," Sov. Phys.-JETP 20, 1433-1442.

s GITTLEMAN, J. I., COHEN, R. W., and HANAK, J. J. (1969) "Fluctuation rounding of the superconducting transition in the three dimensional regime," Phys. Lett. 29A, 56-57.

f GLASSER, M. L. (1969) "Evaluation of the partition functions for some two-dimensional ferroelectric models," Phys. Rev. 184, 539-542.

f ---, ABRAHAM, D. B., and LIEB, E. H. (1972) "Analytic properties of the free energy for the 'ice' models," J. Math. Phys. 13, 887.

l GLAUBER, R. J. (1964) "Time-dependent statistics of the Ising model," Ibid. 4, 294-307.

s GLOVER, R. E. (1967) "Ideal resistive transition of a superconductor," Phys. Lett. 25A, 542-544.

s --- (1971) "Superconducting fluctuation effects above the transition temperature," Physica 55, 3-23.

GLUCK, P. (1968) "Viscosity of liquids near the critical point," Nuovo Cimento 56B, 338-342.

--- (1969) "Impurity effects near the critical point in second-order phase transitions," Lett. Nuovo Cimento 1, 129-130.

--- (1970) "Dynamics of an Ising ferromagnet near T_c," Phys. Stat. Sol. 39, K85.

--- and LUBAN, M. (1969) "Scaling laws and effects of impurities near a critical point," Phys. Lett. 28A, 607-608.

l --- and WOHLMAN, O. (1970) "On the dynamics of the Ising lattice," Phys. Stat. Sol. 40, K85.

d GOBLE, D. F. and TRAINOR, L. E. H. (1968) "Liquid helium and the properties of a Bose-Einstein Gas. II. Effects of finite geometry," Can. J. Phys. 46, 1867-1882.

b GOEL, N. S. (1968) "Relaxation kinetics of the denaturation of DNA," Biopolymers 6, 55-72.

b ---, CAMPBELL, R. D., GORDON, R., ROSEN, MARTINEX, H., and YČAS, M. (1970) "Self-sorting of isotropic cells," J. Theor. Biol. 28, 423-468.

b ---, FUKUDA, N., and REIN, R. (1968) "A semi-empirical analysis of the melting curves of synthetic DNA molecules and the calculation of the stacking and pairing energies and entropies in DNA," Ibid. 18, 350-370.

b --- and MAITRA, S. C. (1969) "On the existence of independently melting subunits in a copolymeric DNA and their statistical distribution," Ibid. 23, 87-98.

b ---, --- and MONTROLL, E. W. (1971) "On the Volterra and other nonlinear models of interacting populations," Rev . Mod. Phys. 43, 231-276.

b --- and MONTROLL, E. W. (1968) "Denaturation and renaturation of DNA. II. Possible use of synethetic periodic copolymers to establish model and parameters," Biopolymers 6, 731-765.

m GOELLNER, G. and MEYER, H. (1971) "Chemical potential of He^3-He^4 solutions near the tricritical point," Phys. Rev. Lett. 26, 1534-1537.

GOLDING, B. (1968) "Ultrasonic propagation in $RbMnF_3$ near the magnetic critical point," Ibid. 20, 5-7.

b --- (1970) "Thermal expansivity and ultrasonic propagation near the structural transition of $SrTiO_3$," Ibid. 25, 1439.

--- (1971) "Critical magnetic thermal expansivity of $RbMnF_3$," Ibid. 27, 1142-1145.

--- and BARMATZ, M. (1969) "Ultrasonic propagation near the magnetic critical point of nickel," Ibid. 23, 223-226.

GOLDRICH, F. E. and B IRMAN, J. L. (1968) "Theory of symmetry change in second-order phase transitions in Perovskite structure," Phys. Rev. 167, 528.

GOLDSTEIN, J. C., SCULLY, M. O., and LEE, P. A. (1971) "Laser-phase transition analogy: II," Phys. Lett. 35A, 317-319.

GOLLUB, J. P., BEASLEY, M. R., NEWBOWER, R. S., and TINKHAM, M. (1969) "Observation of enhanced diamagnetism above T_c in indium due to thermodynamic fluctuations," Phys. Rev. Lett. 22, 1288-1291.

f GONZALO, J. A. (1966) "Critical behavior of ferroelectric triglycine sulfate," Phys. Rev. 144, 662-665.

f --- (1968) "Set of experimental critical exponents for ferroelectric triglycine sulfate," Phys. Rev. Lett. 21, 749-751.

f --- (1970) "Equation of state for the cooperative transition of triglycine sulfate near T_c," Phys. Rev. B1, 3125-3132.

m GOPAL, R. and RICE, O. K. (1955) "Shape of the coexistence curve in the perfluoromethylcyclohexane-carbon tetrachloride system," J. Chem. Phys. 23, 2428-2431.

GORODESTSKY, G. (1968) "Semiempirical expression for the paramagnetic susceptibility of $LuFeO_3$ and $KMnF_3$ by the Padé approximant," Sol. State Commun. 6, 159-162.

---, LUTHI, B., and MORAN, T. J. (1971) "Sound propagation near displacive and magnetic order-disorder phase transitions," Int. J. Magnetism 1, 295-306.

--- SHTRIKMAN, S., and TREVES, D. (1966) "The critical behavior of a weak ferromagnet," Sol. State Commun. 4, 147-151.

GOTTLIEB, A. M. and HELLER, P. (1971) "Nuclear-resonance studies of critical fluctuations in FeF_2 above T_N," Phys. Rev. B3, 3615-3624.

GOULD, H. and WONG, V. K. (1971) "Order-parameter fluctuations in a weakly interacting Bose gas near the superfluid transition," Ibid. A4, 719-734.

GRAHAM, C. D., Jr. (1965) "Magnetic behavior of gadolinium near the Curie point," J. Appl. Phys. 36, 1135-1136.

GREEN, M. S. (1960) "On the theory of the critical point of a simple fluid," J. Chem. Phys. 33, 1403-1409.

r --- (1965) "Critical-point phenomena," Science 150, 229-236.

r --- (1967) "I. Phenomenology of the liquid-vapor phase transition," Proc. 1965 Latin American School of Physics (Gordon & Breach Publ., N.Y.), pp. 923-955.

--- (1969) "Why is C_V less singular than C_P near the critical point?," Phys. Rev. 185, 176-186.

---, COOPER, M. J. and LEVELT SENGERS, J. M. H. (1971) "Extended thermodynamic scaling from a generalized parametric form," Phys. Rev. Lett. 26, 492-495.

---, VICENTINI-MISSONI, M., and LEVELT SENGERS, J. M. H. (1967) "Scaling-law equation of state for gases in the critical region," Ibid. 18, 1113-1117.

GREYTAK, T. J. and BENEDEK, G. B. (1966) "Spectrum of light scattered from thermal fluctuations in gases," Phys. Rev . 17, 179-182.

h GREYWALL, D. S. and AHLERS, G. (1972) "Second-sound velocity, scaling, and universality in HeII under pressure near the superfluid transition," Phys. Rev. Lett. 28, 1251-1254.

g GRIBOV, V. N. and MIGDAL, A. A. (1969) "Strong coupling in the Pomeranchuk pole problem," Sov. Phys.-JETP 28, 784.

m GRIFFIN, A. (1969) "Hydrodynamic modes in ^3He-^4He mixtures at the critical point," Can. J. Phys. 47, 429-434.

GRIFFITHS, R. B. (1963) "Magnetization curve at zero temperature for the antiferromagnetic Heisenberg linear chain," Phys. Rev. 133, 768-775.

--- (1964) "Evidence for exchange-coupled linear chains in $Cu(NH_3)_4SO_4H_2O$," Ibid. 135, A659.

--- (1964) "Peierls proof of spontaneous magnetization in a two-dimensional Ising ferromagnet," Ibid. 136, A437.

--- (1964) "Free energy of the antiferromagnetic linear chain," Ibid. 136, A751.

--- (1964) "A proof that the free energy of a spin system is extensive," J. Math. Phys. 5, 1215-1222.

--- (1965) "Microcanonical ensemble in quantum statistical mechanics," Ibid. 6, 1447.

--- (1965) "Ferromagnets and simple fluids near the critical point: Some thermodynamic inequalities," J. Chem. Phys. 43, 1958-1968.

--- (1965) "Thermodynamic inequality near the critical point for ferromagnets and fluids," Phys. Rev. Lett. 14, 623-624.

--- (1966) "Spontaneous magnetization in idealized ferromagnets," Phys. Rev. 152, 240-246.

--- (1966) "Power-series expansions and specific-heat singularities near the He^4 critical point," Phys. Rev. Lett. 16, 787.

l --- (1967) "Correlations in Ising ferromagnets. III. A mean field bound for binary correlations," Commun. Math. Phys. 6, 121-127.

l --- (1967) "Correlations in Ising ferromagnets. I," J. Math. Phys. 8, 478-483.

l --- (1967) "Correlations in Ising ferromagnets. II. External Magnetic Fields," Ibid. 8, 484-489.

m --- (1967) "First-order phase transitions in spin-one Ising systems," Physica 33, 689-690.

--- (1967) "Thermodynamic functions for fluids and ferromagnets near the critical point," Phys. Rev. 158, 176-187.

--- (1968) "Free energy of interacting magnetic dipoles," Ibid. 176, 655-659.

l --- (1969) "Rigorous results for Ising ferromagnets of arbitrary spin," J. Math. Phys. 10, 1559-1565.

--- (1969) "Ferromagnetic heat capacity in an external magnetic field near the critical point," Phys. Rev. 188, 942-947.

l --- (1969) "Nonanalytic behavior above the critical point in a random Ising ferromagnet," Phys. Rev. Lett. 23, 17-19

--- (1970) "Correlation-function inequality obtained by Yeh," Phys. Rev. B1, 3883.

m --- (1970) "Thermodynamics near the two-fluid critical mixing point in He^3-He^4," Phys. Rev. Lett. 24, 715-717.

u --- (1970) "Dependence of critical indices on a parameter," Ibid. 24, 1479-1482.

u --- (1971) "Critical points dependent on parameters," Critical Phenomena in Alloys, Magnets and Superconductors, R. E. Mills, E. Ascher, and R. I. Jaffee, Eds., McGraw-Hill Book Company, N.Y., pp. 377-391.

r --- (1971) "Phase Transitions," In Statistical Mechanics and Quantum Field Theory, C. DeWitt and R. Stora, Eds. (Gordon and Breach, N.Y.), p. 241.

---, HURST, C. A., SHERMAN, S. (1970) "Concavity of magnetization of an Ising ferromagnet in a positive external field," J. Math. Phys. 11, 790.

--- and LEBOWITZ, J. L. (1968) "Random spin systems: Some rigorous results," Ibid. 9, 1284-1292.

--- and RUELLE, D. (1971) "Strict convexity ('continuity') of the pressure in lattice systems," Commun. Math. Phys. 23, 169-175.

---, WENG, C. Y., and LANGER, J. S. (1966) "Metastable states in the mean-field model of a ferromagnet," Phys. Rev. 149, 301.

m --- and WHEELER, J. C. (1970) "Critical points in multicomponent systems," Ibid. A2, 1047-1064.

h GRILLY, E. R. (1966) "Compressibility of liquid He^4 as a function of pressure," Ibid. 149, 97-101.

f GRINDLAY, J. (1965) "Evidence for logarithmic singularities in the specific heats of the ferroelectric crystals KH_2PO_4 and KH_2AsO_4," Ibid. 139, 1603-1606.

--- (1968) "The second-order phase transition and the Ehrenfest relations," Can. J. Phys. 46, 2253-2258.

d GRUENBERG, L. W. and GUNTHER, L. (1972) "Critical behaviour of one-dimensional superconductors: An exact solution," Phys. Lett. 38A, 463-465.

GUFAN, YU. M. (1971) "Phase transitions characterized by a multicomponent order parameter," Sov. Phys.-Solid State 13, 175-180.

GUGGENHEIM, E. A. (1945) "The principle of corresponding states," J. Chem. Phys. 13, 253-261.

GULLEY, J. E., SILBERNAGEL, B. G., and JACCARINO, V. (1969) "Magnetic resonance lineshapes in exchange coupled paramagnets," Phys. Lett. 29A, 657-658.

d GUNTHER, L. (1967) "Stability of the finite two-dimensional harmonic lattice," Ibid. 25A, 649-650.

--- (1968) "'Boltzmann statistics' without statistics in the classical harmonic lattice," Phys. Rev. 171, 210-212.

--- (1969) "Fluctuation corrected Ginzburg-Landau equation and higher moment relations," Phys. Lett. 29A 102-103.

u ---, BERGMAN, D. J., and IMRY, Y. (1971) "Renormalized critical behavior or first-order phase transitions?," Phys. Rev. Lett. 27, 558-560.

d GUNTON, J. D. (1968) "Finite size effects at the critical point," Phys. Lett. 26A, 406-407.

∞ --- and BUCKINGHAM, M. J. (1968) "Condensation of the ideal Bose gas as a cooperative transition," Phys. Rev. 166, 152-158.

---, --- (1968) "Behavior of the correlation function near the critical point," Phys. Rev. Lett. 20, 143-146.

--- and GREEN, M. S. (1971) "Generalized Ornstein-Zernike approach to critical phenomena III. The critical null space," Phys. Rev. A4, 1282-1299.

l GUPTA, A. K. (1970) "On the critical point behaviour of a model of a 3-d Ising ferromagnet," Phys. Stat. Sol. 38, 837.

GUPTA, R. P. and SEEHRA, M. S. "Critical behavior of the paramagnetic line width in $RbMnF_3$," Phys. Lett. 33A, 347-348.

GUTTMANN, A. J. (1969) "Determination of critical behavior in lattice statistics from series expansions III," J. Phys. C2, 1900-1907.

--- (1972) "Critical behavior of an isotropic spin system," Phys. Lett. 38A, 297-298.

---, NINHAM, B. W., and THOMPSON, C. J. (1968) "Determination of critical behavior in lattice statistics from series expansions. I," Phys. Rev. 172, 554-558.

---, ---, --- (1968) "Determination of critical behaviour in lattice statistics from series expansions," Phys. Lett. 26A, 180-181.

l --- and THOMPSON, C. J. (1969) "Low temperature susceptibility exponent for the three dimensional Ising model," Ibid. 28A, 679-680.

---, ---, and NINHAM, B. W. (1970) "Determination of critical behavior from series expansions in lattice statistics: IV," J. Phys. C3, 1641-1651.

GUTTMAN, L. and SCHNYDERS, H. C. (1969) "Critical scattering of X-rays from Fe_3Al," Phys. Rev. Lett. 22, 520-522.

---, ---, and ARAI, G. J. (1969) "Variation of long-range order in Fe_3Al near its transition temperature," Ibid. 22, 517-518.

L HALLER, I. and LITSTER, J. D. (1970) "Temperature dependence of normal modes in a nematic liquid crystal," Ibid. 25, 1550-1553.

HALPERIN, B. I. and HOHENBERG, P. C. (1967) "Generalization of scaling laws to dynamical properties of a system near its critical point," Ibid. 19, 700-703.

---, --- (1969) "Scaling laws for dynamic phenomena near a critical point," J. Phys. Soc. Japan 26, 131-135.

---, --- (1969) "Scaling laws for dynamic critical phenomena," Phys. Rev. 177, 952-970.

---, --- (1969) "Hydrodynamic theory of spin waves," Ibid. 188, 898-918

HAMAGUCHI, Y. (1968) "Critical magnetic scattering of neutrons in chromium," J. Appl. Phys. 39, 1227-1230.

d HAMILTON, D. C. (1967) "Absence of long-range Overhauser spin-density waves in one or two dimensions," Phys. Rev. 157, 427-428.

HAMMERSLEY, J. M. (1957) "Percolation processes," Proc. Camb. Phil. Soc. 53, 642-645.

---, FEUERVERGER, A., ISENMAN, A., and MAKANI, K. (1969) "Negative finding for the three-dimensional dimer problem," J. Math. Phys. 10, 443-446.

HANDLER, P., MAPOTHER, D. E., and RAYL, M. (1967) "AC measurement of the heat capacity of nickel near its critical point," Phys. Rev. Lett. 19, 356-358.

HANDRICH, K. and KOBE, S. (1970) "On the theory of amorphous and liquid ferromagnets," Acta Phys. Polon. A38, 819.

HANKEY, A. and STANLEY, H. E. (1971) "An alternate formulation of the static scaling hypothesis," Int. J. Quantum Chem. 5S, 593-608.

m ---, --- (1972) "A unified approach to static and dynamic scaling," Proc. 1971 Conf. on Magnetism and Magnetic Materials, Chicago, Ill., AIP Conf. No. 5, p. 1230-1235.

m ---, ---, and CHANG, T. S. (1972) "The geometric predictions of scaling at tricritical points," Phys. Rev. Lett. 29, 278-281.

u ---, --- (1972) "A systematic application of generalized homogeneous functions to static scaling, dynamic scaling, and universality," Phys. Rev. B6, 3515-3542.

HANSEN, J. P. and VERLET, L. (1969) "Phase transitions of the Lennard-Jones system," Ibid. 184, 151.

L HANUS, J. (1969) "Effect of the molecular interaction between anisotropic molecules on the optical Kerr effect. Field-induced phase transitions," Ibid. 178, 420-428.

u HARBUS, F., KRASNOW, R., LIU, L. L., and STANLEY, H. E. (1972) "Evidence supporting scaling with a parameter for the pair correlation function and thermodynamic functions," Phys. Lett. 42A, 65-66.

m --- and STANLEY, H. E. (1972) "Ising model 'metamagnet' and tricritical susceptibility exponent," Phys. Rev. Lett. 29, 58-62.

m ---, --- (1973) "General-R high-temperature series for the susceptibility, second moment, and specific heat of the sc and fcc Ising models with lattice anisotropy," Phys. Rev. B7, 365.

m ---, ---, and CHANG, T. S. (1972) "Tricritical susceptibility in Ising model metamagnets and some remarks concerning tricritical point scaling," Proc. LT 13, Boulder, Colo., August, 1972.

HARKER, Y. D. and BRUGGER, R. M. (1967) "Investigation of the low-temperature phase transition in solid methane by slow neutron inelastic scattering," J. Chem. Phys. 46, 2201-2208.

HARRIS, A. B. (1967) "Bounds for certain thermodynamic averages," J. Math. Phys. 8, 1044-1045.

--- (1968) "Energy width of spin waves in the Heisenberg ferromagnet," Phys. Rev. 175, 674-679.

---, KUMAR, D., HALPERIN, B. I., and HOHENBERG, P. C. (1970) "Dynamics of an antiferromagnet at low temperatures: Spin-wave damping and hydrodynamics," Ibid. B1, 961-1024.

HARTWIG, R. E. and FISHER, M. E. (1969) "Asymptotic behavior of Toeplitz matrices and determinants," Arch. Ratl. Mech. Anal. 32, 190-225.

1 --- and STEPHENSON, J. (1968) "Ising-model spin correlations," J. Math. Phys. 9, 836-848.

1 HATTORI, M. and NAKANO, H. (1968) "Magnetic and thermal properties of an antiferromagnetic decorated Ising lattice in finite magnetic fields," Progr. Theoret. Phys. 40, 958-974.

HAUGE, E. H. and HEMMER, P. C. (1963) "On the Yang-Lee distribution of roots," Physica 29, 1338-1344.

d ---, --- (1971) "The two-dimensional Coulomb gas," Phys. Norveg. 5, 209-218.

HEAP, B. R. (1966) "The enumeration of homeomorphically irreducible star graphs," J. Math. Phys. 7, 1582-1587.

1 HECHT, R. (1967) "Correlation functions for the two-dimensional Ising model," Phys. Rev. 158, 557.

1 HEILMANN, O. J., KLEITMAN, D., LIEB, E. H., and SHERMAN, S. (1971) "Some positive definite functions on sets and their application to the Ising model," Discrete Mathematics 1, 19-27.

---, LIEB, E. H. (1970) "Monomers and dimers," Phys. Rev. Lett. 24, 1412-1414.

---, --- (1971) "Violation of the noncrossing rule: The Hubbard Hamiltonian for benzene," Trans. N.Y. Acad. Sci. 33, 116-149.

---, --- (1972) "Theory of monomer-dimer systems," Commun. Math. Phys. 25, 190-232.

1 HEIMS, S. P. (1965) "Master equation for Ising model," Phys. Rev. A138, 587-590.

HEISENBERG, W. (1928) "Zur Theorie des Ferromagnetismus," Z. Physik 49, 619-636.

r HELFAND, E. (1963) "Statistical mechanics," Ann. Rev. Phys. Chem. 14, 117-139.

--- (1964) "Approach to a phase transition in a one-dimensional system," J. Math. Phys. 5, 1-6.

--- (1964) "Long-range interactions," In The Equilibrium Theory of Classical Fluids, H. L. Frisch and J. L. Lebowitz, Eds. (W. A. Benjamin, Inc., N.Y.).

1 --- (1969) "Ising model: Field-theoretic and functional-integral aspects," Phys. Rev. 180, 600-612.

∞ --- (1969) "Three-dimensional magnetic model with classical spins of high dimensionality," Ibid. 183, 562.

---, FRISCH, H. L., and LEBOWITZ, J. L. (1961) "Theory of two- and one-dimensional rigid sphere fluids," J. Chem. Phys. 34, 1037-1042.

1 --- and LANGER, J. S. (1967) "Critical correlations in the Ising model," Phys. Rev. 160, 437-450.

m --- and STILLINGER, F. H., JR. (1968) "Critical solution behavior in a binary mixture of gaussian molecules. II," J. Chem. Phys. 49, 1232-1242.

b --- and TAGAMI, Y. (1971) "Theory of the interface between immiscible polymers," Polymer Lett. 9, 741-746.

L HELFRICH, W. (1969) "Capillary flow of cholesteric and smectic liquid crystals," Phys. Rev. Lett. 23, 372-374.

HELLEMAN, R. H. (1971) "Fluctuation spectra and quasithermodynamics of a linearized Markov process," Phys. Rev. A3, 1124-1133.

3 HELLER, G. and KRAMERS, H. A. (1934) "Ein klassisches Modell des Ferromagnetikums und seine nachträgliche Quantisierung im Gebiete tiefer Temperaturen," Proc. Roy. Acad. Amsterdam 37, 378-385.

HELLER, P. (1966) "Nuclear-magnetic-resonance studies of critical phenomena in MnF_2 I. Time-average properties," Phys. Rev. 146, 403-422.

r --- (1967) "Experimental investigations of critical phenomena," Rept. Progr. Phys. 30, 731-826.

--- (1970) "An interpretation of the behavior of the longitudinal scattering in MnF_2 below the critical temperature," Int. J. Magnetism 1, 53-59.

--- and BENEDEK, G. B. (1962) "Nuclear magnetic resonance in MnF_2 near the critical point," Phys. Rev. Lett. 8, 428-432.

---, SCHULHOF, M. P., NATHANS, R., and LINZ, A. (1971) "Inelastic neutron scattering studies of critical fluctuations in MnF_2 above and below T_N," J. Appl. Phys. 42, 1258-1264.

1 HELM, M. (1970) "Correlations in the Ising model," Phys. Lett. 33A, 513-514.

1 --- (1972) "The phase transition in the Ising-lattice," Physica 57, 46-56.

3 HEMMANN, J. H. and BROWN, H. A. (1970) "Note on approximate free energies for Heisenberg ferromagnets'," Phys. Rev. B2, 230-231.

HEMMER, P. C. (1964) "On the van der Waals theory of the vapor-liquid equilibrium I.V. The pair correlation function and equation of state for long-range forces," J. Math. Phys. 5, 75-84.

HEMMER, P. C. and HAUGE, E. H. (1964) "Yang-Lee distribution of zeros for a van der Waals gas," Phys. Rev. 133, A1010-1015.

---, ---, and AASEN, J. O. (1966) "Distribution of zeros of the grand partition function," J. Math. Phys. 7, 35-39.

---, KAC, M., and UHLENBECK, G. E. (1964) "On the van der Waals theory of the vapor-liquid equilibrium III. Discussion of the critical region," Ibid. 5, 60-74.

m --- and STELL, G. (1970) "Fluids with several phase transitions," Phys. Rev. Lett. 24, 1284-1287.

HENDERSON, A. J. and ROGERS, R. N. (1966) "Exchange and the 10/3 effect in $K_2CuCl_4 \cdot 2H_2O$ and $(NH_4)_2CuCl_4 \cdot 2H_2O$," Phys. Rev. 152, 218-222.

r HENDERSON, D. (1964) "The theory of liquids and dense gases," Ann. Rev. Phys. Chem. 15, 31-62.

r ---, BARKER, J. A., and KIM, S. (1969) "Theory of the liquid state," Int. J. Quantum Chem. 3S, 265-292.

HENRY, D., SWINNEY, H., CUMMINS, H. (1970) "Rayleigh linewidth in xenon near the critical point," Phys. Rev. Lett. 25, 1170-1173.

HERGET, C. M. (1940) "Ultrasonic velocity in carbon dioxide and ethylene in the critical region," J. Chem. Phys. 8, 537-542.

HERTZ, J. A. (1971) "Critical spin fluctuations in itinerant electron ferromagnets," Int. J. Magnetims 1, 253-269.

--- (1971) "Critical spin fluctuations in itinerant electron ferromagnets II. Sum-rule methods," Ibid. 1, 307-312.

--- (1971) "Critical spin fluctuations in itinerant electron ferromagnets III. Boltzmann equation approach," Ibid. 1, 313-317.

b HERZFELD, J. and STANLEY, H. E. (1972) "A general model of cooperativity and its application to DPG Inhibition of Hemoglobin Oxygenation," Biochemical and Biophysical Research Commun. 48, 307-313.

b ---, --- (1972) "Cooperativity in biological systems: A general quantitative model with specific application to hemoglobin oxygenation," In Biomedical Physics and Biomaterials Science, H. E. Stanley, Ed. (MIT Press, Cambridge), p. 63-83.

b ---, --- (1973) "A general model of cooperativity and its application to the oxygen equilibrium of hemoglobin," Proc. Fourth Internat. Biophys. Congress, Moscow, Aug. 1972.

h HESS, G. B. (1971) "Critical velocities in superfluid helium flow through 10-μm-diameter pinholes," Phys. Rev. Lett. 27, 977-979.

l HILEY, B. J. and JOYCE, G. S. (1965) "The Ising model with long-range interactions," Proc. Phys. Soc. 85, 493-507.

b --- and SYKES, M. F. (1961) "Probability of initial ring closure in the restricted random-walk model a macromolecule," J. Chem. Phys. 34, 1531-1537.

b HILL, T. L. (1959) "Generalization of the one-dimensional Ising model applicable to helix transitions in nucleic acids and proteins," Ibid. 30, 383-387.

b --- (1969) "A proposed common allosteric mechanism for active transport, muscle contraction, and ribosomal translocation," Proc. Natl. Acad. Sci. 64, 267-274.

m HIWATARI, Y. and MATSUDA, H. (1972) "Ideal three-phase model and the melting of molecular crystals and metals," Prog. Theoret. Phys. 47, 741-764.

HO, J. T. and LITSTER, J. D. (1969) "Divergences of the magnetic properties of $CrBr_3$ near the critical point," J. Appl. Phys. 40, 1270-1271.

---, --- (1969) "Magnetic equation of state of $CrBr_3$ near the critical point," Phys. Rev. Lett. 22, 603-606.

---, --- (1970) "Faraday rotation near the ferromagnetic critical temperature of $CrBr_3$," Phys. Rev. A2, 4523-4532.

HOFMANN, J. A., PASKIN, A., TAUER, K. J., and WEISS, R. J. (1956) "Analysis of ferromagnetic and antiferromagnetic second-order transitions," Phys. Chem. Solids 1, 45-60.

d HOHENBERG, P. C. (1967) "Existence of long-range order in one and two dimensions," Phys. Rev. 158, 383-386

s --- (1968) "Critical phenomena and their bearing on the superconducting transition," In Fluctuations and Superconductors, (W. S. Gorce and F. Chilton, eds.) Stanford Research Institute, Menlo Park, p. 305-339.

--- and BARMATZ, M. (1972) "Gravity effects near the gas-liquid critical point," Phys. Rev. A6, 289-313.

h --- and MARTIN, P. C. (1965) "Microscopic theory of superfluid helium," Ann. Phys. 34, 291-359.

HOLMES, L. M. and VAN UITERT, L. G. (1972) "Magnetoelectric effect and metamagnetic transitions in $DyAlO_3$," Phys. Rev. B5, 147-153.

m ---, ---, HECKER, R. R., and HULL, G. W. (1972) "Magnetic behavior of metamagnetic $DyAlO_3$," Ibid. B5, 138-146.

---, ---, and HULL, G. W. (1971) "Magnetoelectric effect and critical behavior in the Ising-like antiferromagnet, $DyAlO_3$," Sol. State Commun. 9, 1373-1376.

3 HONE, D. and PETZINGER, K. G. (1972) "Spin correlations in impure linear Heisenberg chains," Phys. Rev. B6, 245-254.

HOOVER, W. G., YOUNG, D. A., and GROVER, R. (1972) "Statistical mechanics of phase diagrams. I. Inverse power potentials and the close-packed to body-centered cubic transition," J. Chem. Phys. 56, 2207-2211.

HOPKINSON, J. (1890) "Magnetic properties of alloys of nickel and iron," Proc. Roy. Soc. 48, 1-13.

HORIGUCHI, T. (1971) "Lattice Green's function for the simple cubic lattice," J. Phys. Soc. Japan 30, 1261-1272.

HORNER, H. (1967) "Two phonon bound state and phase transition in $SrTiO_3$," Phys. Lett. 25A, 464-465.

--- (1972) "First order phase transition in an anharmonic Ising lattice," Z. Physik 251, 202-213.

HORWITZ, G., BROUT, R., and ENGLERT, F. (1963) "Zero-temperature properties of the many-fermion system," Phys. Rev. 130, 409-419.

1 --- and CALLEN, H. B. (1961) "Diagrammatic expansion for the Ising model with arbitrary spin and range of interaction," Ibid. 124, 1757-1785.

f HOUSTON, G. D. and BOLTON, H. C. (1971) "Quantum field-theoretical treatment of a dynamical model of the ferroelectric KDP," J. Phys. C. 4, 2097-2108.

f ---, --- (1971) "Quantum field theoretical treatment of a dynamical model of the ferroelectric KDP. II," Ibid. 4, 2894-2902.

1 HOUTAPPEL, R. M. F. (1950) "Order-disorder in hexagonal lattices," Physica 16, 425-455.

HOWARD, D. G., DUNLAP, B. D., and DASH, J. G. (1965) "Internal field of Fe^{57} in Ni in the region of the Curie point," Phys. Rev. Lett. 15, 628-632.

HUANG, C. C. and HO, J. T. (1971) "Critical specific-heat scaling in the linear model for fluids," Phys. Lett. 37A, 149-151.

HUANG, J. S. and WEBB, W. W. (1969) "Viscous damping of thermal excitations on the interface of critical fluid mixtures," Phys. Rev. Lett. 23, 160-163.

r HUANG, K. (1963) Statistical Mechanics, Wiley, N.Y.

r --- (1964) "Imperfect Bosegas," in Studies in Statistical Mechanics, J. de Boer and G. E. Uhlenbeck, Eds. (North-Holland, Amsterdam), Vol. 2, p. 1-106.

HUBBARD, J. (1959) "Calculation of partition functions," Phys. Rev. Lett. 3, 77-78.

3 --- (1971) "Spin-correlation functions in the paramagnetic phase of a Heisenberg ferromagnet," J. Phys. C. 4, 53.

1 --- (1972) "Critical behaviour of the Ising model," Phys. Lett. 39A, 365-368.

g --- (1972) "Scaling relations and the Wilson theory," Ibid. 40A, 111-113.

g --- and SCHOFIELD, P. (1972) "Wilson theory of a liquid vapour critical point," Ibid. 40A, 245-246.

3 HUBER, D. L. (1968) "Thermal conductivity of a Heisenberg chain," Progr. Theoret. Phys. 39, 1170-1179.

--- (1968) "Contribution of the diffuse modes to the thermal conductivity of a magnetic lattice near the Curie point," Sol. State Commun. 6, 685-686.

--- (1971) "Critical damping of the uniform mode in anisotropic antiferromagnets: MnF_2," Int. J. Magnetism 2, 405-408.

--- (1971) "Critical dynamics of the staggered magnetization in uniaxial antiferromagnets," Int. J. Quantum Chem. 5, 667-671.

--- (1971) "Spin-lattice relaxation near the critical point: $RbMnF_3$, MnF_2, EuO," Phys. Rev. B3, 836-842.

--- (1972) "Nonlinear effects in the critical dynamics of easy-axis ferro- and antiferromagnets," Ibid. B5, 1980-1986.

--- (1972) "Critical-point anomalies in the electron-paramagnetic-resonance linewidth and in the zero-field relaxation time of antiferromagnets," Ibid. B6, 3180-3186.

--- and SEMURA, J. S. (1969) "Spin and energy transport in anisotropic magnetic chains with S = 1/2," Ibid. 182, 602-603.

---, ---, and WINDSOR, C. G. (1969) "Energy transport in magnetic chains," Ibid. 186, 534-537.

HUISKAMP, W. J. (1966) "Experimental results on specific heat anomalies of magnetic salts below 1°K," Ann. Acad. Sci. Fenn. A VI, 180-193.

HUNT, J. T. and GIRARDEAU, M. D. (1967) "Anisotropic Heisenberg linear chain at nonzero temperature," Phys. Rev. 160, 455-462.

l HUNT, R. A. (1971) "A model of magnetostriction in an Ising system and its application to $PrCl_3$ and $NdCl_3$," J. Phys. C. 4, 2572-2582.

l HUNTER, D. L. (1969) "Padé approximant analysis of the Ising model specific heat above the Curie temperature," Ibid. 2, 941-947.

l HURST, C. A. (1963) "Solution of plane Ising lattices by the Pfaffian method," J. Chem. Phys. 38, 2558-2571.

l --- (1964) "Applicability of the Pfaffian method to combinatorial problems on a lattice," J. Math. Phys. 5, 90-100.

l --- (1965) "Relation between the Onsager and Pfaffian methods for solving the Ising problem. I. The rectangular lattice," Ibid. 6, 11-18.

l --- (1966) "Relation between the Onsager and Pfaffian methods for solving the Ising problem. II. The general lattice," Ibid. 7, 81-87.

l --- (1966) "New approach to the Ising problem," Ibid. 7, 305-310.

HUTCHINGS, M. T., SCHULHOF, M. P., and GUGGENHEIM, H. J. (1972) "Critical magnetic neutron scattering from ferrous fluoride," Phys. Rev. B5, 154-168.

d ---, SHIRANE, G., BIRGENEAU, R. J., and HOLT, S. L. (1972) "Spin dynamics in the one-dimensional antiferromagnet $(CD_3)_4NMnCl_3$," Ibid. B5, 1999-2014.

h ICHIYANAGI, M. (1969) "Attenuation of first sound waves near the λ-point of liquid helium," Progr. Theoret. Phys. 42, 147-157.

IKUSHIMA, A. (1969) "Sound velocity near the Néel point of MnF_2," Phys. Lett. 29A, 364-365.

--- (1969) "Ultrasonic attenuation near the antiferromagnetic critical point of CoO," Ibid. 29A, 417-418.

h IMAI, J. S. and RUDNICK, I. (1969) "Ultrasonic attenuation in liquid helium at 1 GHz," Phys. Rev. Lett. 22, 694-697.

d IMRY, Y. (1969) "Effective long range order and phase transitions in finite, macroscopic one and two dimensional systems," Ann. Phys. 51, 1-27.

d --- (1969) "Do long range quantum phase fluctuations inhibit superfluidity in thin systems?," Phys. Lett. 29A, 82-83.

d --- and BERGMAN, D. (1971) "Critical points and scaling laws for finite systems," Phys. Rev. A3, 1416-1418.

d --- and GUNTHER, L. (1969) "Is there Bragg-scattering off a two-dimensional crystal?," Phys. Lett. 29A, 483-484.

d ---, --- (1971) "Fluctuations and physical properties of the two-dimensional crystal lattice," Phys. Rev. B3, 3939-3945.

m IRANI, N. F. and RICE, O. K. (1967) "Coexistence curve of the cyclohexane+methylene iodide system in the critical region," Trans. Faraday Soc. 63, 2158-2162.

ISHII, K. (1968) "Lattice vibration of a linear chain with long-range interaction," Progr. Theoret. Phys. 39, 593-600.

d ISHIKAWA, T. and OGUCHI, T. (1971) "Critical behavior of the spin system with anisotropic exchange interaction. II. Two-dimensional lattice," J. Phys. Soc. Japan 31, 1021-1025.

ISIHARA, A. (1957) "Defects on a dipole lattice," J. Chem. Phys. 27, 1174-1179.

--- (1968) "Correlation functions in ring-diagram approximation," Phys. Rev. 172, 166-172.

b --- (1968) "Irreversible processes in solutions of chain polymers," Adv. Polymer Sci. 5, 531-567.

--- (1968) "The Gibbs-Bogoliubov inequality," J. Phys. A. 1, 539-548.

--- (1969) "Corrections to the pair distribution function in the ring-diagram approximation," Phys. Rev. 178, 412-419.

l ISING, (1925) "Beitrag zur Theorie des Ferromagnetismus," Z. Physik 31, 253-258.

IZUYAMA, T. (1968) "Spin-spin correlation function at the transition point of itinerant electron ferromagnet," Prog. Chem. Phys. 39, 1365-1367.

--- (1972) "General theorems on ferromagnetism and ferromagnetic spin waves," Phys. Rev. B5, 190-203.

m JACOBS, I. S. and LAWRENCE, P. E. (1967) "Metamagnetic phase transitions and hysteresis in $FeCl_2$," Ibid. 164, 866-878.

m JACOX, M. E., MAC QUEEN, J. T., and RICE, O. K. (1960) "A dilatometric study of the cyclohexane-aniline system near its critical separation temperature," J. Phys. Chem. 64, 972-975.

JACROT, B., KONSTANTINOVIĆ, J., PARETTE, G., and CRIBIER, D. (1963) "Inelastic scattering of neutrons in solids and liquids," IAEA Proc., Vol. II, p. 317.

--- and RISTE, T. (1965) "Magnetic inelastic scattering of neutrons," In Thermal Neutron Scattering, P. A. Egelstaff, Ed. (Academic Press, N.Y.), p. 251-290.

L JÄHNIG, F. and SCHMIDT, H. (1972) "Hydrodynamics of liquid crystals," Ann. Phys. 71, 129-166.

f JAISWAL, V. K. and SHARMA, P. K. (1968) "Variation of Curie temperature with electric field in ferroelectric crystals," Physica 38, 409-414.

l JAMES, H. M. (1971) "On the Ising lattice with randomly distributed bonds of two strengths," Phys. Norveg. 5, 285-290.

d JANCOVICI, B. (1967) "Infinite susceptibility without long-range order: The two-dimensional harmonic 'solid'," Phys. Rev. Lett. 19, 20-22.

d JASNOW, D. and FISHER, M. E. (1969) "Broken symmetry and decay of order in restricted dimensionality," Ibid. 23, 286-288.

d ---, --- (1971) "Decay of order in isotropic systems of restricted dimensionality. I. Bose superfluids," Phys. Rev. B3, 895-907.

l --- and WORTIS, M. (1967) "New derivation of the Ising model cluster expansion," J. Math. Phys. 8, 507-511.

u ---, --- (1968) "High-temperature critical indices for the classical anisotropic Heisenberg model," Phys. Rev. 176, 739-748.

JEEVANANDAM, M. (1971) "Vapor pressure isotope effect near the triple point," J. Chem. Phys. 55, 5735-5741.

JENNINGS, L. D. and HANSEN, W. N. (1965) "Heat capacity of $CrBr_3$ from 14 to 360°K," Phys. Rev. 139, 1694-1697.

JIMBO, T. and ELBAUM, C. (1972) "Ultrasonic studies of antiferromagnetic resonance in $RbMnF_3$ near the Néel temperature," Phys. Rev. Lett. 28, 1393-1396.

JOHANSSON, B. (1971) "Effects of impurities on the Curie temperature," Phys. Lett. 34A, 154-155.

h JOHNSON, D. L. and CROOKS, M. J. (1968) "Second sound velocities in He II near $T\lambda$," Ibid. 27A, 688-689.

h ---, --- (1969) "Second-sound velocities, superfluid hydrodynamics and scaling laws in helium II," Phys. Rev. 185, 253.

2 JOHNSON, J. D. and MC COY, B. M. (1971) "Off-diagonal time-dependent spin correlation functions of the X-Y model," Ibid. A4, 2314-2324.

---, ---, and LAI, C. K. (1972) "On Takahashi's study of the thermodynamics of the Heisenberg-Ising ring for $|\Delta| < 1$," Phys. Lett. 38A, 143-144.

JONES, G. L. (1966) "Complex temperatures and phase transitions, J. Math. Phys. 7, 2000-2005.

JONES, G. J. (1968) "Structure of correlation function near the critical point," Phys. Rev. 171, 243-247.

JOSEPH, R. I. (1967) "High-temperature susceptibility of an exchange-interaction model of ferromagnetism," Ibid. 163, 523-526.

---, COOPER, M. J., AND VICENTINI-MISSONI, M. (1969) "Linear parametrization of the equation of state near the critical point," Phys. Rev. Lett. 23, 70-73.

h JOSEPHSON, B. D. (1966) "Relation between the superfluid density and order parameter for superfluid He near T_c," Phys. Lett. 21, 608-609.

--- (1967) "Inequality for the specific heat I. Derivation," Proc. Phys. Soc. 92, 269-275.

--- (1967) "Inequality for the specific heat II. Application to critical phenomena," Ibid. 92, 276-284.

--- (1969) "Equation of state near the critical point," J. Phys. C. 2, 1113-1115.

JOYCE, G. S. (1966) "Spherical model with long-range ferromagnetic interactions," Phys. Rev. 146, 349-358.

3 --- (1967) "Classical Heisenberg model," Ibid. 155, 478-491.

--- (1967) "Exact results for the one-dimensional, anisotropic classical Heisenberg model," Phys. Rev. Lett. 19, 581-583.

--- (1966) "Absence of ferromagnetism or antiferromagnetism in the isotropic Heisenberg model with long-range interactions," J. Phys. C. 2, 1531-1533.

--- (1971) "Lattice Green function for the anisotropic face center cubic lattice," Ibid. 4, L53-L56.

3 --- and BOWERS, R. G. (1966) "Cluster series for the infinite spin Heisenberg model," Proc. Phys. Soc. 88, 1053-1055.

3 ---, --- (1966) "High-temperature specific heat and susceptibility of the classical Heisenberg model," Ibid. 89, 776-778.

JUDIN, V. M., SHERMAN, A. B., and MYLINIKOVA, I. E. (1966) "Magnetic properties of $YFeO_3$," Phys. Lett. 22, 554-555.

KAC, M. (1959) "On the partition function of a one-dimensional gas," Phys. Fluids 2, 8-12.

r --- (1969) "Some mathematical models in science," Science 166, 695.

--- and HELFAND, E. (1963) "Study of several lattice systems with long range forces," J. Math. Phys. 4, 1078-1088.

--- and THOMPSON, C. J. (1966) "On the mathematical mechanism of phase transition," Proc. Natl. Acad. Sci. U.S. 55, 676-683.

---, --- (1969) "Critical behavior of several lattice models with long-range interaction," J. Math. Phys. 10, 1373-1386.

∞ ---, --- (1971) "Spherical model and the infinite spin dimensionality limit," Phys. Norveg. 5, 163-167.

---, UHLENBECK, G. E., and HEMMER, P. C. (1963) "On the van der Waals theory of the vapor-liquid equilibrium. I. Discussion of a one-dimensional model," J. Math. Phys. 4, 216-228.

l KADANOFF, L. P. (1966) "Spin-spin correlations in the two-dimensional Ising model," Nuovo Cimento 44, 276-305.

--- (1966) "Scaling laws for Ising models near T_c," Physics 2, 263-272.

r --- (1968) "Transport coefficients near critical points," Comments on Sol. State Phys. 1, 5-10.

r --- (1969) "Mode-mode coupling near the critical point," J. Phys. Soc. Japan 26, 122-126.

l --- (1969) "Correlations along a line in the two-dimensional Ising model," Phys. Rev. 188, 859-863.

u --- (1969) "Operator algebra and the determination of critical indices," Phys. Rev. Lett. 23, 1430-1433.

r --- and BAYM, G. (1962) Quantum Statistical Mechanics, W. A. Benjamin, Inc., N.Y.

l --- and CEVA, H. (1971) "Determination of an operator algebra for the two-dimensional Ising model," Phys. Rev. B3, 3918-3939.

r ---, GÖTZE, W., HAMBLEN, D., HECHT, R., LEWIS, E. A. S., PALCIAUSKAS, V. V., RAYL, M., SWIFT, J., ASPNES, D., and KANE, J. (1967) "Static phenomena near critical points: Theory and experiment," Rev. Mod. Phys. 39, 395-431.

s --- and LARAMORE, G. (1968) "Anomalous electrical conductivity above the superconducting transition," Phys. Rev. 175, 579-584.

--- and MARTIN, P. C. (1963) "Hydrodynamic equations and correlation functions," Ann. Phys. 24, 419-469.

--- and SWIFT, J. (1968) "Transport coefficients near the critical point: A master-equation approach," Phys. Rev. 165, 310-322.

---, --- (1968) "Transport coefficients near the liquid-gas critical point," Ibid. 166, 89-101.

f --- and WEGNER, F. J. (1971) "Some critical properties of the eight-vertex model," Ibid. B4, 3989-3993.

KAMENSKIǏ, V. G. (1971) "High-frequency sound near the Curie point of ferromagnets," Sov. Phys.-JETP 32, 1214.

l KANAMORI, J. (1966) "Magnetization process in an Ising spin system," Progr. Theoret. Phys. 35, 16-35.

h KANE, J. W. and KADANOFF, L. P. (1967) "Long-range order in superfluid helium," Phys. Rev. 155, 80-83.

m KANEKO, T. (1971) "Pressure effect on the Curie points of the compounds $CuCr_2X_4$ (X=S, Se, Te) and MPt_3 (M=Cr, Mn, Co)," Sci. Rept. Res. Inst. Tohoku 23, 92.

1 KANO, K. (1966) "Short-range order for the triangular and honeycomb Ising nets in ferro-and anti-ferromagnetic cases," Progr. Theoret. Phys. 35, 1-15.

KAPLAN, D. M. and SUMMERFIELD, G. S. (1969) "Quasiclassical methods in spin dynamics," Phys. Rev. 187, 639-642.

KAPLAN, D. M. M. and WU, F. Y. (1971) "On the eigenvalues of orbital angular momentum," Chinese J. Phys. 9, 31-33.

KAPLAN, J. I. and STENSCHKE, H. (1970) "Absolutely stable and metastable thermodynamic states of a magnetic model exhibiting a first-order phase transition," Am. J. Phys. 38, 1323-1326.

3 KAPLAN, T. A., STANLEY, H. E., DWIGHT, K., and MENYUK, N. (1965) "Determination of magnetic ordering in Heisenberg magnets from high-temperature expansions," J. Appl. Phys. 36, 1129-1130.

d KARIMOV, YU. S. (1972) "Transition of a two-dimensional paramagnet into a magnetically ordered state," Sov. Phys.-JETP Lett. 15, 235-237.

d ---, VOL'PIN, M. E., and NOVIKOV, YU. N. (1971) "Layered compounds of $NiCl_2$ and $CoCl_2$ with graphite as two-dimensional Heisenberg ferromagnets," Ibid. 14, 142-144.

d ---, ZVARYKINA, A. V., and NOVIKOV, YU. N. (1972) "Two-dimensional ferromagnetism in layered compounds of graphite with iron chlorides," Sov. Phys.-Solid State 13, 2388-2391.

b KARLIN, A. (1967) "On the application of 'a plausible model' of allosteric proteins to the receptor for acetylcholine," J. Theoret. Biol. 16, 306-320.

1 KASAI, Y., MIYAZIMA, S., and SYOZI, I. (1969) "Ising lattices with random arrangements of ferromagnetic and antiferromagnetic bonds," Progr. Theoret. Phys. 42, 1-8.

KASHCHEEV, V. N. (1968) "Phonon spectral density in a ferromagnet near the Curie point," Phys. Lett. 26A, 123-124.

--- (1972) "Critical damping coefficient of a phonon in ferromagnets and antiferromagnets," Sov. Phys.-Solid State 13, 2856-2857.

KASTELEYN, P. W. (1963) "Dimer statistics and phase transitions," J. Math. Phys. 4, 287-293.

KASUYA, T. (1956) "Electrical resistance of ferromagnetic metals," Progr. Theoret. Phys. 16, 58-63.

KATSURA, S. (1962) "Statistical mechanics of the anisotropic linear Heisenberg model," Phys. Rev. 127, 1508-1518.

--- (1965) "Two-body problem in the one-dimensional Heisenberg model," Ann. Phys. 31, 325-341.

--- (1967) "Distribution of roots of the partition function in the complex temperature plane," Progr. Theoret. Phys. 38, 1415-1416.

3 --- (1970) "Critical field of the Heisenberg model at zero temperature," J. Phys. Soc. Japan 28, 255.

f ---, ABE, Y., and OHKOHCHI, K. (1970) "Distribution of zeros of the partition function for the Slater model of ferroelectricity," Ibid. 29, 845-850.

--- and HORIGUCHI, T. (1968) "First order Green function theory of ferromagnetism," Ibid. 25, 60-72.

2 ---, ---, and SUZUKI, M. (1970) "Dynamical properties of the isotropic XY model," Physica 46, 67-86.

--- and INAWASHIRO, S. (1965) "An exactly soluable model showing ferromagnetism," J. Math. Phys. 6, 1916-1922.

---, MORITA, T., INAWASHIRO, S., HORIGUCHI, T., and ABE, Y. (1971) "Lattice Green's function. Introduction," Ibid. 12, 892-895.

1 --- and OHMINAMI, M. (1972) "Distribution of zeros of the partition function for the one dimensional Ising model," J. Phys. A. 5, 95.

3 --- and SUZUKI, M. (1970) "Critical field of the Heisenberg model at zero temperature," J. Phys. Soc. Japan 28, 255.

KATYL, R. H. and INGARD, U. (1967) "Line broadening of light scattered from a liquid surface," Phys. Rev. Lett. 19, 64-66.

---, --- (1968) "Scattering of light by thermal ripplons," Ibid. 20, 248-249.

d KATZ, E. I. (1970) "Quasi-two-dimensional nature of the superconductivity of $NbSe_2$," Sov. Phys.-JETP 31, 787.

1 KAUFMAN, B. (1949) "Crystal statistics. II. Partition function evaluated by spinor analysis," Phys. Rev. 76, 1232-1243.

l KAUFMAN, B. and ONSAGER, L. (1949) "Crystal statistics. III. Short-range order in a binary Ising lattice," Phys. Rev. 76, 1244-1252.

d KAWABATA, C. and SUZUKI, M. (1970) "Statistical mechanics of the finite Heisenberg model. I," J. Phys. Soc. Japan 28, 16-28.

d --- (1970) "Statistical mechanics of the finite Heisenberg model. III," Ibid. 28, 1396-1401.

l --- and SUZUKI, M., ONO, S., and KARAKI, Y. (1968) "Zeros of the partition function for the Ising model with higher spin," Phys. Lett. 28A, 113-114.

l KAWAKAMI, A. and OSAWA, T. (1969) "Perturbational approach to the two-dimensional next nearest neighbor Ising problem," Progr. Theoret. Phys. 42, 196-212.

KAWASAKI, K. (1965) "Excitation spectra of linear magnetic chains," Phys. Rev. 142, 164-165.

--- (1966) "Some dynamical properties of linear magnetic chains. I," Ann. Phys. 37, 142-175.

l --- (1966) "Diffusion constants near the critical point for time-dependent Ising models. I," Phys. Rev. 145, 224-230.

l --- (1966) "Diffusion constants near the critical point for time-dependent Ising models. II," Ibid. 148, 375-381.

l --- (1966) "Diffusion constants near the critical point for time-dependent Ising models. III.," Ibid. 150, 285-290.

--- (1966) "Correlation-function approach to the transport coefficients near the critical point. I," Ibid. 150, 291-306.

--- (1967) "Anomalous spin diffusion in ferromagnetic spin systems," J. Phys. Chem. Solids 28, 1277-1283.

--- (1967) "Anomalous spin relaxation near the magnetic transition," Phys. Lett. 25A, 448-449.

3 --- (1967) "Spin pair correlation in the Heisenberg ferromagnet," Progr. Theoret. Phys. 38, 1052-1061.

--- (1968) "Dynamics of critical fluctuations. I," Ibid. 39, 1133-1152.

--- (1968) "Dynamics of critical fluctuations. II," Ibid. 40, 11-35.

--- (1968) "Dynamics of critical fluctuations. III," Ibid. 40, 706-733.

--- (1968) "Dynamics of critical fluctuations. IV," Ibid. 40, 930-941.

--- (1968) "Ultrasonic attenuation and ESR linewdith near magnetic critical points," Phys. Lett. 26A, 543.

--- (1968) "Anomalous spin relaxation near the magnetic transition," Progr. Theoret. Phys. 39, 285-311.

--- (1968) "Asymptotic behaviors of ultrasonic attenuation and the linewdiths of NMR and ESR of the Heisenberg paramagnets near the critical points," Sol. State Commun. 6, 57-59.

--- (1969) "A note on the ultrasonic attenuation near the magnetic critical points," Phys. Lett. 29A, 406-407.

--- (1969) "Transport coefficients of van der Waals fluids and fluid mixtures," Progr. Theoret. Phys. 41, 1190-1210.

d --- (1969) "Magnetic properties of low dimensional spin systems," Progr. Theoret. Phys. 42, 174-195.

--- (1970) "Kinetic equations and time correlation functions of critical fluctuations," Ann. Phys. 61, 1-56.

r --- (1970) "Dynamical aspects of magnetic critical phenomena," J. Appl. Phys. 41, 1311-1316.

--- (1970) "Sound attenuation and dispersion near the liquid-gas critical point," Phys. Rev. A1, 1750-1757.

d --- (1971) "Non-hydrodynamical behavior of two-dimensional fluids," Phys. Lett. 34A, 12-14.

--- and GUNTON, J. D. (1972) "A note on projection operators in non-equilibrium statistical mechanics," Ibid. 40A, 35-36.

--- and IKUSHIMA, A. (1970) "Velocity of sound in MnF_2 near the Néel temperature," Phys. Rev. B1, 3143-3151.

--- and LO, S. M. (1972) "Nonlocal shear viscosity and order-parameter dynamics near the critical point of fluids," Phys. Rev. Lett. 29, 48-51.

m --- and TANAKA, M. (1967) "Correlation function approach to bulk viscosity and sound propagation in critical mixtures," Proc. Phys. Soc. 90, 791-800.

l --- and YAMADA, T. (1968) "Time-dependent Ising model with long range interaction," Progr. Theoret. Phys. 39, 1-25.

KAWATRA, M. P. and BUDNICK, J. I. (1970) "Transport properties of ferromagnetic systems near the critical point: Electrical resistivity," Int. J. Magnetism 1, 61-74.

--- and KIJEWSKI, L. J. (1969) "Lower bound to the free energy of an anisotropic linear Heisenberg chain," Phys. Lett. 28A, 472-473.

KEEN, B. E., LANDAU, D., SCHNEIDER, B., and WOLF, W. P. (1966) "First- and higher-order magnetic phase transitions in dysprosium aluminum garnet," J. Appl. Phys. 37, 1120-1121.

r KELLER, W. E. (1969) Helium-3 and Helium-4, Plenum Press, N. Y.

l KELLY, D. G. and SHERMAN, S. (1968) "General Griffiths' inequalities on correlations in Ising ferromagnets," J. Math. Phys. 9, 466-484.

d KENAN, R. P. (1970) "Comment on 'phase transition in the two-dimensional ferromagnet'," Phys. Rev. B1, 3205-3207.

b KERNER, E. H. (1957) "A statistical mechanics of interacting biological species," Bull. Math. Biophys. 19, 121-146.

f KESSEL', A. R. and KORCHEMKIN, M. A. (1972) "Nature of phase transitions in ferroelectric and antiferroelectric materials," Sov. Phys.-Solid State 13, 2858-2859.

f KESSENIKH, G. G., SHIROKOV, A. M., SHUVALOV, L. A., and SHEHAGINA, N. M. (1971) "Attenuation of sound waves in deuterated Rochelle salt at the phase-transition point," Soviet Phys. Cryst. 15, 1097-1098.

m KHACHATRYAN, YU. M. (1971) "Effect of pressure on the Curie temperature of ferrites," Sov. Phys.-Solid State 13, 795.

KIANG, C. S. (1970) "Use of liquid-droplet model in calculations of the critical exponent δ," Phys. Rev. Lett. 24, 47-50.

h KIERSTEAD, H. A. (1971) "Pressures on the critical isochore of He^4," Phys. Rev. A3, 329-339.

KIM, D. J. (1964) "Electrical resistance in ferromagnetic metals and dilute alloys near the Curie temperature," Progr. Theoret. Phys. 31, 921-923.

KIRKPATRICK, S. (1971) "Classical transport in disordered media: Scaling and effective-medium theories," Phys. Rev. Lett. 27, 1722-1725.

b KIRTLEY, M. E. and KOSHLAND, D. E., Jr. (1967) "Models for cooperative effects in proteins containing subunits," J. Biol. Chem. 242, 4192-4205.

KITTEL, C. and SHORE, H. (1965) "Development of a phase transition for a rigorously solvable many-body system," Phys. Rev. A138, 1165-1169.

KLAMUT, J. (1968) "Thermodynamical properties of the transition to paramagnetic state of uniaxial ferromagnets with domain structures," Bull. Acad. Polon. Sci. Ser. Sci. Math., Astr. et Phys. 16, 515-521.

--- and KOZLOWSKI, G. (1968) "Local transition temperatures for magnetic crystals with domain structures," Acta Phys. Polon. 33, 743-757.

KLEBAN, P. (1969) "Inequalities for the free energy of certain many-body systems," Phys. Rev. Lett. 22, 587-590.

--- and LANGE, R. V. (1969) "Inequalities for the energy and free energy of many-body systems," Ibid. 22, 1045-1047.

KLEIN, M. J. and TISZA, L. (1949) "Theory of critical fluctuations," Phys. Rev. 76, 1861-1868.

l KLEIN, M. W. (1969) "Molecular-field theory of a random Ising-system in the presence of an external magnetic field," Ibid. 188, 933-941.

--- and BROUT, R. (1963) "Statistical mechanics of dilute copper manganese," Ibid. 132, 2412-2426.

KOBE, S. and HANDRICH, K. (1971) "Curie temperature of an amorphous ferromagnet in effective field approximations," Phys. Stat. Sol. (b)44, K53.

KOCIŃSKI, J. (1968) "Critical magnetic moment diffusion in ferromagnets," Phys. Lett. 28A, 128-129.

--- and MRYGOŃ, B. (1967) "Critical magnetic scattering of neutrons and the $|\sin k_2 r|/r$ spin correlation," Ibid. 25A, 600-601.

---, WOJTCZAK, L., and MYRGOŃ, B. (1971) "Correlation of spins in ferromagnets in the immediate vicinity of the Curie point," Ibid. 36A, 171-173.

m KOHLER, F. and RICE, O. K. (1957) "Coexistence curve of the triethylamine-water system," J. Chem. Phys. 26, 1614-1618.

KOMAROV, L. I. and FISHER, I. Z. (1963) "Theory of Rayleigh scattering of light in liquids," Sov. Phys.-JETP 16, 1358-1361.

d KOPPEN, J., HAMERSMA, R., LEBESQUE, J. V., and MIEDEMA, A. R. (1967) "Observation of two transition temperatures in the heat capacity of a two-dimensional ferromagnet," Phys. Lett. 25A, 376-377.

m KOROTKIKH, A. M. and NABUTOVSKII, V. M. (1971) "Critical point of phase transitions of first order but close to second order," JETP Lett. 13, 147-149.

l KORTMAN, P. J. and GRIFFITHS, R. B. (1971) "Density of zeros on the Lee-Yang circle for two Ising ferromagnets," Phys. Rev. Lett. 27, 1439-1442.

b KOSEVICH, A. M. and GALKIN, V. L. (1971) "Phase transition in a double polymer chain in an external field," Sov. Phys.-JETP 33, 444-448.

--- and SHKLOVSKIĬ, V. A. (1971) "Phase transitions and spontaneous magnetization in an Ising lattice containing impurities," Ibid. 33, 588-593.

b KOSHLAND, D. E., Jr., and NEMETHY, G., and FILMER, D. (1966) "Comparison of experimental binding data and theoretical models in proteins containing subunits," Biochemistry 5, 365-385.

KOUVEL, J. S. and COMLY, J. B. (1968) "Magnetic equation of state for nickel near its Curie point," Phys. Rev. Lett. 20, 1237-1239.

--- and FISHER, M. E. (1964) "Detailed magnetic behavior of nickel near its Curie point," Phys. Rev. 136, 1626-1632.

l KOWALSKI, J. M. (1970) "Order parameter and the ground state energy for the Ising model with the perpendicular field," Acta Phys. Polon. A38, 255.

l --- (1972) "The double Ising chain," Ibid. A41, 57-66.

s KOYANAGI, M. and TSUZUKI, T. (1969) "Frequency dependence of the enhanced diamagnetism above the superconducting transition point," Phys. Lett. 30A, 405-406.

KOZAK, J. J., SCHRODT, I. B., and LUKS, L. D. (1972) "Square well potential. II. Some comments on critical point behaviour and scaling laws," J. Chem. Phys. 57, 206-209.

m KOZLOWSKI, G. (1971) "Influence of the field direction on the magnetic phases of a uniaxial two-sublattice antiferromagnet. I. Ground state energies, critical fields and magnetizations," Acta Phys. Polon. A40, 333.

KRAFTMAKHER, Y. A. (1967) "Electrical conductivity of nickel close to the Curie point," Sov. Phys.-Solid State 9, 1199-1200.

KRAMERS, H. A. (1927) "Investigations on the free energy of a mixture of ions," Proc. Amst. Acad. Sci. 30, 145.

--- (1936) "Theory of ferromagnetism," Comm. Kamerlingh Onnes Lab. Leiden Supp. No. 83. [Reprinted in Collected Scientific Papers of H. A. Kramers (North Holland Publ. Co., Amsterdam, 1956), p. 607-628.

l --- and WANNIER, G. H. (1941) "Statistics of the two-dimensional ferromagnet. Part I," Phys. Rev. 60, 252-262.

---, --- (1941) "Statistics of a two-dimensional ferromagnet. Part II," Ibid. 60, 263-276.

u KRASNOW, R., HARBUS, F., LIU, L. L., and STANLEY, H. E. (1973) "Scaling with respect to a parameter for the Gibbs potential and pair correlation function of the S=1/2 Ising model with directional anisotropy," Phys. Rev. B7, 370.

KRÉN, E., KÁDÁR, G., PÁL, L., SÓLYOM, J., SZABÓ, P., and TARNÓCZI, T. (1968) "Magnetic structures and exchange interactions in the Mn-Pt system," Ibid. 171, 574-585.

2 KRINSKY, S. (1972) "Equivalence of the free fermion model to the ground state of the linear XY model," Phys. Lett. 39A, 169-171.

KROÓ, N. and PÁL, L. (1968) "Inelastic scattering of neutrons by virtual magnon states in dilute alloys," J. Appl. Phys. 39, 453-454.

d KRUEGER, D. A. (1967) "Impossibility of Bose condensation or superconductivity in partially finite geometries," Phys. Rev. Lett. 19, 563-565.

d --- (1968) "Finite geometries and ideal Bose gases," Phys. Rev. 172, 211-223.

--- (1971) "Energy transport at finite temperatures in isotropic magnetic chains with S=1/2," Ibid B3, 2348-2351.

--- and HUBER, D. L. (1968) "Theoretical estimate of the spin diffusion constant in iron," Sol. State Commun. 6, 869-871.

---, --- (1970) "Critical spin relaxation in anisotropic ferromagnets," Phys. Rev. B1, 3152-3159.

m KRZEMIŃSKI, S. (1971) "Remark on the field-induced magnetic phases in uniaxial antiferrimagnets with weak anisotropy," Acta Phys. Polon. A39, 201.

KUBO, R. (1957) "Statistical-mechanical theory of irreversible processes. I," J. Phys. Soc. Japan 12, 570-586.

--- (1965) Statistical Mechanics, North-Holland, Amsterdam.

KUMBAR, M. and WINDWER (1971) "Study of self-avoiding walks with excluded first nearest neighbors," J. Chem. Phys. 54, 5051-5058.

KURAMOTO, Y. (1968) "Nonlinear effects on critical fluctuations," Progr. Theoret. Phys. 39, 587-592.

--- (1969) "Modified Landau theory of the second order phase transition," Ibid. 41, 604-618.

KUZMIN, V. L. and KUNI, F. M. (1971) "On an additional relation in the scaling theory," Phys. Lett. 37A, 397-399.

LAI, C. C. and CHEN, S. H. (1972) "Evidence of mode-mode coupling and nonlocal shear viscosity in a binary mixture near the consolute point," Phys. Rev. Lett. 29, 401-404.

--- (1971) "Thermodynamics of fermions in one dimension with a δ-function interaction," Ibid. 26, 1472-1475.

--- and YANG, C. N. (1971) "Ground state energy of a mixture of fermions and bosons in one dimension with a repulsive δ-function interaction," Phys. Rev. A3, 393-399.

u LAMBETH, D. and STANLEY, H. E. (1972) "The planar and Heisenberg models with directional anisotropy: Evidence for a phase transition and susceptibility exponent in the d=2 limit," Phys. Rev. (in press).

m LANDAU, D. P. (1972) "Magnetic tricritical points in Ising antiferromagnets," Ibid. 28, 449-452.

---, KEEN, B. E., SCHNEIDER, B., and WOLF, W. P. (1971) "Magnetic and thermal properties of dysprosium aluminum garnet I. Experimental results for the two-sublattice phases," Phys. Rev. B3, 2310-2343.

LANDAU, L. (1937) "On the theory of phase transitions," reprinted in English translation in D. TerHaar, Collected Works of L. D. Landau (Gordon and Breach, N.Y., 1965).

r --- and LIFSHITZ, E. M. (1969) Statistical Physics, Addison-Wesley, Reading, Mass.

--- and PLACZEK, G. (1934) "Struktur der unverschobenen Streulinie," Phys. Z. Sowjet. 5, 172-173.

b LANG, G. and MARSHALL, W. (1966) "Mössbauer effect in some haemoglobin compounds," Proc. Phys. Soc. 87, 3.

m LANG, J. C., Jr., and FREED, J. H. (1972) "ESR study of Heisenberg spin exchange in a binary liquid solution near the critical point," J. Chem. Phys. 56, 4103.

LANGE, R. V. (1967) "Interaction range, the Goldstone theorem, and long-range order in the Heisenberg ferromagnet," Phys. Rev. 156, 630-631.

∞ LANGER, J. S. (1965) "A modified spherical model of a first-order phase transition," Ibid. A137, 1531-1547.

--- (1967) "Theory of the condensation point," Ann. Phys. 41, 108-157.

h --- (1969) "Coherent states in the theory of superfluidity. II. Fluctuations and irreversible Processes," Phys. Rev. 184, 219.

a --- (1971) "Theory of spinodal decomposition in alloys," Ann. Phys. 65, 53-86.

h --- and FISHER, M. E. (1967) "Intrinsic critical velocity of a superfluid," Phys. Rev. Lett. 19, 560-563.

LANGER, W., PLISCHKE, M., and MATTIS, D. (1969) "Existence of two phase transitions in Hubbard model," Ibid. 23, 1448-1452.

LANGLEY, K. H. and FORD, N. C., Jr. (1969) Attenuation of the Rayleigh component in Brillouin spectroscopy using interferometric filtering," J. Optical Soc. Am. 59, 281-284.

s LARAMORE, G. E. and KADANOFF, L. P. (1969) "Fluctuation enhancement of the electrical conductivity above the superconducting transition," Comments on Solid State Phys. 2, 105-110.

---, --- (1969) "Anomalous ultrasonic attenuation above the magnetic critical point," Phys. Rev. 187, 619-629.

LARSEN, S. Y., MOUNTAIN, R. D., and ZWANZIG, R. (1965) "On the validity of the Lorentz-Lorenz equation near the critical point," J. Chem. Phys. 42, 2187-2190.

a LASSETTRE, E. N. and HOWE, J. P. (1941) "Thermodynamic properties of binary solid solutions on the basis of the nearest neighbor approximation," Ibid. 9,747-754.

LAST, B. J. and THOULESS, D. J. (1971) "Percolation theory and electrical conductivity," Phys. Rev. Lett. 27, 1719-1721.

LASTOVKA, J. B. and BENEDEK, G. B. (1966) "Spectrum of light scattered quasielastically from a normal liquid," Ibid. 17, 1039-1042.

LAU, H. Y., CORLISS, L. M., DELAPALME, A., HASTINGS, J. M., NATHANS, R., and TUCCIARONE, A. (1969) "Test of dynamic scaling by neutron scattering from $RbMnF_3$," Ibid. 23, 1225-1228.

---, ---, ---, ---, ---, --- (1970) "Critical scattering of neutrons from $RbMnF_3$," J. Appl. Phys. 41, 1384-1389.

LAURENCE, G. (1971) "Low-temperature thermal conductivity of $CdCl_2$ and $FeCl_2$ and critical magnetic scattering of phonons in $FeCl_2$," Phys. Lett. 34A, 308-309.

f LAVRENCIC, B. and BLINC, R. (1972) "On the symmetry of the soft modes in the ferroelectric phase transitions," Phys. Stat. Sol. (b) 49, K119.

∞ LAX, M. (1952) "Dipoles on a lattice: The spherical model," J. Chem. Phys. 20, 1351-1359.

∞ --- (1955) "Molecular field in the spherical model," Phys. Rev. 97, 629-640.

--- (1955) "Relation between canonical and microcanonical ensembles," Ibid. 97, 1419-1420.

--- and WINDWER, S. (1971) "Properties of self-avoiding walks not constrained to lattices," J. Chem. Phys. 55, 4167-4174.

s LAYZER, A. and FAY, D. (1971) "The superconducting pairing tendence in nearly-ferromagnetic systems," Int. J. Magnetism 1, 135-141.

d LEAVER, K. D. (1968) "Thin ferromagnetic films," Contemp. Phys. 9, 475-499.

d --- (1968) "Magnetization ripple in ferromagnetic thin films," Thin Solid Films 2, 149-172.

LEBOWITZ, J. L. (1964) "Exact solution of generalized Percus-Yevick equation for a mixture of hard spheres," Phys. Rev. 133, A895-A899.

r --- (1968) "Statistical mechanics-A review of selected rigorous results," Ann. Rev. Phys. Chem. 19, 389-418.

--- (1969) "Analytic properties of systems with Lennard-Jones type potentials," Phys. Lett. 28A, 596-597.

--- (1971) "Griffiths inequalities for anti-ferromagnets," Ibid. 36A, 99-100.

--- (1972) "More inequalities for Ising ferromagnets," Phys. Rev. B5, 2538-2541.

---, BAER, S., and STELL, G. (1966) "Properties of lattice and continuum van der Waals fluids at the critical point," Ibid. 141, 198-199.

m --- and GALLAVOTTI, G. (1971) "Phase transitions in binary lattice gases," J. Math. Phys. 12, 1129-1133.

--- and LIEB, E. H. (1969) "Existence of thermodynamics for real matter with Coulomb forces," Phys. Rev. Lett. 22, 631-634.

---, --- (1972) "Phase transition in a continuum classical system with finite interactions," Phys. Lett. 39A, 98-101.

l --- and MARTIN-LÖF, A. (1972) "On the uniqueness of the equilibrium state for Ising spin systems," Commun. Math. Phys. 25, 276-282.

--- and PENROSE, O. (1964) "Convergence of virial expansions," J. Math. Phys. 5, 841-847.

---, --- (1966) "Rigorous treatment of the van der Waals-Maxwell theory of the liquid-vapor transition," Ibid. 7, 98-113.

---, --- (1968) "Analytic and clustering properties of thermodynamic functions and distribution functions for classical lattice and continuum systems," Commun. Math. Phys. 11, 99-124.

--- and PERCUS, J. K. (1963) "Asymptotic behavior of the radial distribution function," J. Math. Phys. 4, 248-254.

∞ ---, --- (1966) "Mean spherical model for lattice gases with extended hard cores and continuum fluids," Phys. Rev. 144, 251-258.

---, ---, and VERLET, L. (1967) "Ensemble dependence of fluctuations with application to machine computations," Ibid. 153, 250-254.

---, --- (1967) "Exactly solvable one-dimensional system," Ibid. 155, 122-138.

---, ---, and SYKES, J. (1969) "Kinetic-equation approach to time-dependent correlation functions," Ibid. 188, 487-504.

---, STELL, G., and BAER, S. (1965) "Separation of the interaction potential into two parts in treating many-body systems. I. General theory and applications to simple fluids with short-range and long-range forces," J. Math. Phys. 6, 1282-1298.

LEBWOHL, P. A. and LASHER, G. (1972) "Nematic-liquid-crystal order: A Monte Carlo calculation," Phys. Rev. A6, 426-429.

LEE, M. H. (1968) "Some remarks on the theory of interacting many fermions in one dimension," Progr. Theoret. Phys. 40, 990-997.

2 --- (1971) "High-temperature expansion of the spin-1/2 XY model," J. Math. Phys. 12, 61-69.

--- and STANLEY, H. E. (1971) "Critical properties obtained by a conformal transformation method," J. Phys. 32S, 352-353.

3 ---, --- (1971) "The spin-1/2 Heisenberg ferromagnet on cubic lattices: Analysis of critical properties by a transformation method," Phys. Rev. B4, 1613-1625.

LEE, S. P., TSCHARNUTER, W., and CHU, B. (1972) "Kawasaki-Einstein-Stokes formula and dynamical scaling in the critical region of a binary liquid mixture: Isobutyric acid in water," Phys. Rev. Lett. 28, 1509-1512.

1 LEE, T. D. and YANG, C. N. (1952) "Statistical theory of equations of state and phase transitions. II. Lattice gas and Ising model," Phys. Rev. 87, 410-419.

LEFF, H. S. (1970) "Proof of the third law of thermodynamics for Ising ferromagnets," Ibid. A2, 2368-2370.

--- (1971) "Correlation inequalities for coupled oscillators," J. Math. Phys. 12, 569-578.

--- and FLICKER, M. (1968) "Difference-equation solutions for the linear-Ising model and nearest-neighbor fluid," Am. J. Phys. 36, 591-598.

f LEFKOWITZ, I. and HAZONY, Y. (1968) "Elastic and dielectric anomalies and the lattice-dynamical theory of ferroelectric phase transitions," Phys. Rev. 169, 441-442.

LEHMANN-SZWEYKOWSKA, A. (1972) "Magnetic phase transitions in neodymium by phenomenological theory," Phys. Stat. Sol. B51, K139-142.

s LEHOCZKY, S. L. and BRISCOE, C. V. (1969) "Fluctuation effects in the ac conductivity of thin lead films above the superconducting transition temperature," Phys. Rev. 184, 695-697.

s ---, --- (1969) "Fluctuation effects in the ac conductivity of thin lead films above the superconducting transition temperature," Phys. Rev. Lett. 23, 695-697.

s ---, --- (1970) "Fluctuation effects in the ac conductivity of thin lead films below the superconducting transition temperature," Ibid. 24, 880-882.

a LESOILLE, M. R. and GIELEN, P. M. (1970) "The order-disorder transformation in Fe_3Al alloys," Phys. Stat. Sol. 37, 127.

1 LEU, J. A., BETTS, D. D., and ELLIOTT, C. J. (1969) "High-temperature critical properties of the Ising model on a triple of related lattices," Can. J. Phys. 47, 1671-1689.

LEVANYUK, A. P. (1964) "Theory of second-order phase transition," Sov. Phys.-Solid State 5, 1294-1298.

LEVELT-SENGERS, J. M. H. (1970) "Scaling predictions for thermodynamic anomalies near the gas-liquid critical point," Ind. Eng. Chem. Fundam. 9, 470-479.

--- (1971) "Coexistence curves of CO_2, N_2O, and $CClF_3$ in the critical region," J. Chem. Phys. 54, 5034-5051.

--- and CHEN, W. T. (1972) "Vapor pressure, critical isochore, and some metastable states of CO_2," Ibid. 56, 595-608.

---, STRAUB, J., and VICENTINI-MISSONI, M. (1971) "Coexistence curves of CO_2,N_2O, and $CClF_2$ in the critical region," Ibid. 54, 5034-5050.

LEVINSON, L. M. and LUBAN, M. (1968) "Spin relaxation in Mössbauer spectra of magnetically ordered systems," Phys. Rev. 172, 268-276.

LEVY, P. M. (1970) "Paramagnetic Curie temperatures of the rare-earth monophosphides," Ibid. B2, 1429-1431.

f --- and CHEN, H. H. (1971) "Structural and magnetic phase transitions in the rare-earth pnictides," Phys. Rev. Lett. 27, 1385-1388.

LEWIS, E. A. S. (1970) "Heat capacity of gadolinium near the Curie point," Phys. Rev. B1, 4368-4377.

∞ LEWIS, H. W. and WANNIER, G. H. (1952) "Spherical model of a ferromagnet," Ibid. 88, 682-683.

∞ ---, --- (1953) "Spherical model of a ferromagnet (erratum)," Ibid. 90, 1131.

LEY-KOO, M. and GARCIA-COLIN, L. (1970) "Behaviour of the pair correlation function in the critical region," Rev. Mex. Fis. 19, 23-39.

LIE, T. J. (1972) "Mean field theory and infinite dimensionality in Widom-Rowlinson model," J. Chem. Phys. 56, 332-336.

LIEB, E. H. (1963) "New method in the theory of imperfect gases and liquids," J. Math. Phys. 4, 671.

--- (1963) "Exact analysis of an interacting Bose gas. II. The excitation spectrum," Phys. Rev. 130, 1616.

--- (1963) "Simplified approach to the ground state energy of an imperfect Bose gas," Ibid. 130, 2518.

r --- (1965) "The Bose fluid," In Lectures in Theoret. Phys. Vol. VIIC (Univ. Colo. Press, Boulder,), pp. 175-224.

--- (1966) "Quantum-mechanical extension of the Lebowitz-Penrose theorem on the van der Waals theory," J. Math. Phys. 7, 1016-1024.

-- (1967) "Calculation of exchange second virial coefficient of a hard-sphere gas by path integrals," Ibid. 8, 43-52.

--- (1967) "Solution of the dimer problem by the transfer matrix method," Ibid. 8, 2339-2341.

r --- (1967) "Exactly soluble models," In Mathematical Methods in Solid State and Superfluid Theory, (Plenum Press, N.Y.), pp. 286-306.

f --- (1967) "Residual entropy of square ice," Phys. Rev. 162, 162-172.

f --- (1967) "Exact solution of the problem of the entropy of two-dimensional ice," Phys. Rev. Lett. 18, 692-694.

f --- (1967) "Exact solution of the F model of an antiferroelectric," Ibid. 18, 1046-1048.

f --- (1967) "Exact solution of the two-dimensional Slater KDP model of a ferroelectric," Ibid. 19, 108-110.

r --- (1967) "Ice, Ferro- and antiferroelectrics," Proc. Birmingham Conf., in honor of Prof. R. E. Peierls.

--- (1968) "A theorem on Pfaffians," J. Comb. Theory 5, 313-319.

r --- (1969) "Survey of the one-dimensional many body problem and two-dimensional ferroelectric models," In Contemporary Physics: Trieste Symposium (IAEA, Vienna), pp. 163-193.

r --- (1969) "Two-dimensional ferroelectric models," J. Phys. Soc. Japan 26, 94-95.

r --- (1969) "Two-dimensional ice and ferroelectric models," In Lectures in Theoret. Physics, Vol. XI D, (Gordon and Breach, N.Y.), pp. 329-354.

r --- (1969) "Survey of the one-dimensional many-body problem and two-dimensional ferroelectric models," Contemp. Phys. 1, 163-176 [Proc. Trieste Symposium].

r --- (1971) "Models," In Phase Transitions - Proc. 14th Solvay Conf. (Interscience, N.Y.), pp. 45-52.

--- (1972) "Political activism and the future of universities," In Dynamic Aspects of Critical Phenomena," J. I. Budnick and M. P. Kawatra, Eds., Gordon and Breach Publishers, N.Y.

--- and BEYER, W. A. (1969) "Clusters on a thin quadratic lattice (transfer matrix technique), Studies in Appl. Math. 48, 77-90.

--- and LEBOWITZ, J. L. (1969) "The existence of thermodynamics for real matter with Coulomb forces," Phys. Rev. Lett. 22, 631-634.

---, --- (1972) "The constitution of matter," Adv. Math. (in press).

--- and LINIGER, W. (1963) "Exact analysis of an interacting Bose gas. I. The general solution and the ground state," Phys. Rev. 130, 1605.

---, --- (1964) "Simplified approach to the ground state energy of an imperfect Bose gas. III. Application to the one-dimensional model," Ibid. 134, A312.

--- and MATTIS, D. C. (1962) "Ordering energy levels of interacting spin systems," J. Math. Phys. 3, 749.

---, --- (1962) "Theory of ferromagnetism and the ordering of electronic energy levels," Phys. Rev. 125, 164.

r ---, --- (1966) Mathematical Physics in One Dimension, Academic Press, N.Y.

r ---, NIEMEIJER, TH., and VERTOGEN, G. (1971) "Models in statistical mechanics," Proc. 1970 Les Houches School (Gordon and Breach, N.Y.)

1 --- and RUELLE, D. (1971) "A property of zeros of the partition function for Ising spin systems," J. Math. Phys. 13, 781.

--- and SAKAKURA, A. (1964) "Simplified approach to the ground state energy of an imperfect Bose gas. II. The charged Bose gas at high density," Phys. Rev. 133, A899.

2 ---, SCHULTZ, T., and MATTIS, D. (1961) "Two soluble models of an antiferromagnetic chains," Ann. Phys. 16, 407-466.

∞ --- and THOMPSON, C. J. (1969) "Phase transition in zero dimensions: A remark on the spherical model, J. Math. Phys. 10, 1403-1406.

--- and WU, F. Y. (1968) "Absence of Mott transition in an exact solution of the short-range, one-band model in one dimension," Phys. Rev. Lett. 20, 1445-1448.

r ---, --- (1972) "Two-dimensional ferroelectric models," In Phase Transitions and Critical Phenomena, C. Domb and M. S. Green, Eds. (Academic Press, London).

d LIEBERMAN, L. N., FREDKIN, D. R., and SHORE, H. B. (1969) "Two-dimensional 'ferromagnetism' in iron," Phys. Rev. Lett. 22, 539-541.

b LIFSON, S. and ZIMM, B. H. (1963) "Simplified theory of the helix-coil transition in DNA based on a grand partition function," Biopolymers 1, 15-23.

LIM, T. K., SWINNEY, H. L., LANGLEY, K. H., and KACHNOWSKI, T. A. (1971) "Rayleigh linewidth in SF_6 near the critical point," Phys. Rev. Lett. 27, 1776-1780.

LINES, M. E. (1967) "Sensitivity of Curie temperature to crystal-field anisotropy. I. Theory," Phys. Rev. 156, 534-542.

--- (1967) "Sensitivity of Curie temperature to crystal-field anisotropy. II. FeF_2," Ibid. 156, 543-551.

d --- (1967) "Examples of two dimensional ordered magnetic systems," Phys. Lett. 24A, 591-592.

d --- (1967) "Comparative studies of magnetism in $KNiF_3$ and K_2NiF_4," Phys. Rev. 164, 736-748.

d --- (1969) "Magnetism in two dimensions," J. Appl. Phys. 40, 1352-1358.

d --- (1970) "The quadratic-layer antiferromagnet," J. Phys. Chem. Solids 31, 101-116.

--- (1971) "Orbital angular momentum in the theory of paramagnetic clusters," J. Chem. Phys. 55, 2977-2984.

d --- (1971) "New approach to Green's function decoupling in magnetism with specific application to two-dimensional systems," Phys. Rev. B3, 1749-1763.

f --- (1972) "Polarization fluctuations near a ferroelectric phase transition," Ibid. B5, 3690-3702.

f --- (1972) "Nature of the ferroelectric-paraelectric phase transition in lithium tantalate," Sol. State Commun. 10, 793-794.

h LIPA, B. J. and BUCKINGHAM, M. J. (1968) "Critical exponent values at λ-transitions," Phys. Lett. 26A, 643-644.

---, EDWARDS, C., and BUCKINGHAM, M. J. (1970) "Precision measurement of the specific heat of CO_2 near the critical point," Phys. Rev. Lett. 25, 1086-1090.

f LITOV, E. and GARLAND, C. W. (1970) "Ultrasonic investigation of the ferroelectric transition region in KH_2PO_4," Phys. Rev. B2, 4597-4602.

f --- and UEHLING, E. A. (1968) "Polarization relaxation in the ferroelectric transition region of Kd_2PO_4," Phys. Rev. Lett. 21, 809-812.

LIU, L. L. and JOSEPH, R. I. (1971) "Exact solution for a closed chain of classical spins with arbitrary isotropic nearest neighbor exchange," Ibid. 26, 1378-1380.

---, --- (1972) "An exactly soluble model of a system containing arbitrary isotropic bilinear and biquadratic exchange interactions," J. Phys. Chem. Solids 33, 451-456.

---, ---, and STANLEY, H. E. (1971) "Inequalities among the critical-point exponents for the static correlation functions," Proc. 1971 Conf. on Magnetism and Magnetic Materials, Chicago, AIP Conf. No. 5, p. 1235-1240.

---, ---, --- (1972) "New inequalities among the critical-point exponents for the spin-spin and energy-energy correlation functions," Phys. Rev. B6, 1963-1968

u ---, KRASNOW, R., HARBUS, F., and STANLEY, H. E. (1973) "Evidence supporting scaling with a parameter for thermodynamic functions, with application to systems with lower dimensionality," Proc. LT 13, Boulder Colorado, Aug. 1972.

u --- and STANLEY, H. E. (1972) "Some rigorous results concerning the crossover behavior of the Ising model with lattice anisotropy," Phys. Lett. 40A, 272-274.

u ---, --- (1972) "Some results concerning the crossover behavior of 'quasi-two-dimensional' and 'quasi-one-dimensional' systems," Phys. Rev. Lett. 29, 927-930.

u ---, --- (1973) "Rigorous results concerning the lattice anisotropy problem," Phys. Rev. B7

3 LIU, S. H. (1965) "Correlation functions for a Heisenberg ferromagnet," Ibid. 139, 1522-1530.

LO, S. M. and KAWASAKI, K. (1972) "Vortex correction contribution to the decay rate of concentration fluctuations in binary liquid critical mixtures," Ibid. A5, 421-424

h LONDON, F. (1938) "On the Bose-Einstein condensation," Phys. Rev. 54, 947-954.

m LONG, C. and WANG, Y. L. (1971) "Theory of magnetic-moment-jump phase transition with application to UP," Ibid. B3, 1656-1661

LONGWORTH, G. (1968) "Temperature dependence of the ^{57}Fe hfs in the ordered alloys $FePd_3$ near the Curie temperature," Ibid. 172, 572-576.

LORENTZEN, H. L. (1953) "Studies of critical phenomena in carbon dioxide contained in vertical tubes," Acta Chemica Scandinavica 7, 1335-1346.

LOVEN, A. W. and RICE, O. K. (1963) "Coexistence curve of the 2,6-lutidine+water system in the critical region," Trans. Faraday Soc. 59, 2723-2727.

d LOVESEY, S. W. and MESERVE, R. A. (1972) "Dynamic properties of a one-dimensional Heisenberg magnet," Phys. Rev.Lett. 28, 614-617.

LOWDE, R. D. (1958) "Critical magnetic scattering of neutrons by iron," Rev. Mod. Phys. 30, 69-74.

d LOWNDES, D. H., Jr., FINEGOLD, L., ROGERS, R. N., and MOROSIN, B. (1969) "Specific heat of three magnetic linear-chain antiferromagnets," Phys. Rev. 186, 515-521.

LUBAN, M. (1962) "Statistical mechanics of a nonideal boson gas: Pair Hamiltonian model," Ibid. 128, 965-987.

--- (1965) "Quantum hard-sphere gas. I," Ibid. A138, 1028-1032.

--- (1965) "Quantum hard-sphere gas. II," Ibid. A138, 1033-1045.

--- and AMIT, D. J. (1969) "Critical-exponent scaling relations and a generalized Landau theory," J. Phys. Soc. Japan 26, 120-121.

--- and GROBMAN, W. D. (1966) "Bose-Einstein phase transition in an interacting system," Phys. Rev. Lett. 17, 182-185.

h --- and MILLER, S. (1968) "Two critical velocities in superfluid helium?," Phys. Lett. 27A, 501-502.

--- and NOVOGRODSKY, H. (1972) "Statistical mechanics of solids," Phys. Rev. B6, 1130-1134.

--- and REVZEN, M. (1968) "Onset of ODLRO and the phase transition in the ideal boson gas," J. Math. Phys. 9, 347-349.

---, WISER, N., and GREENFIELD, A. J. (1969) "Direct determination of the critical exponent γ for order-disorder phase transitions," Phys. Lett. 29A, 79-80.

L LUBENSKY, T. C. (1972) "Hydrodynamics of cholesteric liquid crystals," Phys. Rev. A6, 452-471.

LUNACEK, J. H. and CANNEL, D. S. (1971) "Long-range correlation length and isothermal compressibility of carbon dioxide near the critical point," Phys. Rev. Lett. 27, 841-844.

LÜTHI, B., MORAN, T. J., and POLLINA, R. J. (1970) "Sound propagation near magnetic phase transitions," J. Phys. Chem. Solids 31, 1741-1758.

---, PAPON, P., and POLLINA, R. J. (1969) "Ultrasonic attenuation of magnetic phase transitions," J. Appl. Phys. 40, 1029-1030.

---, POLLINA, R. J. (1968) "Critical attenuation of sound in gadolinium," Phys. Rev. 167, 488-492.

---, --- (1969) "Sound propagation near the magnetic phase transition in EuO," Phys. Rev. Lett. 22, 717-720.

2 MC COY, B. M. (1968) "Spin correlation functions of the X-Y model," Phys. Rev. 173, 531-541.

1 --- (1968) "Theory of Toeplitz determinants and the spin correlations of the two-dimensional Ising model. V," Ibid. 174, 546-559.

1 --- (1968) "Theory of a two-dimensional Ising model with random impurities. I. Thermodynamics," Ibid. 176, 631-643.

1 --- (1969) "Theory of a two-dimensional Ising model with random impurities. III. Boundary effects," Ibid. 188, 1014-1031.

2 ---, BAROUCH, E., and ABRAHAM, D. B. (1971) "Statistical mechanics of the XY model. IV. Time-dependent spin-correlation functions," Ibid. A4, 2331-2342.

1 ---, and WU, T. T. (1967) "Theory of Toeplitz determinants and the spin correlations of the two-dimensional Ising model. II," Ibid. 155, 438-452.

1 ---, --- (1967) "Theory of Toeplitz determinants and the spin correlations of the two-dimensional Ising model. IV," Ibid. 162, 436-475.

---, --- (1968) "Random impurities as the cause of smooth specific heats near the critical temperature," Phys. Rev. Lett. 21, 549-551.

---, --- (1969) "Theory of a two-dimensional Ising model with random impurities. II. Spin correlation functions," Phys. Rev. 188, 982-1013.

d MC CUMBER, D. E. and HALPERIN, B. I. (1970) "Time scale of intrinsic resistive fluctuations in thin superconducting wires," Ibid. B1, 1054-1070.

3 MC FADDEN, D. G., TAHIR-KHELI, R. A., and TAGGART, G. B. (1969) "Space-time-dependent correlation functiors for one-dimensional Heisenberg spin systems in a Lorentzian-Gaussian approximation," Ibid. 185, 854.

---, --- (1970) "Space-time correlations in anisotropic Heisenberg paramagnets at elevated temperatures," Ibid. B1, 3649-3670.

r MC INTYRE, D. and SENGERS, J. V. (1968) "Study of fluids by light scattering," in Physics of Simple Liquids, p. 449-505.

MC KERRELL, A. and BOWERS, R. G. (1972) "Critical phenomena with many-body interactions: An exactly soluble model system," J. Phys. C5, 1-4.

d MAARSCHALL, E. P., BOTTERMAN, A. C., VEGA, S., and MIEDEMA, A. R. (1969) "Nuclear magnetic resonance in paramagnetic K_2NiF_4," Physica 41, 473-485.

MACCABEE, B. S. and WHITE, J. A. (1971) "Supercritical correlation length of carbon dioxide along the critical isochore," Phys. Rev. Lett. 27, 495-498.

---, --- (1971) "Temperature variation of the correlation length carbon dioxide at its critical density," Phys. Lett. 35A, 187-188.

MACDONALD, J. R., and POWELL, D. R. (1971) "Discrimination between equations of state," J. Res. Natl. Bur. Std. (U.S.) A 75A, 441-453.

m MACQUEEN, J. T. , MEEKS, F. R., and RICE, O. K. (1961) "The effect of an impurity on the phase transition in a binary liquid system as a surface phenomenon," J. Phys. Chem. 65, 1925-1929.

MAJOR, J., MEZEI, F., NAGY, E., SVÁB, E., and TICHY, G. (1971) "Thermal expansion coefficient of nickel near the Curie point," Phys. Lett. 35A, 377-378.

l MAJUMDAR, C. K. (1966) "Analytic properties of the Onsager solution of the Ising model," Phys. Rev. 145, 158-163.

--- (1969) "Problem of two spin deviations in a linear chain with next-nearest-neighbor interactions," J. Math. Phys. 10, 177-180.

s MAKI, K. (1967) "The critical fluctuation of the order parameter in type II superconductors," Progr. Theoret. Phys. 39, 897-906.

s MALASPINAS, A. (1971) "On the order of itinerant antiferromagnetic phase transitions and superconducting phase transitions in an exchange field," Phys. Kond. Mat. 13, 193-203.

MALEEV, S. V. (1971) "On dynamic scaling in the Heisenberg ferromagnet," Phys. Lett. 37A, 406-407.

l MAMADA, H. and TAKANO, F. (1968) "New approximate method for Ising system," J. Phys. Soc. Japan 25, 675-686.

MANNARI, I. (1968) "Anomaly in electrical resistivity of ferromagnetic metals near T_C," Phys. Lett. 26A, 134-135.

--- and KAWABE, T. (1970) "Extended Watson integral for an anisotropic simple cubic lattice," Progr. Theoret. Phys. 44, 359-362.

d MARČELJA, S., MASKER, W. E., and PARKS, R. D. (1969) "Electrical conductivity of a two-dimensional superconductor," Phys. Rev. Lett. 22, 124-217.

l MARINARO, M. and SEWELL, G. L. (1972) "Characterizations of phase transitions in Ising spin systems," Commun. Math. Phys. 24, 310-335.

MARQUARD, C. D. (1967) "High-temperature expansions for magnetic lattices with arbitrary interactions: Application to $GdCl_3$," Proc. Phys. Soc. 92, 650-664.

MARSH, J. S. (1969) "Generalized molecular fields," Phys. Rev. 178, 403-412.

MARSHALL, W. (1966) "Critical scattering of neutrons by ferromagnets," Proc. NBS Conf. on Phenomena in the Neighbourhood of Critical Points, Washington, 1965.

r ---, Ed. (1967) Theory of Magnetism in Transition Metals, Proc. Varenna Summer School (Academic Press, London).

r --- and BEEBY, J. L. (1969) "Collective behavior in classical liquids," Comments Solid State Phys. 2, 167-171.

r --- and LOVESEY, S. W. (1971) Theory of Thermal Neutron Scattering, Oxford Univ. Press, London.

r --- and LOWDE, R. D. (1968) "Magnetic correlations and neutron scattering," Rept. Progr. Phys. 31, 705-775.

MARTIN, J. (1967) "Equations of state," Indust. Engin. Chem. 59, 34-52.

MARTIN, J. L., SYKES, M. F., and HIOE, F. T. (1967) "Probability of initial ring closure for self-avoiding walks on the face-centered cubic and triangular lattices," J. Chem. Phys. 46, 3478-3481.

L MARTIN, P. C., PERSHAN, P. S., SWIFT, J. (1970) "New elastic-hydrodynamic theory of liquid crystals," Phys. Rev. Lett. 25, 844-848.

l MARTIN-LÖF, A. (1972) "On the spontaneous magnetization in the Ising model," Commun. Math. Phys. 24, 253-259.

L MARTINOTY, P., CANDAU, S., and DEBEAUVAIS, F. (1971) "Dynamic properties near the nematic-isotropic transition of a liquid crystal," Phys. Rev. Lett. 27, 1123-1125.

m MARTYNETS, V. G. and MATIZEN, É. V. (1970) "Browian motion near the critical point of the two-phase liquid-liquid equilibrium," Sov. Phys.-JETP 31, 228.

MATHON, J. (1968) "Magnetic and electrical properties of ferromagnetic alloys near the critical concentration," Proc. Roy. Soc. A306, 355-368.

MATLAK, M. and PAWLIKOWSKI, A. (1970) "A simple proof of the linked cluster expansion theorem in the many-body problem," Acta Phys. Polon. A37, 847.

h MATSUBARA, T. and MATSUDA, H. (1956) "A lattice model of liquid helium, I," Progr. Theoret. Phys. 16, 569-582.

h MATSUDA, H. and MATSUBARA, T. (1957) "A lattice model of liquid helium, II," Ibid. 17, 19-29.

l MATSUDAIRA, N. (1967) "Some dynamical properties of the Ising ferromagnet," Can. J. Phys. 45, 2091-2111.

l --- (1967) "Some dynamical properties of the Ising ferromagnet. II. Cubic lattices," J. Phys. Soc. Japan 23, 232-240.

MATSUNO, K. and STANLEY, H. E. (1970) "Self-consistent collective excitation in critical region of second order phase transitions," Phys. Lett. 33A, 425-426.

s ---, --- (1971) "Multiparticle quasi-bound state near the superconducting transition temperature," Proc. 12th International Low-Temperature Physics Conf., p. 311-312 (E. Kanda, Ed., Academic Press).

d MATSUURA, M. (1971) "The magnetization of the linear chain Heisenberg antiferromagnet $CuCl_2 \cdot 2NC_5H_5$, in high magnetic fields," Phys. Lett. 34A, 274-275.

d --- (1971) "The magnetization of the linear chain Heisenberg antiferromagnet $CuCl_2 2NC_5H_5$, in high magnetic fields," Ibid. 34A, 274-275.

---, BLOTE, H. W. J., and HUISKAMP, W. J. (1970) "Heat capacity and magnetic behaviour of cobalt and manganese formate dihydrate," Physica 50, 444-456.

d ---, GILIJAMSE, K., STERKENBURG, J. E. W., and BREED, D. J. (1970) "The effect of anisotropy of the critical temperature of two-dimensional Heisenberg antiferromagnets," Phys. Lett. 33A, 363-364.

2 MATTINGLY, S. R. and BETTS, D. D. (1972) "Dynamical properties of the three-dimensional XY model. II. Singularity structure of the frequency dependent susceptibility," Can. J. Phys. 50, 2415-2420.

r MATTIS, D. C. (1965) The theory of magnetism: An introduction to the study of cooperative phenomena, Harper and Row, N.Y.

--- (1969) "Theory of electronic switching effect as a cooperative phenomenon," Phys. Rev. Lett. 22, 936-939.

--- and LANGER, W. D. (1970) "Role of phonons and band structure in metal-insulator phase transition," Ibid. 25, 376-380.

l --- and WOLF, W. P. (1966) "Soluble extension of the Ising model," Phys. Lett. 16, 899-901.

MATTUCK, R. D. (1968) "Effect of correlations on ferromagnetic transition in δ-function gas," Ibid. 28A, 407.

---, JOHANSSON, B. (1968) "Quantum field theory of phase transitions in Fermi systems," Adv. in Phys. 17, 509-562.

g MAYER, J. E. and STREETER, S. F. (1939) "Phase transitions," J. Chem. Phys. 7, 1019-1025.

∞ MAZO, R. M. (1963) "Antiferromagnetic spherical Ising model and some properties of the associated lattice gas," Ibid. 39, 2196-2200.

MEADEN, G. T., RAO, K. V., and LOO, H. Y. (1969) "Lorenz function enhancement due to inelastic processes near the Neél point of chromium," Phys. Rev. Lett. 23, 475-477.

---, and SZE, N. H. (1969) "Critical exponents and electrical resistivity near the Neél point of chromium," Phys. Lett. 29A, 162-163.

m MEEKS, F. R., GOPAL, R., and RICE, O. K. (1959) "Critical phenomena in the cyclohexane-aniline system: Effect of water at definite activity[1]," J. Phys. Chem. 63, 993-994.

1 MEIJER, P. H. E. and EDWARDS, J. C. (1969) "Dynamic behavior of a near-Ising system near the critical point," J. Appl. Phys. 40, 1543-1545.

3 MENYUK, N., DWIGHT, K., and REED, T. B. (1971) "Critical magnetic properties and exchange interactions in EuO," Phys. Rev. B3, 1689-1698.

---, RACCAH, P. M., and DWIGHT, K. (1968) "Magnetic properties of $La_{0.5}Sr_{0.5}CoO_3$ near its Curie temperature," Ibid. 166, 510-513.

MERMIN, N. D. (1964) "Time-dependent correlations in a solvable ferromagnetic model," Ibid. 134, A112-A125.

d --- (1967) "Absence of ordering in certain classical systems," J. Math. Phys. 8, 1061-1064.

d --- (1968) "Crystalline order in two dimensions," Phys. Rev. 176, 250-254.

--- (1971) "Solvable model of a vapor-liquid transition with a singular coexistence-curve diameter," Phys. Rev. Lett. 26, 169-172.

--- (1971) "Lattice gas with short-range pair interactions and a singular coexistence-curve diameter," Ibid. 26, 957-959.

--- and REHR, J. J. (1971) "Generality of the singular diameter of the liquid-vapor coexistence curve," Ibid. 26, 1155-1156.

---, --- (1971) "Asymmetry in the liquid and vapor density fluctuations at the critical point," Phys. Rev. A4, 2408-2410.

d --- and WAGNER, H. (1966) "Absence of ferromagnetism or antiferromagnetism in one- or two-dimensional isotropic Heisenberg models," Phys. Rev. Lett. 17, 1133-1136.

METCALFE, M. J. and ROSENBERG, H. M. (1972) "The magnetothermal resistivity of antiferromagnetic crystals at low temperatures: I. $DyPO_4$, a nearly ideal Ising system," J. Phys. C. 5, 450.

3 ---, --- (1972) "The magnetothermal resistivity of antiferromagnetic crystals at low temperatures: II. $GdVO_4$, a nearly ideal Heisenberg system," Ibid. C. 5, 459.

---, --- (1972) "The magnetothermal resistivity of antiferromagnetic crystals at low temperatures: III. $ErVO_4$, a new antiferromagnet," Ibid. C. 5, 474.

MEYER, H., WEINHAUS, F., MARAVIGLIA, B., and MILLS, R. D. (1972) "Orientational order parameter in cubic D_2," Phys. Rev. B6, 1112-1121.

MICHELS, A., BLAISSE, B., and MICHELS, C. (1937) "The isotherms of CO_2 in the neighbourhood of the critical point and round the coexistence line," Proc. Roy. Soc. A160, 358-375.

---, LEVELT, J. M., and DE GRAAFF, W. (1958) "Compressibility isotherms of argon at temperatures between -25°C and 155°C, and at densities up to 640 Amagat (pressures up to 1050 atmospheres)," Physica 24, 659-671.

---, SENGERS, J. V., and VAN DER GULIK, P. S. (1962) "The thermal conductivity of carbon dioxide in the critical region," Ibid. 28, 1201-1264.

--- and STRIJLAND, J. (1952) "The specific heat at constant volume of compressed carbon dioxide," Ibid. 18, 613-628.

MIEDEMA, A. R., VAN KEMPEN, H., and HUISKAMP, W. J. (1963) "Experimental study of the simple Heisenberg ferromagnetism in CuK_2Cl_4-$2H_2O$ and $Cu(NH_4)_2Cl_4 \cdot 2H_2O$," Ibid. 29, 1-15.

---, WIELINGA, R. F., and HUISKAMP, W. J. (1965) "Experimental study of the heat capacity of antiferromagnetic $MnCl_2 \cdot 4H_2O$ and $MnBr_2 \cdot 4H_2O$," Ibid. 31, 835-844.

---, ---, --- (1965) "Experimental study of the body-centered-cubic Heisenberg ferromagnet," Ibid. 31, 1585-1598.

g MIGDAL, A. A. (1969) "A diagram technique near the Curie point and the second order phase transition in a Bose liquid," Sov. Phys.-JETP 28, 1036-1044.

g --- (1971) "Conformal invariance and bootstrap," Phys. Lett. 37B, 386-388.

g --- (1971) "Correlation functions in the theory of phase transitions: Violation of the scaling laws," Sov. Phys.-JETP 32, 552-560.

d MIKESKA, H. J. and SCHMIDT, H. (1970) "Phase transition without long-range order in two dimensions," J. Low Temp. Phys. 2, 371-381.

MIKULINSKIĬ, M. A. (1970) "The behavior of high-order correlation functions near second-order phase-transition points," Sov. Phys.-JETP 31, 991-993.

--- (1971) "Effect of small perturbations on the shift of the critical point of a second-order phase transition," Ibid. 33, 782-785.

MILLS, D. L. (1971) "Surface effects in magnetic crystals near the ordering temperature," Phys. Rev. B3, 3887-3895.

MILLS, R. E. (1969) "Generalized molecular field calculation of critical properties," Phys. Lett. 29A, 184-185.

u MILOŠEVIĆ, S., KARO, D., KRASNOW, R. A. C., and STANLEY, H. E. (1972) "Magnetic equations of state in the critical region," Invited talk at LT 13, Boulder, Colo., August, 1972 (in press).

---, MATSUNO, K., and STANLEY, H. E. (1970) "A montonicity relation for the two-spin correlation function," Phys. Stat. Sol. 42, K163-K167.

∞ --- and STANLEY, H. E. (1971) "Series expansions for the spherical and Ising models with large lattice dimensionality," J. Phys. 32S, 346-348.

u ---, --- (1972) "A method of calculating the scaling function directly from high-temperature series expansions," Proc. 1971 Conf. on Magnetism and Magnetic Materials, Chicago, Ill., AIP Conf. No. 5, p. 1225-1230 (1972).

u ---, --- (1972) "Calculation of the scaling function for the Heisenberg model," Phys. Rev. B5, 2526-2529.

u ---, --- (1972) "Equation of state near the critical point I. Calculation of the scaling function for the S=1/2 and S=∞ Heisenberg models using high-temperature expansions," Ibid. B6, 986-1001.

u ---, --- (1972) "Equation of state near the critical point. II. Comparison with experiment and possible universality with respect to lattice structure and spin quantum number," Ibid. B6, 1002-1009.

u ---, --- (1973) "Calculations of the equation of state near the critical point for the Heisenberg model using Padé approximants," Proc. Intern. Conf. on Padé Approximants, Canterbury, July, 1972.

MINKIEWICZ, V. J., COLLINS, M. F., NATHANS, R., and SHIRANE, G. (1969) "Critical and spin-wave fluctuations in nickel by neutron scattering," Phys. Rev. 182, 624-631.

d MINOT, M. J. and PERLSTEIN, J. H. (1971) "Mixed-valence square planar complexes: A new class of solids with high electrical conductivity in one dimension," Phys. Rev. Lett. 26, 371-373.

MIRACLE-SOLE, S. and ROBINSON, D. W. (1970) "Statistical mechanics of quantum mechanical particles with hard cores II. The equilibrium states," Commun. Math. Phys. 19, 204-219.

m MISTURA, L. (1971) "Effect of gravity near a critical point in binary mixtures," J. Chem. Phys. 55, 2375-2377.

m --- (1972) "Thermodynamic structure of correlation functions near a critical point in multicomponent systems," Phys. Rev. A6, 471-477.

--- and SETTE, D. (1966) "Shape of the coexistence curve in the critical region," Phys. Rev. Lett. 16, 268-270.

---, --- (1968) "Specific heat near the critical point," J. Chem. Phys. 49, 1419-1423.

d MITSEK, A. I., SEMYANNIKOV, S. S., and NOSKOV, A. S. (1971) "Critical state of thin ferromagnetic films," Sov. Phys.-Solid State 12, 2053.

MIYAZIMA, S. (1972) "On critical behavior of dilute ferromagnet with four-spin interaction," Progr. Theoret. Phys. 47, 2146-2148.

MOHR, R. and LANGLEY, K. H. (1972) "Light scattering from SF_6 in the vicinity of the critical point," J. de Physique 33, 97-103.

h MOLDOVER, M. R. (1969) "Scaling of the specific-heat singularity of He^4 near its critical point," Phys. Rev. 182, 342-351.

h --- and LITTLE, W. A. (1965) "Specific heat of He^3 and He^4 in the neighborhood of their critical points," Phys. Rev. Lett. 15, 54-56.

---, SJOLANDER, G., and WEYHMANN, W. (1971) "Second-order nature of the spin-reorientation phase transitions in $YbFeO_3$," Ibid. 26, 1257-1259.

b MONOD, J., CHANGEUX, J., and JACOB, F. (1963) "Allosteric proteins and cellular control systems," J. Mol. Biol. 6, 306-329.

b ---, WYMAN, J., and CHANGEUX, J. (1965) "On the nature of allosteric transitions: A plausible model," Ibid. 12, 88-118.

l MONTROLL, E. W. (1941) "Statistical mechanics of nearest neighbor systems," J. Chem. Phys. 9, 706-721.

l --- (1949) "Continuum models of cooperative phenomenon," Nuovo Cimento 6, 264-278.

l --- (1968) "Lectures on the Ising model of phase transitions," In Statistical Physics, Phase Transitions and Superfluidity, Vol. 2 (Eds. M. Chrétien, E. P. Gross, and S. Deser) Gordon and Breach, N.Y., p. 197-267.

l --- and BERLIN, T. H. (1951) "An analytical approach to the Ising problem," Commun. Pure Appl. Math. 4, 23-30.

---, ---, and HART, R. W. (1952) "Fonctions delta et intégrales gaussiennes en mécanique statistique," Extrait des Comptes Rendus de la 2e Réunion de Chimie Physique, 211-223.

b --- and GOEL, N. S. (1966) "Denaturation and renaturation of DNA. I. Equilibrium statistics of copolymeric DNA," Biopolymers 4, 855-878.

1 ---, POTTS, R. B., and WARD, J. C. (1963) "Correlations and spontaneous magnetization of the two-dimensional Ising model," J. Math. Phys. 4, 308-322.

--- and WEISS, G. H. (1965) "Random walks on lattices. II," Ibid. 6, 167-181.

h MOORE, M. A. (1967) "On an objection to the Patashinskii-Pokrovskii theory of the λ-transition in liquid ^{4}He," Phys. Lett. 25A, 499-500.

d --- (1969) "Additional evidence for a phase transition in the plane-rotator and classical Heisenberg models for two-dimensional lattices," Phys. Rev. Lett. 23, 861-863.

g --- (1972) "Renormalization and phase transitions," Lett. Nuovo Cimento 3, 275.

1 ---, JASNOW, D., and WORTIS, M. (1969) "Spin-spin correlation function of the three-dimensional Ising ferromagnet above the Curie temperature," Phys. Rev. Lett. 22, 940-943.

1 --- and TRAININ, J. E. T. (1972) "Spin dependence of critical indices in the two-dimensional Ising model," J. Phys. C. 5, L9-L12.

3 MOORJANI, K. and TANAKA, T. (1969) "Critical properties of a spin 1/2 Heisenberg ferromagnet," Phys. Lett. 29A, 188.

MORAN, T. J. and LUTHI, B. (1969) "Critical changes in sound velocity near a magnetic phase transition," Ibid. 29A, 665-666.

---, --- (1971) "High-frequency sound propagation near magnetic phase transitions," Phys. Rev. B4, 122-132.

MORGAN, D. J. and RUSHBROOKE, G. S. (1961) "On the magnetically dilute Heisenberg and Ising ferromagnetics. II. High-temperature expansions," Mol. Phys. 4, 291-303.

---, --- (1963) "On the magnetically dilute Heisenberg and Ising ferromagnetics. III. Concentration expansions for the Heisenberg model," Ibid. 6, 477-488.

g MORI, H. (1966) "Relaxation phenomena near the critical point," 1965 Tokyo Summer Lectures in Theoret. Phys. 1, 17-35.

--- and FUKUYAMA, Y. (1968) "Anomalous transport in critical gases," Phys. Lett. 27A, 214-215.

--- and KAWASAKI, K. (1961) "On the critical scattering of neutrons," Progr. Theoret. Phys. 25, 723-725.

--- and OKAMOTO, H. (1968) "Collective modes of ferromagnetic spins," Phys. Lett. 26A, 249-250.

---, --- (1968) "Dynamic critical phenomena in magnetic systems. I," Progr. Theoret. Phys. 40, 1287-1311.

---, ---, and ISA, S. (1972) "A simplified theory of liquid-solid transitions. I," Ibid. 47, 1087-1109.

MORITA, T. (1969) "Asymptotic behavior of the two-position correlation functions of classical systems," J. Phys. Soc. Japan 27, 19-25.

--- (1972) "Spin diffusion in the Heisenberg magnet at infinite temperature," Phys. Rev. B6, 3385-3393.

---, HORIGUCHI, T., and KATSURA, S. (1970) "Tetradics formulation of the two-time Green's function method and its application to the Heisenberg Ferromagnet," J. Phys. Soc. Japan 29, 84-89.

---, --- (1971) "Formulas for the lattice Green's functions for the cubic lattices in terms of the complete elliptic integral," J. Phys. Soc. Japan 30, 957-964.

---, --- (1972) "Analytic properties of the lattice Green function," J. Phys. A5, 67-77.

1 ---, --- (1972) "Spin orderings of the one-dimensional Ising magnet with the nearest and next nearest neighbor interaction," Phys. Lett. 38A, 223-225.

--- and KATSURA, S. (1969) "Calculation of the isotherm susceptibility by the Kubo formula," J. Phys. C. 2, 1030-1036.

MORIYA, T. (1956) "Nuclear magnetic relaxation in antiferromagnetics," Progr. Theoret. Phys. 16, 23-44.

--- (1956) "Nuclear magnetic relaxation in antiferromagnetics, II," Ibid. 16, 641-657.

--- (1962) "Nuclear magnetic relaxation near the Curie temperature," Ibid. 28, 371-400.

MOROZOV, V. G. (1972) "Phase transitions in a uniaxial antiferromagnet at low temperatures," Phys. Stat. Sol. 50, 479-490.

MORRIS, R. G. and CARSON, J. L., Jr. (1968) "Thermal conductivity of europium oxide (EuO) across the Curie temperature," Helv. Phys. Acta 41, 1045-1051.

r MOTT, N. F. and ZINAMON, Z. (1970) "The metal-nonmetal transition," Rept. Progr. Phys. 33, 881.

MOUNTAIN, R. D. (1966) "Thermal relaxation and Brillouin scattering in liquids," J. Res. Natl. Bur. Std. 70A, 207-220.

r --- (1966) "Spectral distribution of scattered light in a simple fluid," Rev. Mod. Phys. 38, 205-214.

--- (1969) "Dynamical model for Brillouin scattering near the critical point of a fluid," J. Res. Natl. Bur. Std. 73A, 593-598.

r --- (1970) "Liquids: dynamics of liquid structure," Chem. Rubber Co. 1, 5-46.

--- and LITOVITZ, T. A. (1967) "Negative dispersion and Brillouin scattering," J. Acoust. Soc. Am. 42, 516-517.

--- and ZWANZIG, R. (1968) "Anomalous transport properties of a van der Waals gas," J. Chem. Phys. 48, 1451-1458.

MUBAYI, V. (1969) "Spontaneous magnetization in isotropic ferromagnets," Phys. Lett. 29A, 663-664.

d --- and LANGE, R. V. (1969) "Phase transition in the two-dimensional Heisenberg ferromagnet," Phys. Rev. 176, 882-894.

L MULLEN, M. E., LÜTHI, B., and STEPHEN, M. J. (1972) "Sound velocity in a nematic liquid crystal," Phys. Rev. Lett. 28, 790-800.

f MÜLLER, K. A. and BERLINGER, W. (1971) "Static critical exponents at structural phase transitions," Ibid. 26, 13-16.

f ---, ---, and SLONCZEWSKI, J. C. (1970) "Order parameter and phase transitions of stressed $SrTiO_3$," Ibid. 25, 734-737.

f ---, ---, and WALDNER, F. (1968) "Characteristic structural phase transition in Perovskite-type compounds," Ibid. 21, 814-817.

MÜNSTER, A. (1960) "Sur la theorie des fluctuations critiques," J. Chimie Phys. 492-499.

r --- (1966) "Critical fluctuations," In Fluctuation Phenomena in Solids, R. E. Burgess, Ed. (Academic Press, N.Y.), p. 179-266.

--- and SAGEL, K. (1958) "Kritische Opaleszenz fester Lösungen," Mol. Phys. 1, 23-43.

a ---, --- (1958) "Entmischungskurve und kritischer Punkt des Systems Gold-Nickel," Z. Physik. Chem. (Neue Folge) 14, 296-305.

a ---, --- (1960) "Entmischungskurve und kritischer Punkt des Systems Gold-Platin," Ibid. 23, 416-425.

--- and SCHNEEWEISS, CH. (1963) "Kritische Schwankungen und kritische Opaleszenz binarer Systeme," Ibid. 37, 24-368.

MURRAY, F. E. and MASON, S. G. (1952) "Coexistence phenomena in the critical region, I. The gravity effect in ethane from light scattering," Can. J. Phys. 30, 550-561

f MURZIN, V. N., PASYNKOV, R. E., and SOLOV'EV, S. P. (1968) "Ferroelectricity and crystal-lattice dynamics," Sov. Phys.-Usp. 10, 453-484.

l MUSHINSKII, S. D. and NABUTOVSKII, V. M. (1972) "Point, one-dimensional and two-dimensional defects in the Ising model near a transition point," Sov. Phys.-Solid State 13, 2059-2061.

NABUTOVSKII, V. M. and PATASHINSKII, A. Z. (1969) "The resistance anomaly of ferromagnetic and anti-ferromagnetic metals near the magnetic ordering point," Ibid. 10, 2462-2464.

--- and PEISAKHOVICH, YU. G. (1972) "Anomalous light scattering in magnets near the critical point," Ibid. 13, 2724-2728.

f NAGLE, J. F. (1966) "Lattice statistics of hydrogen bonded crystals. I. The residual entropy of ice," J. Math. Phys. 7, 1484-1491.

f --- (1966) "Lattice statistics of hydrogen bonded cyrstals. II. The Slater KDP model and the Rys F-model," Ibid. 7, 1492-1496.

--- (1966) "On ordering and identifying undirected linear graphs," Ibid. 7, 1588-1592.

--- (1966) "New series-expansion method for the dimer problem," Phys. Rev. 152, 190-197.

f --- (1968) "The one-dimensional KDP model in statistical mechanics," Am. J. Phys. 36, 1114-1117.

--- (1968) "Weak-graph method for obtaining formal series expansions for lattice statistical problems," J. Math. Phys. 9, 1007-1019.

f --- (1969) "Proof of the first order phase transition in the Slater KDP model," Commun. Math. Phys. 13, 62-67.

f --- (1969) "Study of the F model using low-temperature series," J. Chem. Phys. 50, 2813-2818.

--- (1969) "Exact configuration entropy of copper formate tetrahydrate model," Phys. Rev. 186, 594.

m --- (1970) "Ising chain with competing interactions," Ibid. A2, 2124-2128.

f --- and ALLEN, G. R. (1971) "Models for the order-disorder transition in $NaH_3(SeO_3)_2$," J. Chem. Phys. 55, 2708-2714.

--- and BONNER, J. C. (1970) "Numerical studies of the Ising chain with long-range ferromagnetic interactions," J. Phys. C. 3, 352-366.

m ---, --- (1971) "Ising chain with competing interactions in a staggered field," J. Chem. Phys. 54, 729-734.

--- and TEMPERLEY, H. N. V. (1968) "Combinatorial theorem for graphs on a lattice," J. Math. Phys. 9, 1020-1026.

NAGY, I. and PAL, L. (1970) "Electrical resistivity and thermoelectric power of Ni near the Curie point," Phys. Rev. Lett. 24, 894-896.

NAKANO, H. (1968) "Ordering in certain statistical systems of Ising spins," Progr. Theoret. Phys. 39, 1121-1132.

l --- (1968) "Existence of three transition temperatures in decorated triangular and square Ising lattices with anisotropic couplings," Ibid. 40, 231-236.

NAKATANI, N. (1972) "The dependence of the coercive field of tri-glycine sulfate on frequency, amplitude, and temperature," J. Phys. Soc. Japan 32, 1556-1559.

NATHANS, R., MENZINGER, F., and PICKART, S. J. (1968) "Inelastic magnetic scattering from $RbMnF_3$ in the neighborhood of its Néel point," J. Appl. Phys. 39, 1237-1238.

s NAUGLE, D. G. and GLOVER, R. E., III (1968) "Thickness dependence of the resistive transition of superconducting films," Phys. Lett. 28A, 110-111.

l NAYA, S. (1954) "On the spontaneous magnetizations of honeycomb and kagome Ising lattices," Progr. Theoret. Phys. 11, 53-62.

m NEECE, G. A. (1967) "Extensions of the Widom model for two- and three-component liquid mixtures," J. Chem. Phys. 47, 4112.

r --- and WIDOM, B. (1969) "Theories of liquids," Ann. Rev. Phys. Chem. 20, 167-190.

NEIGHBOURS, J. R. and MOSS, R. W. (1968) "Ultrasonic attenuation near the critical point of MnF_2," Phys. Rev. 173, 542-546.

r NETTLETON, R. E. (1971) "Ferroelectric phase transitions: A review of theory and experiment," Ferroelectrics 3, 77-92.

r NEWELL, G. F. and MONTROLL, E. W. (1953) "On the theory of the Ising model of ferromagnetism," Rev. Mod. Phys. 25, 353-389.

NEWMAN, D. J. (1972) "Superexchange models: I. General theory," J. Phys. C. 5, 1089.

NICOLIS, G., WALLENBORN, J., and VELARDE, M. G. (1969) "On the validity of Gibbs' entropy law in strongly coupled systems," Physica 43, 263-276.

NIEMEIJER, TH. (1967) "Some exact calculations on a chain of spins 1/2," Ibid. 36, 377-419.

NILSEN, T. S. and HEMMER, P. C. (1969) "Note on the quantum displacement of the critical point," J. Stat. Phys. 1, 175-178.

NINHAM, B. W. (1963) "Asymptotic form of the coefficients of some Ising-model series," J. Math. Phys. 4, 679-685.

f NISHIKAWA, K. (1967) "Contribution to the theory of paraelectric relaxation near the Curie point," Progr. Theoret. Phys. 38, 305-321.

a NIX, F. C. and SHOCKLEY, W. (1938) "Order-disorder transformations in alloys," Rev. Mod. Phys. 10, 1-71.

b NJUS, D. L. and STANLEY, H. E. (1971) "Present and future problems in critical phenomena: Cooperativity in biological systems," In Proceedings of the Fordham Conference on the Dynamical Aspects of Critical Phenomena (Eds., J. I. Budnick and M. P. Kawatra) Gordon and Breach, N.Y., pp. 586-613.

NOAKES, J. E. and ARROTT, A. (1967) "Surface of magnetization, field, and temperature for nickel near its Curie temperature," J. Appl. Phys. 38, 973-974.

---, --- (1968) "Magnetization of nickel near its critical temperature," Ibid. 39, 1235-1236.

NOAKES, J. E., TORNBERG, N. E., and ARROTT, A. (1966) "Initial susceptibility of iron and iron alloys just above their Curiė temperatures," Ibid. 37, 1264-1265.

NOBLE, J. D. and BLOOM, M. (1965) "Self-diffusion in ethane near the critical point," Phys. Rev. Lett. 14, 250-251.

NORVELL, J. C., WOLF, W. P., CORLISS, L. M., HASTINGS, J. M., and NATHANS, R. (1969) "Magnetic neutron scattering in dysprosium aluminium garnet. I. Long-range order," Phys. Rev. 186, 557-566.

---, ---, ---, ---, --- (1969) "Magnetic neutron scattering in dysprosium aluminum garnet. II. Short-range order and critical scattering," Ibid. 186, 567-596.

NOWIK, I. (1969) "Ratio of first to second neighbour exchange in ferromagnets," Phys. Lett. 29A, 718-719.

l OBOKATA, T. (1969) "Time-dependent one-dimensional Ising model with spin S=1," J. Phys. Soc. Japan 26, 895-900.

--- and OGUCHI, T. (1968) "One-dimensional Ising model with general spin," Ibid. 25, 322-330.

---, ONO, I., and OGUCHI, T. (1967) "Padé approximation to ferromagnet with anisotropic exchange interaction," Ibid. 23, 516-521.

l OGITA, N., UEDA, A., MATSUBARA, T., MATSUDA, H., and YONEZAWA, F. (1969) "Computer simulation of order-disorder phenomena," Ibid. 26, 145-149.

OGUCHI, T. (1951) "Statistics of the three-dimensional ferromagnet. II," Ibid. 6, 27-35.

--- (1971) "Critical behavior of the spin system with anisotropic exchange interactions. I," Ibid. 30, 988-994.

--- and ONO, I. (1966) "Theory of critical magnetic scattering of neutrons by ferromagnet and antiferromagnet," Ibid. 21, 2178-2193.

---, --- (1968) "Time-dependent spin-correlation function in ferromagnet and antiferromagnet," J. Appl. Phys. 39, 1353-1354.

2 ---, --- (1969) "Analytical methods and computer calculations of XY model with impurity spin," J. Phys. Soc. Japan 26, 1378-1384.

OHBAYASHI, K. and IIDA, S. (1968) "Magnetic equation of state of YIG near tne Curie point," Ibid. 25, 1187.

m OITMAA, J. (1971) "High temperature series expansions for a lattice model of critical behaviour in multicomponent systems," J. Phys. C. 4, 2466-2474.

u --- (1972) "Critical behaviour of a generalized Ising model," Ibid. C. 5, 435.

u --- and ENTING, I. G. (1971) "Critical behaviour of the anisotropic Ising model," Phys. Lett. 36A, 91-92.

u ---, --- (1972) "Critical behaviour of the anisotropic Ising model," J. Phys. C. 5, 231-244.

∞ OKAMOTO, H. (1970) "A spherical model of ferromagnet with a uniaxial anisotropic interaction," Phys. Lett. 32A, 315-316.

OKASAKI, A., STEVENSON, R. W. H., and TURBERFIELD, K. C. (1964) "Critical magnetic scattering of neutrons in MnF_2," Proc. Int. Conf. Magnetism, Nottingham, pp. 92-94.

O'LEARY, G. P. and WHEELER, R. G. (1970) "Phase transitions and soft librational modes in cubic crystals," Phys. Rev. B1, 4409-4439.

ONO, I. (1972) "Ground state energies for a finite linear Heisenberg chain with nearest and next-nearest neighbor interactions," Phys. Lett. 38A, 327-329.

d ONO, S., KARAKI, Y., SUZUKI, M., and KAWABATA, C. (1968) "Statistical thermodynamics of finite Ising model. I," J. Phys. Soc. Japan 25, 54-59.

l ONSAGER, L. (1944) "Crystal statistics. I. A two-dimensional model with an order-disorder transition," Phys. Rev. 65, 117-149.

OPECHOWSKI, W. (1937) "On the exchange interaction in magnetic crystals," Physica 4, 181-199.

ORBACH, R. (1958) "Linear antiferromagnetic chain with anisotropic coupling," Phys. Rev. 112, 309-316.

ORNSTEIN, L. and ZERNIKE, F. (1914) "Accidental deviations of density and opalescence at the critical point of a single substance," Proc. Amst. Acad. Sci. 17, 793-806.

---, --- (1918) "Die linearen Dimensionen der Dichteschwankungen," Phys. Z. 19, 134-137.

---, --- (1926) "Die Molekularzerstreuung des Lichtes beim kritischen Zustande," Ibid. 27, 761-763.

L ORSAY LIQUID CRYSTAL GROUP (1969) "Quasielastic Rayleigh scattering in nematic liquid crystals," Phys. Rev. Lett. 22, 1361-1363.

s OSTENSON, J. E. and FINNEMORE, D. K. (1969) "Critical phenomena in sheath superconductivity of Nb," Ibid. 22, 188-190.

f OTNES, K., RISTE, T., SHIRANE, G., and FEDER, J. (1971) "Temperature dependence of the soft mode in $SrTiO_3$ above the 105°K transition," Sol. State Commun. 9, 1103-1106.

b OZAKI, M., TANAKA, M., and TERAMOTO, E. (1963) "Dependence of the transition temperatures of DNA molecules upon their base compositions," J. Phys. Soc. Japan 18, 551-557.

d PADMORE, T. C. (1972) "Superfluid helium in restricted geometries," Phys. Rev. Lett. 28, 1512-1515.

r PÁL, L. (1969) "First-order magnetic phase transitions," Acta Phys. Acad. Sci. Hung. 27, 47-85.

---, KRÉN, E., KÁDÁR, G., SZABÓ, P., and TARNÓCZI, T. (1968) "Magnetic structures and phase transformations in Mn-based CuAu-I type alloys," J. Appl. Phys. 39, 538-544.

PAPOULAR, M. (1967) "Diffusion magnétique critique de phonons," J. de Physique 28, 140-147.

L --- (1969) "On the behavior of viscosity at the nematic-isotropic transition," Phys. Lett. 30A, 5-6.

PARETTE, G. and USHA DENIZ, K. (1968) "Magnetic scattering of neutrons in MnF_2 near the critical point," J. Appl. Phys. 39, 1232-1234.

--- and KAHN, R. (1971) "Study of the critical neutron scattering by iron in 'hydrodynamic region' and 'quasi-hydrodynamic region'," J. de Physique 32, 447-460.

PARK, D. (1956) "A summation method for crystal statistics," Physica 12, 932-940.

r PARKS, R. D. (1971) "Thermodynamic fluctuations in superconductors: experimental situation," Proc. LT-12, Kyoto, p. 217-223.

PARLANGE, J. Y. (1968) "Phase transition and surface tension in the quasichemical approximation," J. Chem. Phys. 48, 169-173.

PASSELL, L., BLINOWSKI, K., BRUN, T., and NIELSEN, P. (1965) "Critical magnetic scattering of neutrons in iron," Phys. Rev. 139, A1866-A1876.

PATASHINSKIǏ, A. Z. and POKROVSKIǏ, V. L. (1964) "Second order phase transitions in a Bose fluid," Sov. Phys.-JETP 19, 677-691.

---, --- (1966) "Behavior of ordered systems near the transition point," Ibid. 23, 292-297.

d PATHRIA, R. K. (1972) "Bose-Einstein condensation in thin films," Phys. Rev. A5, 1451.

PATTERSON, H. S., CRIPPS, R. S., and WHYTLAW-GRAY, R. (1912) "The critical constants and orthobaric densities of xenon," Proc. Roy. Soc. A86, 579-590.

d PATTON, B. R. (1971) "Fluctuation theory of the superconducting transition in restricted dimensionality," Phys. Rev. Lett. 27, 1273-1276.

PAUKOV, I. E. and RAKHMENKULOV, F. S. (1972) "Phase transitions in rubidium superoxide," Sov. Phys.-Solid State 13, 1846-1848.

b PAUL, E. and MAZO, R. M. (1969) "Hydrodynamic properties of a plane-polygonal polymer, according to Kirkwood-Riseman theory," J. Chem.Phys. 51, 1102-1107.

PAUL, G., LAMBETH, D., and STANLEY, H. E. (1973) "The problem of antiferromagnetic structures," Phys. Rev. B7 (submitted for publication).

3 --- and STANLEY, H. E. (1971) "Theorems for series expansions for the generalized Heisenberg model," J. de Physique 32S, 350-352.

u ---, --- (1971) "Do critical-point exponents depend upon the strength of second neighbor interactions?," Phys. Lett. 37A, 328-330.

u ---, --- (1971) "Universality of critical-point exponents with respect to lattice anisotropy," Ibid. 37A, 347-349.

u ---, --- (1972) "Partial test of the universality hypothesis: The case of different coupling strengths in different lattice directions," Phys. Rev. B5, 2578-2599.

u ---, --- (1972) "A partial test of the universality hypothesis. The case of next-nearest neighbor interactions," Ibid. B5, 3715-3726.

h PEARCE, C. J., LIPA, J. A., and BUCKINGHAM, M. J. (1968) "Velocity of second sound near the λ point of helium," Phys. Rev. Lett. 20, 1471-1473.

b PEARLSTEIN, R. M. (1968) "Possibility of limited pairon condensation in double-stranded DNA," Ibid. 20, 594-596.

m PEARSON, F. J. and RUSHBROOKE, G. S. (1957) "On the theory of binary fluid mixtures," Proc. Roy. Soc. (Edinburgh) 64, 305-317.

b PECORA, R. (1964) "Doppler shifts in light scattering from pure liquids and polymer solutions," J. Chem. Phys. 40, 1604-1614.

--- (1969) "Dispersion of the electrically induced refractive-index anisotropy in nonpolar fluids," J. Chem. Phys. 50, 2650-2656.

PEIERLS, R. (1936) "Statistical theory of superlattices with unequal concentrations of the components," Proc. Roy. Soc. A154, 207-222.

m PEKALSKI, A. (1971) "The field-dependent magnetic phases of a uniaxial two-sublattice antiferromagnet of Néel type. II. Magnetizations, magnetic susceptibilities and phase transitions," Acta Phys. Polon. A40, 189.

PENROSE, O. (1963) "Convergence of fugacity expansions for fluids and lattice gases," J. Math. Phys. 4, 1312-1320.

--- (1964) "Two inequalities for classical and quantum systems of particles with hard cores," Phys. Lett. 2, 224-226.

--- and ELVEY, J. S. N. (1968) "The Yang-Lee distribution of zeros for a classical one-dimensional fluid," J. Phys. A. 1, 661-674.

--- and LEBOWITZ, J. L. (1971) "Rigorous treatment of metastable states in the van der Waals-Maxwell theory," J. Stat. Phys. 3, 211-236.

---, --- (1972) "A functional equation in the theory of fluids," J. Math. Phys. 13, 604.

h --- and ONSAGER, L. (1956) "Bose-Einstein condensation and liquid helium," Phys. Rev. 104, 576-584.

PEREL'MAN, M. E. (1971) "Phase transitions caused by the opening of new channels in electron-photon interactions," Phys. Lett. 37A, 411.

PERL, R. and FERRELL, R. A. (1972) "Critical viscosity and diffusion in the binary-liquid phase transition," Phys. Rev. Lett. 29, 51-55.

b PERUTZ, M. F. (1970) "Sterochemistry of cooperative effects in haemoglobin," Nature 228, 726-739.

PETER, H., KORPIUN, P., and LÜSCHER, E. (1968) "Measurement of the longitudinal sound velocity in solid krypton at 4.2°K, 77°K, and 90°K," Phys. Lett. 26A, 207.

PETERSON, R. L. (1968) "Short-range magnetic order in a modified Weiss molecular-field theory," Phys. Rev. 171, 586-590.

--- (1968) "Ising chain as the basic cluster in effective-field theories of magnetism," Phys. Lett. 27A, 177-178.

PETRESCU, V., REZLESCU, N., and CUCIUREANU (1970) "Influence of cation distribution on the Curie temperature of copper containing ferrites," Rev. Roum. Phys. 15, 965-972.

m PFEIFFER, H. (1971) "Green-functions theory of ferro-paramagnetic phase transitions in uniaxial spin-1/2 ferromagnets with external magnetic field," Acta. Phys. Polon. A39, 213.

l PFEUTY, P. and ELLIOTT, R. J. (1971) "The Ising model with a transverse field - II. Ground state properties," J. Phys. C. 4, 2370-2385.

∞ PHILHOURS, J. (1969) "Generalization of the spherical model," Phys. Rev. 177, 976-980.

l --- (1971) "Ising model with first-, second-, and third-neighbor interactions," Ibid. B4, 929-937.

a --- and HALL, G. L. (1967) "Ordering conditions. I. For alloys represented by generalized Ising models," Ibid. 163, 460-465.

∞ ---, --- (1968) "Comparison of the spherical model and the Clapp-Moss equations with the Ising model," Ibid. 170, 496-502.

PHILIP, J. W., GONANO, R., and ADAMS, E. D. (1969) "Critical-region thermal expansion in $MnCl_2 4H_2O$ and $MnBr_2 \cdot 4H_2O$," Ibid. 188, 973-981.

l PIASECKI, J. (1967) "Dynamics of the Ising chain II," Bull. Acad. Pol. Sci. 15, 357-361.

d PIERCE, R. D. and FRIEDBERG, S. A. (1967) "Magnetic ordering in $Mn(HCOO)_2 \cdot 2H_2O$ related compounds," J. Appl. Phys. 38, 1462-1463.

f PIETRASS, B. (1971) "Analysis of the elastic anomalies at the structural phase transition of $SrTiO_3$ near 105 K," Phys. Stat. Sol. (b)47, 495.

PIKIN, S. A. and TSUKERNIK, V. M. (1966) "The thermodynamics of linear spin chains in a transverse magnetic field," Sov. Phys.-JETP 23, 914-916.

m PINGS, C. J. (1967) "Simplification and symmetrization of the Fixman relation for viscosity of critical mixtures," J. Chem. Phys. 46, 2013-2014.

--- and TEAGUE, R. K. (1968) "Experimental study of the shape of the coexistence curve of argon near the critical state," Phys. Lett. 26A, 496-497.

l PINK, D. A. (1969) "Row correlation functions of the two-dimensional Ising model," Phys. Rev. 188, 1032-1037.

3 PIRNIE, K. and WOOD, P. J. (1965) "The Heisenberg ferromagnet with second neighbour interactions for general spin," Phys. Lett. 17, 241-243.

3 ---, ---, and EVE, J. (1966) "On high temperature susceptibilities of Heisenberg model ferromagnetics and antiferromagnetics," Mol. Phys. 11, 551-577.

m PLISCHKE, M. and MATTIS, D. (1971) "Critical curves and thermodynamic phases of lattice fluids and antiferromagnets with structured interactions," Phys. Rev. A3, 2092-2099.

a ---, --- (1971) "A short-range versus long-range order in a model binary alloy," Phys. Rev. Lett. 27, 42-45.

d PLUMIER, R. (1963) "Etude par diffraction des neutrons du desordre magnetique d'empilement dans l'antiferromagnetique K_2NiF_4," J. Phys. Radium 24, 741-745.

d --- (1964) "Neutron diffraction study of magnetic stacking faults in antiferromagnetic K_2NiF_4," J. Appl. Phys. 35, 950-951.

g POKROVSKII, V. L. (1968) "Similarity hypothesis in the theory of phase transitions," Sov. Phys.-Usp. 11, 66-74.

b POLAND, D. and SCHERAGA, H. A. (1966) "Phase transitions in one dimension and the helix-coil transition in polyamino acids," J. Chem . Phys. 45, 1456-1463.

r ---, --- (1970) Theory of Helix-Coil Transitions in Biopolymers, Academic Press, N.Y.

m POLGAR, L. G. and FRIEDBERG, S. A. (1971) "Low-temperature heat capacity of the metamagnet $Ni(NO_3)_2 \cdot 2H_2O$," Phys. Rev. B4, 3110-3115.

g POLYAKOV, A. M. (1969) "Microscopic description of critical phenomena," Sov. Phys.-JETP 28, 533-539.

g --- (1970) "Conformal symmetry of critical fluctuations," JETP Lett. 12, 381-383.

g --- (1970) "Properties of long and short range correlations in the critical region," Sov. Phys.-JETP 30, 151.

g --- (1971) "A similarity hypothesis in the strong interactions. I. Multiple hadron production in e^+e^- annihilation," Ibid. 32, 296.

f POMPE, G. and HEGENBARTH, E. (1970) "Measurement of the specific heat of KH_2PO_4 by a dynamic method," Ibid. 12, 357.

POPOVICI, M. (1971) "Wavelength-dependent susceptibility of iron in the critical region," Phys. Lett. 34A, 319-321.

3 POTAPKOV, N. A. and SHELKOV, N. V. (1971) "On the calculation of the complex susceptibility of a Heisenberg ferromagnet in the vicinity of the Curie point," Phys. Stat. Sol. (b)43, 35.

POTTER, H. H. (1934) "The magneto-caloric effect and other magnetic phenomena in iron," Proc. Roy. Soc. A146, 362-387.

1 POTTS, R. B. (1952) "Spontaneous magnetization of a triangular Ising lattice," Phys. Rev. 88, 352.

POULIS, N. J. and HARDEMAN, G. E. G. (1952) "Behavior of a single crystal of $CuCl_2 \cdot 2H_2O$ near the Néel temperature," Physica 18, 429-432.

PRASAD RAO, A. D., KATIYAR, R. S., and PORTO, S. P. S. (1972) "Relation between phonon structure and phase transition in $NaClO_3$," Phys. Rev. Lett. 28, 665-668.

PRESSING, J. (1971) "New equation for surface tension near the critical point," J. Chem. Phys. 54, 3485-3487.

∞ PRESSMAN, W. and KELLER, J. B. (1960) "Equation of state and phase transition of the spherical lattice gas," Ibid. 120, 22-32.

PRESTON, R. S. (1968) "Temperature dependence of isomer shift and hyperfine field near the Curie point in iron," J. Appl. Phys. 39, 1231.

L PRIEST, R. G. (1971) "Comments on the lattice model of liquid crystals," Phys. Rev. Lett. 26, 423-425.

PUGLIELLI, V. and FORD, N. C., JR. (1970) "Turbidity measurements in SF_6 near its critical point," Ibid. 25, 143-147.

PURIĆ, J., LABAT, J., ĆIRKOVIĆ, L. J., and KONJEVIĆ, N. (1970) "Experimental study of Stark broadening of neutral helium line 5876Å in a plasma," Fizika 2, 67-72.

m PUSEY, P. N. and GOLDBERG, W. I. (1968) "Temperature dependence of concentration fluctuations in a binary liquid mixture using a photon correlation method," Appl. Phys. Lett. 13, 321-323.

m ---, --- (1969) "Light-scattering measurement of concentration fluctuations in phenol-water near its critical point," Phys. Rev. Lett. 23, 67-70.

f PYTTE, E. (1970) "Theory of the structural transition in NbSn and V_3Si," Ibid. 25, 1176-1180.

PYTTE, E. (1972) "Spurious first-order phase transitions in the self-consistent phonon approximation," Ibid. 28, 895-897.

--- and BENNETT, H. S. (1967) "Ultrasonic attenuation in the Heisenberg paramagnet. II. Antiferromagnets," Phys. Rev. 164, 712-715.

f --- and FEDER, J. (1969) "Theory of a structural transition in Perovskite-type crystals," Ibid. 187, 1077-1088.

f --- and THOMAS, H. (1969) "Soft modes, critical fluctuations, and optical properties for a two-valley model of Gunn-instability semiconductors," Ibid. 179, 431-443.

RADO, G. T. (1970) "Magnetoelectric studies of critical behavior in the Ising-like antiferromagnet $DyPO_4$," Sol. State Commun. 8, 1349-1352.

RAMANATHAN, G. V. (1966) "A new cluster scheme in statistical physics," J. Math. Phys. 7, 1507-1514.

RAMÍREZ, R. and KIWI, M. (1972) "Electronic phase transitions in $Ce_{1-x}La_x$ alloys," Phys. Rev. Lett. 28, 344-346.

l RAPAPORT, D. C. (1971) "Critical curves for the Ising model with quenched impurities," Phys. Lett. 37A, 15-16.

u --- (1971) "On the critical behavior of the anisotropic Ising model," Ibid. 37A, 407-408.

l --- (1971) "On the Ising model with impurities," J. Phys. C. 4, L322-L324.

∞ --- (1972) "Perturbation theory, scaling and the spherical model," Ibid. C. 5, 933-955.

--- (1972) "The Ising ferromagnet with impurities: a series expansion approach. I," Ibid. C. 5, 1830-1858.

u --- and DOMB, C. (1971) "The smoothness postulate and the Ising antiferromagnet," Ibid. C. 4, 2684-2694.

l --- and FRANKEL, N. E. (1968) "On the existence of a phase transition on an Ising chain with a long-range interaction," Phys. Lett. 28A, 405-406.

RASAIAH, J. C. and FRIEDMAN, H. J. (1969) "Integral equation computations for aqueous 1-1 electrolytes. Accuracy of the method," J. Chem. Phys. 50, 3965-3976.

--- and STELL, G. (1970) "Upper bounds on free energies in terms of hard-sphere results," Mol. Phys. 18, 249-260.

RATHJEN, W., STAUFFER, D., and KIANG, C. S. (1972) "Equation of state from triple point to critical point," Phys. Lett. 40A, 345-346.

l RATNER, I. M. (1968) "Ising lattice at negative absolute temperatures," Sov. Phys.-Solid State 9, 1652-1653.

RAYL, M. and WOJTOWICZ, P. J. (1968) "Critical behavior of the spontaneous magnetization of an isotropic ferromagnet," Phys. Lett. 28A, 142-143.

d REATTO, L. (1968) "Low temperature transition of a two-dimensional Bose system," Ibid. 26A, 400-401.

m --- (1970) "Droplet model for tricritical points: metamagnetic transition," Phys. Rev. B5, 204-221.

--- (1970) "On the droplet model for critical phenomena," Phys. Lett. 33A, 519-520.

m --- (1972) "Droplet model for tricritical points: metamagnetic transition," Phys. Rev. B5, 204.

RECHTIN, M. D. and AVERBACH, B. L. (1971) "Critical magnetic neutron scattering from CoO," Phys. Rev. Lett. 26, 1480-1483.

---, --- (1971) "Tetragonal elongation in CoO near the Néel point," Ibid. 26, 1483-1485.

---, MOSS, S. C., and AVERBACH, B. L. (1970) "Influence of lattice contraction on long-range order in CoO near T_N," Ibid. 24, 1485-1489.

REE, F. H. and CHAY, T. R. (1972) "Density fluctuations in microscopic volumes of fluid argon around its critical point," Phys. Rev. A6, 796-799.

f REESE, W. and MAY, L. F. (1977) "Critical phenomena in order-disorder ferroelectrics. I. Calorimetric studies of KH_2PO_4," Ibid. 162, 510-518.

f ---, --- (1968) "Studies of phase transition in order-disorder ferroelectrics. II. Calorimetric investigations of KD_2PO_4," Ibid. 167, 504-510.

REITER, G. F. (1970) "Kinetic theory calculation of the spectral density for $RbMnF_3$," J. Appl. Phys. 41, 1368-1369.

3 --- (1972) "Spin fluctuations in Heisenberg paramagnets. I. Diagrammatic expansion for the moments of the spectral density at finite temperature," Phys. Rev. B5, 222-235.

l RENARD, R. and GARLAND, C. W. (1966) "Order-disorder phenomena. II. Elastic constants of a two-dimensional Ising model," J. Chem. Phys. 44, 1125-1129.

---, --- (1966) "Order-disorder phenomena. V. Pippard equations and the phase diagram for ammonium chloride," Ibid. 45, 763-766.

RENNINGER, A., MOSS, S. C., and AVERBACH, B. L. (1966) "Local antiferromagnetic order in single-crystal MnO above the Néel temperature," Phys. Rev. 147, 418-422.

RENO, R. C. and HOHENEMSER, C. (1970) "Perturbed angular correlation measurement on ^{100}Rh in a Ni host: Critical exponent β for Ni," Phys. Rev. Lett. 25, 1007-1011

h REPPY, J. D. (1965) "Application of a superfluid gyroscope to the study of critical velocities in liquid helium near the λ transition," Ibid. 14, 733-735.

3 RÉSIBOIS, P. and DEWEL, G. (1972) "Dynamical scaling and kinetic equations for Heisenberg spin systems just below the critical temperature," Ann. Phys. 69, 299-320.

3 --- and DE LEENER, M. (1967) "Critical time-dependent fluctuations in Heisenberg spin systems," Phys. Lett. 25A, 65-66.

e ---, --- (1969) "Irreversibility in Heisenberg spin systems. III. Kinetic equations for the auto-correlation function at finite temperature," Phys. Rev. 178, 806-818.

--- and PIETTE, C. (1970) "Temperature dependence of the linewidth in critical spin fluctuation," Phys. Rev. Lett. 24, 514-516.

--- and VELARDE, M. G. (1971) "Non-analytic density expansion of transport coefficients in the quantum Lorentz gas and weak-coupling approximation," Physica 51, 541-560.

h REVZEN, M. (1969) "Fluctuations and the onset of superfluidity," Phys. Rev. 185, 337.

RICCI, F. P. and SCAFE, E. (1969) "Orthobaric density of CH_4 in the critical region," Phys. Lett. 29A, 650-651.

RICE, O. K. (1947) "On the behavior of pure substances near the critical point," J. Chem. Phys. 15, 314-332.

r --- (1950) "Introduction to the Symposium on Critical Phenomena," J. Phys. Chem. 54, 1293-1305.

--- (1955) "Relation between isotherms and coexistence curve in the critical region," J. Chem. Phys. 23, 169-173.

--- (1960) "The thermodynamics of non-uniform systems, and the interfacial tension near a critical point," J. Phys. Chem. 64, 976-984.

h --- (1967) "Statistical thermodynamics of λ transitions, especially of liquid helium," Phys. Rev. 153, 275-279.

m --- (1967) "Possible relation between phase separation and the λ transition in ^{3}He-^{4}He mixtures," Phys. Rev. Lett. 19, 295-297.

m --- and CHANG, D.-R. (1972) "Thermodynamic relationships at the tricritical point in ^{3}He-^{4}He mixtures," Phys. Rev. A5, 1419.

m --- and MAC QUEEN, J. T. (1962) "The effect of an impurity on the phase transition in a binary liquid system II," J. Phys. Chem. 66, 625-631.

d RICE, T. M. (1965) "Superconductivity in one and two dimensions," Phys. Rev. 140, A1889-A1891.

--- and BRINKMAN, W. F. (1972) "Effects of impurities on the metal-insulator transition," Ibid. B5, 4350-4357.

d RICHARDS, P. M. (1971) "Spin-wave theory and static correlations in a one-dimensional Heisenberg antiferromagnet," Phys. Rev. Lett. 27, 1800-1803.

--- and CARBONI, F. (1972) "Spin waves in finite spin-1/2 Heisenberg chains," Phys. Rev. B5, 2014-2018.

u RIEDEL, E. K. (1971) "Dynamic scaling theory for anisotropic magnetic systems," J. Appl. Phys. 42, 1383-1389.

m --- (1972) "Scaling approach to tricritical phase transitions," Phys. Rev. Lett. 28, 675-678.

--- and WEGNER, F. (1969) "Anomaly of the ferromagnetic susceptibility χ_q near T_c," Phys. Lett. 29A, 77.

u ---, --- (1969) "Scaling approadch to anisotropic magnetic systems. Statics," Z. Physik 225, 195-215.

u ---, --- (1970) "Dynamic scaling theory for anisotropic magnetic systems," Phys. Rev. Lett. 24, 730-733.

g ---, --- (1972) "Tricritical exponents and scaling fields," Ibid. 29, 349-353.

RIEDER, Z., LEBOWITZ, J. L., and LIEB, E. (1967) "Properties of a harmonic crystal in a stationary nonequilibrium state," J. Math. Phys. 8, 1073-1078.

u RIMMER, D. E., DALTON, N. W., and WOOD, D. W. (1971) "The high temperature susceptibility of the anisotropic Heisenberg model," J. Phys. C. 4, L4-L5.

RISTE, T. and WANIC, A. (1965) "The range of the critical fluctuations of the sublattice magnetization in Cr_2O_3 and α-Fe_2O_3 and its temperature variation," Phys. Lett. 16, 231-232.

3 RITCHIE, D. S. and FISHER, M. E. (1972) "Theory of critical-point scattering and correlations. II. Heisenberg models," Phys. Rev. B5, 2668-2692.

ROACH, P. (1968) "Pressure-density-temperature surface of He^4 near the critical point," Ibid. 170, 213-223.

h ROBERTS, P. H. and DONNELLY, R. J. (1970) "Theory of the onset of superflow," Phys. Rev. Lett. 24, 367-371.

ROCKER, W. and KOHLHAAS, R. (1968) "Uber die magnetische Zustandsgleichung des Kobalts in der Umgebung seiner Curie-Temperatur," Z. Angew. Physik 25, 343-346.

ROKNI, M. and WALL, L. S. (1971) "Soft-mode study of the stress-induced phase transitions near T_o in $SrTiO_3$," J. Chem. Phys. 55, 435-438.

ROSEN, M. (1968) "Elastic moduli and ultrasonic attenuation of gadolinium, terbium, dysprosium, holmium, and erbium from 4.2 to 300°K," Phys. Rev. 174, 505-514.

--- and KLIMKER, H. (1969) "Elastic constants and ultrasonic attenuation of a β-tungsten-structure V_3Ge crystal from 4.2 to 300°K," Ibid. 184, 466-471.

b ROTHSCHILD, K. J. and STANLEY, H. E. (1972) "Globular membrane proteins as functional units of ionic transport," In Biomedical Physics and Biomaterials Science, H. E. Stanley, Ed., MIT Press, Cambridge, Mass., p. 3-24.

---, --- (1972) "The molecular organization and function of biological membranes: A possible microscopic picture of ionic permeation," In Problems of the Biochemical Dialogue at the Cell Membrane Surface, Immunopathological Mechanism, and Aspects as They Relate to Cancer, Transplantation, and Cellular Engineering, S. B. Day and R. H. Good, Eds., Academic Press, N.Y., p. 1-38.

b ---, --- (1973) "All-or-none transitions in membranes without directly-interacting functional units of ionic transport," Proc. Fourth Internat. Biophys. Congress, Moscow, Aug. 1972.

m ROWDEN, R. W. and RICE, O. K. (1951) "Critical phenomena in the cyclohexane-aniline system," J. Chem. Phys. 19, 1-2.

ROWE, J. E. and TRACY, J. C. (1971) "Critical behavior of d-band photoelectrons near the Curie temperature of nickel," Phys. Rev. Lett. 27, 799-802.

r ROWLINSON, J. S. (1970) Liquids and Liquid Mixtures, 2nd Edition, Butterworths, London.

RUDD, W. G. and FRISCH, H. L. (1970) "Critical behavior in percolation processes," Phys. Rev. B2, 162-164.

m RUDD, D. F. P. and WIDOM, B. (1960) "Critical solution phenomenon in two-component liquid systems. The system water-ethylene glycol mono-isobutyl ether," J. Chem. Phys. 33, 1816-1819.

1 RUDOY, Y. G. (1967) "Green's function method in the Ising model," Sov. Phys. Doklady 12, 437-440.

3 --- (1967) "High-temperature expansion for Heisenberg spin 1/2 ferromagnet with an impurity atom," Phys. Lett. 24A, 488-490.

RUELLE, D. (1963) "Correlation functions of classical gases," Ann. Phys. 25, 109-120.

--- (1963) "Statistical mechanics of quantum systems of particles," Physica 36, 789-799.

--- (1967) "A variational formulation of equilibrium statistical mechanics and the Gibbs phase rule," Commun. Math. Phys. 5, 324-329.

--- (1967) "States of classical statistical mechanics," J. Math. Phys. 8, 1657-1668.

r --- (1969) Statistical Mechanics: Rigorous Results, W. A. Benjamin, Inc., N.Y.

--- (1970) "Superstable interactions in classical statistical mechanics," Commun. Math. Phys. 18, 127-159.

--- (1971) "Extension of the Lee-Yang circle theorem," Phys. Rev. Lett. 26, 303-304.

--- (1971) "Existence of a phase transition in a continuous classical system," Ibid. 27, 1040-1041.

RUNNELS, L. K. and COLVIN, C. (1970) "Nature of the rigid-rod mesophase," J. Chem. Phys. 53, 4219-4226.

---, CRAIG, J. R., and STREIFFER, H. R. (1971) "Exact finite method of lattice statistics. V. The thermodynamic phases of a triangular lattice gas," Ibid. 54, 2004-2013.

--- and HUBBARD, J. B. (1972) "Applications of the Yang-Lee-Ruelle theory to hard-core lattice gases," J. Stat. Phys. 6, 1-20.

RUSHBROOKE, G. S. (1938) "A note on an assumption in the theory of cooperative phenomena," Proc. Camb. Phil. Soc. 34, 424-428.

--- (1938) "A note on Guggenheim's theory of strictly regular binary liquid mixtures," Proc. Roy. Soc. 166, 296-315.

--- (1940) "Note on the composition of alloys with composition near Al_2Cu," Proc. Phys. Soc. 52, 701-706.

r --- (1949) Statistical Mechanics, Oxford Univ. Press, London.

--- (1949) "On the theory of regular solutions," Nuovo Cimento 51, 1-10.

--- (1952) "Statistique des Cristaux," Changements de Phases, 177-191.

--- (1963) "On the thermodynamics of the critical region for the Ising problem," J. Chem. Phys. 39, 842-843.

--- (1964) "On the theory of randomly dilute Ising and Heisenberg ferromagnets," J. Math. Phys. 5, 1106-1116.

--- (1965) "On the Griffiths inequality at a critical point," J. Chem. Phys. 43, 3439-3441.

--- (1971) "Randomly dilute Ising and Heisenberg systems," In Critical Phenomena in Alloys, Magnets, and Superconductors, R. E. Mills, E. Ascher, and R. I. Jaffee, Eds. (McGraw-Hill Book Co., N.Y.), pp. 155-163.

r ---, BAKER, G. A., JR., and WOOD, P. J. (1973) "High-temperature expansions: Heisenberg model," In Phase Transitions and Critical Phenomena, Vol. 3, C. Domb and M. S. Green, Eds. (Academic Press, London).

--- and EVE, J. (1959) "On noncrossing lattice polygons," J. Chem. Phys. 31, 1333-1334.

l ---, --- (1962) "High-temperature Ising partition function and related noncrossing polygons for the simple cubic lattice," J. Math. Phys. 3, 185-189.

--- and MORGAN, D. J. (1961) "On the magnetically dilute Heisenberg and Ising ferromagnetics," Mol. Phys. 4, 1-15.

l --- and SCOINS, H. I. (1955) "On the Ising problem and Mayer's cluster sums," Proc. Roy. Soc. 230, 74-90.

l ---, --- (1962) "Cluster sums for the Ising model," J. Math. Phys. 3, 176-184.

--- and URSELL, H. D. (1948) "On one-dimensional regular assemblies," Proc. Camb. Phil. Soc. 44, 263-271.

3 --- and WOOD, P. J. (1955) "On the high-temperature susceptibility for the Heisenberg model of a ferromagnet," Proc. Phys. Soc. 68, 1161-1169.

3 ---, --- (1958) "On the Curie points and high temperature susceptibilities of Heisenberg model ferromagnetics," Mol. Phys. 1, 257-283.

---, --- (1963) "On the high temperature staggered susceptibility of Heisenberg model antiferromagnetics," Ibid. 6, 409-421.

l RYAZANOV, G. V. (1966) "Correlation asymptotics for a plane Ising lattice," Sov. Phys--JETP 22, 789-795.

h --- (1971) "Influence of violation of electroneutrality on a phase transition in a ferroelectric with hydrogen bonding," Ibid. 32, 544-546.

m SAAM, W. F. (1970) "Thermodynamics of binary systems near the liquid-gas critical point," Phys. Rev. A2, 1461-1466.

SALAMON, M. B. (1970) "Specific heat of CoO near T_N: Anisotropy effects," Ibid. B2, 214-220.

d --- and HATTA, I, (1971) "Specific heat of the two-dimensional antiferromagnet K_2NiF_4," Phys. Lett. 36A, 85-87.

d SALZANO, F. J. and STRONGIN, M. (1967) "Dimensionality of superconductivity in graphite lamellar compounds," Phys. Rev. 153, 533-534.

f SAMULIONIS, V. I. and KUNIGELIS, V. F. (1971) "Anomalous absorption of sound near the ferroelectric phase transition," JETP Lett. 13, 207-208.

SARKIES, K. W., NINHAM, B. W., and RICHMOND, P. (1971) "Van der Waals forces and critical phenomena," J. Phys. C. 4, L235-L237.

SATO, H. (1961) "Remarks on the magnetic properties of Au-Mn system," J. Phys. Chem. Solids 19, 54-60.

SAUER, G. E. and BORST, L. B. (1968) "Lambda transition in liquid sulfur," Science 158, 1567-1569.

m SAUL, D. M. and WORTIS, M. (1972) "Series analysis of tricritical behavior in an Ising-like magnetic model," In Proc. of the 1971 Magnetism Conf. (AIP Publication No. 5) p. 349-351.

f SAWADA, A., ISHIBASHI, Y., and TAKAGI, Y. (1971) "Dielectric study of critical behavior of ferroelectric triglycine sulfate by a digital technique," J. Phys. Soc. Japan 31, 823-827.

--- and TAKAGI, Y. (1971) "Superstructure in the ferroelectric phase of ammonium Rochelle salt," Ibid. 31, 952.

SCALAPINO, D. J., SEARS, M., and FERRELL, R. A. (1972) "Statistical mechanics of one-dimensional Ginzburg-Landau fields," Phys. Rev. B6, 3409-3416.

SCESNEY, P. E. (1970) "Mobile-electron Ising ferromagnet," Ibid. B1, 2274-2288.

SCHAEFER, D. W., BENEDEK, G. B., SCHOFIELD, P., and BRADFORD, E. (1971) "Spectrum of light quasi-elastically scattered from tobacco mosaic virus," J. Chem. Phys. 55, 3884-3895.

SCHELLENG, J. H. and FRIEDBERG, S. A. (1969) "Thermal behavior of the antiferromagnet $MnBr_2 \cdot 4H_2O$ in applied magnetic fields," Phys. Rev. 185, 728-734.

SCHICK, M. and ZILSEL, P. R. (1969) "Order parameter, mean-field theory, and the ideal Bose gas," Ibid. 188, 522-525.

s SCHMIDT, H. (1968) "The onset of superconductivity in the time dependent Ginzburg-Landau theory," Z. Physik 216, 336-345.

--- (1971) "Critical region scaled equation-of-state calculations and gravity effects," J. Chem. Phys. 54, 3610-3621.

m ---, JURA, G., and HILDEBRAND, J. H. (1959) "The heat capacity of the system carbon tetrachloride-perfluoromethycyclohexane through the critical region," J. Phys. Chem. 63, 297-299.

---, OPDYCKE, J., and GRAY, C. F. (1967) "Heat capacity in the critical region of xenon," Phys. Rev. Lett. 19, 887-890.

SCHMIDT, P. W. and HIGHT, R., JR. (1960) "Slit height corrections in small angle X-ray scattering," Acta Cryst. 13, 480-483.

SCHMIDT, V. A. and FRIEDBERG, S. A. (1967) "Low-temperature magnetic phase transitions in $CoCl_2 \cdot 6H_2O$ and $MnBr_2 \cdot 4H_2O$," J. Appl. Phys. 38, 5319-5326.

m ---, --- (1970) "Metamagnetism of $Ni(NO_3)_2 \cdot 2H_2O$," Phys. Rev. B5, 2250-2256.

SCHNEIDER, T. (1971) "Theory of the liquid-solid phase transition," Phys. Rev. A3, 2145-2148.

---, BROUT, R., THOMAS, H., and FEDER, J. (1970) "Dynamics of the liquid-solid transition," Phys. Rev. Lett. 25, 1423-1426.

h --- and ENZ, C. P. (1971) "Theory of the superfluid-solid transition of ^4He," Ibid. 27, 1186-1188.

f ---, SRINIVASAN, G., and ENZ, C. P. (1972) "Phase transitions and soft modes," Phys. Rev. A5, 1528-1537.

SCHNEIDER, W. G. (1950) "Sound velocity and sound absorption in the critical temperature region," Can. J. Chem. 29, 243-252.

SCHNEIDER, W. R. and STRÄSSLER, S. (1972) "Entropy of mixing of nonideal solutions," Phys. Rev. A6, 333-341.

SCHOFIELD, P. (1966) "Wavelength-dependent fluctuations in classical fluids," Proc. Phys. Soc. 88, 149-170.

--- (1969) "Parametric representation of the equation of state near a critical point," Phys. Rev. Lett. 22, 606-608.

---, LITSTER, J. D., and HO, J. T. (1969) "Correlation between critical coefficients and critical exponents," Ibid. 23, 1098-1102.

SCHREIBER, J. and HANDRICH, K. (1971) "Curie temperature and susceptibility of an amorphous Heisenberg ferromagnet (high-temperature expansion)," JETP Lett. 14, 37-39.

SCHULHOF, M. P., HELLER, P., NATHANS, R., and LINZ, A. (1970) "Inelastic neutron scattering from MnF_2 in the critical region," Phys. Rev. Lett. 24, 1184-1187.

---, ---, ---, --- (1970) "Critical magnetic scattering in manganese fluoride," Phys. Rev. B1, 2304-2311.

---, NATHANS, R., HELLER, P., and LINZ, A. (1971) "Inelastic neutron scattering from MnF_2 in the critical region," Ibid. B4, 2254-2276.

l SCHULTZ, T. D., MATTIS, D. C., and LIEB, E. H. (1964) "Two-dimensional Ising model as a soluble problem of many fermions," Rev. Mod. Phys. 36, 856-871.

m SCHÜRMANN, H. K. and PARKS, R. D. (1971) "Critical exponent of phase separation in metallic binary liquids," Phys. Rev. Lett. 26, 367-370.

---, --- (1971) "Gravitational separation and resistance anomalies in liquid binary alloys with miscibility gaps," Ibid. 26, 835-837.

---, --- (1971) "Paraconductivity in binary metallic liquids above the critical point," Ibid. 27, 1790-1793.

SCHWABL, F. (1971) "On the critical dynamics of isotropic ferromagnets below the Curie point," Z. Physik 246, 13-28.

f --- (1972) "Critical dynamics of $SrTiO_3$ for $T \geq T_c$," Phys. Rev. Lett. 28, 500-503.

3 --- and MICHEL, K. H. (1970) "Hydrodynamics of Heisenberg ferromagnets," Phys. Rev. B2, 189-205.

SCHWARTZ, P. (1971) "Order-disorder transition in NH_4Cl. III. Specific heat," Ibid. B4, 920-928.

SCOTT, H. L., JR. (1970) "Phenomenological model for ^3He near the critical point," Ibid. A2, 1099-1101.

SEARS, V. F. (1967) "Temperature dependence of paramagnetic neutron scattering," Can. J. Phys. 45, 2923-2929.

d SEEHRA, M. S. (1969) "Two-dimensional magnetic behavior of copper formate tetrahydrate," Phys. Lett. 28A, 754-755.

r SENGERS, J. V. (1968) "Transport properties of compressed gases," Recent Advances in Eng. Sci. 3, 153-196.

r --- (1971) "Triple collision effects in the transport properties for a gas of hard spheres," In Kinetic Equations, R. L. Loboff and N. Rostoker, Eds., (Gordon and Breach, N.Y.) pp. 137-193.

r --- (1972) "Transport processes near the critical point of gases and binary liquids in the hydrodynamic regime," Ber. der Bunsen-Gesellschaft 76, 234-249.

--- and KEYES, P. H. (1971) "Scaling of the thermal conductivity near the gas-liquid critical point," Phys. Rev. Lett. 26, 70-73.

r --- and LEVELT-SENGERS, J. M. H. (1968) "The critical region," Chem. Eng. News 46, 104-118.

SENTURIA, S. D. and BENEDEK, G. B. (1966) "Nuclear resonance in ferromagnetic chromium tribromide from 4.2°K to the Curie point," Phys. Rev. Lett. 17, 475-478.

SHAFER, M. W. and MC GUIRE, T. R. (1968) "Studies of Curie-point increases in EuO," J. Appl. Phys. 39, 588-590.

r SHANTE, V. K. S. and KIRKPATRICK, S. (1971) "An introduction to percolation theory," Adv. Phys. 11, 325-355.

m SHAPIRA, Y. (1969) "Ultrasonic behavior near the spin-flop transitions of hematite," Phys. Rev. 184, 589-600.

m --- (1969) "Ultrasonic behavior near the spin-flop transition of Cr_2O_3," Ibid. 187, 734-736.

m --- (1970) "Paramagnetic-to-antiferromagnetic phase boundaries of FeF_2 from ultrasonic measurements," Ibid. B2, 2725-2734.

--- (1971) "Observation of antiferromagnetic phase transitions by ultrasonic techniques," J. Appl. Phys. 42, 1588-1594.

m --- and FONER, S. (1970) "Magnetic phase diagram of MnF_2 from ultrasonic and differential magnetization measurements," Phys. Rev. B1, 3083-3096.

---, ---, REED, T. B., BIRECKI, H., and STANLEY, H. E. (1972) "Dependence of the insulator metal transition in EuO on magnetic order," Phys. Lett. 41A, 471-472.

m --- and ZAK, J. (1968) "Ultrasonic attenuation near and above the spin-flop transition of MnF_2," Phys. Rev. 170, 503-512.

SHAPIRO, S. M. and CUMMINS, H. Z. (1968) "Critical opalescence in quartz," Phys. Rev. Lett. 21, 1578-1581.

h SHERMAN, R. H. (1965) "Behavior of He^3 in the critical region," Ibid. 15, 141-142.

--- and HAMMEL, E. F. (1965) "Application of the quantum theory of corresponding states to the problem of critical-point exponents," Ibid. 15, 9-11.

SHERMAN, S. (1960) "Combinatorial aspects of the Ising model for ferromagnetism. I. A conjecture of Feynman on paths and graphs," J. Math. Phys. 1, 202-217

--- (1969) "Cosets and ferromagnetic correlation inequalities," Commun. Math. Phys. 14, 1-4.

SHINOZAKI, S. S. and ARROTT, A. (1967) "Correlation of electronic specific heats and Curie temperatures in dilute alloys of iron," J. Appl. Phys. 38, 1241-1242.

r SHIRANE, G. (1952) Ferroelectrics, Pergamon Press, N.Y.

f --- and AXE, J. D. (1971) "Acoustic-phonon instability and critical scattering in Nb_3Sn," Phys. Rev. Lett. 27, 1803-1806.

f ---, --- (1971) "Neutron scattering study of the lattice-dynamical phase transition in Nb_3Sn," Phys. Rev. B4, 2957-2963.

f ---, ---, and BIRGENEAU, R. J. (1971) "Neutron scattering study of the lattice dynamical phase transition in V_3Si," Sol. State Commun. 9, 397-400.

1 SHKLOVSKIǏ, V. A. (1971) "Effect of impurities on the spontaneous magnetization of a transformed Ising lattice," Sov. Phys.-JETP 32, 979.

b SHUSKUS, A. J. (1964) "Doppler shifts in light scattering from pure liquids and polymer solutions," J. Chem. Phys. 40, 1604-1614.

r SIEGERT, A. J. F. (1971) "From the mean field approximation to the method of random fields," In Statistical Mechanics at the Turn of the Decade, E. G. D. Cohen, Ed. (Marcel Dekker, N.Y.) p. 145-174.

h SIKORA, P. T. and MOHLING, F. (1971) "Constant-pressure heat capacity of a Bose fluid immediately above the λ point," J. Low Temp. Phys. 5, 537.

SILVER, H., FRANKEL, N. E., and NINHAM, B. W. (1972) "A class of mean field models," J. Math. Phys. 13, 468-473.

s SILVERT, W. and SINGH, A. (1972) "Superconducting colloids with multiple transition temperatures," Phys. Rev. Lett. 28, 222-224.

a SIMONS, D. S. and SALAMON, M. B. (1971) "Specific heat and resistivity near the order-disorder transition in β-brass," Ibid. 26, 750-752.

SIVARDIERE, J. and BLUME, M. (1972) "Dipolar and quadrupolar ordering in S=3/2 Ising systems," Phys. Rev. B5, 1126-1134.

SKALYO, J., JR., COHEN, A. F., FRIEDBERG, S. A., and GRIFFITHS, R. B. (1967) "Antiferromagnetic transition in $CoCl_2 \cdot 6H_2O$ and Fisher's relation," Ibid. 164, 705.

--- and FRIEDBERG, S. A. (1964) "Heat capacity of the antiferromagnet $CoCl_2 \cdot 6H_2O$ near its Néel point," Ibid. 13, 133-135.

d ---, SHIRANE, G., and FRIEDBERG, S. A. (1969) "Two-dimensional antiferromagnetism in $Mn(HCOO)_2 \cdot 2H_2O$," Ibid. 188, 1037-1041.

d ---, ---, BIRGENEAU, R., and GUGGENHEIM, H. (1969) "Magnons at low and high temperatures in the planar antiferromagnet K_2NiF_4," Phys. Rev. Lett. 23, 1394.

d ---, ---, FRIEDBERG, S. A., and KOBAYASHI, H. (1970) "Neutron scattering in the linear chain antiferromagnet $CsMnCl_3 \cdot 2H_2O$," Phys. Rev. B2, 1310-1317.

SLICHTER, C. P., SEIDEL, H., SCHWARTZ, P., and FREDERICKS, G. (1971) "Order-disorder transition in NH_4Cl. I. Phenomenological theory," Ibid. B4, 907-911.

f SLONCZEWSKI, J. C. and THOMAS, H. (1970) "Interaction of elastic strain with the structural transition of strontium titanate," Ibid. B1, 3599-3608.

r SMART, J. S. (1966) Effective field theories of magnetism, Saunders, Philadelphia.

2 SMITH, E. R. (1969) "One dimensional X-Y model with random interaction constants," Phys. Lett. 29A, 460-461.

2 --- (1970) "One-dimensional X-Y model with random coupling constants. I. Thermodynamics," J. Phys. C. 3, 1419-1432.

SMITH, I. W., GIGLIO, M., and BENEDEK, G. B. (1971) "Correlation range and compressibility of xenon near the critical point," Phys. Rev. Lett. 27, 1556-1560.

SMOLUCHOWSKI, M. V. (1908) "Molekula-kinetische Theorie der Opaleszenz von Gasen im kritischen Zustande, sowie einiger verwandter Erscheinungen," Ann. Physik 25, 205-226.

SNIDER, N. S. (1966) "Hard core model of liquids," J. Chem. Phys. 45, 378.

--- (1971) "Remarks concerning the scaling relations for critical exponents," Ibid. 54, 4587-4590.

m --- and HERRINGTON, T. M. (1967) "Hard-sphere model of binary liquid mixtures," Ibid. 47, 2248.

h SOBYANIN, A. A. (1971) "Surface tension of He II near the λ point and the boundary condition of the ordering parameter on the free surface," Sov. Phys.-JETP 34, 229.

r SOMMERVILLE, D. M. Y. (1958) An introduction to the geometry of n dimensions," Dover Publ., N.Y.

SPEAR, R. R., ROBINSON, R. L., and CHAO, K. C. (1971) "Critical states of ternary mixtures and equations of state," Ind. & Eng. Chem. Fundamentals 10, 588-592.

SPENCER, E. G., BERGER, S. B., LINARES, R. C., and LENZO, P. V. (1963) "Sodium iron fluoride, a transparent ferrimagnet," Phys. Rev. Lett. 10, 236-239.

SPOONER, S. and AVERBACH, B. L. (1966) "Spin correlations in iron," Phys. Rev. 142, 291-299.

SQUIRES, G. L. (1954) "The scattering of slow neutrons by ferromagnetic crystals," Proc. Phys. Soc. 67, 248-253.

SRIVASTAVA, K. G. (1963) "A mangetic study of some compounds having the K_2NiF_4 structure," Phys. Lett. 4, 55-56.

STACEY, L. M., PASS, B., CARR, H. Y. (1969) "NMR measurements of the xenon coexistence curve near the critical point," Phys. Rev. Lett. 23, 1424-1426.

3 STANLEY, H. E. (1967) "High-temperature expansions for the classical Heisenberg model. I. Spin correlation function," Phys. Rev. 158, 537-545.

3 --- (1967) "High-temperature expansions for the classical Heisenberg model. II. Zero-field susceptibility," Ibid. 158, 546-551.

--- (1967) "New expansion for the classical Heisenberg model and its similarity to the S = 1/2 Ising model," Ibid. 164, 709-711.

d --- (1968) "Critical properties of isotropically interacting classical spins constrained to a plane," Phys. Rev. Lett. 20, 150-154.

u --- (1968) "Dependence of critical properties on dimensionality of spins," Ibid. 20, 589-592.

∞ --- (1968) "Spherical model as the limit of infinite spin dimensionality," Phys. Rev. 176, 718-722.

r --- (1968) "Critical phenomena in Heisenberg models of magnetism," In Solid State Physics, Nuclear Physics, and Particle Physics: The Ninth Latin American School of Physics, I. Saavedra, Ed. (W. A. Benjamin, Inc., New York), p. 833-844.

∞ --- (1968) "Observations on the Kac-Baker model and the 'mathematical mechanism' of phase transitions," MIT Lincoln Lab. Solid State Research Report Vol. 4.

u --- (1969) "Critical indices for a system of spins of arbitrary dimensionality situated on a lattice of Arbitrary dimensionality," J. Appl. Phys. 40, 1272-1274.

d --- (1969) "Some critical properties of quantum-mechanical Heisenberg ferro- and antiferromagnets," Ibid. 40, 1546-1548.

u --- (1969) "Relation among Ising, Heisenberg, and spherical models: properties of isotropically-interacting spins of arbitrary dimensionality," J. Phys. Soc. Japan 26, 102-104.

u --- (1969) "Exact solution for a linear chain of isotropically interacting classical spins of arbitrary dimensionality," Phys. Rev. 179, 570-577.

u --- (1971) "Series expansions and the universality hypothesis," In Critical Phenomena in Alloys, Magnets, and Superconductors," (Eds., R. E. Mills, E. Ascher, and R. I. Jaffee), McGraw-Hill Book Co., N.Y., 203-221.

r --- (1971) Introduction to Phase Transitions and Critical Phenomena, Oxford University Press, London and New York.

--- (1971) "The static scaling hypothesis and its implications for the planar spin model of strontium titanate," In Structural Phase Transitions and Soft Modes (Eds. E. J. Samuelsen, E. Andersen, and J. Feder) Universitetsforlaget, Oslo, pp. 271-289.

--- (1972) 相転移と臨界現象 translated by Koichiro Matsuno (Tokyo-Tosho Publishing Company, Tokyo, 1972).

r --- (1972) Biomedical Physics and Biomaterials Science, M.I.T. Press, Cambridge.

r --- (1972) "Scaling laws and universality--or--statistical mechanics is not dead!," In Statistical Mechanics and Field Theory, R. Sen and C. Weil, Eds. (Keter Publishing House, Jerusalem), pp. 225-267.

--- (1972) фазовые переходы и критические явления, translated by S. V. Vonsovsky (Mir, Moscow, 1972).

--- (1973) "Cooperative phenomena in physical and biological systems," Scientific American.

--- and BETTS, D. D. (1973) "Dependence of critical-point exponents on model parameters: a bilinear form hypothesis," Phys. Rev. B7.

---, BLUME, M., MATSUNO, K., and MILOŠEVIĆ, S. (1970) "Eigenvalue degeneracy as a possible 'mathematical mechanism' for phase transitions," J. Appl. Phys. 41, 1278-1280.

r ---, GORDON, J. M., LEE, M. H., and MILOŠEVIĆ, S. (1972) "Cooperative phenomena in some magnetic and biological systems: homogeneous functions, scaling laws, and molecular pharmacology," In Proceedings of the NATO Summer Institute on Magnetism (Ed. S. Foner), Gordon and Breach, N.Y.

r ---, HANKEY, A., and LEE, M. H. (1972) "Series expansions, transformation methods, scaling and universality," In Proc. Varenna Summer School on Critical Phenomena, Ed. M. S. Green (Academic Press, N.Y.).

3 --- and KAPLAN, T. A. (1966) "High-temperature expansions--the classical Heisenberg model," Phys. Rev. Lett. 16, 981-983.

d ---, --- (1966) "Possibility of a phase transition for the two-dimensional Heisenberg Ferromagnet," Ibid. 17, 913-916.

d ---, --- (1967) "On the possible phase transition in two-dimensional Heisenberg models," J. Appl. Phys. 38, 975-977.

---, --- (1967) "Dependence of critical properties of Heisenberg magnets upon spin and lattice," Ibid. 38, 977-979.

--- and LEE, M. H. (1970) "Diagrammatic representation of the two-spin correlation function for the generalized Heisenberg model," Int. J. Quantum Chem. 4S, 407-416.

---, MATSUNO, K., and PAUL, G. (1971) "The dynamic scaling hypothesis, the mode-mode coupling approximation, and a dynamic cluster theory of time-dependent critical phenomena," In Structural Phase Transitions and Soft Modes (Eds. E. J. Samuelsen, E. Andersen, and J. Feder) Universitesforlaget, Oslo, pp. 289-329.

r ---, PAUL, G., and MILOŠEVIĆ, S. (1971) "Dynamic critical phenomena in fluid systems," In The Liquid State, Vol. 8B of a 10-volume Treatise on Phyiscal Chemistry. (H. Eyring, D. Henderson, and W. Jost, Eds.) Academic Press, N.Y., pp. 795-878.

STARK, J. P. (1967) "Density fluctuations at critical points," Physica 37, 125-128.

m STAUFFER, D. (1972) "Another droplet model for the tricritical point in the Blume-Capel magnet," Phys. Rev. B6, 1839-1847.

u ---, FERER, M., and WORTIS, M. (1972) "Universality of second-order phase transitions: the scale factor for the correlation length," Phys. Rev. Lett. 29, 345-349.

---, KIANG, C. S., EGGINGTON, A., PATTERSON, E. M., PURI, O. P., WALTER, G. H., and WISE, J. D., JR. (1972) "Heterogeneous nucleation and Fisher's droplet picture," Phys. Rev. B6, 2780-2783.

---, ---, and WALTER, G. H. (1971) "Possible symmetrization of Fisher's droplet picture neat T_c: A bubble-droplet formula," J. Stat. Phys. 3, 323.

---, --- (1971) "Droplet model near T_c: droplet-droplet interaction and correlation function," Phys. Rev. Lett. 27, 1783-1786.

---, ---, WALKER, G. H., PURI, O. P., WISE, J. D., JR., and PATTERSON, E. M. (1971) "Corrections to asymptotic scaling laws in a modified liquid droplet model," Phys. Lett. 35A, 172-173.

STEIN, W. A. (1972) "The critical point of pure fluid materials," Chem. Eng. Sci. 27, 257-273.

STELL, G. (1968) "Extension of the Ornstein-Zernike theory of the critical region," Phys. Rev. Lett. 20, 533-536.

--- (1968) "Homogeneity of the correlation length in the critical region," Phys. Rev. 173, 314-317.

--- (1968) "The two-point correlation function and the specific heat," Phys. Lett. 27, 550-551.

--- (1969) "Free-energy density near the critical point," J. Chem. Phys. 51, 2037-2039.

--- (1969) "Self-consistent equations for the radial distribution function," Mol. Phys. 16, 209-]15.

--- (1969) "Some critical properties of Ornstein-Zernike systems," Phys. Rev. 184, 135-144.

--- (1970) "Upper bounds on the Helmholtz free energy," Chem. Phys. Lett. 4, 651-652.

--- (1970) "Extension of the Ornstein-Zernike theory of the critical region. II," Phys. Rev. B1, 2265-2270.

--- (1970) "Weak-scaling theory," Phys. Rev. Lett. 24, 1343-1345.

--- (1971) "Relation between ordering and the mode expansion," J. Chem . Phys. 55, 1485-1486.

--- (1972) "Scaling theory of the critical region for systems with long-range forces," Phys. Rev. B5, 981-985.

m --- and HEMMER, P. C. (1972) "Phase transitions due to softness of the potential core," J. Chem. Phys. 56, 4274.

---, LEBOWITZ, J. L., BAER, S., and THEUMANN, W. (1966) "Separation of the interaction potential into two parts in statistical mechanics. II. Graph theory for lattice gases and spin systems with application to systems with long-range potentials," J. Math. Phys. 7, 1532-1547.

---, --- (1968) "Equilibrium properties of a system of charged particles," J. Chem. Phys. 49, 3706-3717.

---, NARANG, H., and HALL, C. K. (1971) "Simple lattice gas with realistic phase changes," Phys. Rev. Lett. 28, 292-294.

--- and PENROSE, O. (1971) "Bounds on the thermodynamic behavior of systems with generalized Coulomb interactions," Phys. Rev. A4, 1567-1569.

m --- and THEUMANN, W. K. (1969) "Critical correlations of certain lattice systems with long-range forces," Ibid. 186, 581-587.

STENSCHKE, H. (1968) "The ferromagnetic transition of iron according to a Landau-theory of phase transitions of second order," Z. Physik 216, 456-458.

--- (1969) "Restriction of the values of the critical indices by the condition of thermodynamic stability," Ibid. 221, 469.

--- and FALK, G. (1968) "On the logarithmic singularity of the specific heat of He II," Ibid. 210, 11-112.

---, --- (1968) "Thermal properties of He II according to a Landau-theory of phase transitions of second order," Ibid. 212, 308-314.

L STEPHEN, M. J. (1970) "Hydrodynamics of liquid crystals," Phys. Rev. A2, 1558-1562.

l STEPHENSON, J. (1964) "Ising-model spin correlations on the triangular lattice," J. Math. Phys. 5, 1009-1024.

l --- (1966) "Ising model spin correlations on the triangular lattice. II. Fourth-order correlations," Ibid. 7, 1123-1132.

l --- (1969) "Close-packed anisotropic antiferromagnetic Ising lattices. I," Can. J. Phys. 47, 2621-2631.

l --- (1970) "Two one-dimensional Ising models with disorder points," Ibid. 48, 1724-1734.

--- (1970) "Range of order in antiferromagnets with next-nearest neighbor coupling," Ibid. 48, 2118-2122.

l --- (1970) "Ising-model spin correlations on the triangular lattice. IV. Anisotropic ferromagnetic and antiferromagnetic lattices," J. Math. Phys. 11, 420-431.

l --- (1970) "Ising model with antiferromagnetic next-nearest-neighbor coupling: spin correlations and disorder points," Phys. Rev. B1, 4405-4409.

--- (1971) "On the critical region of a simple fluid. I. Two elementary index inequalities," J. Chem. Phys. 54, 890-894.

--- (1971) "On the critical region of a simple fluid. II. Scaling-law equation of state," Ibid. 54, 895-897.

r --- (1971) "Critical phenomena: static aspects," In Physical Chemistry Vol. 8B, 717-793 (Ed. D. Henderson), Academic Press, N.Y.

l --- and BETTS, D. D. (1970) "Ising model with antiferromagnetic next-nearest-neighbor coupling. II. Ground states and phase diagrams," Phys. Rev. B2, 2702-2706.

3 STEPHENSON, R. L., PIRNIE, K., WOOD, P. J., and EVE, J. (1968) "On the high temperature susceptibility and specific heat of the Heisenberg ferromagnet for general spin," Phys. Lett. 27A, 2-3.

3 --- and WOOD, P. J. (1968) "Free energy of the classical Heisenberg model," Phys. Rev. 173, 475-480.

3 ---, --- (1970) "On the spontaneous magnetization of the classical Heisenberg ferromagnet," J. Phys. C. 3, 90-93.

l STILLINGER, F. H., JR. (1964) "Ising model reformulation. I. Fundamentals," Phys. Rev. 135, A1646-A1661.

l --- (1965) "Ising-model reformulation. II. Relation to the random-external-field method," Ibid. 138, A1174-A1181.

l --- (1966) "Ising-model reformulation. III. Quadruplet spin averages," Ibid. 146, 209-221.

--- (1971) "Pair distribution in the classical rigid disk and sphere systems," J. Computational Phys. 7, 367-384.

--- and COTTER, M. A. (1971) "Free energy in the presence of constraint surfaces," J. Chem. Phys. 55, 3449-3458.

--- and FRISCH, H. L. (1961) "Critique of the cluster theory of critical point density fluctuations," Physica 27, 751-752.

m --- and HELFAND, E. (1964) "Critical solution behavior in a binary mixture of gaussian molecules," J. Chem. Phys. 41, 2495-2502.

3 STINCHCOMBE, R. B., HORWITZ, G., ENGLERT, F., and BROUT, R. (1963) "Thermodynamic behavior of the Heisenberg ferromagnet," Phys. Rev. 130, 155-176.

L STINSON, T. W., III, and LITSTER, J. D. (1970) "Pretransitional phenomena in the isotropic phase of a nematic liquid crystal," Phys. Rev. Lett. 25, 503-506.

STOLL, E., BINDER, K., and SCHNEIDER, T. (1972) "Evidence for Fisher's droplet model in simulated two-dimensional cluster distributions," Phys. Rev. B6, 2777-2780.

1 --- and SCHNEIDER, T. (1972) "Computer simulation of critical properties and metastable states in a finite square Ising system," Ibid. A6, 429-433.

STORMARK, O. (1969) "The Yang-Lee pictures of phase transitions in two variables," Phys. Lett. 29A, 566-568.

STOUT, J. W. (1961) "Magnetic transitions at low temperatures," Pure and Appl. Chem. 2, 287-296.

--- and CATALANO, E. (1955) "Heat capacity of zinc fluoride from 11 to 300°K. Thermodynamic functions of zinc fluoride. Entropy and heat capacity associated with the antiferromagnetic ordering of manganous fluoride, ferrous fluoride, cobaltous fluoride and nickelous fluoride," J. Chem. Phys. 23, 2013-2022.

--- and CHISHOLM, R. C. (1962) "Heat capacity and entropy of $CuCl_2$ and $CrCl_2$ from 11° to 300°K. Magnetic ordering in linear chain crystals," Ibid. 36, 979-991.

b STRÄSSLER, S. (1967) "Theory of the helix-coil transition in DNA considered as a periodic copolymer," Ibid. 46, 1037-1042.

STRATONVICH, R. L. (1958) "On a method of calculating quantum distribution functions," Sov. Phys. Doklady 2, 416-419.

m STRINGFELLOW, G. B. (1972) "Calculation of ternary phase diagrams of III-V systems," J. Phys. Chem. Solids 33, 665-678.

f STRUJOV, B. A., KORZHUEV, M. A., BADDUR, A., and KOPTSIK, V. A. (1972) "Spontaneous polarization of a KH_2PO_4 crystal near the Curie point," Sov. Phys.-Solid State 13, 1569.

STUMP, N. and MAIER, G. (1969) "Temperature shift and crystalline perfection in critical magnetic scattering," Phys. Lett. 29A, 75-76.

SUEZAKI, Y. and MORI, H. (1969) "Dynamic critical phenomena in magnetic systems. II," Progr. Theoret. Phys. 41, 1177-1189.

SUKIENNICKI, A. and ZAGORSKI, A. (1969) "On anisotropic one-dimensional Heisenberg chain with impurities," Phys. Lett. 29A, 684-689.

f SUTHERLAND, B. (1967) "Exact solution of a two-dimensional model for hydrogen-bonded crystals," Phys. Rev. Lett. 19, 103-104.

f --- (1968) "Further results for the many-body problem in one dimension," Ibid. 20, 98-100.

f --- (1968) "Correlation functions for two-dimensional ferroelectrics," Phys. Lett. 26A, 532-533.

f --- (1970) "Two-dimensional hydrogen bonded crystals without the ice rule," J. Math. Phys. 11, 3183-3186.

SUZUKI, M. (1965) "A theorem on the exact solution of a spin system with a finite magnetic field," Phy . Lett. 18, 233.

--- (1966) "Exactly soluble models of quantum systems showing the second order phase transitions," J. Phys. Soc. Japan 21, 2140-2146.

3 --- (1966) "Pair-product model of Heisenberg ferromagnets," Ibid. 21, 2274-2290.

1 --- (1967) "One-dimensional Ising model with general spin," J. Math. Phys. 8, 124-130.

--- (1967) "A semi-phenomenological theory of the second order phase transitions in spin systems. I," J. Phys. Soc. Japan 22, 757-761.

--- (1967) "A new method for determining singularities in phase transitions," Phys. Lett. 24A, 470-471.

--- (1967) "Solution of Potts model for phase transition," Progr. Theoret. Phys. 37, 770-771.

--- (1967) "A theory on the critical behaviour of ferromagnets," Ibid. 38, 238-290.

--- (1967) "A theory of the second order phase transitions in spin systems. II. Complex magnetic field," Ibid. 38, 1225-1242.

--- (1967) "Note on singularity of specific heat in the second order phase transition," Ibid. 38, 1243-1251.

1 --- (1968) "Theorems on the Ising model with general spin and phase transition," J. Math. Phys. 9, 2064-2068.

--- (1968) "A theory of the second order phase transitions in spin systems. III," Progr. Theoret. Phys. 39, 349-364.

1 --- (1968) "Theorems on extended Ising model with applications to dilute ferromagnetism," Ibid. 40, 1246-1256.

3 --- (1969) "On the distribution of zeros for the Heisenberg model," Ibid. 41, 1438-1449.

SUZUKI, M. (1969) "Singularity of nonlinear response near the critical field. I. Static case," Ibid. 42, 1076-1085.

d --- (1969) "Long-range order in ideal ferromagnets," Ibid. 42, 1086-1097.

--- (1970) "On the singularity of dynamical response and critical slowing down," Ibid. 43, 882-906.

--- (1971) "Relationship among exactly soluble models of critical phenomena. I," Ibid. 46, 1337-1359.

--- (1971) "Ergodicity, constants of motion, and bounds for susceptibilities," Physica 51, 277-291.

--- (1971) "Equilvalence of the two-dimensional Ising model to the ground state of the linear XY-model," Phys. Lett. 34A, 94-95.

2 --- (1971) "The dimer problem and the generalized XY-model," Ibid. 34A, 338-339.

u --- (1971) "Critical exponents of scaling with a parameter in spin systems," Ibid. 35A, 23-24.

u --- (1971) "Scaling with a parameter in spin systems near the critical point. I," Progr. Theoret. Phys. 46, 1054-1070.

--- (1971) "Relationship among exactly soluble models of critical phenomena. I. Ising model, dimer problem, and the generalized XY model," Ibid. 46, 1337.

--- (1971) "Upper bounds of magnetization and broken symmetry in spin systems," Phys. Lett. 36A, 115-116.

--- (1971) "Theory of critical slowing down," In Critical Phenomena in Alloys, Magnets, and Superconductors, R. E. Mills, E. Ascher, and R. I. Jaffee, Eds. (McGraw-Hill Book Co., N.Y.) p. 231-240.

--- (1971) "Recent developments of the Lee-Yang circle theorem and analyticity of the free energy," Ibid. p. 613-621.

--- (1971) "Nonlinear critical slowing down in ergodic and in non-ergodic systems," Int. J. Magnetism 1, 123-134.

u --- (1972) "Solution and critical behavior of some 'three-dimensional' Ising models with a four-spin interaction," Phys. Rev. Lett. 28, 507-510.

u --- (1972) "Dependence of critical exponents upon symmetry, dimensionality, potential-range and strength of interaction," Phys. Lett. 38A, 23-25.

--- (1972) "On the temperature-dependence of effective critical exponents and confluent singularities," Progr. Theoret. Phys. 47, 722-723.

g --- (1972) "Critical exponents for long-range interactions. I. Dimensionality, symmetry, and potential-range," Ibid. 48,

--- and FISHER, M. E. (1971) "Zeros of the partition function for the Heisenberg, ferroelectric, and general Ising models," J. Math. Phys. 12, 235-246.

1 ---, IKARI, H., and KUBO, R. (1969) "Dynamics of the Ising model near the critical point," J. Phys. Soc. Japan 26, 153-156.

3 --- and KAWABATA, C. (1966) "An application of Padé approximants to Heisenberg ferromagnetism and pair-product model," Ibid. 21, 1063-1068.

d ---, ---, ONO, S., KARAKI, Y., and IKEDA, M. (1970) "Statistical thermodynamics of finite Ising model, II," Ibid. 29, 837-844.

1 --- and KUBO, R. (1968) "Dynamics of the Ising model near the critical point I," Ibid. 24, 51-60

1 ---, TSUJIYAMA, B., and KATSURA, S. (1967) "One-dimensional Ising model with general spin," J. Math. Phys. 8, 124-130.

m SWIFT, J. (1968) "Transport coefficients near the consolute temperature of a binary liquid mixture," Phys. Rev. 173, 257-260.

h --- and KADANOFF, L. P. (1968) "Transport coefficients near the λ-transition of helium," Ann. Phys. 50, 312-321.

SWINNEY, H. L. and CUMMINS, H. Z. (1968) "Thermal diffusivity of CO_2 in the critical region," Phys. Rev. 171, 152-160.

a SYKES, C. and WILKINSON, H. (1937) "The transformation in the β brasses," J. Inst. Met. 61, 223-240.

SYKES, J. and BROOKER, G. A. (1970) "The transport coefficients of a Fermi liquid," Ann. Phys. 56, 1-39.

SYKES, M. F. (1962) "Some counting theorems in the theory of the Ising model and the excluded volume problem," J. Math. Phys. 2, 52-62.

--- (1963) "Self-avoiding walks on the simple cubic lattice," J. Chem. Phys. 39, 410-412.

--- and ESSAM, J. (1963) "Some exact critical percolation probabilities for bond and site problems in two dimensions," Phys. Rev. Lett. 10, 3-4.

---, --- (1964) "Exact critical percolation probabilities for site and bond problems in two dimensions," J. Math. Phys. 5, 1117-1126.

---, --- (1964) "Critical percolation probabilities by series methods," Phys. Rev. 133, A310-A315.

1 ---, ---, and GAUNT, D. S. (1965) "Derivation of low-temperature expansions for the Ising model of a ferromagnet and an antiferromagnet," J. Math. Phys. 6, 283-298.

---, ---, HEAP, B. R., and HILEY, B. J. (1966) "Lattice constant systems and graph theory," Ibid. 7, 1557-1572.

1 --- and FISHER, M. E. (1958) "The susceptibility of the Ising model of an antiferromagnet," Phys. Rev. Lett. 1, 321.

1 ---, --- (1962) "Antiferromagnetic susceptibility of the plane square and honeycomb Ising lattice," Physica 28, 919-938.

1 ---, GAUNT, D. S., ROBERTS, P. D., and WYLES, J. A. (1972) "High temperature series for the susceptibility of the Ising model. I. Two dimensional lattices," J. Phys. A5, 624-639.

1 ---, ---, ---, --- (1972) "High temperature series for the susceptibility of the Ising model. II. Three dimensional lattices," Ibid. A5, 640-652.

---, GUTTMANN, A. J., WATTS, M. G., and ROBERTS, P. D. (1972) "The asymptotic behaviour of self-avoiding walks and returns on a lattice," Ibid. A5, 653-660.

1 ---, HUNTER, D. L., MC KENZIE, D. S., and HEAP, B. R. (1972) "Specific heat of a three dimensional Ising ferromagnet above the Curie temperature II," Ibid. A5, 667-673.

1 ---, MARTIN, J. L., and HUNTER, D. L. (1967) "Specific heat of a three-dimensional Ising ferromagnet above the Curie temperature," Proc. Phys. Soc. 91, 671-677.

---, MC KENZIE, D. S., WATTS, M. G., and MARTIN, J. L. (1972) "The number of self-avoiding rings on a lattice," J. Phys. A5, 661-666.

1 --- and ZUCKER, I, J. (1961) "Antiferromagnetic susceptibility of the plane triangular Ising lattice," Phys. Rev. 124, 410.

1 SYOZI, I. (1968) "A decorated Ising lattice with three transition temperatures," Progr. Theoret. Phys. 39, 1367-1368.

SYOZI, I. and SIYAZIMA, S. (1966) "A statistical model for the dilute ferromagnet," Ibid. 36, 1083-1094.

b SZILARD, L. (1964) "On memory and recall," Proc. Natl. Acad. Sci. U.S. 51, 1092-1099.

SZNAJD, J. (1971) "Landau theory of the second-order magnetic phase transitions in uniaxial ferromagnets with external field," Acta Phys. Polon. A40, 687.

TAHIR-KHELI, R. A. (1963) "Interpolation approach to the Green function theory of ferromagnetism," Phys. Rev. 132, 689-701.

1 --- (1967) "Odd-order correlations in disordered Ising systems," Phys. Lett. 25A, 641-642.

--- (1969) "Spin diffusion and propagating modes in anisotropic paramagnet," J. Appl. Phys. 40, 1550-1552.

--- (1972) "Spatially random Heisenberg spins at very low temperatures. I. Dilute ferromagnet," Phys. Rev. B6, 2808-2826.

--- (1972) "Spatially random Heisenberg spins at very low temperatures. II. Dilute antiferromagnet with nearest-neighbor substitutional short-range order," Ibid. B6, 2826-2838.

--- (1972) "Spatially random Heisenberg spins at very low temperatures. III. Multicomponent ferromagnetic alloy with substitutional short-range order," Ibid. B6, 2838-2858.

---, CALLEN, H. B., and JARRETT, H. (1966) "Magnetic ordering in cubic crystals with first and second neighbor exchange," J. Phys. Chem. Solids 27, 23-32.

--- and JARRETT, H. S. (1964) "Ferromagnetic Curie temperature in cubic lattices with next-nearest-neighbor interaction," Phys. Rev. 135, A1096-A1098.

3 --- and MC FADDEN, D. G. (1969) "Frequency-dependent self-correlation function for the Heisenberg spin system in one dimension," Ibid. 178, 800-803.

---, --- (1969) "Space-time correlations in exchange-coupled paramagnets at elevated temperatures," Ibid. 182, 604-613.

--- and TER HAAR, D. (1962) "Use of Green functions in the theory of ferromagnetism. I. General discussion of the spin-S case," Ibid. 127, 88-94.

s TAKADA, S. and IZUYAMA, T. (1969) "Superconductivity in a molecular field. I," Progr. Theoret. Phys. 41, 635-666.

TAKAGI, S. (1972) "Lattice model for quantum liquid mixtures," Ibid. 47, 22-36.

TAKAHASHI, M. (1971) "Thermodynamics of the Heisenberg-Ising model for $|\Delta| < 1$ in one dimension," Phys. Lett. 36A, 325-327.

3 --- (1971) "One-dimensional Heisenberg model at finite temperature," Progr. Theoret. Phys. 46, 401-415.

--- (1972) "One-dimensional Hubbard model at finite temperature," Ibid. 47, 69-82.

--- and SUZUKI, M. (1972) "A reply to the comments of Johnson et al. on the thermodynamics of the Heisenberg-Ising ring for $|\Delta| < 1$," Phys. Lett. 41A, 81-83.

d TAKEDA, K., HASEDA, T., and MATSUURA, M. (1971) "A study of magnetic phase transition in a two-dimensional lattice by heat capacity measurements," Physica 52, 225-236.

d --- and KAWASAKI, K. (1971) "Magnetism and phase transition in two-dimensional lattices; $M(HCOO)_2 \cdot 2H_2O$ (M; Mn, Fe, Ni, Co)," J. Phys. Soc. Japan 31, 1026.

d --- and MATSUKAWA, S. (1971) "Magnetic susceptibility of antiferromagnetic $Co(HCOO)_2 \cdot 2H_2O$," Ibid. 30, 887.

d ---, ---, and HASEDA, T. (1971) "Thermal and magnetic properties of linear ferro- and antiferro-magnetic substances, Mcl_2 $2NC_5H_5$(M; Co, Cu)," Ibid. 30, 1330-1336.

d ---, ---, --- (1970) "Magnetic phase transition for dilute Heisenberg spin system in two-dimensional lattice," Ibid. 28, 29-35.

d ---, ---, --- (1970) "Magnetic phase transition in $Ni_2Co_{(1-x)}Cl_2 \cdot 6H_2)$," Ibid. 29, 885-889.

TANAKA, M. (1968) "Molecular theory of scattering of light from liquids," Progr. Theoret. Phys. 40, 975-989.

TANAKA, T., KATSUMORI, H., and TOSHIMA, S. (1951) "On tne theory of cooperative phenomena," Ibid. 6, 17-26.

---, MEIJER, P. H. E., and BARRY, J. H. (1962) "Theory of relaxation phenomena near the second-order phase-transition point," J. Chem. Phys. 37, 1397-1402.

TANI, K. (1966) "Ultrasonic attenuation in magnetics at low temperatures," Progr. Theoret. Phys. 36, 848-849.

--- and MORI, H. (1968) "Ultrasonic attenuation near the magnetic critical point," Ibid. 39, 876-896.

f --- and TAKEMURA, M. (1968) "Coupled mode of sound with soft mode in displacive-type ferroelectrics," Phys. Lett. 27A, 1-2.

m --- and TANAKA, H. (1968) "Anomalous behavior of sound near the critical point in Heisenberg Ferro- and antiferromagnets with uniaxial anisotropy," Ibid. 27A, 25-26.

f --- and TSUDA, N. (1967) "Anomalous behaviour of sound near the Curie points in displacive-type ferroelectrics," Ibid. 25A, 529-530.

L TAYLOR, T. R., FERGASON, J. L., and ARORA, S. L. (1970) "Biaxial liquid crystals," Phys. Rev. Lett. 24, 359-362.

TEANEY, D. T. (1965) "Specific-heat singularity in MnF_2," Ibid. 14, 898-900.

---, MORUZZI, V. L., and ARGYLE, B. E. (1966) "Critical point of the cubic antiferromagnet $RbMnF_3$," J. Appl. Phys. 37, 1122-1123.

---, VAN DER HOEVEN, B. J. C., JR., and MORUZZI, V. L. (1968) "Singular behavior of a ferromagnet in nonzero field," Phys. Rev. Lett. 20, 722-724.

TEMPERLEY, H. N. V. (1950) "Statistical mechanics of the two-dimensional assembly," Proc. Roy. Soc. 202-A, 202-207.

r --- (1956) Changes of state, Cleaver-Hume Press, London.

--- and FISHER, M. E. (1961) "Dimer problem in statistical mechanics--an exact result," Phil. Mag. 6, 1061-1063.

--- and LIEB, E. H. (1971) "Relations between the 'percolation' and 'colouring' problem and other graph-theoretical problems associated with regular planar lattices: some exact results for the 'percolation' problem," Proc. Roy. Soc. A322, 251-280.

---, ROWLINSON, J. S., and RUSHBROOKE, G. S. (1968) Physics of Simple Liquids, Wiley Interscience, N.Y.

TERAUCHI, H., NODA, Y., and YAMADA, Y. (1972) "X-ray critical scattering in NH_4Br," J. Phys. Soc. Japan 32, 1560-1564.

∞ THEUMANN, W. K. (1970) "Some critical properties of Ornstein-Zernike systems and spherical models," Phys. Rev. B2, 1396-1405.

--- (1970) "Mean-field aspect of Ornstein-Zernike systems," Phys. Lett. 32A, 1-2.

--- (1972) "Phenomenological weak-scaling theory," Phys. Rev. B6, 281-286.

m --- and HYE, J. W. (1971) "Ising chain with several phase transitions," J. Chem. Phys. 55, 4159-4166.

THOEN, J., VANGEEL, E., and VAN DAEL, W. (1971) "Experimental investigation of the sound velocity in the critical region of argon," Physica 52, 205-224.

d THOMAS, G. A. and PARKS, R. D. (1971) "Fluctuation-induced conductivity of a one-dimensional, nearly ideal BCS superconductor," Phys. Rev. Lett. 27, 1276-1279.

m THOMAS, H. (1969) "Phase transitions in a uniaxial ferromagnet," Phys. Rev. 187, 630-637.

f --- and MÜLLER, K. A. (1968) "Structural phase transitions in Perovskite-type crystals," Phys. Rev. Lett. 21, 1256-1259.

f ---, --- (1972) "Theory of a structural phase transitions induced by the Jahn-Teller effect," Ibid. 28, 820-823.

THOMAS, J. E. and SCHMIDT, P. W. (1963) "X-ray study of critical opalescence in argon," J. Chem. Phys. 39, 2506-2516.

l THOMPSON, C. J. (1965) "Algebraic derivation of the partition function of a two-dimensional Ising model," J. Math. Phys. 6, 1392-1395.

3 --- (1965) "Heisenberg model with long range interaction," J. Phys. Chem. Solids 26, 1977-1981.

l --- (1966) "Phase transition of a two-dimensional continuum Ising model," J. Math. Phys. 7, 531-534.

--- (1966) "Strong-coupling limit in dilute alloys," Phys. Rev. 141, 479-482.

b --- (1968) "Models for hemoglobin and allosteric enzymes," Biopolymers 6, 1101-1118.

l --- (1968) "Infinite-spin Ising model in one dimension," J. Math. Phys. 9, 241-245.

∞ --- (1968) "Spherical model as an instance of eigenvalue degeneracy," Ibid. 9, 1059-1062.

l --- (1971) "Upper bounds for Ising model correlation functions," Commun. Math. Phys. 24, 61-66.

r --- (1972) *Mathematical Statistical Mechanics*, Macmillan, N.Y. See especially the excellent treatment of the Ising model presented herein.

---, GUTTMANN, A. J., and NINHAM, B. W. (1969) "Determination of critical behavior in lattice statistics from series expansions II," J. Phys. C2, 1889-1899.

THOMPSON, D. R. and RICE, O. K. (1964) "Shape of the coexistence curve in the perfluoromethylcyclohexane-carbon tetrachloride system II. Measurements accurate to 0.0001^{o1}," J. Am. Chem. Soc. 86, 3547-3553.

THORPE, M. F. and BLUME, F. (1972) "Soluble model of interacting classical quadrupoles in one dimension," Phys. Rev. B5, 1961-1965.

l THOULESS, D. J. (1969) "Critical region for the Ising model with a long-range interaction," Ibid. 181, 954-968.

l --- (1969) "Long-range order in one-dimensional Ising systems," Ibid. 187, 732-733.

m TIMOSHCHUK, V. I. and FAKIDOV, I. G. (1971) "Metamagnetism in Mn_3B_4," Sov. Phys.-Solid State 13, 85-87.

m TINKHAM, M. (1969) "Microscopic dynamics of metamagnetic transitions in an approximately Ising system: $CoCl_2 \cdot 2H_2O$," Phys. Rev. 188, 967-973.

h TISZA, L. (1938) "Transport phenomena in helium II," Nature 141, 913.

--- (1951) "On the general theory of phase transitions," In *Phase Transformations in Solids*, ed. R. Smoluchowski, J. E. Mayer, and W. A. Weyl (John Wiley, N.Y.).

TISZA, L. (1961) "The thermodynamics of phase equilibrium," Ann. Phys. 13, 1-92.

r --- (1966) *Generalized thermodynamics*, MIT Press, Cambridge, Mass.

--- (1971) "The thermodynamics of phase equilibrium: from the phase rule to the scaling laws," Proc. Intern. Conf. on Thermodynamics, Cardiff, 1970.

h --- and CHASE, C. E. (1965) "Equation of state of ^4He in the critical region," Phys. Rev. Lett. 15, 4-6.

TOMITA, K. and KAWASAKI, T. (1971) "Critical magnetic relaxation," Progr. Theoret. Phys. 45, 1-24.

--- and TOMITA, H. (1971) "A dynamic approach to phase transition based on moments," Ibid. 45, 1407-1436.

TORRIE, B. H. (1966) "Transverse magnetic fluctuations in MnF_2 near the critical point," Proc. Phys. Soc. 89, 77-85.

TRELIN, Y. S. and SHELUDYAKOV, E. P. (1966) "Experimental determination of the speed of sound in the critical region of carbon dioxide," JETP Lett. 3, 63-64.

r TREVES, D. (1965) "Studies on orthoferrites at the Weizmann Institute of Science," J. Appl. Phys. 36, 1033-1039.

TRIEZENBERG, D. G. and ZWANZIG, R. (1972) "Fluctuation theory of surface tension," Phys. Rev. Lett. 28, 1183-1185.

m TRUNOV, V. A., YAGUD, A. Z., EGOROV, A. I., DMITRIEV, R. P., and UL'YANOV, V. A. (1971) "Investigation of metamagnetic transition in $FeCl_2$ with the aid of polarized neutrons," JETP Lett. 14, 146-147.

b TSCHARNUTER, W., LEE, S. P., CHU, B., and KUWAHARA, N. (1972) "Dynamics of concentration fluctuations of a macromolecular solution very near its critical point," Phys. Lett. 39A, 257-259.

m ---, THIEL, D., and CHU, B. (1972) "Spectrum of scattered light in a critical binary mixture," Ibid. 38A, 299-300.

m ---, ---, --- (1972) "Experimental evidence of non-local shear viscosity correction of the mode-mode coupling theory of Kawasaki in a binary liquid mixture," Ibid. 40A, 275-276.

TSUEI, C. C., LONGWORTH, G., and LIN, S. C. H. (1968) "Temperature dependence of the magnetization of an amorphous ferromagnet," Phys. Rev. 170, 603-606.

h TSUZUKI, T. (1969) "Critical anomaly of the first sound in a Bose liquid above the phase transition point," Progr. Theoret. Phys. 41, 1387-1394.

TUCCIARONE, A., CORLISS, L. M., and HASTINGS, J. M. (1971) "Neutron investigation of the spin dynamics in paramagnetic $RbMnF_3$," J. Appl. Phys. 42, 1378-1380.

---, ---, --- (1971) "Theoretical and experimental spin-relaxation functions in paramagnetic $RbMnF_3$," Phys. Rev. Lett. 26, 257-261.

---, LAU, H. Y., CORLISS, L. M., DELAPALME, A., and HASTINGS, J. M. (1971) "Quantitative analysis of inelastic scattering in two-crystal and three-crystal neutron spectrometry; critical scattering from $RbMnF_3$," Phys. Rev. B4, 3206-3245.

d TUCKER, J. R. and HALPERIN, B. I. (1971) "Onset of superconductivity in one-dimensional systems," Ibid. B3, 3768-3782.

h TYSON, J. A. (1968) "Superfluid density of He II in the critical region," Ibid. 166, 166-176.

h --- (1968) "Critical-region second-sound damping in He II," Phys. Rev. Lett. 21, 1235-1237.

h --- (1969) "Superfluid exponents ζ from self-consistent Landau-Ginzburg theory," Phys. Lett. 28A, 526-527.

h --- and DOUGLASS, D. H. (1966) "Superfluid density and scaling laws for liquid helium near T_λ," Phys. Rev. Lett. 17, 472-474.

h ---, --- (1968) "Critical-region second-sound velocity in He II," Ibid. 21, 1308-1310.

h UEYAMA, H. (1971) "Phenomenological theory on critical phenomena in liquid helium," Progr. Theoret. Phys. 45, 25-35.

UHLENBECK, G. E. (1971) "Some questions about quantum statistical mechanics," Phys. Norveg. 5, 139-144.

---, HEMMER, P. C., and KAC, M. (1963) "On the van der Waals theory of the vapor-liquid equilibrium, II. Discussion of the distribution functions," J. Math. Phys. 4, 229-247.

L UHRICH, D. L., WILSON, J. M., and RESCH, W. A. (1970) "Mossbauer investigation of the smectic liquid crystalline state," Phys. Rev. Lett. 24, 355-359.

∞ URYU, N. (1954) "Application of the spherical model to the antiferromagnetism I, II. On the approximate nature of the spherical model," Mem. Fac. Sci., Kyushu Univ., Ser B. 1, 131-149.

h VAKS, V. G. (1970) "Correlation effects in displacement phase transitions in ferroelectrics," Sov. Phys.-JETP 31, 161.

---, GALITSKII, V. M., and LARKIN, A. I. (1967) "Collective excitations near second-order phase-transition points," Ibid. 24, 1071-1108.

--- and GEĬLIKMAN, M. B. (1971) "Low-temperature and virtual phase transitions," Ibid. 33, 179-184.

2 --- and LARKIN, A. I. (1966) "On phase transitions of second order," Ibid. 22, 678-687.

---, ---, and OVCHINNIKOV, Y. N. (1966) "Ising model with interaction between non-nearest neighbors," Ibid. 22, 820-826.

---, ---, and PIKIN, S. A. (1967) "Self-consistent field method for the description of phase transitions," Ibid. 24, 240-249.

---, ---, --- (1968) "Thermodynamics of an ideal ferromagnetic substance," Ibid. 26, 188-199.

VAN CRAEN, J. (1970) "Statistical mechanics of rectilinear trimers on the square lattice," Physica 49, 558-564.

--- and BELLEMANS, A. (1972) "Series expansion for the monomer-trimer problem. I. General formulation and application to the square lattice," J. Chem. Phys. 56, 2041-2049.

VAN DER HOEVEN, B. J. C., JR., TEANEY, D. T., and MORUZZI, V. L. (1968) "Magnetic equation of state and specific heat of EuS near the Curie point," Phys. Rev. Lett. 20, 719-722.

VAN HOVE, L. (1954) "Correlations in space and time and Born approximation scattering in systems of interacting particles," Phys. Rev. 95, 249-262.

--- (1954) "Time-dependent correlations between spins and neutron scattering in ferromagnetic crystals," Pbid. 95, 1374-1384.

VAN KAMPEN, N. G. (1964) "Condensation of a classical gas with long-range attraction," Ibid. 135, A362-A369.

∞ VAN LEEUWEN, J. M. J. and GUNTON, J. D. (1972) "Dynamical properties of the spherical model in the low-temperature and critical regions," Ibid. B6, 231-245.

m VAN TOL, M. W., HENKENS, L. S. J. M., and POULIS, N. J. (1971) "High-field magnetic phase transition in $Cu(NO_3)_2 \cdot 2\text{-}1/2H_2O$," Phys. Rev. Lett. 27, 739-741.

VAN VLECK, J. H. (1937) "The influence of dipole-dipole coupling on the specific heat and susceptibility of a paramagnetic salt," J. Chem . Phys. 5, 320-337.

VAN VLIET, K. M. (1971) "Markov approach to density fluctuations due to transport and scattering. I. Mathematical formalism," J. Math. Phys. 12, 1981-2012.

VAN DER WAERDEN, V. B. L. (1941) "Die lange Reichweite der regelmäßigen Antomanordnung in Mischkristallen," Z. Physik 118, 473-488.

f VAVREK, A. F., EPIFANOV, A. S., and SHAPOVAL, E. A. (1972) "Phase transitions in the Slater model with impurities," Sov. Phys.-JETP 34, 428.

VICENTINI-MISSIONI, M., JOSEPH, R. I., GREEN, M. S., and LEVELT-SENGERS, J. M. H. (1970) "Scaled equation of state and critical exponents in magnets and fluids," Phys. Rev. B1, 2312-2331.

--- and LEVELT-SENGERS, J. M. H. (1969) "Thermodynamic anomalies of CO_2, Xe, and He^4 in the critical region," Phys. Rev. Lett. 22, 389-393.

---, --- (1969) "Scaling analysis of thermodynamic properties in the critical region of fluids," J. Res. Matl. Bur. Std. (U.S.) 73A, 563-583.

f VIELAND, L. J., COHEN, R. W., and REHWALD, W. (1971) "Evidence for a first-order structural transformation in Nb_3Sn," Phys. Rev. Lett. 26, 373-376.

VILLAIN, J. (1968) "Un modele pour la relaxation ferromagnetique critique," J. de Physique 29, 321-328.

--- (1968) "Relaxation critique au-dessus du point de Néel dans les antiferromagnetiques et les helimagnetiques," Ibid. 29, 687-693.

--- (1968) "Critical exponents in the dynamics of an isotropic magnetic system near its transition temperature," Phys. Stat. Sol. 26, 501-507.

--- (1971) "Theorie de la relaxation magnetique critique," J. de Physique 32, 310-317.

--- (1971) "Lois d'echelle dans un ferromagnetique anisotrope," Ibid. 32, 646-647.

f --- (1972) "The 3-dimensional eight vertex model and the proton-proton correlation functions in ice," Solid State Commun. 10, 967-970.

VOL'KENSHTEIN, M. V. Molecules and Life (Plenum Press, N.Y., 1969).

b --- (1970) "Physics of muscle contraction," Usp. Fiz. Nuak 100, 681-717.

m VOLOCHINE, B., BERGÉ, P., and LAGUES, I. (1970) "Experimental study of the relaxation of concentration fluctuations in the 'critical region' of a binary mixture," Phys. Rev. Lett. 25, 1414-1416.

f VON WALDKIRCH, TH., MÜLLER, K. A., BERLINGER, W., and THOMAS, H. (1972) "Fluctuations and correlations in $SrTiO_3$ for $T \geq T_c$," Ibid. 28, 503-506.

VORONEL', A. V., CHASHKIN, Y. R., POPOV, V. A., and SIMKIN, V. G. (1964) "Measurement of the specific heat C_V of oxygen near the critical point," Sov. Phys.-JETP 18, 568-569.

--- and GITERMAN, M. SH. (1965) "Hydrostatic effect at the critical point of a binary mixture," Ibid. 21, 958-961.

---, GORBUNOVA, V. G., CHASHKIN, YU. R., and SHCHEKOCHIKHINA, V. V. (1966), Sov. Phys.-JETP 23, 597-601

---, SNIGIREV, V. G., and CHASHKIN, YU. R. (1965) "Behavior of the specific heat C_V of pure substances near the critical point," Ibid. 21, 653-655.

VORONTSOV-VELIAMINOV, P. N., ELIASHEVICH, A. M., RASAIAH, J. C., and FRIEDMAN, H. L. (1970) "Comparison of hypernetted chain equation and Monte Carlo results for a system of charged hard spheres," J. Chem. Phys. 52, 1013-1014.

l VUL, E. G. and SINAĬ, YA. G. (1970) "On one form of singularities in models of the Ising type," Sov. Phys.-JETP 31, 1144.

WAGNER, H. (1966) "Long-wavelength excitations and the Goldstone theorem in many-particle systems with 'broken symmetries,'" Z. Physik 195, 273-299.

--- (1970) "Phase transition in a compressible Ising ferromagnet," Phys. Rev. Lett. 25, 31-34.

--- (1970) "On dynamic scaling for isotropic magnets in the paramagnetic critical region," Phys. Lett. 33A, 58-59.

WAISMAN, E. and LEBOWITZ, J. L. (1970) "Exact solution of an integral equation for the structure of a primitive model of electrolytes," J. Chem. Phys. 52, 4307-4309.

∞ ---, --- (1972) "Mean spherical model integral equation for charged hard spheres. I. Method of solution," Ibid. 56, 3086-3093.

∞ ---, --- (1972) "Mean spherical model integral equation for charged hard spheres. II. Results," Ibid. 56, 3093-3100.

WAKEFIELD, A. J. (1951) "Statistics of the simple cubic lattice," Proc. Camb. Phil. Soc. 47, 419-435.

WALDO, G. V. and DORFMAN, J. R. (1972) "Bounds on the critical temperature of Ising ferromagnets," Phys. Lett. 40A, 333-334.

a WALKER, C. B. and CHIPMAN, D. R. (1971) "Cowley theory of long-range order in β-CuZn," Phys. Rev. B4, 3104-3106.

d WALKER, M. B. (1968) "Nonexistence of excitonic insulators in one and two dimensions," Can. J. Phys. 46, 817-821.

d --- and RUIJGROK, TH. W. (1968) "Absence of magnetic ordering in one and two dimensions in a many-band model for interacting electrons in a metal," Phys. Rev. 171, 513-515.

b WALL, F. T. and ERPENBECK, J. J. (1959) "New method for the statistical computation of polymer dimensions," J. Chem. Phys. 30, 634-637.

m WALLACE, B., JR., HARRIS, J., and MEYER, H. (1972) "Boiling and dew curves of He^3-He^4 mixtures," Phys. Rev. A5, 964-967.

h --- and MEYER, H. (1970) "Equation of state of He^3 close to the critical point," Ibid. A2, 1563-1575.

h ---, --- (1970) "Critical isotherm of He^3," Ibid. A2, 1610-1612.

m ---, --- (1972) "Pressure-density-temperature relations of He^3-He^4 mixtures near the liquid-vapor critical point," Ibid. A5, 953-964.

d WALSTEDT, R. E., DE WIJN, H. W., and GUGGENHEIM, H. J. (1970) "Observation of zero-point spin reduction in quadratic layer antiferromagnets," Phys. Rev. Lett. 25, 1119-1122.

WANG, C. H. and WRIGHT, R. B. (1972) "Raman scattering study of the disorder-order phase transition in NH_4Cl," J. Chem. Phys. 56, 2124-2130.

WANG, Y. L. and KHAJEHPOUR, M. R. H. (1972) "Magnetic-field-induced phase transitions in uniaxial ferromagnets," Phys. Rev. B6, 1778-1787.

--- and COOPER, B. R. (1968) "Collective excitations and magnetic ordering in materials with singlet crystal-field ground state," Ibid. 172, 539-551.

r WANNIER, G. H. (1945) "The statistical problem in cooperative phenomena," Rev. Mod. Phys. 17, 50-60.

r --- (1966) Statistical physics, Wiley, N.Y.

∞ WATSON, G. N. (1939) "Three triple integrals," Quart. J. Math. 10, 266-276.

WATSON, P. G. (1968) "Critical behaviour of boundary susceptibility and boundary tension," J. Phys. C. 1, 268-269.

1 --- (1968) "Impurities and defects in Ising lattices," Ibid. 1, 575-579.

--- (1969) "Critical behavior of inhomogeneous lattices," Ibid. 2, 948-958.

--- (1969) "Formation of invariants from critical amplitudes of ferromagnets," Ibid. 2, 1883-1885.

--- (1969) "Further critical invariants of ferromagnets," Ibid. 2, 2158-2159.

--- (1970) "Critical behavior of inhomogeneous lattices II," Ibid. 3, L25-L28.

--- (1970) "The effect of nearest-neighbour interactions on excluded-volume lattice problems," Ibid. 3, L28-L30.

3 WATSON, R. E., BLUME, M., and VINEYARD, G. H. (1969) "Spin motions in a classical ferromagnet," Phys. Rev. 181, 811-823.

d ---, ---, --- (1970) "Classical Heisenberg magnet in two dimensions," Ibid. B2, 684-690.

WEBER, R. (1969) "Spin wave impurity states in linear chain ferromagnets and antiferromagnets," Z. Phys. 223, 229-337.

d WEGNER, F. (1967) "Magnetic ordering in one and two dimensional systems," Phys. Lett. 24A, 131-132.

d --- (1967) "Spin-ordering in a planar classical Heisenberg model," Z. Physik 206, 465-470.

3 --- (1968) "On the Heisenberg model in the paramagnetic region and at the critical point," Ibid. 216, 433-455.

3 --- (1969) "On the dynamics of the Heisenberg antiferromagnet at T_N," Ibid. 218, 260-265.

f --- (1972) "Duality relation between the Ashkin-Teller and the eight-vertex model," J. Phys. C. 5, L131-L132.

g --- (1972) "Corrections to scaling laws," Phys. Rev. B5, 4529-4536.

g --- (1972) "Critical exponents in isotropic spin systems," Ibid. B6, 1891-1893.

g --- and RIEDEL, E. K. (1972) "Logarithmic corrections to the molecular field behavior of critical and tricritical systems," Ibid. B6 (Dec.).

WEINBERGER, M. A. and SCHNEIDER, W. G. (1952) "On the liquid-vapor coexistence curve of xenon in the region of the critical temperature," Can. J. Chem. 30, 422-437.

WEINER, B. B. and GARLAND, C. W. (1972) "Order-disorder phenomena. VII. Critical variations in the length of NH_4Cl single crystals at high pressures," J. Chem. Phys. 56, 155-166.

WENG, C. Y., GRIFFITHS, R. B., and FISHER, M. E. (1967) "Critical temperature of anisotropic Ising lattices. I. Lower bounds," Phys. Rev. 162, 475-479.

WERTHEIM, G. K. and BUCHANAN, D. N. E. (1967) "Temperature dependence of the Fe^{57} hfs in FeF_2 below the Néel temperature," Ibid. 161, 478-482.

---, ---, and GUGGENHEIM, H. J. (1970) "Thermal shift of a Mössbauer gamma ray at a magnetic phase transition," Ibid. B2, 1392-1395.

---, GUGGENHEIM, H. J., WILLIAMS, H. J., and BUCHANAN, D. N. E. (1967) "Mössbauer effect in a cubic antiferromagnet near the Néel point," Ibid. 158, 446-450.

---, ---, BUCHANAN, D. N. E. (1968) "Sublattice magnetization in FeF_3 near the critical point," Ibid. 169, 465-470.

d ---, ---, LEVINSTEIN, H. J., BUCHANAN, D. N. E., and SHERWOOD, R. C. (1968) "Magnetic properties of the planar antiferromagnet Rb_2FeF_4," Ibid. 173, 614-616.

---, ---, and BUCHANAN, D. N. E. (1968) "Critical-point magnetization of an impurity in an antiferromagnet," Phys. Rev. Lett. 20, 1158-1161.

∞ --- (1971) "Exact solution of the mean spherical model for fluids of hard spheres with permanent electric dipole moments," J. Chem. Phys. 55, 4291-4299.

m WHEELER, J. C. (1972) "Behavior of a solute near the critical point of an almost pure solvent," Berichter der Bunsen-Gesellschaft 76, 308-318.

--- and CHANDLER, D. (1971) "Catastrophe in the random-phase approximation: critique of a theory of phase transitions," J. Chem. Phys. 55, 1645-1654.

--- and GRIFFITHS, R. B. (1968) "Thermodynamic bounds on constant-volume heat capacities and adiabatic compressibilities," Phys. Rev. 170, 249-256.

m --- and WIDOM, B. (1968) "Phase transitions and critical points in a model three-component system," J. Am. Chem. Soc. 90, 3064-3071.

m ---, --- (1970) "Phase equilibrium and critical behavior in a two-component Bethe-lattice gas or three-component Bethe-lattice solution," J. Chem. Phys. 52, 5334-5343.

WHITE, J. A. and MACCABEE, B. S. (1971) "Temperature dependence of critical opalescence in carbon dioxide," Phys. Rev. Lett. 26, 1468-1471.

d WIDOM, A. (1968) "Superfluid phase transitions in one and two dimensions," Phys. Rev. 176, 254-257.

WIDOM, B. (1955) "On the structure of the configuration integral in the statistical mechanics of pure fluids," J. Chem. Phys. 23, 560.

--- (1957) "Statistical mechanics of liquid-vapor equilibrium," Ibid. 26, 887.

--- (1960) "Rotational relaxation of rough spheres," Ibid. 32, 913.

--- (1962) "Relation between the compressibility and the coexistence curve near the critical point," Ibid. 37, 2703-2704.

r --- (1963) "Collision theory of chemical reaction rates," Adv. Chem. Phys. 5, 353.

--- (1963) "Some topics in the theory of fluids," J. Chem. Phys. 39, 2808.

--- (1964) "Degree of the critical isotherm," Ibid. 41, 1633-1634.

--- (1965) "Surface tension and molecular correlations near the critical point," Ibid. 43, 3892-3897.

--- (1965) "Equation of state in the neighborhood of the critical point," Ibid. 43, 3898-3905.

m --- (1967) "Plait points in two- and three-component liquid mixtures," Ibid. 46, 3324-3333.

r --- (1967) "Intermolecular forces and the nature of the liquid state," Science 157, 375-382.

--- (1971) "Geometrical aspects of the penetrable-sphere model," J. Chem.Phys. 54, 3950.

--- (1971) "Reaction kinetics in stochastic models," Ibid. 55, 44.

r --- (1972) "Critical phenomena in the surface tension of fluids," In C. Domb and M. S. Green, Phase Transitions and Critical Phenomena, Academic Press, N.Y.

--- and RICE, O. K. (1955) "Critical isotherm and the equation of state of liquid-vapor systems," J. Chem. Phys. 23, 1250-1255.

--- and ROWLINSON, J. S. (1970) "New model for the study of liquid-vapor phase transitions," Ibid. 52, 1670-1684.

WIECHERT, H. and MEINHOLD-HEERLEIN, L. (1969) "A possible determination of all coefficients of viscosity and the thermal conductivity by sound measurements in liquid helium," Phys. Lett. 29A, 41-42.

WIEGEL, F. W. and JALICKEE, J. B. (1972) "The Bogoliubov theory of Bose condensation as a mean field theory," Physica 57, 317-333.

r WIELINGA, R. F. (1969) "Critical behaviour in magnetic crystals," Progr. Low Temp. Phys. 6, 333-373.

---, LUBBERS, J., and HUISKAMP, W. J. (1967) "Heat capacity singularities in two gadolinium salts below 1°K," Physica 37, 375-392.

m WIGNALL, G. D. and EGELSTAFF, P. A. (1968) "Critical opalescence in binary liquid metal mixtures. I. Temperature dependence," J. Phys. C. 1, 1088-1096.

WILCOX, L. R. and BALZARINI, D. (1968) "Interferometric determination of near-critical isotherms of xenon in the earth's field," J. Chem. Phys. 48, 753-763.

WILCOX, R. M. (1968) "Bounds for the isothermal, adiabatic, and isolated static susceptibility tensors," Phys. Rev. 174, 624-629.

WILKINSON, M. K. and SHULL, C. G. (1956) "Neutron diffraction studies on iron at high temperatures," Ibid. 103, 516-524.

r WILKS, J. (1967) The Properties of Liquid and Solid Helium, Oxford University Press, London.

h WILLIAMS, R., BEAVER, S. E. A., FRASER, J. C., KAGIWADA, R. S., and RUDNICK, I. (1969) "The velocity of second sound near T_λ," Phys. Lett. 29A, 279-280.

h --- and RUDNICK, I. (1970) "Attenuation of first sound near the lambda transition of liquid helium," Phys. Rev. Lett. 25, 276-280.

h WILLIAMSON, R. C. and CHASE, C. E. (1968) "Velocity of sound at 1 MH_z near the He^4 critical point," Phys. Rev. 176, 285-294.

g WILSON, K. G. (1970) "Model of coupling-constant renormalization," Phys. Rev. D2, 1438.

--- (1971) "Renormalization group and strong interactions," Ibid. D3, 1818.

g --- (1971) "Renormalization group and critical phenomena. I. Renormalization group and the Kadanoff scaling picture," Ibid. B4, 3174-3183.

g --- (1971) "Renormalization group and critical phenomena. II. Phase-space cell analysis of critical behavior," Ibid. B4, 3184-3205.

g --- (1972) "Feynman graph expansion for critical exponents," Phys. Rev. Lett. 28, 548-551.

g --- and FISHER, M. E. (1972) "Critical exponents in 3.99 dimensions," Ibid. 28, 240-243.

g --- and KOGUT, J. (1972) "The renormalization group and the ε expansion," Princeton Univ. Lecture Notes, July 1972.

WISER, N. and COHEN, M. H. (1969) "Phase transitions of the electron fluid," J. Phys. C. 2, 193-199.

m WISSEL, C. (1972) "A model for the phase diagram of $Fe(Pd_xPt_{1-x})_3$ showing a quadruple point," Phys. Stat. Sol. 41, 669.

WOJTCZAK, L. and KOCIŃSKI, J. (1971) "Critical scattering cross section in ferromagnetic polycrystals," Phys. Lett. 34A, 306-308.

WOJTOWICZ, P. J. (1959) "Theoretical model for tetragonal-to-cubic phase transformations in transition metal spinels," Phys. Rev. 116, 32-45.

--- (1960) "High-temperature susceptibility of ferrimagnetic spinels," J. Appl. Phys. 31, 265-266.

--- (1962) "High temperature susceptibility of garnets: exchange interactions in YIG and LuIB, Ibid. 33, 1257-1258.

--- (1963) "Diagram enumeration in the theory of magnetically dilute Heisenberg ferromagnets," Mol. Phys. 6, 157-160.

--- (1964) "High-temperature heat capacity of ferromagnets having 1st- and 2nd-neighbor exchange: application to EuS," J. Appl. Phys. 35, 991-993.

--- (1965) "High-temperature susceptibility of Heisenberg ferrimagnets," Phys. Rev. 155, 492-495.

3 --- and JOSEPH, R. I. (1964) "High-temperature susceptibility of Heisenberg ferromagnets having first- and second-neighbor interactions," Ibid. 135, A1314-A1320.

--- and RAYL, M. (1968) "Phase transitions of an isotropic ferromagnet in an external magnetic field," Phys. Rev. Lett. 20, 1489-1491.

WOLF, W. P., SCHENIDER, B., LANDAU, D. P., and KEEN, B. E. (1972) "Magnetic and thermal properties of dysprosium aluminum garnet. II. Characteristic parameters of an Ising antiferromagnet," Phys. Rev. B5, 4472-4496.

h WONG, V. K. (1969) "Critical order-parameter relaxation and hypersonic attenuation in helium," Phys. Lett. 29A, 441-442.

d --- (1969) "Unique set of critical exponents and order-parameter theories for 1, 2, 3-dimensional superfluids," Ibid. 30A, 38-39.

f WOOD, D. W. (1972) "A self-dual relation for a three-dimensional assembly," J. Phys. C. 5, L181-L182.

--- and DALTON, N. W. (1966) "Determination of exchange in $Cu(NH_4)_2$, $Cl_4 \cdot 2H_2O$, and $CuK_2Cl_4 \cdot 2H_2O$," Proc. Phys. Soc. 87, 755-765.

---, --- (1972) "High-temperature series expansions for the anisotropic Heisenberg model," J. Phys. C. 5, 1675-1687.

3 WOOD, P. J. and RUSHBROOKE, G. S. (1957) "On the high temperature susceptibility for the Heisenberg model of a ferromagnetic," Proc. Phys. Soc. 70, 765-768.

3 ---, --- (1966) "Classical Heisenberg ferromagnet," Phys. Rev. Lett. 17, 307-308.

WORTIS, M., JASNOW, D., and MOORE, M. A. (1969) "Renormalization of the linked cluster expansion for a classical magnet," Phys. Rev. 185, 805-816.

WRIGHT, P. G. (1969) "A simple alternative derivation of Pippard's thermodynamic relations for λ-transitions," J. Phys. C. 2, 226-228.

--- (1969) "Note on departures form Garland's relation between the elastic constants of a cubic crystal near λ-transition," Ibid. 2, 1352-1353.

m --- (1972) "Revised thermodynamic relations for λ transitions in mixtures. I. Examination in terms of properties generally relevant," J. Phys. A 5, 1004-1014.

m --- (1972) "Revised thermodynamic relations for λ transitions in mixtures. II. Special relations for Curie points and Néel points," Ibid. 5, 1015.1017.

m --- (1972) "Revised thermodynamic relations for λ transitions in mixtures. III. The saturated vapour of solutions of ^{3}He in liquid ^{4}He below their λ temperatures," Ibid. 5, 1018-1024.

--- (1972) "Pippard's relations applied to certain λ transitions in dielectric and magnetic materials," Ibid. 5, 1206-1220.

f WU, F. Y. (1967) "Remarks on the modified potassium dihydrogen phosphate model of a ferroelectric," Phys. Rev. 168, 539-543.

f --- (1967) "Exactly soluble model of the ferroelectric phase transition in two dimensions," Phys. Rev. Lett. 18, 605-607.

f --- (1969) "Exact solution of a model of an antiferroelectric transition," Phys. Rev. 183, 604-607.

f --- (1969) "Critical behavior of two-dimensional hydrogen-bonded antiferroelectrics," Phys. Rev. Lett. 22, 1174-1176.

f --- (1970) "Critical behavior of hydrogen-bonded ferroelectrics," Ibid. 24, 1476-1478.

--- (1971) "Multiple density correlations in a many-particle system," J. Math. Phys. 12, 1923-1929.

f --- (1971) "Ising model with four-spin interactions," Phys. Rev. B4, 2312-2314.

f --- (1971) "Modified potassium dihydrogen phosphate model in a staggered field," Ibid. B3, 3895-3900.

f --- (1972) "Solution of an Ising model with two- and four-spin interactions," Phys. Lett. 38A, 77-78.

--- (1972) "Exact results on a general lattice statistical model," Sol. State Commun. 10, 115-117.

f --- (1972) "Phase transition in a sixteen-vertex lattice model," Phys. Rev. B6, 1810-1814.

--- and CHIEN, M. K. (1970) "Convolution approximation for the n-particle distribution function," J. Math. Phys. 11, 1912-1916.

d ---, LAI, H. W., and WOO, C. W. (1970) "Communications on a microscopic theory of helium submonolayers," J. Low Temp. Phys. 3, 331-333.

l WU, T. T. "Theory of Toeplitz determinants and the spin correlations of the two-dimensional Ising model. I.," Phys. Rev. 149, 380-401.

WYN N, P. (1960) "The rational approximation of functions which are formally defined by a power series expansions," Math. Comput. 15, 147-186.

2 YAHATA, H. (1971) "Critical slowing down of the XY model along the critical field," Phys. Lett. 35A, 366-368.

l --- and SUZUKI, M. (1969) "Critical slowing down in the kinetic Ising model," J. Phys. Soc. Japan 27, 1421-1438.

YAMADA, T. (1969) "Fermi-liquid theory of linear antiferromagnet chains," Progr. Theoret. Phys. 41, 880-890.

---, MORI, M., and NODA, Y. (1972) "A microscopic theory of the phase transition in NH_4Br--an Ising spin phonon coupled system," J. Phys. Soc. Japan 32, 1565-1577.

d YAMAGATA, K., HAYAMA, M., and ODAKA, T. (1971) "Proton resonance study of sublattice rotations and spin deviation in $Cu(HCOO)_2 \cdot 4D_2O$," Ibid. 31, 1279.

m YAMASHITA, N. (1972) "Field induced phase transitions in uniaxial antiferromagnets," Ibid. 32, 610-615.

∞ YAN, C. C. and WANNIER, G. H. (1965) "Observations on the spherical model of a ferromagnet," J. Math. Phys. 6, 1833-1838.

L YANG, C. C. (1972) "Light-scattering study of the dynamical behavior of ordering just above the phase transition to a cholesteric liquid crystal," Phys. Rev. Lett. 28, 955-958.

l YANG, C. N. (1952) "The spontaneous magnetization of a two-dimensional Ising model," Phys. Rev. 85, 808-816.

h --- (1962) "Concept of off-diagonal long-range order and the quantum phases of liquid He and of superconductors," Rev. Mod. Phys. 34, 694-704.

--- and LEE, T. D. (1952) "Statistical theory of equations of state and phase transitions I. Theory of condensation," Phys. Rev. 87, 404-409.

--- and YANG, C. P. (1964) "Critical point in liquid-gas transitions," Phys. Rev. Lett. 13, 303-305.

---, --- (1969) "Thermodynamics of a one-dimensional system of Bosons with repulsive delta-function interaction," J. Math. Phys. 10, 1115-1122.

YANG, C. P. (1970) "One-dimensional system of Bosons with repulsive δ-function interactions," Phys. Rev. A2, 154-157.

YEA, R. and ISIHARA, A. (1969) "Correlation and distribution of segments in a chain molecule," J. Chem. Phys. 51, 1215-1221.

s YEE, B. G. W. and DEATON, B. C. (1969) "Change in velocity of sound between normal and superconducting tin for ql>1," Phys. Rev. Lett. 23, 1438-1440.

YEH, R. H. (1968) "On the validity of recent phenomenological theories of the Ising model and superfluidity," Phys. Lett. 28A, 345-346.

YEH, Y. (1967) "Observation of the long-range correlations effect in the Rayleigh linewidth near the critical point of Xe," Phys. Rev. Lett. 18, 1043-1046.

m YELON, W. B. and BIRGENEAU, R. . (1972) "Magnetic properties of $FeCl_2$ in zero field. II. Long-range order," Phys. Rev. B5, 2615-2621.

f ---, COCHRAN, W., and SHIRANE, G. (1971) "Neutron scattering study of the soft modes in cubic potassium tantalate niobate," Ferroelectrics 2, 261-270.

YUVAL, G. and ANDERSON, P. W. (1970) "Exact results for the Kondo problem: One-body theory and extension to finite temperature," Phys. Rev. B1, 1522-1528.

ZAGORSKI, A. and SUKIENNICKI, A. (1970) "On the anisotropic one-dimensional Heisenberg chain with defects," Acta Phys. Polon. A37 429.

ZEIGER, H. J., KAPLAN, T. A., and RACCAH, P. M. (1971) "Semiconductor-metal transition in Ti_2O_3," Phys. Rev. Lett. 26, 1328-1331.

ZERNIKE, F. (1916) "The clustering-tendency of the molecules in the critical state and the extinction of light caused thereby," Proc. Acad Sci. Amsterdam 18, 1520-1539.

--- (1940) "The propagation of order in co-operative phenomena," Physica 7, 565-585.

m ZIMM, B. H. (1950) "Opalescence of a two-component liquid system near the critical mixing point," J. Phys. Chem. 54, 1306-1317.

b --- and BRAGG, J. K. (1959) "Theory of the phase transition between helix and random coil in polypeptide chains," J. Chem. Phys. 31, 526-535.

b --- and RICE, S. A. (1960) "The helix-coil transition in charged macromolecules," Mol. Phys. 3, 391.

h ZIMMERMAN, G. O. and CHASE, C. E. (1967) "Orthobaric density of ^3He in the critical region," Phys. Rev. Lett. 19, 151-154.

ZIMMERMAN, N. J., BASTMEIJER, J. D., and VAN DEN HANDEL, J. (1972) "Antiferromagnetic resonance and critical paramagnetic linewidth in $LiCuCl_3 \cdot 2H2O$," Phys. Lett. 40A, 259-261.

r ZINN-JUSTIN, J. (1971) "Strong interactions dynamics with Padé approximants," Physics Reports 1, 55-102.

ZITTARTZ, J. (1964) "Funktionalimittelwerte in der statistischen Theorie von quantenmechanischen und klassischen Systemen vieler Teilchen," Z. Physik 180, 219-254.

l --- (1965) "Microscopic approach to interfacial structure in Ising-like ferromagnets," Phys. Rev. 154, 529-534.

m ZOLLWEG, J. A. (1971) "Shape of the coexistence curve near a Plait point in a three-component system," J. Chem. Phys. 55, 1430-1436.

---, HAWKINS, G., and BENEDEK, G. B. (1971) "Surface tension and viscosity of xenon near its critical point," Phys. Rev. Lett. 27, 1182-1185.

---, ---, SMITH, I. W., GIGLIO, M., and BENEDEK, G. B. (1972) "The spectrum and intensity of light scattered from the liquid-vapor interface of xenon near its critical point," J. de Physique 33S, 135-139.

ZUMSTEG, F. C., CADIEU, F. J., MARCELJA, S., and PARKS, R. D. (1970) "Effect of spin-lattice coupling on the critical resistivity of a ferromagnet," Ibid. 25, 1204-1207.

--- and PARKS, R. D. (1970) "Electrical resistivity of nickel near the Curie point," Ibid. 24, 520-524.

r ZWANZIG, R. (1965) "Time-correlation functions and transport coefficients in statistical mechanics," Ann. Rev. Phys. Chem. 16, 67-102.

b --- and LAURITZEN, J. I., JR. (1968) "Exact calculation of the partition function for a model of two-dimensional polymer crystallization by chain folding," J. Chem. Phys. 48, 3351-3360.

ADDENDA:

f ABRAHAM, D. B., LIEB, E. H., OGUCHI, T., and YAMAMOTO, T. (1970) "On the anomolous specific heat of sodium trihydrogen selenite," Progr. Theoret. Phys. 44, 1114-1115.

h AHLERS, G. (1967) "Order of the He^4 II-He^4 I transition under rotation," Phys. Rev. 164, 259-262.

h --- (1969) "Critical heat flow in a thick He II film near the superfluid transition," J. Low Temp. Phys. 1, 159-172.

h --- (1969) "Mutual friction in He II near the superfluid transition," Phys. Rev. Lett. 22, 54-56.

h --- (1972) "Temperature derivative of the pressure of ^{4}He at the superfluid transition," J. Low Temp. Phys. 7, 361-365.

d BINDER, K. and HOHENBERG, P. C. (1972) "Phase transitions and static spin correlations in Ising models with free surfaces," Phys. Rev. B6, 3461-3488.

--- and STAUFFER, D. (1972) "Monte Carlo study of the surface area of liquid droplets," J. Stat. Phys. 6, 49.

BORTZ, A. B. and GRIFFITHS (1972) "Phase transitions in anisotropic Heisenberg ferromagnets," Commun. Math. Phys. 26, 102.

BROWN, G. R. and MEYER, H. (1972) "Study of the specific heat of a He^3-He^4 mixture near its gas-liquid critical point," Phys. Rev. A6, 1578-1588.

---, --- (1972) "Study of the specific-heat singularity of He^3 near its critical point," Phys. Rev. A6, 364-377.

CAMP, W. J. and FISHER, M. E. (1972) "Decay of order in classical many-body systems. I. Introduction and formal theory," Ibid. B6, 946-959.

--- (1972) "Decay of order in classical many-body system. II. Ising model at high temperatures," Ibid. B6, 960-979.

L CHENG, J. and MEYER, R. B. (1972) "Pretransitional optical rotation in the isotropic phase of cholesteric liquid crystals," Phys. Rev. Lett. 29, 1240-1243.

COLLINS, M. F. and MARSHALL, W. (1967) "Neutron scattering from paramagnets," Proc. Phys. Soc. 92, 390.

---, MINKIEWICZ, V. J., NATHANS, R., PASSELL, L., and SHIRANE, G. (1969) "High resolution studies of critical scattering of neutrons from iron," J. Phys. Soc. Japan 26, 169-173.

f FAN, C. (1972) "Symmetry properties of the Ashkin-Teller model and the eight-vertex model," Phys. Rev. B6, 902-910.

f --- (1972) "Remarks on the eight-vertex model and the Ashkin-Teller model of lattice statistics," Phys. Rev. Lett. 29, 158-160.

FELDERHOF, B. U. (1972) "Time-dependent statistics of binary linear lattices," J. Stat. Phys. 6, 21-38.

l FERER, M. and WORTIS, M. (1972) "High-temperature series and critical amplitudes for the spin-spin correlations of the three-dimensional Ising ferromagnet," Phys. Rev. B6, 3426-3444.

r FERRELL, R. A. (1969) "Field theory of phase transitions," In Contemporary Physics: Trieste Symposium (IAEA, Vienna), p. 129-156.

g FISHER, M. E., MA, S., and NICKEL, B. G. (1972) "Critical exponents for long-range interactions," Phys. Rev. Lett. 29, 917-920.

l GAUNT, D. S. and SYKES, M. F. (1972) "Re-analysis of the critical isotherm of the Ising ferromagnet," J. Phys. C. 5, 1429-1445.

GERSCH, H. A., SHULL, C. G., and WILKINSON, M. K. (1956) "Critical magnetic scattering of neutrons by iron," Phys. Rev. 103, 525-534.

r GLOVER, R. E., III (1968) "Superconductivity above the transition temperature," Prog. Low Temp. Phys. 6, 291-332.

g GROVER, M. K. (1972) "Critical exponents for the X-Y model," Phys. Rev. B6, 3546-3547.

g ---, KADANOFF, L. P., and WEGNER, F. J. (1972) "Critical exponents for the Heisenberg model," Ibid. B6 311-313.

GUTTMANN, A. J. and JOYCE, G. S. (1972) "On a new method of series analysis in lattice statistics," J. Phys. A5, L81-L84.

u HARBUS, F., KRASNOW, R., LAMBETH, D., LIU, L. L., and STANLEY, H. E. (1973) "Ising, planar, and Heisenberg models with directional anisotropy," Proc. Intern. Conf. on Padé Approximants, Canterbury, July, 1972.

HO, J. T. (1971) "Correlation between magnetization and field-dependent specific-heat data near the Curie Point," Phys. Rev. Lett. 26, 1485-1487.

f HUBER, D. L. (1972) "Critical dynamics of the order-disorder transformation in ferroelectrics," Phy . Rev. B6, 3379-3385.

f JOHNSON, J. D., KRINSKY, S., and MC COY, B. M. (1972) "Critical index ν of the vertical-arrow correlation length in the eight-vertex model," Phys. Rev. Lett. 29, 492-494.

r KITTEL, C. (1969) Thermal Physics, Wiley, N.Y.

LANGER, J. S. (1969) "Statistical theory of the decay of metastable states," Ann. Phys. 54, 258-275.

II
EXPERIMENTAL ARTICLES

EXPERIMENTAL TESTS OF SCALING LAWS NEAR THE SUPERFLUID TRANSITION IN He^4 UNDER PRESSURE

G. AHLERS

Measurements of the heat capacity, the pressure coefficient $(\partial P/\partial T)_v$, and the thermal conductivity of liquid helium near the superfluid transition under pressure are presented and analyzed in terms of Scaling Law predictions. Several deviations from theory are revealed by the data. Neither the heat capacity at constant pressure nor the thermal conductivity fully reflect the symmetry of the transition which is implied by Scaling.

1. Introduction

The superfluid transition in liquid helium offers a unique opportunity to study in detail singularities both in equilibrium and transport properties. Primarily because of the high homogeneity of the samples, and because of relatively short thermal relaxation times even in He I, one can obtain very precise measurements near this transition; and theoretical predictions pertaining to critical phenomena can be put to a more detailed test here than has been possible for critical points or "λ-lines" in other systems. In this paper I would like to summarize our current knowledge about the heat capacity at constant pressure C_p of He I and He II, and of the thermal conductivity κ of He I, all in the immediate vicinity of T_λ. Some of the work which I will discuss, particularly that at pressures larger than the vapor pressure, is still being pursued, and a number of interesting unanswered questions remain. Nonetheless, sufficient information is now available to make it reasonably certain that at least some of the predictions based upon current theories of critical phenomenon are not quantitatively correct near the superfluid transition, and it seems likely that further measurements will confirm additional suspected failures of predictions. However, it might be hoped that a careful study of the dependence of the deviations from theory upon such supposedly inert parameters as the pressure will provide the kind of insight that may be required to appropriately modify the theory.

2. Theoretical Predictions

For the discussion of properties with singularities at the transition temperature T_λ it is convenient to use the dimensionless parameter

$$\epsilon \equiv 1 - T/T_\lambda(P) \qquad (1)$$

where $T_\lambda(P)$ is a function of the pressure P. Since almost any singular function $f(\epsilon)$ can be described asymptotically as a power law[1], i.e. $f(\epsilon) \sim \epsilon^{-z}$, we write, for He I, C_p as[1]

$$C_p \sim (A/\alpha)\ \{|\epsilon|^{-\alpha}-1\}\ + B, \qquad (2)$$

the thermal conductivity as

$$\kappa \sim |\epsilon|^{-x}. \qquad (3)$$

and the coherence length ξ for fluctuations in the order parameter as

$$\xi \sim |\epsilon|^{-\nu}. \qquad (4)$$

These relations may be regarded as definitions of the parameters A, B, α, x, and ν. For He II, analogous definitions apply for the parameters of C_p and ξ, but by convention these parameters are identified as A', B', α', and ν'. The parameters for C_p were defined deliberately by the at first sight overly complicated form of Eq. 2 because, as α vanishes, Eq. 2 becomes[1]

$$C_p \sim -A \ln|\epsilon| + B \text{ if } \alpha = 0. \qquad (5)$$

In addition to the asymptotic contributions given by Eqs. 2 to 5, it must be expected that there are other, higher order, terms contributing[1] when $|\epsilon| > 0$. These terms may also be singular at the transition; but in most cases they can be expected to vanish at T_λ. There are usually no theoretical predictions about higher order terms, and when they are appreciable in the entire experimentally accessible temperature range, it becomes extremely difficult to extract the asymptotic behavior from the experimental measurements.

Current theories of critical phenomena do not predict any of the parameters in Eqs. 2 to 5. They do, however, predict relations, called scaling laws, between these parameters, and we shall attempt to test these relations by comparing them with experimental measurements. On the basis of very general assumptions about the functional form of the free energy of the system, it can be predicted[2-4] that

$$\alpha = \alpha'. \qquad (6)$$

In addition,[2-4]

$$A = A' \text{ if } \alpha = \alpha' = 0. \qquad (7)$$

If further assumptions similar to those made about the free energy are made also about the correlation function for order parameter fluctuations, then it follows[2,3] that

$$\nu = (2-\alpha)/3, \qquad (8)$$

$$\nu' = (2-\alpha')/3, \qquad (9)$$

and, by virtue of Eq. 6,

$$\nu = \nu'. \qquad (10)$$

The exponent ν' can be evaluated from measurements of the superfluid density ρ_s and the relation[5,6]

$$\xi \sim \rho_s^{-1}. \qquad (11)$$

Unfortunately, ν is not accessible to direct measurement in helium. However, Eq. 8 or 10, in combination with another prediction, Eq. 15 below, can be tested by examining the thermal conductivity of He I. For κ, dynamic scaling predicts,[5,6] on the basis of some very general assumptions about the critical frequency which describes the time dependence of order parameter fluctuations, that

$$\kappa/[\rho C_p^+(\xi^+)^2] \sim u_2/\xi^- \text{ with } \xi^+ = \xi^-, \qquad (12)$$

where κ, $C_p{}^+$, and u_2 are functions of ξ. Here ρ is the density, u_2 the second sound velocity which is given by the hydrodynamic relation

$$u_2^2 = S^2 T\rho_s/(\rho_n C_p^-), \qquad (13)$$

$C_p{}^+$ and ξ^+ pertain to He I, and $C_p{}^-$ and ξ^- to He II. We emphasize that κ, $C_p{}^+$, and u_2 in Eq. 12 should be evaluated at those pairs of values of ϵ for which the coherence lengths ξ^+ and ξ^- are equal to each other. From Eqs. 11 to 13 one obtains for κ

$$\kappa \sim [C_p^+/(C_p^-)^{1/2}]\,(\xi^+)^2/(\xi^-)^{3/2}. \qquad (14)$$

If ξ is symmetric in ϵ, then $\xi^+=\xi^-$ at the same value of $|\epsilon|$ above and below T_λ, and

$$x = 2\nu - \frac{3}{2}\nu' + \alpha - \tfrac{1}{2}\alpha' = \tfrac{1}{2}(\nu' + \alpha'). \qquad (15)$$

If α and α' are zero, then $x = \frac{1}{2}\nu'$; however, since in any experimentally accessible temperature region C_p is dependent upon ϵ even if $\alpha = \alpha' = 0$, it is necessary in the determination of x from experimental data to include C_p in the analysis.

If Eq. 10 should fail, then ξ is no longer symmetric in ϵ, and Eq. 15 is not valid. In this case, one obtains, from Eq. 12 with $\xi^+ = \xi^-$,

$$x = \nu/2 + \alpha - \nu\alpha'/(2\nu'). \qquad (15a)$$

When $\alpha = \alpha' = 0$, then one can predict a relation between the amplitudes A and A' of C_p and the exponents ν and ν' of ξ. This follows from the relation between C_p and the energy density - energy density correlation function,[7] and is given by

$$A/A' = \nu/\nu' \text{ if } \alpha = \alpha' = 0. \qquad (16)$$

Eq. 16 obviously is consistent with Eqs. 7 and 10. It is, however, not dependent upon all the usual scaling hypotheses, and Eq. 16 might be valid even if the prediction Eq. 7 should fail. Thus Eq. 16 may be considered as a link between the result Eq. 7 of thermodynamic scaling and the result Eq. 10 of correlation function scaling which predicts how Eq. 10 should fail if we know from experiment how Eq. 7 fails.

There is one additional point of considerable practical importance to be considered. It has been recognized recently that the detailed nature of singularities in various properties may depend upon the path which is followed in approaching a critical point.[8-10] Thus, for instance, κ along an isochore may depend upon $T_\lambda(V)-T$ in a manner different in detail from the dependence of κ along an isobar upon $T_\lambda(P)-T$. It appears from various model calculations that the divergences which characterize the "ideal" transition will be observed only when, in our specific case, the variable under consideration is examined along an isobar as a function of $T_\lambda(P)-T$.[9]

We are now ready to proceed to the experimental results, and we shall atempt to test particularly Eqs. 6, 7, 9, 10, 15, 15a, and 16. Within experimental error, we shall find that, in agreement with Eq. 6, $\alpha = \alpha' = 0$ if the pressure is not too large; that, contrary to Eq. 7, $A \neq A'$; that, in agreement with Eq. 9, $\nu' = (2-\alpha')/3$; and that, contrary to Eqs. 8, 10, and 15, $x \neq \nu'/2$. It also appears that a determination of ν from Eq. 15a is inconsistent with Eq. 16. At sufficiently high pressure, there is some indication that $\alpha > 0$ although $\alpha' = 0$, in disagreement with Eq. 6; however, we do not regard the available experimental results as entirely definitive on this point, and additional measurements are in progress.

3. The Heat Capacity and $(\partial P/\partial T)_v$

Although we would like to know C_p along isobars, it is experimentally difficult to measure this variable along this thermodynamic path. More readily measurable quantities are the heat capacity at saturated vapor pressure C_s, and the heat capacity at constant volume C_v along isochores. It can be shown that near T_λ C_s differs negligibly from C_p along the isobar $P = 0.05$ bars. However, C_v along isochores, particularly at the higher pressures, differs considerably from C_p along the same isochores. Furthermore, C_p along isochores is not the same function of $T_\lambda(V)-T$ as C_p along isobars is of $T_\lambda(P)-T$. However, these practical difficulties are not unsurmountable, for it is possible to obtain enough thermodynamic information to calculate C_p along isobars reliably from the readily measurable quantities.

Along a particular path we have

$$C_p = C_v - T(\partial P/\partial T)_v^2\,(\partial V/\partial P)_T. \qquad (17)$$

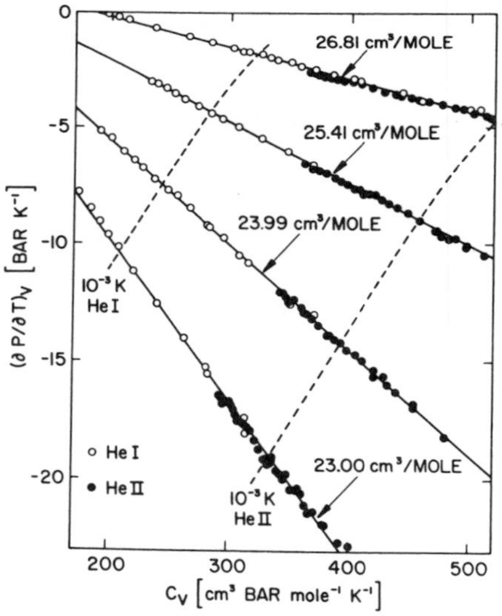

Fig. 1. $(\partial P/\partial T)_v$ as a function of C_v. The dashed lines show where $|T_\lambda(V)-T| = 10^{-3} K$.

Most of the dependence upon $T_\lambda - T$ of $C_p - C_v$ is contained in $(\partial P/\partial T)_v$, since the compressibility, although divergent at T_λ, varies little over the experimental temperature range. Therefore $(\partial P/\partial T)_v$ was determined experimentally in addition to C_v. Knowing these two variables has several further advantages. Near the transition, $(\partial P/\partial T)_v$ asymptotically becomes a linear function of C_v.[10,11] One has

$$C_v = T(\partial S/\partial T)_t - T(\partial V/\partial T)_t(\partial P/\partial T)_v \qquad (18)$$

where

$$t = T_\lambda(V) - T. \qquad (19)$$

Along isochores, the parameter $(\partial V/\partial T)_t = (\partial V/\partial T)_\lambda$ and is a property of the λ-line only, and $(\partial S/\partial T)_t$, although singular at T_λ, is finite and varies almost negligibly for $t \lesssim 10^{-3} K$.[10,11] We therefore can test the consistency of the two measurements by examining $(\partial P/\partial T)_v$ as a function of C_v. The data should yield, at least asymptotically, a unique straight line for both phases. They are shown in Fig. 1, and are seen to be thermodynamically consistent within the scatter. In addition the parameters of the line are some of those quantities which are required to calculate $(\partial V/\partial P)_T$ which we need in Eq. 17. We have[10-11]

$$\left(\frac{\partial V}{\partial P}\right)_T = \left(\frac{\partial S}{\partial T}\right)_t \left(\frac{\partial P}{\partial T}\right)_t^2 + \left(\frac{\partial V}{\partial T}\right)_\lambda \left(\frac{\partial P}{\partial T}\right)_t - \frac{C_p}{T}\left(\frac{\partial P}{\partial T}\right)_t^2 \qquad (20)$$

where $(\partial P/\partial T)_t$ again is a singular, but slowly varying function of t. At T_λ, $(\partial P/\partial T)_\lambda$ is known accurately from independent measurements.[12-13] The very small dependence upon t of $(\partial S/\partial T)_t$ and $(\partial P/\partial T)_t$ was calculated from the volume dependence of C_v by a method[10] upon which we shall not elaborate here. In order to obtain C_p from Eq. 18, we shall first assume in Eq. 20 that $C_p \cong C_v$. The resulting C_p can then be used to give a better estimate of $(\partial V/\partial P)_T$ and C_p. About 4 iterations are sufficient to yield C_p. We now have C_p along isochores, i.e. as a function of t. However, we would like C_p along isobars, i.e. as a function[14] of

$$\theta \equiv T_\lambda(P) - T = T_\lambda(P)\epsilon. \qquad (21)$$

The difference between θ and t is illustrated schematically in the insert in Fig. 2. Again we can use our measured $(\partial P/\partial T)_v$, and obtain θ from

$$\theta = t - \left[\int_0^t (\partial P/\partial T)_v dt'\right] \Big/ (\partial P/\partial T)_\lambda. \qquad (22)$$

We show in Fig. 2 the experimental results for C_v as a function of $\log_{10} |t|$ for the four molar volumes which have been studied. Although C_v is not expected to vary as $\ln |t|$, it is apparent that the data can be approximated by the straight lines which are shown in the figure. Close examination reveals, however, that the measured C_v for $t \gtrsim 10^{-3}$K systematically fall below the lines. This is true even for He I, where higher order contributions are expected to yield positive deviations for C_p from the asymptotic behavior. When the results are, somewhat arbitrarily, fitted to a power law $C_v \sim t^{-z}$, then one

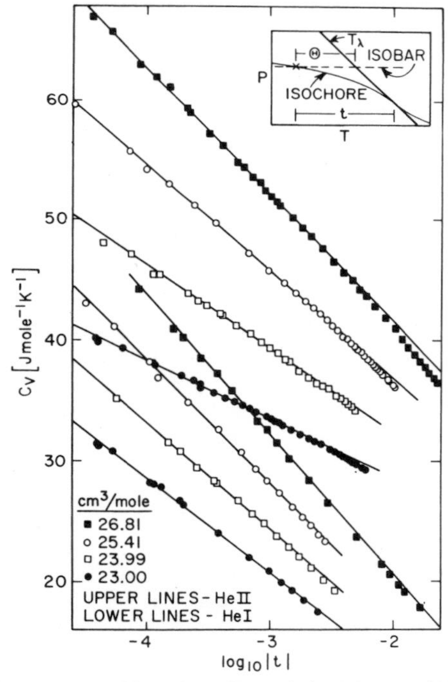

Fig. 2. C_v as a function of $\log_{10} |t|$, with $t = T_\lambda(V) - T$ in K. The insert shows schematically the difference between t and $\theta = T_\lambda(P) - T$.

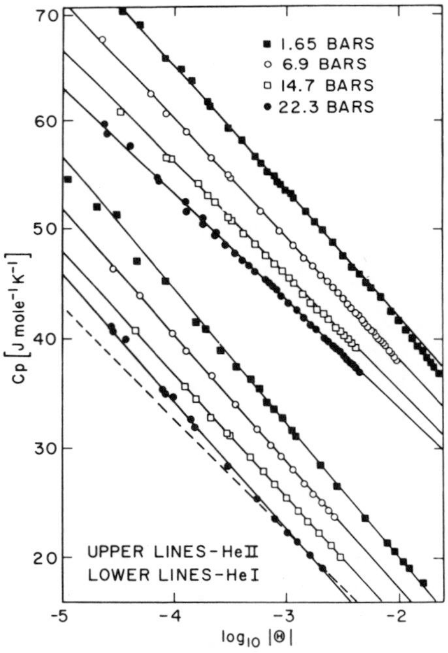

Fig. 3. C_p as a function of $\log_{10} |\theta|$, with θ in K. The dashed line has 1.05 times the slope of the solid line through the He II data at 22.3 bars.

obtains $z < 0$, and at 24 cm^3/mole $z \cong z' \cong -0.05$. The value of this power law at T_λ is only about a factor of 2 smaller than the finite value of C_v which can be calculated[11] from properties of the λ-line.

In Fig. 3 we show C_p as a function of $\log_{10} |\theta|$ for the four isobars which meet the λ-line at the same temperatures as the isochores of Fig. 2. We have not shown the results at saturated vapor pressure; but these lie only slightly above the data for 1.65 bars. They have been published elsewhere,[15] and they are in good agreement with, but more precise than, measurements by others.[11] In the case of C_p, we believe that the straight lines drawn through the data represent the asymptotic behavior. Close inspection of the measurements at 1.65 bars reveals negative deviations from the lines for He II and positive deviations for He I at large $|\theta|$, and we attribute this to higher order contributions. These deviations are even larger for He II at vapor pressure, and they made it difficult to determine the asymptotic behavior.[15] However, the results at vapor pressure and at 1.65 bars are consistent with a logarithmic divergence of C_p both in He I and He II, and higher order singular terms of the form $D\epsilon \ln |\epsilon| + E\epsilon$.[15] At the higher pressures, higher order terms are far less important, and not noticeable in the data for $|\theta| \lesssim 6 \times 10^{-3}$K.

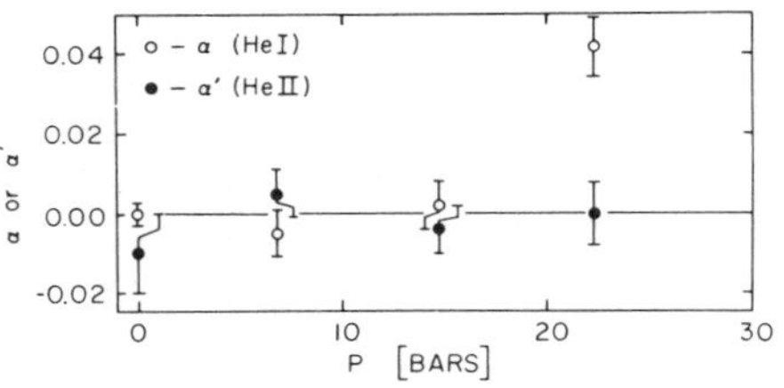

Fig. 4. α and α' as a function of P.

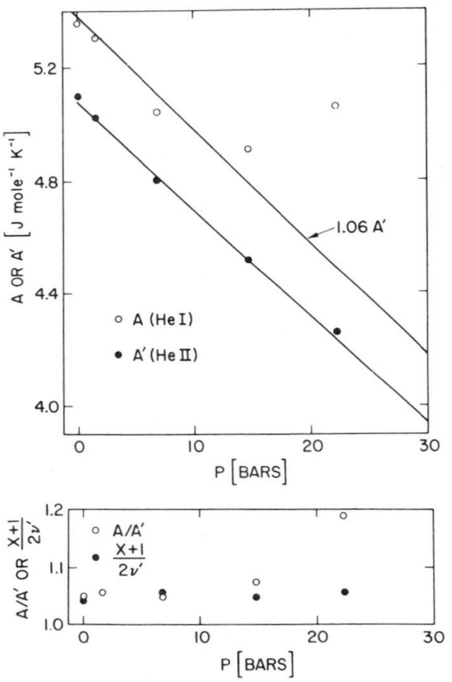

Fig. 5. Upper half: A and A' for Eq. 5 as a function of P. The straight lines are provided only for reference, and are not meant to imply a particular functional form. Lower half: A/A' and $(x + 1)/(2\nu')$ as a function of P. See text.

The results were fitted to a power law, i.e. Eq. 2, by a least squares procedure, and the resulting values of α and α' are shown in Fig. 4. We see that, to within ±0.005, α and α' are equal to zero, except for He I at 22.3 bars. At this highest pressure, $\alpha = 0.042$ whereas $\alpha' = 0.000$. Although there is no known possible systematic error in the experimental measurement or in the data processing which could account for the large deviation of α from zero, it appears wise to reserve judgment about whether the effect is real until further work has been done between 20 and 30 bars.

We see from Fig. 4 that Eq. 6 is obeyed rather well for $P \lesssim 15$ bars. Since in this range $\alpha = \alpha' = 0$, we expect Eq. 7 to hold also. Therefore we have set $\alpha = \alpha' = 0$, and determined A and A' from Eq. 5 and the data. The results are shown in the top half of Fig. 5. We see that A' is almost a linear function of P. At none of the pressures is A equal to A'. It is evident that Eq. 7 fails in this system. In the lower half of Fig. 5 we show A/A',[16] which, according to Eq. 16, we expect to be equal to ν/ν'

We now proceed to examine Eqs. 8 to 10 which involve the exponents ν and ν' for the coherence length. From measurements of the superfluid density at vapor pressure[17–19] we know that $\nu' = 2/3$ within 1% of its value. This, and $\alpha' = 0$, satisfies Eq. 9 rather well. At higher pressures, there are no equally accurate determinations of ν'; but measurements of ρ_s at rather large ϵ[20] indicate that ρ_s/ρ is to within about 1% a universal function of ϵ, independent of P for $P \lesssim 20$ bars. Therefore it is likely that ν' is at least approximately independent of P. Thus, all available information about ν' is consistent with $\alpha' = 0$ and Eq. 9 even at high pressure.

If we believe Eq. 16, then $\nu > \nu'$, although $\alpha = \alpha'$ up to about 15 bars. Clearly this is inconsistent with Eqs. 8 and 10, and either Eqs. 8 and 10 or Eq. 16 must be incorrect at all pressures.

4. The Thermal Conductivity of He I

The results of the thermal conductivity measurements are shown as a function of $\log_{10}(\epsilon)$ in Fig. 6. Most of the data were obtained in a celi of length $L = 0.1$ cm. At saturated vapor pressure, some results with a cell length of 1 cm which have been reported previously[21] are shown also. Data obtained with the two cell lengths agree within experimental error. However, with the greater length, extremely small power densities had to be used and thermal time constants were large. For these reasons the precision is only 1 or 2% for large ϵ, and perhaps 5% for $\epsilon \lesssim 2 \times 10^{-5}$. For $L = 0.1$ cm, higher power densities could be used, and thermal time constants were shorter. Here the precision is limited by the temperature resolution, and varies between 1% at 10^{-7} W/cm^2 and 0.1% for power densities greater than about 10^{-6} W/cm^2. Unfortunately the shorter cell has the disadvantage that corrections for the thermal boundary resistance are larger; and this increases considerably possible systematic errors in the derived exponents.

For κ, it is extremely difficult to obtain the asymptotic behavior. This is most apparent from the data for 22.3 bars. The lower dashed line in Fig. 6 represents what we now believe to be a good representation of Eq. 14. It is evident that there are appreciable contributions to κ from higher order terms for $\epsilon \gtrsim 10^{-5}$, and if one wants to attribute all of the measured κ to the asymptotic behavior, then one must confine the data analysis to $\epsilon \lesssim 10^{-5}$. Clearly such a procedure would lead to rather large errors for the exponent x in Eq. 3. Therefore we have made the assumption that

$$\kappa = \kappa_1 + \Delta\kappa \qquad (23)$$

where κ_1 is given by Eq. 14, and where

$$\Delta\kappa = a\epsilon^{-y};\ y < x \qquad (24)$$

adequately describes the higher order terms *over an appreciable accessible range of* ϵ. We would certainly hope that the inclusion of $\Delta\kappa$ in the analysis would extend the range of ϵ over which the data can be fitted by a decade, and that data up to $\epsilon \cong 10^{-4}$ may be included in the analysis.

We have subtracted Eq. 14 with several trial values of x[22] from the measured κ, and plotted the differences

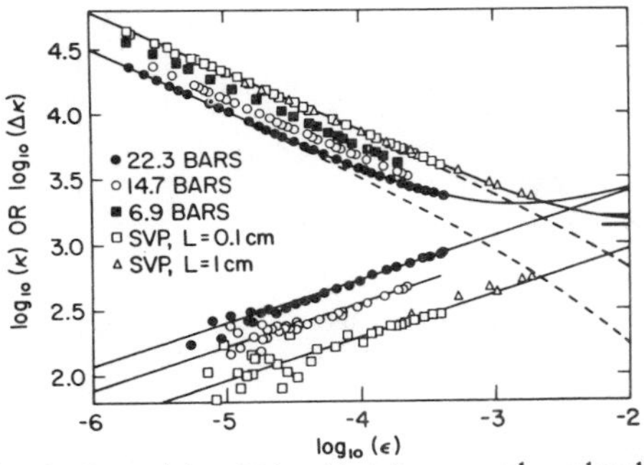

Fig. 6. $\log_{10}(\kappa)$ and $\log_{10}(\Delta\kappa)$ in erg sec^{-1} cm^{-1} K^{-1} as a function of $\log_{10} |\epsilon|$. The upper sets of data pertain to κ, and the lower ones to $\Delta\kappa$. The dashed lines are an extrapolation of the asymptotic behavior. The difference between the dashed lines and the data for κ is equal to $\Delta\kappa$. The solid lines through $\Delta\kappa$ have a slope of 1/3, and the solid lines through κ correspond to the sum of the solid lines through $\Delta\kappa$ and the dashed lines. The bars at the right hand margin are from Ref. 24, upper for 22 bars and lower for vapor pressure.

$\Delta\kappa$ against ϵ on logarithmic scales. Only for a well defined value of x did this procedure result in a linear relation between $\ln(\Delta\kappa)$ and $\ln(\epsilon)$ for $\epsilon \lesssim 10^{-4}$. If x is changed by 0.005 from its best value, then $\ln(\Delta\kappa)$ will deviate appreciably at small ϵ from a straight line. The lower three sets of points in Fig. 6 are $\Delta\kappa$ obtained with the best x. This procedure results in values of x which vary from 0.39 to 0.41 as the pressure is changed from vapor pressure to 22 bars. We estimate that systematic errors in x, due primarily to uncertainties in the boundary resistance, probably are not larger than 0.02.

If all of static and dynamic scaling were correct, then x should be equal to 1/3 if $\alpha = \alpha' = 0$. We do not know of any possible systematic errors in the measurements that could be as large as the difference between 0.39 and 1/3, and we conclude at least tentatively that $x > 1/3$. This conclusion is different from the one at which we arrived on the basis of the earlier measurements in the 1 cm cell.[21] The difference is attributable to the fact that only the more precise measurements with $L = 0.1$ cm, and particularly those at higher pressure, revealed the importance and functional form of the higher order terms. The open triangles in Fig. 6 within their errors can be described well up to $\epsilon \cong 10^{-3}$ by $x = 1/3$ and Eq. 14, without higher order terms.[21]

Since the prediction $x = 1/3$ depends upon the validity of several scaling laws,[23] we can now inquire whether the available experimental evidence contains any hints about the origin of the departure of κ from scaling. Particularly, one would like to know whether it is necessary to invoke failure of frequency scaling, i.e. Eq. 15a, or whether our information about the size of departures from static correlation function scaling, particularly Eq. 10, is sufficient to explain the experimental values of x. We find from Eq. 15a that ν varies from 0.78 to 0.82 between vapor pressure and 22.3 bars, with possible systematic errors presently estimated at ±0.04. This yields $\nu/\nu' > 1.11$, which is inconsistent with Eq. 16 and the measured A/A'. Although a final conclusion will have to be postponed until additional work on the boundary resistance is completed, it appears that either Eq. 16 or Eq. 15a fail.

Although on theoretical grounds it is most reasonable to evaluate Eq. 14 at the same ξ above and below the transition, we shall also explore how the experimental data compare with predictions when Eq. 14 is evaluated at the same $|\epsilon|$. In this case, the left equality of Eq. 15 applies, and $(x + 1)/(2\nu')$ should be equal to ν/ν' if $\alpha = \alpha' = 0$ and $\nu' = 2/3$. From Eq. 16, we would expect this ratio to be equal to A/A' if both Eq. 15 and 16 are correct. These results are compared with A/A' in Fig. 5. We find up to 15 bars that within experimental error $A/A' = (x + 1)/(2\nu')$, but regard this as an empirical observation.

Although there is no theory about the exponent y of $\Delta\kappa$, we note that at all pressures y is about equal to $-1/3$.

The solid lines through $\Delta\kappa$ in Fig. 6 have a slope of 1/3, and the solid extrapolations of the data for κ correspond to Eq. 23 and 24 with $y = -1/3$. We know from a comparison of the pressure dependence of the present data with the pressure dependence observed by Kerrisk and Keller[24] at large ϵ that extrapolations of κ at different pressures should cross each other, as indeed they do in Fig. 6. The results of Kerrisk and Keller[24] at vapor pressure and 22 bars near $\epsilon \cong 10^{-2}$ are shown near the right hand margin as two solid bars. It is seen that the extrapolation at vapor pressure is about correct; but at 22 bars it appears that Eqs. 23 and 24 do not fully account for κ up to $\epsilon \cong 10^{-2}$.

Summary and Conclusion

We have seen that within the rather small uncertainty of ± 0.005, $\alpha = \alpha' = 0$, at least if the pressure is not too high. At the highest pressure, there is some indication that $\alpha > 0$ although $\alpha' = 0$; but this has to be confirmed by additional work. To within about 1%, $\nu' = (2-\alpha')/3$, probably at all pressures. Except possibly at the highest pressure, these results agree with scaling predictions. However, $A \neq A'$, and therefore it is likely that $\nu \neq \nu'$ and $\nu \neq (2-\alpha)/3$. The thermal conductivity exponent x is not equal to its predicted value of 1/3. It appears at the present that x also does not agree with the dynamic scaling prediction $x = \nu/2$ when ν/ν' is set equal to A/A'; but additional work to reduce possible systematic errors in x is required to make this latter conclusion more firm. The exponent x does, however, agree with Eq. 15 when $\nu/\nu' = A/A'$ is used. But we must regard this as an empirical observation unless the use of Eq. 15 can be theoretically justified even when Eq. 10 fails. We see that all the presently known departures from scaling can be attributed to a breakdown of correlation function and frequency scaling in the high temperature phase, He I. We hasten to add, however, that such a breakdown creates considerable difficulties for thermodynamic scaling, because it results in a departure of A/A' from unity.

Within the present errors, it is apparent that α, α', A/A', and x are all independent of pressure. A possible exception to this is of course the behavior of α for $P >$ 15 bars; but we shall discount this until additional measurements are made. This suggests that, even though there is a breakdown of scaling in helium, a certain principle of universality may indeed apply to this system. Such universality hypotheses have been advanced by several theorists and particularly by Kadanoff.[25] In order to test universality in more detail, it will be especially interesting to see if additional measurements confirm $\alpha > 0$ at high pressure; for a pressure dependence of α appears to constitute a clear violation of this hypothesis.

Acknowledgment

I am grateful to B. I. Halperin and P. C. Hohenberg for many stimulating discussions throughout this work, and to L. P. Kadanoff for calling to my attention the relation between A/A' and ν/ν'.

References

1 See for instance, M.E. Fisher, Rept. Progr. Phys.,**30** (1967) 615.
2 B. Widom, J. Chem. Phys.,**43** (1965) 3892, 3898.
3 L.P. Kadanoff, Physics,**2** (1966) 263.
4 R.B. Griffiths, Phys. Rev.,**158** (1967) 176.
5 R.A. Ferrell, N. Menyhàrd, H. Schmidt, F. Schwabl, and P. Szépfalusy, Phys. Rev. Letters, **18** (1967) 891; Phys. Letters,**24B** (1967) 493; Ann. Phys.(N.Y.), **47** (1968) 565.
6 B.I. Halperin and P.C. Hohenberg, Phys. Rev. Letters, **19** (1967) 700; and Phys. Rev., **177** (1969) 952.
7 See for instance L.P. Kadanoff, W. Götze, D. Hamblen, R. Hecht, E.A.S. Lewis, V.V. Palciauskas, M. Rayl, and J. Swift, Rev. Mod. Phys.,**39** (1967) 395.
8 B.J. Lipa and M.J. Buckingham, Phys. Letters,**26A** (1968) 643.
9 M.E. Fisher, Phys. Rev.,**176** (1968) 257.
10 G. Ahlers, Phys. Rev.,**182** (1969) 352.
11 M.J. Buckingham and W.M. Fairbank, in Progress in Low Temperature Physics, ed. by C.J. Gorter, North Holland Pub. Co. (1961) 80; and C.F. Kellers, PhD Thesis, Duke Univ., Durham, N. C. (1960).
12 H.A. Kierstead, Phys. Rev.,**153** (1967) 258.
13 D. L. Elwell and H. Meyer, Phys. Rev.,**164** (1967) 245.
14 We have neglected the very small change in C_p at constant θ due to changes of P along the isochore.
15 G. Ahlers, Phys. Rev. Letters,**23** (1969) 464 and Phys. Rev. (to be published).
16 Elwell and Meyer (Ref. 13) found $A/A' < 1$ from thermal expansion measurements. The present results do not agree with theirs.
17 J.P. Clow and J.D. Reppy, Phys. Rev. Letters,**16** (1969) 887.
18 J.A. Tyson and D.H. Douglass, Jr., Phys. Rev. Letters,**17** (1966) 472; and J.A. Tyson, Phys. Rev.,**166** (1968) 166.
19 M. Kriss and I. Rudnick, to be published.

[20] R.H. Romer and R.J. Duffy, Phys. Rev., **186** (1969) 255.
[21] G. Ahlers, Phys. Rev. Letters, **21** (1968) 1159.
[22] For this purpose, we have assumed that C_p^+ and C_p^- may be evaluated at the same value of $|\epsilon|$.
[23] G. Ahlers, Phys. Letters, **28A** (1969) 507.
[24] J.F. Kerrisk and M.E. Keller, Phys. Rev., **177** (1969) 341.
[25] L.P. Kadanoff, Varenna Lectures on Critical Phenomena (1970) to be published.

PHYSICAL REVIEW VOLUME 185, NUMBER 2 10 SEPTEMBER 1969

Investigation of Scaling Laws by Critical Neutron Scattering from Beta-Brass

J. ALS-NIELSEN
The Danish Atomic Energy Commission, Research Establishment Risö, Roskilde, Denmark
(Received 4 April 1969)

Using a Cu^{65}-Zn β-brass crystal, the critical scattering of neutrons has been studied, both above and below T_c. The staggered susceptibilities χ vary as $C_+(T/T_c-1)^{-\gamma}$ and $C_-(1-T/T_c)^{-\gamma'}$, respectively. We find that $\gamma=\gamma'$ within an accuracy of 3%, in agreement with the scaling hypothesis of static critical phenomena; and that $C_+/C_-=5.46\pm0.05$, in excellent agreement with the recent parametric representation theory of Schofield and in fair agreement with the results of series expansions by Essam and Hunter. For fixed q, a flat maximum is observed in the wave-vector-dependent susceptibility $\chi(q,T)$ at temperatures which agree well with the predictions of Fisher and Burford.

IN theoretical investigations of phase transitions, the Ising magnet has been studied extensively because it is the simplest model which displays the essential features of critical phenomena. Only very few magnetic materials can be described adequately by the Ising model, so the possibilities for experimental verification of the theories may appear to be rather limited. However, the configurational energy in an AB alloy such as β-brass is formally equivalent to the Ising Hamiltonian, and, indeed, our previous investigations of the order-disorder transition in β-brass[1] have provided some accurate verifications of the theoretical predictions for the Ising model.[2]

The critical scattering cross section is proportional to the wave-vector-dependent susceptibility $\chi(\mathbf{q},t)$, where $\mathbf{q}$ is the difference between the scattering vector and a point in reciprocal space with an odd sum of indices, e.g., (1,0,0). The temperature dependence of the susceptibility is given in terms of $t=|T-T_c|/T_c$. We have, in particular, been interested in determining the temperature dependence of $\chi(0,t)$ above and below T_c, in order to compare the critical exponents γ and γ' given by:

$$\chi_+(0,t_+)=C_+t_+^{-\gamma},\quad \chi_-(0,t_-)=C_-t_-^{-\gamma'}. \tag{1}$$

The indices + and − indicate temperatures above and below T_c, respectively. The scaling hypothesis of static critical phenomena implies $\gamma=\gamma'$.[3]

The long-range-order Bragg scattering occurring below T_c obscures the critical scattering for $\mathbf{q}$ near zero, so $\chi_-(0,t_-)$ can only be determined by correctly extrapolating the data for $\mathbf{q}\neq0$. The low critical scattering intensity in the region with no Bragg scattering contamination has previously prevented accurate determination of the susceptibility below T_c. However, the cross section is enhanced by a factor of 7 in a β-brass crystal isotopically enriched in Cu^{65}. We have produced a Cu^{65}-Zn crystal of 100 g, and the results described are derived from this crystal.

[1] J. Als-Nielsen and O. W. Dietrich, Phys. Rev. **153**, 706 (1967); O. W. Dietrich and J. Als-Nielsen, *ibid.* **153**, 711 (1967); J. Als-Nielsen and O. W. Dietrich, *ibid.* **153**, 717 (1967).
[2] M. E. Fisher and R. Burford, Phys. Rev. **156**, 583 (1967).
[3] L. Kadanoff *et al.*, Rev. Mod. Phys. **39**, 395 (1967).

Unfortunately, although the q dependence of $\chi_+(\mathbf{q},t)$ has been calculated very accurately,[2] theoretical knowledge of $\chi_-(\mathbf{q},t)$ is very limited, so the optimum method of extrapolating to $q=0$ must be determined empirically. An appropriate expansion of $\chi(\mathbf{q},t)$ for small q/κ is

$$\chi_\pm(q,t_+)=\chi_\pm(0,t_\pm)/[1+b_\pm(q/\kappa_\pm)^2+\cdots]. \tag{2}$$

κ is the true correlation range as defined by Fisher and Burford.[2] A similar expansion with $b=1$ defines the effective correlation range κ_1, and $b\equiv(\kappa/\kappa_1)^2$. In general, $b\neq1$, but for the bcc lattice $b_+\simeq1$. The temperature dependence of κ near T_c is given by:

$$\kappa_+=F_+t_+^{\nu},\quad \kappa_-=F_-t_-^{\nu'}. \tag{3}$$

For a fixed q, we may calculate the ratio of the temperatures t_+' and t_-' for which $\chi_+(q,t_+')=\chi_-(q,t_-')$. For small q, the leading term in an expansion of t_+'/t_-' will be the constant $(C_+/C_-)^{1/\gamma}$. The expansion to first order in $(\gamma'-\gamma)/\gamma$ and $(q/\kappa)^2$ is easily derived from (1)–(3):

$$\frac{t_+'}{t_-'}\simeq\left(\frac{C_+}{C_-}\right)^{1/\gamma}\left[1+\frac{\gamma-\gamma'}{\gamma}\ln t_-'-\frac{b_+}{\gamma}\left(\frac{q}{\kappa_+}\right)^2\epsilon\right], \tag{4}$$

with

$$\epsilon=1-b_-\kappa_+^2/b_+\kappa_-^2\simeq1-b_-F_+^2C_+/b_+F_-^2C_-. \tag{5}$$

We shall discuss the experimental data in terms of Eqs. (4) and (5).

In Fig. 1 the intensity of critical scattering in a temperature scan at fixed $q=0.04$ Å^{-1} is shown. The high-temperature data extending up to $t_+=0.035$ are not shown. Similar measurements were made at $q=0.02$ Å^{-1} and $q=0.06$ Å^{-1}. The intensities were corrected for Bragg scattering in the following way: The ratio $R(q)$ between Bragg scattering at wave vector q and $q=0$ was measured far below T_c, where the critical scattering is negligible. $R(q)$ was found to be independent of temperature in the interval $0.07<t_-<0.23$. Assuming that $R(q)$ remains constant up to the critical temperature, the Bragg scattering correction at a certain temperature is $R(q)$ times the Bragg peak

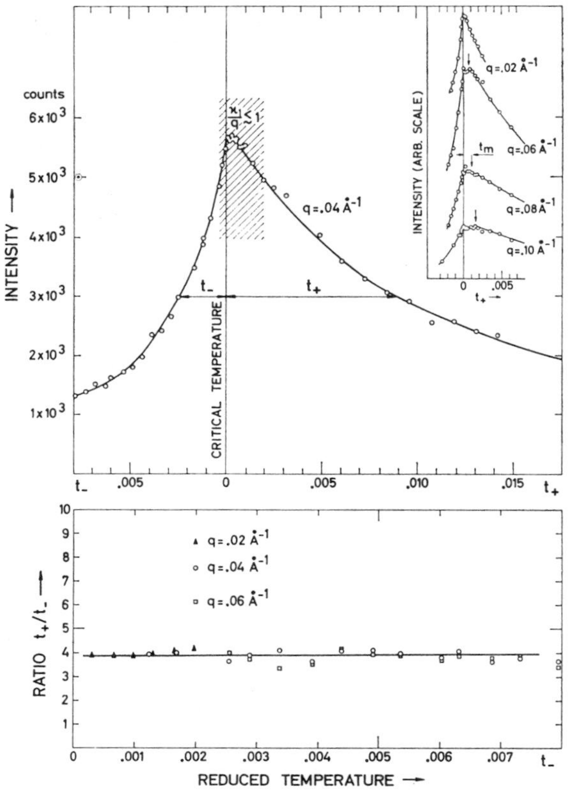

FIG. 1. Critical scattering intensities at fixed $q=0.04$ Å^{-1}. The ratio of equi-intensity temperatures outside the shaded region is constant, independent of q (lower part). Insert: The flat maximum above T_c is located near t_m, given theoretically by $\kappa^2(t_m)/q^2 \simeq 0.023$.

intensity at that temperature. At $q=0.02$ Å^{-1}, one obtains $R(q)=0.18\%$, implying a 10% correction of the critical scattering at $t_-=0.002$. As the correction increased very rapidly for q below 0.02 Å^{-1}, it was not possible to determine $\chi_-(q,t)$ at smaller wave vectors.

We now consider the temperatures t_+ and t_- at which the scattered intensities above and below T_c are equal for a fixed value of $\mathbf{q}=\mathbf{q}_0$. The intensity at a spectrometer setting $\mathbf{q}_0$ is a folding of the cross section and the instrumental resolution. Therefore, it is not possible to conclude immediately that equal intensities correspond to equal cross sections $\chi_\pm(\mathbf{q}_0,t_\pm)$. However, if the intensities at t_- and t_+ are equal within a region in $\mathbf{q}$ space which is considerably greater than the extension of the resolution function, we may conclude unambiguously that $\chi_-(\mathbf{q}_0,t_-)=\chi_+(\mathbf{q}_0,t_+)$. Provided that $q_0<\kappa_+(t_+)$, the experimental data satisfy this condition and the ratio t_+/t_- is shown for various values of q_0 in the lower part of Fig. 1. Under these circumstances $t_+/t_-=t_+'/t_-'$ and, as the ratio is independent of q, the last term of Eq. (4) must be negligible. Since $(q/\kappa_+)^2$ varies from about 1 to less than 0.01, this implies that ϵ must be of the order of 0.05 or less, i.e., $b_-\simeq(F_-/F_+)^2/(C_+/C_-)$. In the mean-field approximation $F_-/F_+=\sqrt{2}$ and $C_+/C_-=2$. For the two-dimensional Ising magnet $F_-/F_+=2$ and $C_+/C_-=37$. For the three-dimensional Ising magnet $C_+/C_-=5.1$,[4] and we estimate that $F_+/F_-\simeq1.6$–1.8. This estimate and the experimental result that $\epsilon\simeq0$ indicate that $b_-\simeq0.4$–0.7, i.e., there is probably a rather large difference between the true and the effective correlation ranges below T_c.

The difference between γ and γ' can now be found from the variation of t_+/t_- with t_-. With a conservative estimate of systematic errors, we conclude from the data in the lower part in Fig. 1 that

$$(\gamma-\gamma')/\gamma=0.01\pm0.02.$$

This result is significant in providing the first accurate experimental evidence for that aspect of the scaling hypothesis which implies $\gamma=\gamma'$. A weighted average of t_+/t_- yields

$$C_+/C_-=5.46\pm0.05.$$

This result can be compared to the parametric representation theory of the equation of state by Schofield[5] which implies that $C_+/C_-=[(1-2\beta)^\gamma/2(\gamma-1)\beta]^{\gamma-1}\gamma/\beta$. Here β is the critical exponent for the magnetization near T_c. Inserting the experimental values for β and γ [1] gives $C_+/C_-=5.5$, whereas the theoretical values for β and γ give $C_+/C_-=5.3$. The value for C_+/C_- has also been determined from the series expansions of Essam and Hunter[4]; they find $C_+/C_-=5.1$ for the bcc Ising magnet.

The data for small values of t_+ display a flat maximum at $t_+=t_m$, and t_m increases with increasing q, as seen from the insert of Fig. 1. The occurrence of a flat maximum above T_c was anticipated by Fisher and Burford.[2] They predicted that t_m is determined by $\kappa_+^2(t_+=t_m)/q^2=0.023$ for the bcc Ising model. The arrows in Fig. 1 indicate the theoretical values of t_m and the agreement with the experimental data is satisfactory.

It may be noted that a temperature shift of the maximum scattered intensity has been observed in Fe, Ni, and Co.[6–8] However, as pointed out by Kocinski and Mrygon,[9] an interpretation of these temperature shifts in terms of the theory of Fisher and Burford seems unjustified, since the observed shift in Fe is zero for $q\lesssim0.10$ Å^{-1} and then increases as q^4. Furthermore, Stump and Springer (private communication) have found that the temperature shift in Ni depends strongly on the perfection of the crystal. We

[4] J. W. Essam and D. L. Hunter, Proc. Phys. Soc. (London) [J. Phys. **C1**, 392 (1968)].

[5] P. Schofield, Phys. Rev. Letters **22**, 606 (1969).
[6] K. Blinowski and R. Ciszewski, Phys. Letters **28A**, 389 (1968).
[7] N. Stump and G. Maier, Phys. Letters **12**, 625 (1967).
[8] D. Bally, B. Grabcev, M. Popovic, M. Totia, and A. A. Lungu, J. Appl. Phys. **39**, 459 (1968).
[9] J. Kocinski and B. Mrygon, Phys. Letters **28A**, 386 (1968).

believe, therefore, that the observed temperature shift in β-brass is the first relevant experimental evidence for the predictions of Fisher and Burford.

I am grateful to M. E. Fisher and A. R. Mackintosh for discussions on the interpretation of the reported experimental results, and to L. Passell of Brookhaven National Laboratory for help in procuring the separated Cu^{65} used in the crystal.

EXPERIMENTAL ARTICLES 100

XVIII. The Bakerian Lecture.—*On the Continuity of the Gaseous and Liquid States of Matter.* By Thomas Andrews, *M.D., F.R.S., Vice-President of Queen's College, Belfast.*

Received June 14,—Read June 17, 1869.

In 1822 M. Cagniard de la Tour observed that certain liquids, such as ether, alcohol, and water, when heated in hermetically sealed glass tubes, became apparently reduced to vapour in a space from twice to four times the original volume of the liquid. He also made a few numerical determinations of the pressures exerted in these experiments*. In the following year Faraday succeeded in liquefying, by the aid of pressure alone, chlorine and several other bodies known before only in the gaseous form†. A few years later Thilorier obtained solid carbonic acid, and observed that the coefficient of expansion of the liquid for heat is greater than that of any aëriform body‡. A second memoir by Faraday, published in 1826, greatly extended our knowledge of the effects of cold and pressure on gases§. Regnault has examined with care the absolute change of volume in a few gases when exposed to a pressure of twenty atmospheres, and Pouillet has made some observations on the same subject. The experiments of Natterer have carried this inquiry to the enormous pressure of 2790 atmospheres; and although his method is not altogether free from objection, the results he obtained are valuable and deserve more attention than they have hitherto received‖.

In 1861 a brief notice appeared of some of my early experiments in this direction. Oxygen, hydrogen, nitrogen, carbonic oxide, and nitric oxide were submitted to greater pressures than had previously been attained in glass tubes, and while under these pressures they were exposed to the cold of the carbonic acid and ether-bath. None of the gases exhibited any appearance of liquefaction, although reduced to less than $\frac{1}{500}$ of their ordinary volume by the combined action of cold and pressure¶. In the third edition of Miller's 'Chemical Physics,' published in 1863, a short account, derived from a private letter addressed by me to Dr. Miller, appeared of some new results I had obtained, under certain fixed conditions of pressure and temperature, with carbonic acid. As these results constitute the foundation of the present investigation and have never been published in a separate form, I may perhaps be permitted to make the following extract from my original communication to Dr. Miller. "On partially liquefying carbonic acid by pressure alone, and gradually raising at the same time the temperature to

* Annales de Chimie, 2ème série, xxi. pp. 127 and 178; also xxii. p. 410.

† Philosophical Transactions for 1823, pp. 160–189.

‡ Annales de Chimie, 2ème série, lx. pp. 427, 432.

§ Philosophical Transactions for 1845, p. 155.

‖ Poggendorff's 'Annalen,' xciv. p. 436.

¶ Report of the British Association for 1861. Transactions of Sections, p. 76.

88° Fahr., the surface of demarcation between the liquid and gas became fainter, lost its curvature, and at last disappeared. The space was then occupied by a homogeneous fluid, which exhibited, when the pressure was suddenly diminished or the temperature slightly lowered, a peculiar appearance of moving or flickering striæ throughout its entire mass. At temperatures above 88° no apparent liquefaction of carbonic acid, or separation into two distinct forms of matter, could be effected, even when a pressure of 300 or 400 atmospheres was applied. Nitrous oxide gave analogous results"*.

The apparatus employed in this investigation is represented in Plate LXIII. It is shown in the simple form in which one gas only is exposed to pressure in figures 1 & 2. In figure 3 a section of the apparatus is given, and in figure 4 another section, with the arrangement for exposing the compressed gas to low degrees of cold *in vacuo*. In figures 5 and 6 a compound form of the same apparatus is represented, by means of which two gases may be simultaneously exposed to the same pressure. The gas to be compressed is introduced into a tube (fa) having a capillary bore from a to b, a diameter of about 2·5 millimetres from b to c, and of 1·25 millimetre from c to f. The gas carefully dried is passed for several hours through the tube open at both ends, as represented below.

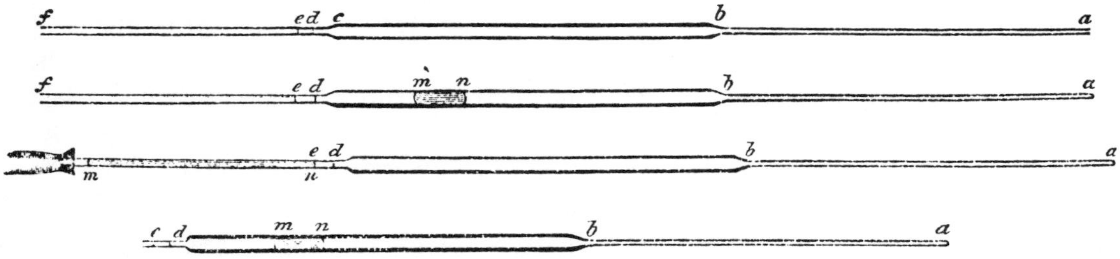

The presence of a column of water of two metres in height was necessary to maintain a moderate stream of gas through the fine capillary tube. In the case of carbonic acid, the gas, after passing through the apparatus, was made to bubble by means of a connecting-tube through mercury, and a portion was collected from time to time, in order to ascertain its purity. The current was continued till the residual air, after the action of caustic potash, was reduced to a constant minimum. In repeated trials I found that in the complicated arrangements I had to adopt, the residual air could not be reduced to less than from $\frac{1}{500}$ to $\frac{1}{1000}$ of the entire volume of the carbonic acid. Even after continuing the current for twenty-four hours this residue appeared; and in discussing some of the results obtained by exposing the gas to high pressures, the presence of this small quantity of air must be carefully taken into account. The capillary end at a was then sealed, and the other end was also closed, and afterwards introduced under a surface of pure mercury contained in a glass capsule. The lower end, while under the surface of the mercury, was opened, and heat applied so as to expel a little of the gas. On cooling contraction occurred, and a short column of mercury entered. The capsule and lower end of the tube were then placed under the receiver of an air-pump, and a partial

* MILLER's 'Chemical Physics.' 3rd edition, p. 328.

vacuum was formed till about one-fourth of the gas was removed. On restoring the pressure, a column of mercury entered and occupied the place of the expelled gas. By withdrawing the end of the tube from below the surface of the mercury in the capsule, and again exhausting cautiously, the column of mercury could be reduced to any required length. The tube, when thus filled, had the form shown (figure, p. 576).

Two file-marks had been made, one at *d*, the other at *e*, in the narrow part of the tube, about 10 millims. distant from each other, and the capacity of the tube from a mark near *a* to *d*, and also from the same mark to *e*, had been determined by filling it with mercury at a known temperature and weighing the mercury. The tube was now placed accurately in a horizontal position and connected by an air-tight junction with one limb of a long U-tube filled with mercury. Each limb of the U-tube was 600 millims. long, and 11 millims. in diameter. By removing mercury from the outer limb of the U-tube, a partial vacuum was obtained, and the column of mercury ($m\,n$) was drawn into the narrow tube ($d\,f$). From the difference of capacity of this part of the tube, the column of mercury was now about four times longer than before. It was easy with a little care so to adjust the pressure that the inner end of the mercurial column coincided with the mark *e*. When this was accomplished, the difference of level of the mercury in the two limbs of the U-tube was accurately read by means of a cathetometer, and the height of the barometer as well as the temperature were carefully noted. Similar observations were made with the gas expanded to the mark *d*. Two independent sets of data were thus obtained for calculating the volume of the gas at 0° C. and 760 millims., and the results usually agreed to less than $\frac{1}{1000}$ part. The tube, after being disconnected with the U-tube, was cut across a little beyond *e*, as shown (figure, p. 576), and was now ready to be introduced into the pressure apparatus.

The capillary tubes were calibrated with great care, and their mean capacity was determined by weighing a column of mercury whose length and position in the tube were accurately observed. One millim. of the air-tube used in these experiments had an average capacity of 0·00002477 cub. centim., and 1 millim. of the carbonic-acid-tube of 0·00003376 cub. centim. A table was constructed showing the corrected capacity of each capillary tube from the sealed end for every millimetre of its length. An allowance of 0·5 millim. was made for the cone formed in sealing the tube.

For the sake of clearness I have described these operations as if they were performed in the detached tube. In actual practice, the tube was in the brass end-piece before it was filled with gas (Plate LXIII. fig. 7).

The construction of the apparatus employed in these experiments will be readily understood from figures 3 and 4, Plate LXIII., which exhibit a section of the simple form. Two massive brass flanges are firmly attached round the ends of a cold-drawn copper tube of great strength, and by means of these flanges two brass end-pieces can be securely bolted to the ends of the copper tube, and the connexions made air-tight by the insertion of leather washers. The lower end-piece (fig. 7) carries a steel screw, 180 millims. long, 4 millims. in diameter, and with an interval of 0·5 millim. between each thread. The screw is packed with care, and readily holds a pressure of 400 atmospheres or more. A similar end-piece attached to the upper flange carries the glass tube containing the gas to be compressed (fig. 7). The apparatus, before being screwed up, is filled with water, and the pressure is obtained by screwing the steel screw into the water*.

In the compound apparatus (figs. 5 & 6) the internal arrangements are the same as in the simple form. A communication is established between the two sides of the apparatus through *a b*. It is indifferent which of the steel screws below is turned, as the pressure is immediately diffused through the interior of both copper tubes, and is applied through the moveable columns of mercury to the two gases to be compressed. Two screws are employed for the purpose of giving a greater command of pressure. In fig. 5 the apparatus is represented without any accessories. In fig. 6 the same apparatus is shown with the arrangements for maintaining the capillary tubes and the body of the apparatus itself at fixed temperatures. A rectangular brass case, closed before and behind with plate glass, surrounds each capillary tube, and allows it to be maintained at any required temperature by the flow of a stream of water. In the figure, the arrangement for obtaining a current of heated water in the case of the carbonic-acid tube is shown. The body of the apparatus itself, as is shown in the figure, is enclosed in an external vessel of copper, which is filled with water at the temperature of the apartment. This latter arrrangement is essential when accurate observations are made.

The temperature of the water surrounding the air-tube was made to coincide, as closely as possible, with that of the apartment, while the temperature of the water surrounding the carbonic-acid tube varied in different experiments from 13° C. to 48° C. In the experiments to be described in this communication, the mercury did not come into view in the capillary part of the air-tube till the pressure amounted to about forty atmospheres. The volumes of the air and of the carbonic acid were carefully read by a cathetometer, and the results could be relied on with certainty to less than 0·05 millim. The temperature of the water around the carbonic-acid tube was ascertained by a thermometer carefully graduated by myself according to an arbitrary scale. This thermometer was one of a set of four, which I constructed some years ago, and which all agreed so closely in their indications, that the differences were found to be altogether insignificant when their readings were reduced to degrees.

I have not attempted to deduce the actual pressure from the observed changes in the volume of the air in the air-tube. For this purpose it would be necessary to know with precision the deviations from the law of MARIOTTE exhibited by atmospheric air within the range of pressure employed in these experiments, and also the change of capacity in the capillary tube from internal pressure. In a future communication I hope to have an opportunity of considering this problem, which must be resolved rather by indirect than direct experiments. As regards the deviation of air from MARIOTTE'S law,

* The first apparatus was constructed for me by Mr. J. CUMINE, of Belfast, to whose rare mechanical skill and valuable suggestions I have been greatly indebted in the whole course of this difficult investigation.

it corresponds, according to the experiments of REGNAULT, to an apparent error of a little more than one-fourth of an atmosphere at a pressure of twenty atmospheres, and according to those of NATTERER, to an approximate error of one atmosphere when the pressure attains 107 atmospheres. These data are manifestly insufficient, and I have therefore not attempted to deduce the true pressure from the observed change of volume in the air-tube. It will be easy to apply hereafter the corrections for true pressure when they are ascertained, and for the purposes of this paper they are not required. The general form of the curves representing the changes of volume in carbonic acid will hardly undergo any sensible change from the irregularities in the air-tube; nor will any of the general conclusions at which I have arrived be affected by them. It must, however, always be understood that, when the pressures are occasionally spoken of, as indicated by the apparent contraction of the air in the air-gauge, the approximate pressures only are meant.

To obtain the capacity in cubic centimetres from the weight in grammes of the mercury which filled any part of a glass tube, the following formula was used,

$$C=W\,.\,\frac{1+0\cdot000154\,.\,t}{13\cdot596}\,.\,1\cdot00012,$$

where C is the capacity in cubic centims., W the weight of the mercury which filled the tube at the temperature t, 0·000154 the coefficient of apparent expansion of mercury in glass, 13·596 the density of mercury at 0°, and 1·00012 the density of water at 4°.

The volume of the gas V, at 0° and 760 millims. of pressure, was deduced from the double observations, as follows,

$$V=C\,.\,\frac{1}{1+\alpha t}\,.\,\frac{h-d}{760},$$

where C is the capacity of the tube (figure, page 576) from a to d, or from a to e, t the temperature, α the coefficient of expansion of the gas for heat (0·00366 in the case of air, 0·0037 in that of carbonic acid), h the height of the barometer reduced to 0° and to the latitude of 45°, d the difference of the mercurial columns in the U-tube similarly reduced.

Having thus ascertained the volumes of the air and of the carbonic acid before compression, at 0° and 760 millims., it was easy to calculate their volumes, under the same pressure of 760 millims., at the temperatures at which the measurements were made when the gases were compressed, and thence to deduce the values of the fractions representing the diminution of volume. But the fractions thus obtained would not give results directly comparable for air and carbonic acid. Although the capillary glass tubes in the apparatus (fig. 6) communicated with the same reservoir, the pressure on the contained gases was not quite equal, in consequence of the mercurial columns, which confined the air and carbonic acid, being of different heights. The column always stood higher in the carbonic-acid-tube than in the air-tube, so that the pressure in the latter was a little greater than in the former. The difference in the lengths of the mercurial columns rarely exceeded 200 millims., or about one-fourth of an atmosphere. This correction was always applied, as was also a trifling correction of 7 millims. for a difference of capillary depression in the two tubes.

In order to show more clearly the methods of reduction, I will give the details of one experiment.

Volume of air at 0° and 760 millims. calculated from the observations when the air was expanded to *a e*, 0·3124 cub. centim.

Volume of same air calculated from the observations when the air was expanded to *a d*, 0·3122 cub. centim.

Mean volume of air at 0° and 760 millims., 0·3123 cub. centim.

The volumes of the carbonic acid, deduced in like manner from two independent observations, were 0·3096 cub. centim. and 0·3094 cub. centim. Mean 0·3095 cub. centim.

The length of the column of air after compression, at 10°·76, in the capillary air-tube was 272·9 millims., corresponding to 0·006757 cub. centim. Hence we have

$$\delta' = \frac{0{\cdot}006757}{0{\cdot}3123 \times 1{\cdot}0394} = \frac{1}{48{\cdot}04}.$$

But as the difference in the heights of the mercurial columns in the air-tube and carbonic-acid-tube, after allowing for the difference of capillary depression, was 178 millims., this result requires a further correction ($\frac{178}{760}$ of an atmosphere), in order to render it comparable with the compression in the carbonic-acid-tube. The final value for δ, the fraction representing the ratio of the volume of the compressed air at the temperature of the experiment to its volume at the same temperature and under the pressure of one atmosphere, will be

$$\delta = \frac{1}{47{\cdot}81}.$$

The corresponding length of the carbonic acid at 13°·22, in its capillary tube, was 124·6 millims., equivalent to 0·004211 cub. centim., from which we deduce for the corresponding fraction for the carbonic acid

$$\varepsilon = \frac{0{\cdot}004211}{0{\cdot}3095 \times 1{\cdot}0489} = \frac{1}{77{\cdot}09}.$$

Hence it follows that the same pressure, which reduced a given volume of air at 10°·76 to $\frac{1}{47{\cdot}81}$ of its volume at the same temperature under one atmosphere, reduced carbonic acid at 13°·22 to $\frac{1}{77{\cdot}09}$ of its volume at the temperature of 13°·22, and under a pressure of one atmosphere. Or assuming the compression of the air to be approximately a measure of the pressure, we may state that under a pressure of about 47·8 atmospheres carbonic acid at 13°·22 contracts to $\frac{1}{77{\cdot}09}$ of its volume under one atmosphere.

In the following Tables, δ is the fraction representing the ratio of the volumes of the air after and before compression to one another, ε the corresponding fraction for the carbonic acid, t and t' the temperatures of the air and carbonic acid respectively, l the number of volumes which 17,000 volumes of carbonic acid, measured at 0° and 760 millims., would occupy at the temperature at which the observation was made under the pressure indicated by the air in the air-tube. The values of l are the ordinates of the curve lines shown in the figure, page 583*.

TABLE I.—Carbonic Acid at 13°·1.

δ.	t.	ϵ.	t'.	l.
$\frac{1}{47·50}$	10°·75	$\frac{1}{76·16}$	13°·18	234·1
$\frac{1}{48·76}$	10·86	$\frac{1}{80·43}$	13·18	221·7
$\frac{1}{48·89}$	10·86	$\frac{1}{80·96}$	13·09	220·3
$\frac{1}{49·00}$	10·86	$\frac{1}{105·9}$	13·09	168·2
$\frac{1}{49·08}$	10·86	$\frac{1}{142·0}$	13·09	125·5
$\frac{1}{49·15}$	10·86	$\frac{1}{192·3}$	13·09	92·7
$\frac{1}{49·28}$	10·86	$\frac{1}{268·8}$	13·09	66·3
$\frac{1}{49·45}$	10·86	$\frac{1}{342·8}$	13·09	52·0
$\frac{1}{49·63}$	10·86	$\frac{1}{384·9}$	13·09	46·3
$\frac{1}{50·15}$	10·86	$\frac{1}{462·9}$	13·09	38·5
$\frac{1}{50·38}$	10·86	$\frac{1}{471·5}$	13·09	37·8
$\frac{1}{54·56}$	10·86	$\frac{1}{480·4}$	13·09	37·1
$\frac{1}{75·61}$	10·86	$\frac{1}{500·7}$	13·09	35·6
$\frac{1}{90·43}$	10·86	$\frac{1}{510·7}$	13·09	34·9

It will be observed that at the pressure of 48·89 atmospheres, as measured by the contraction of the air in the air-tube, liquefaction began. This point could not be fixed by direct observation, inasmuch as the smallest visible quantity of liquid represented a column of gas at least 2 or 3 millims. in length. It was, however, determined indirectly by observing the volume of the gas 0°·2 or 0°·3 above the point of liquefaction, and calculating the contraction the gas would sustain in cooling down to the temperature at which liquefaction began. A slight increase of pressure, it will be seen, was required even in the early stages to carry on the process. Thus the air-guage, after all reductions

* As l is the entire volume to which the carbonic acid is reduced, it does not always refer to homogeneous matter, but sometimes to a mixture of gas and liquid. Its value in the example given in the text is obtained as follows:—

$$l=17000 . \frac{0·004211}{0·3095}=231·3.$$

When l is homogeneous, $\frac{1}{\epsilon}$ represents the density of the carbonic acid referred to carbonic acid gas, at the temperature t', and under a pressure of one atmosphere.

MDCCCLXIX.

were made, indicated an increase of pressure of about one-fourth of an atmosphere (from 48·89 to 49·15 atmospheres) during the condensation of the first and second thirds of the carbonic acid. According to theory no change of volume ought to have occurred. This apparent anomaly is explained by the presence of the trace of air (about $\frac{1}{500}$ part) in the carbonic acid to which I before referred. It is easy to see that the increase of pressure shown in these experiments is explained by the presence of this small quantity of air. If a given volume of carbonic acid contain $\frac{1}{500}$ of air, that air will be diffused through a space 500 times greater than if the same quantity of air were in a separate state. Compress the mixture till 50 atmospheres of pressure have been applied, and the air will now occupy, or be diffused through, ten times the space it would occupy if alone and under the pressure of one atmosphere; or it will be diffused through the space it would occupy, if alone and under the pressure of $\frac{1}{10}$ of an atmosphere. While the carbonic acid is liquefying, pressure must be applied in order to condense this air; and to reduce it to one-half its volume, an increase of $\frac{1}{10}$ of an atmosphere is required. The actual results obtained by experiment approximate to this calculation. From similar considerations, it follows that if a mixture of air and carbonic acid be taken, for example in equal volumes, the pressure, after liquefaction has begun, must be augmented by several atmospheres, in order to liquefy the whole of the carbonic acid. Direct experiments have shown this conclusion to be true.

The small quantity of air in the carbonic acid disturbed the liquefaction in a marked manner, when nearly the whole of the carbonic acid was liquefied, and when its volume relatively to that of the uncondensed carbonic acid was considerable. It resisted for some time absorption by the liquid, but on raising the pressure to 50·4 atmospheres it was entirely absorbed. If the carbonic acid had been absolutely pure, the part of the curve for 13°·1 (figure, page 583) representing the fall from the gaseous to the liquid state, would doubtless have been straight throughout its entire course, and parallel to the lines of equal pressure.

TABLE II.—Carbonic Acid at 21°·5.

δ.	t.	ϵ.	t'.	l.
$\frac{1}{46·70}$	8°·63	$\frac{1}{67·26}$	21°·46	272·9
$\frac{1}{60·05}$	8·70	$\frac{1}{114·7}$	21·46	160·0
$\frac{1}{60·29}$	8·70	$\frac{1}{174·8}$	21·46	105·0
$\frac{1}{60·55}$	8·70	$\frac{1}{240·5}$	21·46	76·3
$\frac{1}{61·00}$	8·70	$\frac{1}{367·7}$	21·46	49·9
$\frac{1}{62·21}$	8·70	$\frac{1}{440·0}$	21·46	41·7
$\frac{1}{62·50}$	8·70	$\frac{1}{443·3}$	21·46	41·4

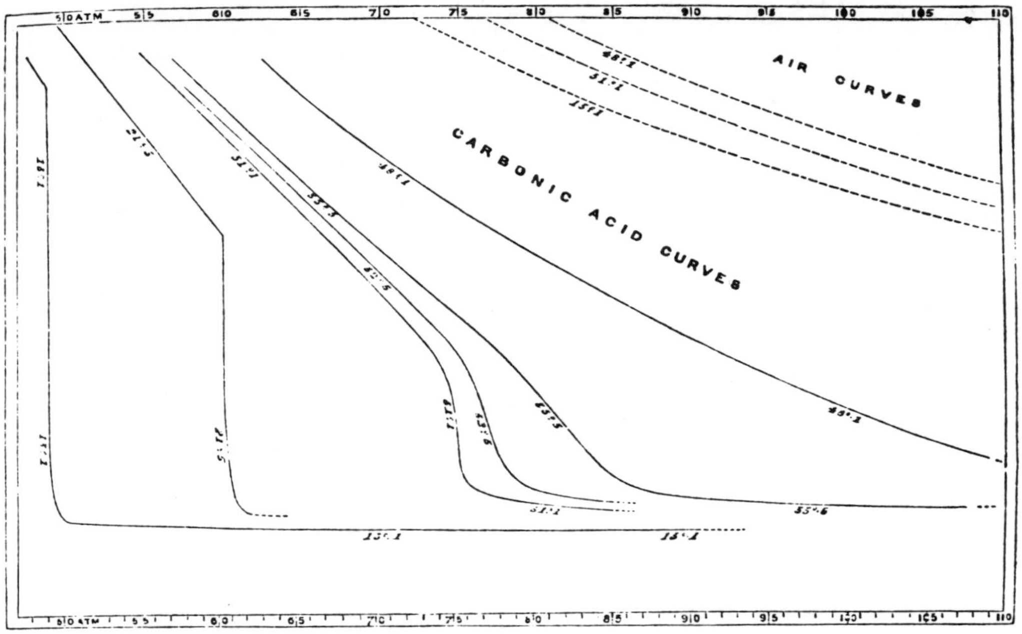

The curve representing the results at 21°·5 agrees in general form with that for 13°·1, as shown in the above figure. At 13°·1, under a pressure of about 49 atmospheres, the volume of carbonic acid is little more than three-fifths of that which a perfect gas would occupy under the same conditions. After liquefaction carbonic acid yields to pressure much more than ordinary liquids; and the compressibility appears to diminish as the pressure increases. The high rate of expansion by heat of liquid carbonic acid, first noticed by Thilorier, is fully confirmed by this investigation.

The next series of experiments was made at the temperature of 31°·1, or 0°·2 above the point at which, by compression alone, carbonic acid is capable of assuming visibly the liquid form. Since I first announced this fact in 1863, I have made careful experiments to fix precisely the temperature of this critical point in the case of carbonic acid. It was found in three trials to be 30°·92 C., or 87°·7 Fahr. Although for a few degrees above this temperature a rapid fall takes place from increase of pressure, when the gas is reduced to the volume at which it might be expected to liquefy, no separation of the carbonic acid into two distinct conditions of matter occurs, so far as any indication of such a separation is afforded by the action of light. By varying the pressure or temperature, but always keeping the latter above 30°·92, the great changes of density which occur about this point produce the flickering movements I formerly described, resembling in an exaggerated form the appearances exhibited during the mixture of liquids of different densities, or when columns of heated air ascend through colder strata. It is easy so to adjust the pressure that one-half of the tube shall be filled with uncondensed gas and one-half with the condensed liquid. Below the critical temperature this distinction is easily seen to have taken place, from the visible surface of demarcation between the liquid and gas, and from the shifting at the same surface of the image of any perpendicular line placed behind the tube. But above 30°·92 no such appearances are seen, and the most careful examination fails to discover any heterogeneity in the carbonic acid, as it exists in the tube.

TABLE III.—Carbonic Acid at 31°·1.

δ.	t.	ϵ.	t'.	l.
$\frac{1}{54·79}$	11°·59	$\frac{1}{80·55}$	31°·17	235·4
$\frac{1}{55·96}$	11·59	$\frac{1}{83·39}$	31·22	227·4
$\frac{1}{57·18}$	11·58	$\frac{1}{86·58}$	31·15	219·0
$\frac{1}{58·46}$	11·55	$\frac{1}{90·04}$	31·19	210·6
$\frac{1}{59·77}$	11·41	$\frac{1}{93·86}$	31·18	202·0
$\frac{1}{61·18}$	11·40	$\frac{1}{98·07}$	31·20	193·3
$\frac{1}{62·67}$	11·44	$\frac{1}{103·1}$	31·19	183·9
$\frac{1}{64·27}$	11·76	$\frac{1}{109·6}$	31·13	173·0
$\frac{1}{65·90}$	11·73	$\frac{1}{116·2}$	31·19	163·2
$\frac{1}{67·60}$	11·63	$\frac{1}{124·4}$	31·15	152·4
$\frac{1}{69·39}$	11·55	$\frac{1}{134·5}$	31·03	140·9
$\frac{1}{71·25}$	11·40	$\frac{1}{147·8}$	31·06	128·2
$\frac{1}{73·26}$	11·45	$\frac{1}{169·0}$	31·09	112·2
$\frac{1}{73·83}$	13·00	$\frac{1}{174·4}$	31·08	108·7
$\frac{1}{75·40}$	11·62	$\frac{1}{311·1}$	31·06	60·9
$\frac{1}{77·64}$	11·65	$\frac{1}{369·1}$	31·06	51·3
$\frac{1}{79·92}$	11·16	$\frac{1}{383·0}$	31·10	49·4
$\frac{1}{82·44}$	11·23	$\frac{1}{395·7}$	31·07	47·9
$\frac{1}{85·19}$	11·45	$\frac{1}{405·5}$	31·05	46·7

The graphical representation of these experiments, as shown in the preceding page, exhibits some marked differences from the curves for lower temperatures. The dotted lines in the figure represent a portion of the curves of a perfect gas (assumed to have the same volume at 0° and under one atmosphere as the carbonic acid) for the temperatures of 13°·1, 31°·1, and 48°·1. The volume of the carbonic acid at 31°·1, it will be observed, diminishes with tolerable regularity, but much faster than according to the law of MARIOTTE, till a pressure of about 73 atmospheres is attained. The diminution of volume then goes on very rapidly, a reduction to nearly one-half taking place, when the pressure is increased from 73 to 75 atmospheres, or only by $\frac{1}{37}$ of the whole pressure. The fall is not, however, abrupt as in the case of the formation of the liquid at lower temperatures, but a steady increase of pressure is required to carry it through. During this fall, as has already been stated, there is no indication at any stage of the process of two conditions of matter being present in the tube. Beyond 77 atmospheres carbonic acid at 31°·1 yielded much less than before to pressure, its volume having become reduced nearly to that which it ought to occupy as a liquid at the temperature at which the observations were made.

TABLE IV.—Carbonic Acid at 32°·5.

δ.	t.	ε.	t'.	l.
$\frac{1}{57\cdot38}$	12°·10	$\frac{1}{85\cdot90}$	32°·50	221·7
$\frac{1}{71\cdot52}$	12·15	$\frac{1}{140\cdot3}$	32·34	135·6
$\frac{1}{73\cdot60}$	12·30	$\frac{1}{156\cdot0}$	32·45	122·0
$\frac{1}{74\cdot02}$	12·30	$\frac{1}{159\cdot9}$	32·46	119·1
$\frac{1}{76\cdot25}$	12·40	$\frac{1}{191\cdot7}$	32·38	99·3
$\frac{1}{78\cdot52}$	12·50	$\frac{1}{311\cdot8}$	32·48	61·1
$\frac{1}{79\cdot77}$	12·35	$\frac{1}{351\cdot3}$	32·54	54·2
$\frac{1}{84\cdot90}$	12·35	$\frac{1}{387\cdot8}$	32·75	49·1

TABLE V.—Carbonic Acid at 35°·5.

δ.	t.	ε.	t'.	l.
$\frac{1}{56\cdot80}$	15°·68	$\frac{1}{82\cdot72}$	35°·49	232·5
$\frac{1}{59\cdot34}$	15·70	$\frac{1}{88\cdot94}$	35·54	216·2
$\frac{1}{62\cdot15}$	15·66	$\frac{1}{96\cdot41}$	35·52	199·5
$\frac{1}{65\cdot23}$	15·66	$\frac{1}{106\cdot0}$	35·51	181·4
$\frac{1}{68\cdot66}$	15·75	$\frac{1}{118\cdot4}$	35·47	162·4
$\frac{1}{72\cdot45}$	15·79	$\frac{1}{135\cdot1}$	35·48	142·3
$\frac{1}{76\cdot58}$	15·52	$\frac{1}{161\cdot2}$	35·55	119·3
$\frac{1}{81\cdot28}$	15·61	$\frac{1}{228\cdot0}$	35·55	84·4
$\frac{1}{86\cdot60}$	15·67	$\frac{1}{351\cdot9}$	35·48	54·6
$\frac{1}{89\cdot52}$	15·67	$\frac{1}{373\cdot7}$	35·50	51·5
$\frac{1}{92\cdot64}$	15·64	$\frac{1}{387\cdot9}$	35·61	49·6
$\frac{1}{99\cdot57}$	15·61	$\frac{1}{411\cdot0}$	35·55	46·8
$\frac{1}{107\cdot6}$	15·47	$\frac{1}{430\cdot2}$	35·53	44·7

TABLE VI.—Carbonic Acid at 48°·1.

δ.	t.	ϵ.	t'.	l.
$\frac{1}{62 \cdot 60}$	15°·67	$\frac{1}{86 \cdot 45}$	47·95	231·5
$\frac{1}{68 \cdot 46}$	15·79	$\frac{1}{99 \cdot 39}$	48·05	201·4
$\frac{1}{75 \cdot 58}$	15·87	$\frac{1}{117 \cdot 8}$	48·12	170·0
$\frac{1}{84 \cdot 35}$	15·91	$\frac{1}{140 \cdot 8}$	48·25	136·5
$\frac{1}{95 \cdot 19}$	15·83	$\frac{1}{198 \cdot 5}$	48·13	100·8
$\frac{1}{109 \cdot 4}$	16·23	$\frac{1}{298 \cdot 4}$	48·25	67·2

The curve for 32°·5 (page 583) resembles closely that for 31°·1. The fall is, however, less abrupt than at the latter temperature. The range of pressure in the experiments at 35°·5 extends from 57 to above 107 atmospheres. The fall is here greatly diminished, and it has nearly lost its abrupt character. It is most considerable from 76 to 87 atmospheres, where an increase of one-seventh in the pressure produces a reduction of volume to one-half. At 107 atmospheres the volume of the carbonic acid has come almost into conformity with that which it should occupy, if it were derived directly from liquid carbonic acid, according to the law of the expansion of that body for heat.

The curve for 48°·1 is very interesting. The fall shown in the curves for lower temperatures has almost, if not altogether, disappeared, and the curve itself approximates to that which would represent the change of volume in a perfect gas. At the same time the contraction is much greater than it would have been if the law of MARIOTTE had held good at this temperature. Under a pressure of 109 atmospheres, the carbonic acid is rapidly approaching to the volume it would occupy if derived from the expansion of the liquid; and if the experiment had not been interrupted by the bursting of one of the tubes, it would doubtless have fallen into position at a pressure of 120 or 130 atmospheres.

I have not made any measurements at higher temperatures than 48°·1; but it is clear that, as the temperature rises, the curve would continue to approach to that representing the change of volume of a perfect gas.

I have frequently exposed carbonic acid, without making precise measurements, to much higher pressures than any marked in the Tables, and have made it pass, without break or interruption from what is regarded by every one as the gaseous state, to what is, in like manner, universally regarded as the liquid state. Take, for example, a given volume of carbonic acid gas at 50° C., or at a higher temperature, and expose it to increasing pressure till 150 atmospheres have been reached. In this process its volume will steadily diminish as the pressure augments, and no sudden diminution of volume, without the application of external pressure, will occur at any stage of it. When the full pressure has been applied, let the temperature be allowed to fall till the carbonic acid has reached the ordinary temperature of the atmosphere. During the whole of this operation no breach of continuity has occurred. It begins with a gas, and by a series of gradual changes, presenting nowhere any abrupt alteration of volume or sudden evolution of heat, it ends with a liquid. The closest observation fails to discover anywhere indications of a change of condition in the carbonic acid, or evidence, at any period of the process, of part of it being in one physical state and part in another. That the gas has actually changed into a liquid would, indeed, never have been suspected, had it not shown itself to be so changed by entering into ebullition on the removal of the pressure. For convenience this process has been divided into two stages, the compression of the carbonic acid and its subsequent cooling; but these operations might have been performed simultaneously, if care were taken so to arrange the application of the pressure and the rate of cooling, that the pressure should not be less than 76 atmospheres when the carbonic acid had cooled to 31°.

We are now prepared for the consideration of the following important question. What is the condition of carbonic acid when it passes, at temperatures above 31°, from the gaseous state down to the volume of the liquid, without giving evidence at any part of the process of liquefaction having occurred? Does it continue in the gaseous state, or does it liquefy, or have we to deal with a new condition of matter? If the experiment were made at 100°, or at a higher temperature, when all indications of a fall had disappeared, the probable answer which would be given to this question is that the gas preserves its gaseous condition during the compression; and few would hesitate to declare this statement to be true, if the pressure, as in NATTERER'S experiments, were applied to such gases as hydrogen or nitrogen. On the other hand, when the experiment is made with carbonic acid at temperatures a little above 31°, the great fall which occurs at one period of the process would lead to the conjecture that liquefaction had actually taken place, although optical tests carefully applied failed at any time to discover the presence of a liquid in contact with a gas. But against this view it may be urged with great force, that the fact of additional pressure being always required for a further diminution of volume, is opposed to the known laws which hold in the change of bodies from the gaseous to the liquid state. Besides, the higher the temperature at which the gas is compressed, the less the fall becomes, and at last it disappears.

The answer to the foregoing question, according to what appears to me to be the true interpretation of the experiments already described, is to be found in the close and intimate relations which subsist between the gaseous and liquid states of matter. The ordinary gaseous and ordinary liquid states are, in short, only widely separated forms of the same condition of matter, and may be made to pass into one another by a series of gradations so gentle that the passage shall nowhere present any interruption or breach of continuity. From carbonic acid as a perfect gas to carbonic acid as a perfect liquid, the transition we have seen may be accomplished by a continuous process, and the gas and liquid are only distant stages of a long series of continuous physical changes. Under certain conditions of temperature and pressure, carbonic acid finds itself, it is true, in what may be described as a state of instability, and suddenly passes, with the evolution of heat, and without the application of additional pressure or change of temperature, to the volume, which by the continuous process can only be reached through a long and circuitous route. In the abrupt change which here occurs, a marked difference is exhibited, while the process is going on, in the optical and other physical properties of the carbonic acid which has collapsed into the smaller volume, and of the carbonic acid not yet altered. There is no difficulty here, therefore, in distinguishing between the liquid and the gas. But in other cases the distinction cannot be made; and under many of the conditions I have described it would be vain to attempt to assign carbonic acid to the liquid rather than the gaseous state. Carbonic acid, at the temperature of 35°·5, and under a pressure of 108 atmospheres, is reduced to $\frac{1}{430}$ of the volume it occupied under a pressure of one atmosphere; but if any one ask whether it is now in the gaseous or liquid state, the question does not, I believe, admit of a positive reply. Carbonic acid at 35°·5, and under 108 atmospheres of pressure, stands nearly midway between the gas and the liquid; and we have no valid grounds for assigning it to the one form of matter any more than to the other. The same observation would apply with even greater force to the state in which carbonic acid exists at higher temperatures and under greater pressures than those just mentioned. In the original experiment of Cagniard de la Tour, that distinguished physicist inferred that the liquid had disappeared, and had changed into a gas. A slight modification of the conditions of his experiment would have led him to the opposite conclusion, that what had been before a gas was changed into a liquid. These conditions are, in short, the intermediate states which matter assumes in passing, without sudden change of volume, or abrupt evolution of heat, from the ordinary liquid to the ordinary gaseous state.

In the foregoing observations I have avoided all reference to the molecular forces brought into play in these experiments. The resistance of liquids and gases to external pressure tending to produce a diminution of volume, proves the existence of an internal force of an expansive or resisting character. On the other hand, the sudden diminution of volume, without the application of additional pressure externally, which occurs when a gas is compressed, at any temperature below the critical point, to the volume at which liquefaction begins, can scarcely be explained without assuming that a molecular force of great attractive power comes here into operation, and overcomes the resistance to diminution of volume, which commonly requires the application of external force. When the passage from the gaseous to the liquid state is effected by the continuous process described in the foregoing pages, these molecular forces are so modified as to be unable at any stage of the process to overcome alone the resistance of the fluid to change of volume.

The properties described in this communication, as exhibited by carbonic acid, are not peculiar to it, but are generally true of all bodies which can be obtained as gases and liquids. Nitrous oxide, hydrochloric acid, ammonia, sulphuric ether, and sulphuret of carbon, all exhibited, at fixed pressures and temperatures, critical points, and rapid changes of volume with flickering movements when the temperature or pressure was changed in the neighbourhood of those points. The critical points of some of these bodies were above 100°; and in order to make the observations, it was necessary to bend the capillary tube before the commencement of the experiment, and to heat it in a bath of paraffin or oil of vitriol.

The distinction between a gas and vapour has hitherto been founded on principles which are altogether arbitrary. Ether in the state of gas is called a vapour, while sulphurous acid in the same state is called a gas; yet they are both vapours, the one derived from a liquid boiling at 35°, the other from a liquid boiling at −10°. The distinction is thus determined by the trivial condition of the boiling-point of the liquid, under the ordinary pressure of the atmosphere, being higher or lower than the ordinary temperature of the atmosphere. Such a distinction may have some advantages for practical reference, but it has no scientific value. The critical point of temperature affords a criterion for distinguishing a vapour from a gas, if it be considered important to maintain the distinction at all. Many of the properties of vapours depend on the gas and liquid being present in contact with one another; and this, we have seen, can only occur at temperatures below the critical point. We may accordingly define a vapour to be a gas at any temperature under its critical point. According to this definition, a vapour may, by pressure alone, be changed into a liquid, and may therefore exist in presence of its own liquid; while a gas cannot be liquefied by pressure—that is, so changed by pressure as to become a visible liquid distinguished by a surface of demarcation from the gas. If this definition be accepted, carbonic acid will be a vapour below 31°, a gas above that temperature; ether a vapour below 200°, a gas above that temperature.

We have seen that the gaseous and liquid states are only distant stages of the same condition of matter, and are capable of passing into one another by a process of continuous change. A problem of far greater difficulty yet remains to be solved, the possible continuity of the liquid and solid states of matter. The fine discovery made some years ago by James Thomson, of the influence of pressure on the temperature at which liquefaction occurs, and verified experimentally by Sir W. Thomson, points, as it appears to me, to the direction this inquiry must take; and in the case at least of those bodies which expand in liquefying, and whose melting-points are raised by pressure, the transition may possibly be effected. But this must be a subject for future investigation; and for the present I will not venture to go beyond the conclusion I have already drawn from direct experiment, that the gaseous and liquid forms of matter may be transformed into one another by a series of continuous and unbroken changes.

Appendix.

The following experiments, made at temperatures differing from any of the foregoing series, are added, as they may hereafter be useful for reference.

δ.	t.	ϵ.	t'.
$\frac{1}{48\cdot15}$	12°·42	$\frac{1}{75\cdot00}$	15·76
$\frac{1}{53\cdot04}$	11·13	$\frac{1}{92\cdot53}$	16·45
$\frac{1}{47\cdot45}$	11·50	$\frac{1}{64\cdot14}$	31·91
$\frac{1}{71\cdot75}$	13·10	$\frac{1}{148\cdot5}$	31·65
$\frac{1}{73\cdot88}$	13·20	$\frac{1}{170\cdot5}$	31·71
$\frac{1}{73\cdot92}$	13·20	$\frac{1}{157\cdot9}$	33·15
$\frac{1}{73\cdot77}$	12·74	$\frac{1}{152\cdot3}$	33·58
$\frac{1}{73\cdot89}$	13·14	$\frac{1}{144\cdot5}$	35·00
$\frac{1}{73\cdot89}$	13·21	$\frac{1}{140\cdot0}$	36·03
$\frac{1}{76\cdot05}$	13·27	$\frac{1}{153\cdot4}$	36·05
$\frac{1}{78\cdot35}$	13·38	$\frac{1}{171\cdot1}$	36·11
$\frac{1}{80\cdot74}$	13·40	$\frac{1}{197\cdot8}$	36·22
$\frac{1}{83\cdot31}$	13·45	$\frac{1}{251\cdot4}$	36·20
$\frac{1}{86\cdot01}$	13·50	$\frac{1}{323\cdot6}$	36·08
$\frac{1}{88\cdot92}$	13·53	$\frac{1}{358\cdot1}$	36·18
$\overline{92\cdot06}$	13·55	$\frac{1}{377\cdot8}$	36·22

Fig. 1.

Fig. 2.

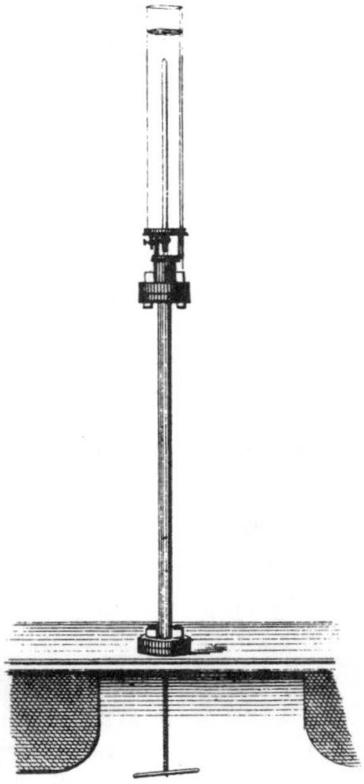

Fig. 3.

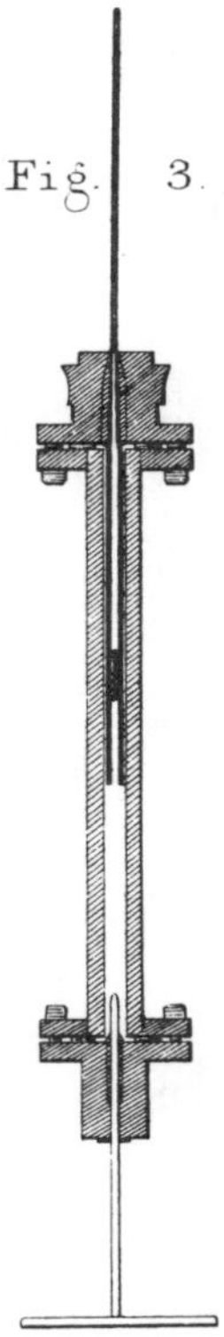

Fig. 4.

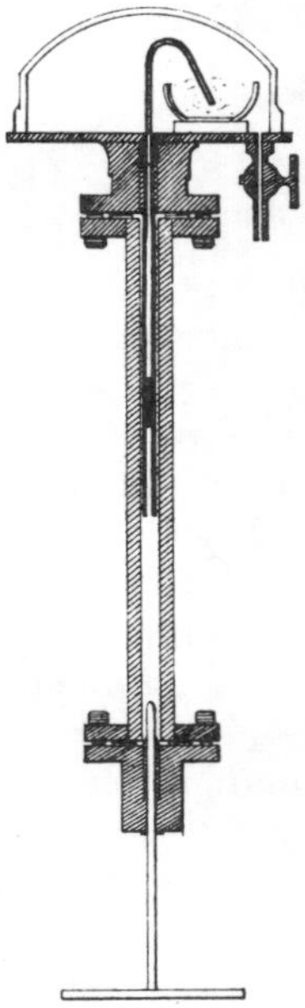

Fig. 6.

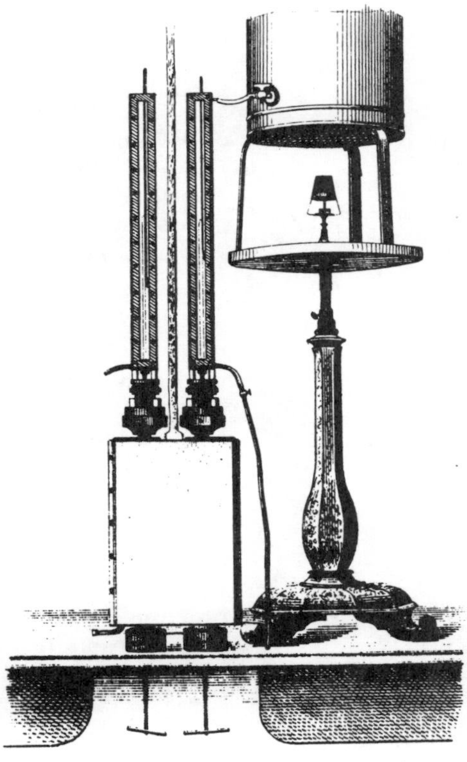

Fig. 5.

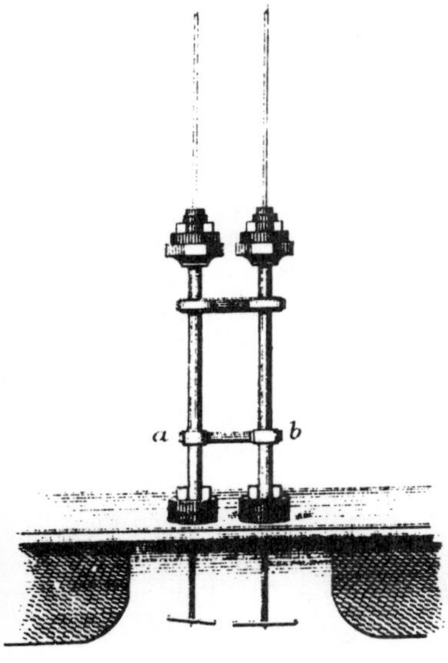

Fig. 7.

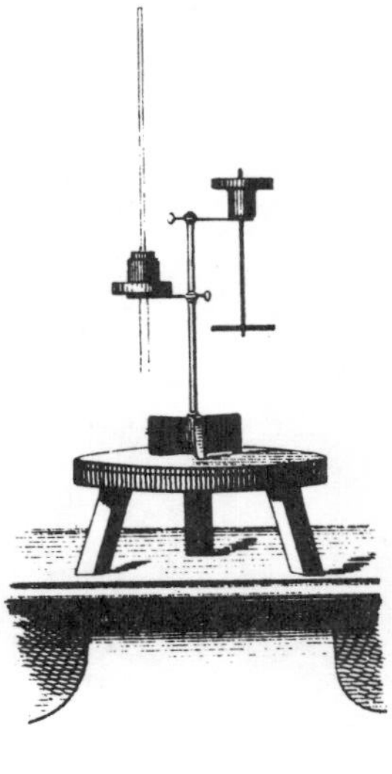

MEASUREMENT OF THE SPECIFIC HEAT C_V OF ARGON IN THE IMMEDIATE VICINITY OF THE CRITICAL POINT

M. I. BAGATSKIĬ, A. V. VORONEL', and V. G. GUSAK

Scientific-Research Institute for Physico-technical and Radiotechnical Measurements

Submitted to JETP editor June 1, 1962

J. Exptl. Theoret. Phys. (U.S.S.R.) **43**, 728-729 (August, 1962)

USING a previously-developed measuring technique,[1,2] we have investigated the temperature dependence of the specific heat of argon in the vicinity of its transformation from a two-phase (liquid-vapor) system into a homogeneous one, at a constant overall volume near the critical value. According to current theory[3,4] a finite discontinuity in the specific heat should be observed at such a transition.

As a result of specific heat measurements at intervals of $\sim 0.05°C$, for a density $\rho = 0.521$ g/cm^3, a curve was obtained (cf. Fig. 1) whose form suggests that the specific heat may rise to an infinite value at the critical point. In Fig. 2 the same data are plotted on a semi-logarithmic scale. The logarithmic temperature difference $\log|T - T_c|$ is plotted along the x axis. On this scale the λ-shaped curve is transformed into a pair of straight lines, of which the higher refers to temperatures $T < T_c$, and the lower, to $T > T_c$. It is evident that the nature of the rise in the specific heat both on the right and the left is identical, since the lines in Fig. 2 are parallel. This fact makes it possible in the present case to retain the concept of a discontinuity in the specific heat at the critical point, determined by the constant distance between the lines in Fig. 2, or by the limit $\lim_{T \to T_c} (C_{het} - C_{hom}) = \Delta C$, where C_{het} and C_{hom} are the specific heats of the material in the heterogeneous and homogeneous regions, respectively. In our experiment, the value of this quantity was found to be $\Delta C = 20$ cal/mole °C.

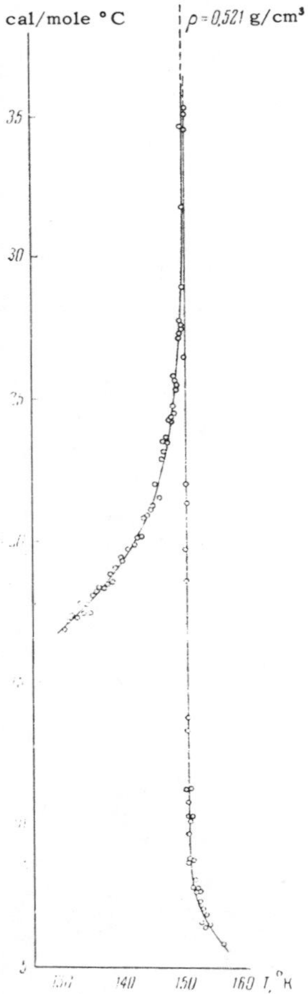

FIG. 1. Variation of C_v of argon with T along an isochore $V \approx V_c$.

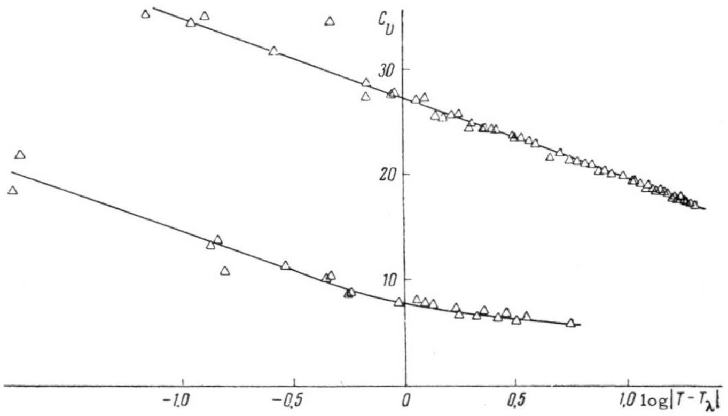

FIG. 2. Variation of C_v of argon with $\log|T - T_c|$. ($T_c = 150.5°K$).

It is thus clear from Fig. 2 that near the critical point the specific heat C_V of argon tends to infinity like $C_V \sim \ln|T - T_c|$. This result does not follow from the present theory, according to which the thermodynamic potential, expressed as a function of V and T, possesses no mathematical singularity at the critical point. The similarity of these results, however, to those of Fairbank et al, who observed a logarithmic singularity in the specific heat of He near the λ-point,[5] is striking. We should note that the point investigated by these authors was not an ordinary λ-point, but rather a critical point for second-order transitions.

The smallest temperature interval over which we were able to measure the specific heat with reasonable accuracy was 0.04–0.05°K. For lesser intervals the errors of measurement were found to exceed 20%.

The purity of the argon sample studied was of the order of 99.99%. In a calorimetric experiment near the critical point, an important factor is the time required for establishment of equilibrium. In our experiments, we allowed 5—6 hours for equilibrium to be reached, continuously stirring the liquid in the calorimeter. Failure to do this leads to non-reproducibility of the data. That we obtained such a clear-cut result is to be attributed to the fact that we carried our measurements closer by an order of magnitude to the critical point than was possible in previous work,[6-8] owing to the advantages offered by our methods; it is also a consequence of the particular attention which we paid to the necessity of waiting for equilibrium to be reached before making the measurements.

From our measurements, the temperature of the transition was found to be $T_c = 150.5°K$, which is close to the critical temperature of argon (150.7°K[9]).

In conclusion, it should be noted that this work required uninterrupted measurements over a period of several days, in which, besides the authors, V. V. Shchekochikhina, T. S. Salikova, Yu. P. Chashkin and others took part. The authors take this opportunity to express their gratitude to these individuals.

[1] P. G. Strelkov, et al., ZhFKh (J. of Phys. Chemistry) **28**, 650 (1954).

[2] A. V. Voronel' and P. G. Strelkov, PTÉ **6**, 111 (1960).

[3] L. D. Landau and E. M. Lifshitz, Statisticheskaya fizika (Statistical Physics) Gostekhizdat, 1951.

[4] I. R. Krivskiĭ and N. E. Khazanova, ZhFKh **29**, 1087 (1955).

[5] Fairbank, Buckingham, and Kellers, Proc. V Internat. Conf. on Low Temp. Phys. and Chem., Univ. of Wisconsin Press, Madison, 1958; p. 50.

[6] A. Michels and J. Strijland, Physica **18**, 613 (1952).

[7] Pall, Broughton, and Maas, Can. J. Res. **B16**, 230 (1938).

[8] Kh. I. Amirkhanov and N. K. Kerimov, Teploénergetika (Heat Power Engineering) No. 9, 68 (1957).

[9] Michels, Levelt, and DeGraaff, Physica **24**, 659 (1958).

Translated by S. D. Elliott
123

DYNAMICS OF CONCENTRATION FLUCTUATIONS IN A BINARY MIXTURE IN THE HYDRODYNAMICAL AND NONHYDRODYNAMICAL REGIMES

P. Bergé, P. Calmettes, C. Laj, M. Tournarie, and B. Volochine
Service de Physique du Solide et de Résonance Magnétique, Centre d'Etudes Nucléaires de Saclay, 91 Gif-sur-Yvette, France
(Received 20 April 1970)

The experimental results obtained by optical-mixing spectroscopy on a critical binary mixture are compared with Kawasaki's theory.

By means of self-beating optical-mixing spectroscopy we have completed a series of detailed experiments on a binary mixture at critical concentration (53% cyclohexane, 47% aniline weight concentration) for different temperatures and scattering angles. Some of the results have already been reported elsewhere.[1-4] In these previous papers it was shown that the spectra of the scattered light are Lorentzian (except maybe in the immediate vicinity of the critical temperature where a slight deviation of the spectra from a Lorentzian shape was observed). It was in any case possible to obtain the half-width of the spectra as a function of $(T-T_c)/T_c$ and of the scattering angle θ.

Our results were interpreted in terms of three distinct regions according to the value of ξK, where K is the scattering vector defined by $|\vec{K}| = (4\pi/\lambda_0)n \sin\frac{1}{2}\theta$, and where ξ is the long-range correlation length which is defined by the asymptotic form $(r\to\infty)$ of the modified Ornstein-Zernike formula,

$$g(r) = e^{-r/\xi}/r^{1-\eta/2}.$$

The three distinct regions are defined as follows:

(1) $\xi K \ll 1$ (hydrodynamical region). In this region our results agree with the Landau-Placzek equation $\Gamma = DK^2$ and we obtain

$$D \sim [(T-T_c)/T_c]^{0.61\pm0.07},$$

D being the mass diffusion coefficient.

(2) ξK small but not negligible compared with unity. In this region, which we may call the nonlocal hydrodynamical region, our results agree with the Fixman equation,

$$\Gamma = DK^2(1+K^2\xi'^2)^{1-\eta/2},$$

with $\eta \simeq 0$; and by fitting our results to it, we obtain

$$\xi'^2 = \xi_0'^2[(T-T_c)/T_c]^{-1.21\pm0.05},$$

where $\xi_0' = 1.65 \pm 0.07$ Å. (The prime symbol is used to distinguish Fixman's from Kawasaki's definitions.)

(3) Finally, $\xi K \geq 1$ (critical region). We have shown experimentally that the half-width of our spectra is well represented by the formula $\Gamma = AK^3$, where A is a coefficient whose value in the critical region was found to be $A = 1.52\times10^{-13}$ $\text{cm}^3\ \text{sec}^{-1}$.[5] This result was in agreement with the theoretical predictions of Halperin and Hohenberg.[6,7]

Quite recently Kawasaki[8,9] has developed a theory according to which the spectral half-width Γ is given by

$$\Gamma = AK^3F(1/K\xi),$$

where $A = \frac{3}{8}\pi D\xi = k_BT/16\eta^*$ (η^* being the high-frequency part of the viscosity) and $F(1/\xi K)$ is an analytical function of the form

$$F(x) = 2\pi^{-1}[x + x^3 + (1-x^4)\arctan(x^{-1})].$$

In the hydrodynamical region $K\xi \ll 1$, so that

$$F(x) = 8x/3\pi,$$

and therefore

$$\Gamma = (8A/3\pi\xi)K^2 \equiv DK^2.$$

In the nonlocal hydrodynamical region $K\xi < 1$, and one obtains

$$\Gamma = DK^2(1 + \tfrac{3}{5}K^2\xi^2).$$

This expression differs from Fixman's equation because of the factor $\frac{3}{5}$. Finally, in the critical region $\xi K \gtrsim 1$, $F(x) \to 1$ and one gets

$$\Gamma = AK^3.$$

Thus Kawasaki's theory accounts continuously for the three otherwise distinct regions. It should be noted however that Kawasaki's result is only a first-order one. Furthermore, the critical exponent η is taken equal to zero and the high-frequency part η^* of the viscosity is supposed temperature independent.

We have tried to fit all our results, obtained on the cyclohexane-aniline mixture under different temperatures and scattering angles, to Kawasaki's expression for Γ. This fit was made using a statistical refining program developed by Tournarie.[10] This program gives a plot of Γ/AK^3 as a function of $[K\xi_0(T/T_c-1)^{-\nu}]^{-1}$, where Γ, K, and $(T-T_c)/T_c$ were known experimentally and A, ξ_0, and ν were determined by the computer in order to get the best fit to Kawasaki's function $F(1/\xi K)$.

Figure 1 shows the results obtained. The 262 experimental points reported on this figure were obtained using different scattering cells and apparatus and under different and quite various conditions. The solid line represents the best fit to these points of a function having Kawasaki's form. This best-fit curve is obtained for the following values of A, ξ_0, and ν:

$$A = (1.51 \pm 0.02) \times 10^{-13} \text{ cm}^3 \text{ sec}^{-1},$$

$$\xi_0 = (2.11 \pm 0.08) \times 10^{-8} \text{ cm},$$

$$\nu = 0.588 \pm 0.06.$$

Thus, except perhaps for ν, this set of values is in agreement with the values obtained by considering three distinct regions which are, respectively,

$$A = (1.52 \pm 0.03) \times 10^{-13} \text{ cm}^3 \text{ sec}^{-1},$$

$$\xi_0' = (1.65 \pm 0.07) \times 10^{-8} \text{ cm},$$

$$\nu = 0.61 \pm 0.07.$$

There is no doubt that Kawasaki's expression for Γ represents fairly well the overall results of our experiments and accounts, with a single analytical expression, for the otherwise distinct regions. However, one might wonder if a fit of the experimental points by this expression is the best way to attain the exact value of the critical parameters and particularly of ν. Indeed the computer program used in the fit gives, together with the results, a "quality factor," defined as the contribution of the statistical fluctuation to the total error.[10] The "quality factor" obtained in this fit is only 0.601, which indicates that there is a slight distortion in the fit. (As a comparison, the fit of a Lorentzian spectrum usually obtained has a quality factor $\geqslant 0.95$, and unity represents an undistorted fit.) This indicates that different portions of the curve do not represent equally well the experimental points.

One may try to localize which portion of the curve (if any) is mostly responsible for this low

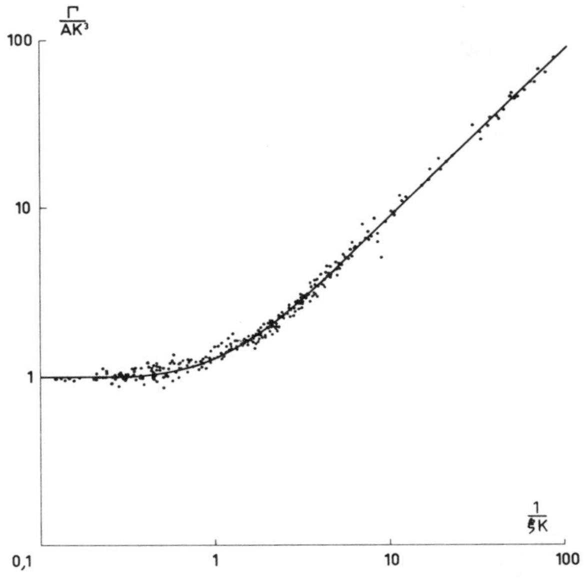

FIG. 1. Plot of Γ/AK^3 vs $1/\xi K$. The solid line represents the best fit of a function having Kawasaki's form.

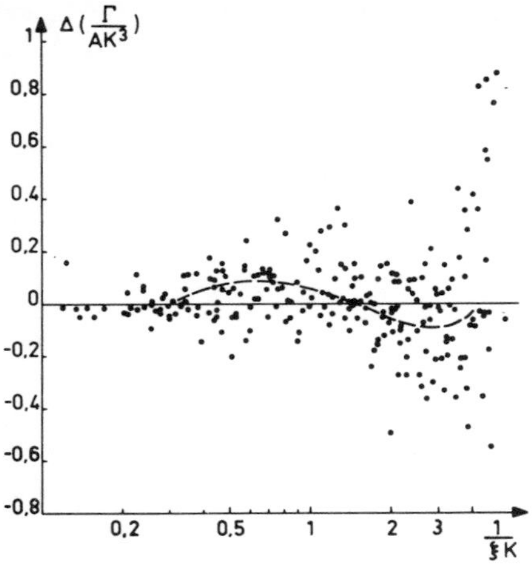

FIG. 2. Deviation of the experimental points from the best-fit curve in the range $0.5 \leq 1/\xi K \leq 4$.

quality factor. This can be done by studying the deviation of each experimental point from the "best fit" theoretical curve as a function of $1/\xi K$. The experimental points should be statistically distributed on both sides of a "true" theoretical curve. On the other hand we find that the experimental points are not statistically distributed around the theoretical curve in the region $0.5 \leq 1/\xi K \leq 4$. This is shown in Fig. 2 where schematically the dotted line shows the behavior of the experimental points around the calculated curve. The fact that this region is mostly responsible for the low quality factor is confirmed by the fact that a fit using only the points outside and on both sides of the region $0.5 \leq 1/\xi K \leq 4$ gives a quality factor of 0.876.

As a final remark we note that, using Kawasaki's expression $A = k_B T/16\eta^*$ and our experimental value $A = 1.51 \times 10^{-13}\ \text{cm}^3\ \text{sec}^{-1}$, one gets

$$\eta^* = 1.97 \times 10^{-2}\ \text{stokes},$$

which is in excellent agreement with the static value determined by Arcovito et al.[11] using a capillary-flow viscosimeter ($\eta \simeq 1.9 \times 10^{-2}$ stokes).

Thus we may conclude that Kawasaki's expression gives a complete and fairly accurate description of the behavior of the spectrum of the light scattered by a binary mixture in the hydrodynamical and nonhydrodynamical regimes. There is, however, at least in a limited region, a small but significant discrepancy between theory and experiment. We feel that, because of their great number and overall accuracy, our experiments could be compared with an even more refined theory; for instance, one which would consider $\eta \neq 0$, $\eta^* = \eta^*(\xi, K)$, or second-order terms.

[1] P. Berge and B. Volochine, Phys. Letters 26A, 267 (1968).

[2] P. Berge, P. Calmettes, and B. Volochine, Phys. Letters 27A, 637 (1968).

[3] P. Berge, P. Calmettes, B. Volochine, and C. Laj, Phys. Letters 30A, 7 (1969).

[4] P. Berge, P. Calmettes, C. Laj, and B. Volochine, Phys. Rev. Letters 23, 693 (1969).

[5] A numerical error was made in the value published in Ref. 4.

[6] B. Halperin and P. Hohenberg, Phys. Rev. Letters 19, 700 (1967).

[7] B. Halperin and P. Hohenberg, Phys. Rev. 177, 952 (1969).

[8] K. Kawasaki, to be published.

[9] K. Kawasaki, Phys. Letters 30A, 325 (1969).

[10] M. Tournarie, J. Phys. (Paris) 30, 737 (1969).

[11] G. Arcovito, C. Faloci, M. Roberti, and L. Mistura, Phys. Rev. Letters 22, 1040 (1969).

Spin Correlations in a One-Dimensional Heisenberg Antiferromagnet

R. J. Birgeneau* and R. Dingle
Bell Laboratories, Murray Hill, New Jersey 07974

and

M. T. Hutchings† and G. Shirane
Brookhaven National Laboratory,‡ Upton, New York 11973

and

S. L. Holt
University of Wyoming, Laramie, Wyoming 82070

(Received 13 January 1971)

Quasielastic magnetic neutron scattering from the linear-chain antiferromagnet $(CD_3)_4NMnCl_3$ is reported. The system is found to exhibit *planes* of critical scattering perpendicular to the $MnCl_3$ chains from >40°K down to 1.1°K. Both the spatial and thermal variation of the scattering can be quantitatively accounted for at all temperatures using Fisher's theory for the classical Heisenberg linear chain.

Over the past several decades considerable attention has been devoted by theoretical physicists to the problem of the one-dimensional ([1]) antiferromagnet with interaction Hamiltonians varying between the Ising, XY, and Heisenberg limits.[1] For the physically important Heisenberg case this has led to a few exact results such as the determination of the ground-state[2] and first-excited-state[3] energies at $T=0$°K for $S=\frac{1}{2}$ and to the instantaneous correlations at all temperatures[4] for $S=\infty$, but there are still many unanswered questions, particularly with regard to the dynamics. It has also been evident that there are a variety of materials in nature which for structural reasons could approximate well over a certain range of reduced temperature to the Heisenberg linear antiferromagnet. However, in all cases which have been reported to date, the [3] aspects of the system seem to manifest themselves at a relatively early stage so that it has not proven possible to study the static and dynamic behavior of a [1] system with very long-range [1] correlations in the absence of [3] effects.[5] In this Letter we present measurements of the instantaneous correlations in the linear-chain antiferromagnet, $(CD_3)_4NMnCl_3$, which is found to be purely [1] at all temperatures. Direct comparison is then possible with Fisher's exact solution of the classical [1] Heisenberg antiferromagnet. As we shall see, the agreement is excellent at all temperatures. This work then represents the first experimental illustration of an exactly soluble model in the phase-transition problem.

The structure and magnetic properties of $(CD_3)_4NMnCl_3$ (hereafter denoted as TMMC) have been discussed extemsively elsewhere.[6,7] For our purposes here it is sufficient to note that TMMC consists of a hexagonal array of antiferromagnetic $MnCl_3$ chains (Mn^{++}, $S=\frac{5}{2}$) which are magnetically insulated from each other by intervening $(CD_3)_4N^+$ ions. The quasielastic neutron scattering cross section[8,9] for such a chain may be derived simply from Fisher's exact solution for the correlations $\langle \vec{S}_i \cdot \vec{S}_{i+n}\rangle = u^{|n|}S(S+1)$, where u is defined below. This assumes classical spins interacting with a nearest-neighbor Heisenberg exchange $-2J_{nn}\vec{S}_i\cdot\vec{S}_{i+1}$. The result is

$$\frac{d\sigma}{d\Omega_f} \propto S(\vec{Q}) = \frac{B}{\kappa^2+(2/\pi^2)(\cos\pi Q^z+1)}, \tag{1}$$

with

$$B=(u^2-1/\pi^2u)S(S+1), \quad \kappa=(1+u)/\pi\sqrt{-u},$$

where $u=\coth K - K^{-1}$, $K=2J_{nn}S(S+1)/kT$. In the above we have written Q^z in units of $2\pi/2a$ with a the nearest-neighbor separation along the chain. Inspection of Eq. (1) shows two important features. Firstly, for an antiferromagnet where $-1 \leq u < 0$, the cross section has minima at even integer Q^z and maxima at odd integer Q^z. Secondly, as is obviously required for a [1] system, $S(\vec{Q})$ is independent of the two momentum coordinates Q^x and Q^y perpendicular to the chains so that these maxima at odd Q^z will have the form of *planes* of scattering. This is just the [1] analog of the ridges observed in the [2] antiferromagnet K_2NiF_4.[10] It should also be noted that for $Q^z \sim n$, where n is odd, Eq. (1) reduces to a simple Lorentzian in $q^z = Q^z - n$.

The experiments were performed on a double-axis spectrometer at the Brookhaven high-flux beam reactor using neutrons of wavelength 1.029 Å; 10-min collimation before and after the scattering was employed. We used different crystals, both with nominally 99% deuteration to minimize the large hydrogen incoherent scattering. The crystals had volumes of 0.7 and 0.4 cm^3, respectively, with mosaic spreads of less than 10 min each; they were oriented such that $(h, 0, l)$ and (h, h, l) wave vectors, respectively, could be surveyed.[11] Scans were carried out both along and across the anticipated planes of diffuse scattering. Typical scans across the planes for various temperatures and positions are shown in Fig. 1. From the figure it may be seen that the scattering does indeed peak at integer values of l as anticipated. Scans with l fixed at 1, 2, 3, etc. in the $(h, 0, l)$ and (h, h, l) directions show that the scattering is independent of h thus confirming that it is *planar*, or, in Fourier transform, that the fluctuations giving rise to the scattering are purely one-dimensional.

Somewhat surprisingly, however, the spectra exhibit maxima at *both* even- and odd-integer Q^z (denoted by l in Fig. 1) contrary to our expectations based on Eq. (1). At 22°K at the even-l position a sharp, resolution-limited peak with essentially a flat background is observed. At the odd-l position at 22°K the scattering clearly has two distinct components: a sharp peak at the center similar to that observed at the even position together with a much broader Lorentzian peak with an integrated intensity considerably larger than that of the central component. Studies of the temperature dependence between 1.1 and 30°K show that both the even-l peak and the central peak at odd l are temperature independent, whereas the residual scattering at odd l varies rapidly in peak intensity and width with temperature. It is evident that the temperature-independent scattering observed at both even and odd l has a unique origin; it must be nonmagnetic since it does not follow the Mn^{++} form factor. The probable explanation is that it arises from pseudo-one-dimensional correlated motion of the $(CD_3)_4N^+$ ions. The remaining scattering at the odd-l positions then may be identified as the anticipated [1] magnetic scattering.

The scans across the $l = 1$ planes in the $(h, 0, l)$, (h, h, l) crystals were analyzed as follows. Data taken above 8°K were fitted by Eq. (1) with the central points omitted so that the nuclear and magnetic contributions could be separated out.

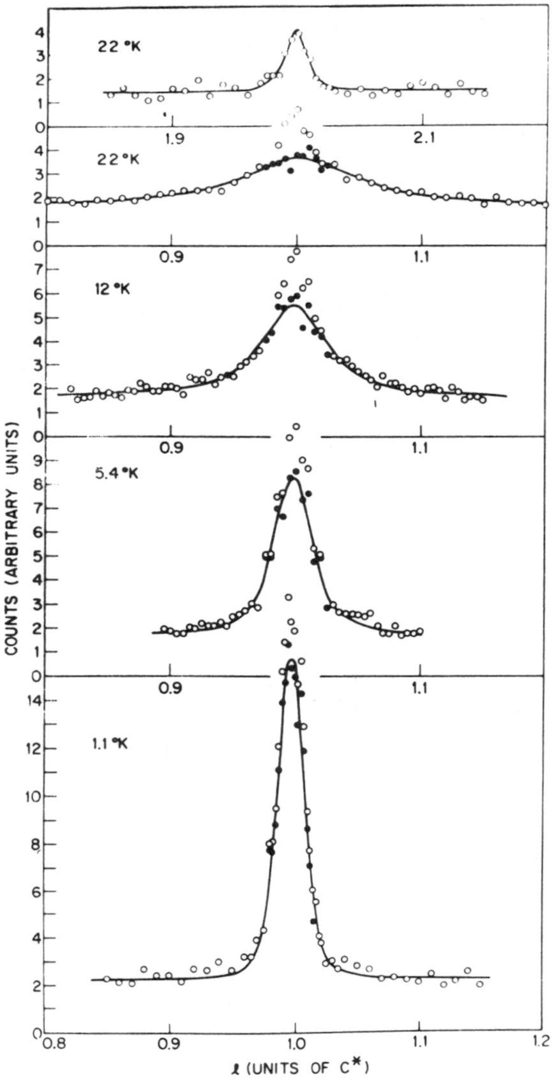

FIG. 1. Quasielastic scans across the planes in TMMC. The upper four scans were made in the $(h, 0, l)$ crystal with h fixed at 0.28 reciprocal-lattice units. The lowest scan was made in the (h, h, l) crystal with $h = 0.1$. The open circles are the actual experimental data; the closed circles give the residual magnetic scattering after the nuclear contribution has been subtracted off. The solid lines in the lowest four scans are theoretical fits as described in the text.

As noted above, the nonmagnetic scattering was found to be temperature independent up to 30°K so that it could be uniformly subtracted off from all of the data. This is necessary at the lower temperatures since the two components are not resolved. The magnetic scattering at all temperatures was then fitted by Eq. (1) convoluted with

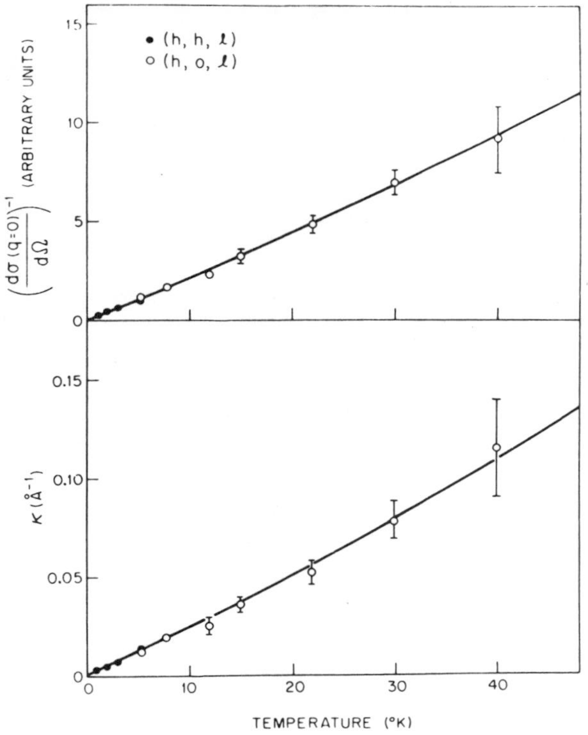

FIG. 2. Results obtained from the fits of the cross section, Eq. (3), to the magnetic scattering. The solid lines are fits by the classical theory as described in the text.

the instrumental resolution function. The solid lines in Fig. 1 are the curves generated by these fits. From the figure it may be seen that the fits are excellent ($\chi^2 \sim 1$) at all temperatures.

Figure 2 gives the deconvoluted inverse peak intensity and the inverse correlation length as a function of temperature. Both are observed to extrapolate to zero at 0°K as expected for the [1] system. This behavior is quite analogous to a normal second-order phase transition in a [3] antiferromagnet with the notable difference that in this [1] system $T_N^{(1)} = 0$°K. In three dimensions $d\sigma(q=0)/d\Omega$ is proportional to the staggered susceptibility.

It is of interest to compare our results quantitatively with the predictions of Fisher's classical model. From Eq. (1) it may be seen that the theoretical expression for κ involves only the one adjustable constant J_{nn}. The solid line in Fig. 2 for κ represents the best fit of the classical expression for κ, as given in Eq. (1), to our experimental results with $J_{nn} = -7.7 \pm 0.3$°K. This value for J_{nn} is somewhat higher than that determined from the bulk susceptibility, $J_{nn} \sim -6.5$°K, but probably lies within the uncertainty associated with the neglect of further neighbor interactions. The corresponding prediction of the classical model for $d\sigma(q^z=0)/d\Omega$ is shown as the solid line in the upper part of the figure. Here one overall scaling factor is adjusted to give the best fit; again the agreement is excellent. These results may be stated as follows: *Fisher's classical nearest-neighbor Heisenberg model properly predicts both the spatial and thermal variation of the instantaneous correlations in TMMC at all temperatures between 40 and 1.1°K.* This is a remarkable result especially when one considers the tremendous labor involved in the corresponding theory for [2] and [3] systems.

There are several additional points of note. Firstly, we have searched for evidence of [3] magnetic correlations at temperatures down to 1.1°K without success. Our search was by no means exhaustive, but it does suggest that the [3] ordering temperature may be somewhat lower than the 0.84°K inferred by Dingle, Lines, and Holt.[6] Secondly, it is clear that the dynamics in a [1] Heisenberg antiferromagnet are of considerable interest. Inelastic measurements show that at low temperatures there are long-lived spin waves over almost the entire [1] Brillouin zone in spite of the absence of long-range order. This will be the subject of a detailed report to be published elsewhere.[12] Finally, our experimental results pose a number of interesting theoretical problems. Although the Mn^{++} spin is large ($S=\frac{5}{2}$), we (at least) find it surprising that quantum mechanical effects do not seem important at 1.1°K. To our knowledge no theoretical calculations have as yet been reported for the linear chain for finite S other than $S=\frac{1}{2}$. In addition, TMMC would seem to offer an ideal test case for the theories of spin dynamics in paramagnetic systems. So far the only published work on [1] Heisenberg systems at finite temperatures is that of McLean and Blume.[13] Clearly, further theoretical work is required.

We should like to acknowledge stimulating conversations with M. Blume, W. F. Brinkman, M. E. Fisher, B. I. Halperin, P. C. Hohenberg, M. E. Lines, J. Skalyo, Jr., and L. R. Walker.

*Guest scientist at Brookhaven National Laboratory.

†Present address: Atomic Energy Research Establishment, Harwell, Didcot, Berkshire, England.

‡Work performed under the auspices of the U. S. Atomic Energy Commission.

[1]For a review see E. H. Lieb and D. C. Mattis, *Mathematical Physics in One Dimension* (Academic, New

York, 1966).
[2]H. A. Bethe, Z. Phys. 71, 205 (1931); L. Hulthén, Arkiv. Mat. Astron. Fys. 26A, No. 11 (1938).
[3]J. des Cloizeaux and J. J. Pearson, Phys. Rev. 128, 2131 (1962).
[4]M. E. Fisher, Amer. J. Phys. 32, 343 (1964).
[5]The most thoroughly studied system of this type is $CsMnCl_3 \cdot 2H_2O$. See J. Skalyo, Jr., G. Shirane, S. A. Friedberg, and H. Kobayashi, Phys. Rev. B 2, 1310, 4632 (1970).
[6]R. Dingle, M. E. Lines, and S. L. Holt, Phys. Rev. 187, 643 (1969).
[7]B. Morosin and E. J. Graeber, Acta Crystallogr. 23, 766 (1967).
[8]For a review see W. Marshall and R. D. Lowde, Rep. Progr. Phys. 31, 705 (1968).
[9]For $\vec{k}_f$ perpendicular to the chains, Eq. (1) is nearly exact. See R. J. Birgeneau, J. Skalyo, Jr., and G. Shirane, to be published.
[10]R. J. Birgeneau, H. J. Guggenheim, and G. Shirane, Phys. Rev. Lett. 22, 720 (1969).
[11]In this paper, for convenience, all momenta are referred to the room-temperature hexagonal axes. There is a small monoclinic distortion at 128°K, but this is not of importance here.
[12]M. T. Hutchings, G. Shirane, R. J. Birgeneau, R. Dingle, and S. L. Holt, to be published.
[13]F. B. McLean and M. Blume, to be published.

NEUTRON SCATTERING FROM K_2NiF_4: A TWO-DIMENSIONAL HEISENBERG ANTIFERROMAGNET*

R. J. Birgeneau
Bell Telephone Laboratories, Murray Hill, New Jersey 07974,
and Brookhaven National Laboratory, Upton, New York 11973

and

H. J. Guggenheim
Bell Telephone Laboratories, Murray Hill, New Jersey 07974

and

G. Shirane
Brookhaven National Laboratory, Upton, New York 11973
(Received 7 February 1969)

The quasielastic magnetic scattering from K_2NiF_4 over the temperature range from 97.2°K to at least 200°K is found to correspond to reciprocal lattice rods rather than points thus giving the first concrete evidence of the two-dimensional character of K_2NiF_4. At 97.1°K the crystal undergoes an extremely sharp phase transition to long-range order in three dimensions. From 97.0°K to 5°K the sublattice magnetization follows a $(97.1-T)^\beta$ law with $\beta = 0.15$.

Recently considerable theoretical attention has been directed towards two-dimensional magnetic systems.[1,2] Such systems are expected to exhibit cooperative properties, particularly in the neighborhood of the critical point, which differ appreciably from their three-dimensional counterparts.[3,4] It is therefore of considerable interest to find in nature systems which are truly two dimensional in their behavior. The most promising candidates at the present time seem to be the family of planar compounds with the K_2NiF_4 structure.[5] Early neutron scattering experiments on powdered K_2NiF_4 by Plumier[6] led him to postulate that long-range order (LRO) set in within the perovskite NiF_2 planes at 180°K but that even at 4.2°K there was no true LRO between the planes.

In this Letter we report results of quasielastic magnetic scattering from a large single crystal of K_2NiF_4. It is shown that at high temperatures Plumier's model is approximately correct although the actual details are rather more complicated. From at least 200°K down to 97.2°K the magnetic scattering indicates that there are very long-range correlations within the antiferromagnetic perovskite NiF_2 planes with no measurable correlations between the planes. At $T_N = 97.1$°K the crystal undergoes an extremely sharp phase transition to LRO in three dimensions. Below T_N the three-dimensional Bragg peaks are accompanied by "critical" scattering which is completely two dimensional in form. In addition, the sublattice magnetization from 97.0°K to 5°K is found to vary as $(97.1-T)^{\beta}$ with $\beta=0.15$. The latter two results indicate that even in the ordered phase the magnetic behavior is essentially two dimensional in character.

The compound K_2NiF_4 was prepared by reacting reagent-grade potassium fluoride with nickel fluoride. The nickel fluoride was made from 99.999%-pure nickel-metal powder and reagent-grade aqueous hydrogen fluoride. Single crystals were grown by a horizontal-zone-melting method using a platinum container in a dry hydrogen fluoride atmosphere. The crystal used in these experiments was approximately 0.5 cm^3 in volume and had a mosaic spread (full width at half-maximum) of 0.5 deg. The experiments were performed on a double-axis spectrometer at the Brookhaven high-flux-beam reactor using neutrons of wavelength 1.029 Å; 20′ collimation before and after the scattering was employed.

The crystal structure and magnetic order of K_2NiF_4 at low temperatures are shown in Fig. 1. Lines[7] has argued that the interactions within the NiF_2 planes may be adequately represented by a Heisenberg exchange Hamiltonian with[7,8] $J_{nn} \simeq 100$°K ($H=J\vec{S}_1\cdot\vec{S}_2$, $S=1$) together with an anisotropy term which is the order of 1°K. The coupling between the planes, however, is almost certainly significantly smaller since, as may be

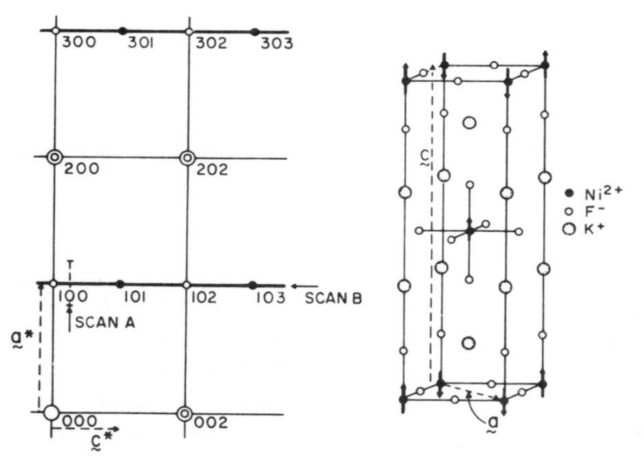

FIG. 1. Right-hand side: Crystal structure of K_2NiF_4 showing the antiferromagnetic arrangement of the nickel spins as exhibited at low temperatures; the figure corresponds to domain 1; domain 2 is produced by inverting the middle spin. Left-hand side: (0, 1, 0) reciprocal lattice (rl) plane. The double circles refer to nuclear rl points; the open and filled circles refer to domain-1 and -2 magnetic rl points, respectively. The thick lines refer to magnetic rl rods in the two-dimensional phase.

seen in Fig. 1, by symmetry there is no net interaction between adjacent planes in the Néel state. Thus any LRO in the C direction must be established by the coupling between the next nearest-neighbor layers. This is expected to be smaller than the intraplane exchange by at least three to four orders of magnitude.[3] It seems plausible therefore that K_2NiF_4 should indeed approximate well to a two-dimensional Heisenberg antiferromagnet over a wide range of temperatures. The two-dimensional nature of the system should manifest itself experimentally in a rather simple way. If LRO exists in three dimensions, then Bragg scattering will be observed at the magnetic reciprocal lattice points in the usual fashion. However, it may readily be seen that if LRO exists only within the layers, then the reciprocal lattice points become extended in the $\vec{C}^*$ direction resulting in the reciprocal lattice rods; for example, in Fig. 1, (100), (101), $\cdots \to (1, 0, l)$. Thus in a two-dimensional system we expect "Bragg ridges" rather than "Bragg peaks." Qualitatively similar scattering will also be observed if there are long-range correlations within the planes without the necessity of true LRO. In the following discussion for simplicity we use nomenclature which is only strictly correct for true LRO to describe both situations.

The experimental results at 99 and 95°K are given in Fig. 2. From the figure it may be seen that at 99°K the reciprocal lattice rod does indeed exist. Scan B along $(1, 0, l)$, that is, along the top of the ridge, gives a constant value far above the background; the apparent decrease in intensity at large l is due to the form factor. Scan A along $(h, 0, 0.25)$, that is, perpendicular to the ridge, shows a sharp peak with a linewidth determined by the instrumental resolution. The lack of concavity in scan B together with the sharpness of scan A shows unambiguously that at 99°K, K_2NiF_4 behaves as a pure two-dimensional antiferromagnet with very long-range correlations within the planes (>100 Å) and no measurable correlations between the planes. We believe that this is the first time that such two-dimensional scattering has been observed in a magnetic system.

The scattering remains essentially identical in form down to 97.2°K. However, in cooling from 97.2 to 97.0°K a rather unusual phase transition is observed. Sharp Bragg peaks appear on top of the ridge at the magnetic reciprocal lattice points (1, 0, 0), (1, 0, 1), etc.[9] The intensity of the (100) peak at 97.0°K is about 13% of its value at 4.2°K

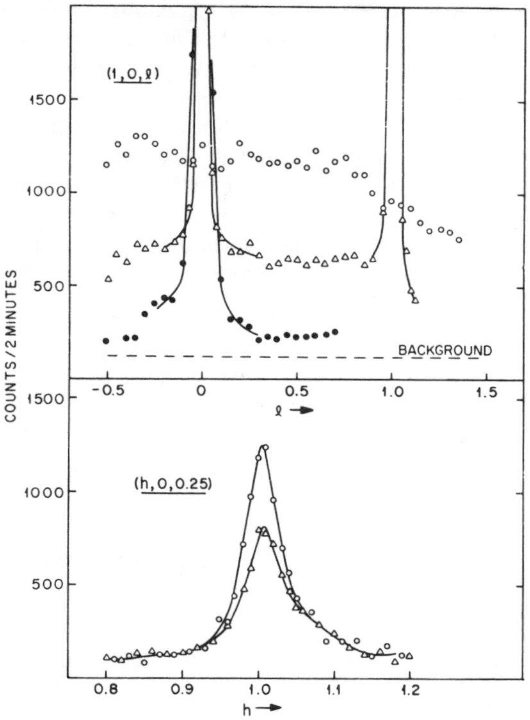

FIG. 2. The upper set of curves correspond to scan B along the top of the ridge as shown in Fig. 1; the lower set of curves correspond to scan A across the ridge. Open circles, data taken at 99°K; open triangles, 95°K; filled circles, 5°K

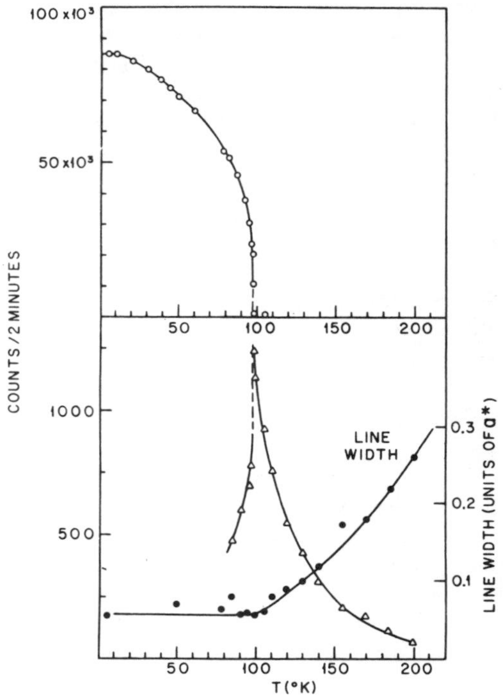

FIG. 3. Scattering intensity in counts/2 min at peak intensity (1, 0, 0), top, and (1, 0, 0.25), bottom, together with the linewidth in rl units (full width at half-maximum) for scan A in Fig. 1 as a function of temperature.

and the linewidth is just the mosaic spread of the crystal indicating that true LRO in all three dimensions has been established. The scattering at 95°K is shown in Fig. 2. From the figure it may be seen that at 95°K the Bragg ridge has remained essentially identical to that at 99°K except that it has decreased in intensity by about $\frac{1}{3}$. In addition, it has developed sharp peaks at the (100) and (101) reciprocal lattice positions. Approximate integration indicates that the intensity in the Bragg peak is just that lost by the Bragg ridge.

The (1, 0, 0) and (1, 0, 0.25) peak intensities together with the $(1, 0, l)$ ridge linewidths over the range from 5 to 200°K are shown in Fig. 3. We consider first the behavior for $T > 97.1$°K. A series of scans of types A, B show that ridge remains well defined up to at least 200°K. The behavior in the range from 97.2 to 200°K is analogous to the critical region in a three-dimensional system except that the temperature scale is greatly expanded. For comparison purposes it is interesting to note that in $KMnF_3$[10] at $T/T_C = 1.1$ the correlation range has decreased to 12 Å, whereas here at $T/T_C = 2$ the range as estimated from the inverse linewidth is 23 Å. The $(1, 0, l)$ ridge reaches its limiting linewidth at T_N. A proper analysis of the linewidth measurements requires further investigation involving precise resolution corrections. As a temperature "calibration" we may note that T_C (molecular field) = 270°K, and that the maximum in the susceptibility[11] (which is nearly isotropic down to 110°K) occurs at 250°K.

The scattering observed below $T_N = 97.1$°K is rather more complicated. Firstly, as shown in Fig. 3, the ridge decreases rapidly in intensity with decreasing temperature. This diffuse scattering may be thought of as the $T < T_N$ counterpart of the critical scattering observed above T_N. The fact that it retains the form of a ridge indicates that even below T_N the critical behavior of K_2NiF_4 is two dimensional in character. The (1, 0, 0) peak intensity, on the other hand, increases extremely rapidly with decreasing temperature. Scans at several temperatures show that the (100) line shape is temperature independent; in addition, extinction effects are found to be negligible for this sample. Under these conditions the peak intensity is directly proportional to the square of the sublattice magnetization; so we expect that in the critical region $I \propto (97.1 - T)^{2\beta}$. A fit to the data in Fig. 3 shows that over the complete temperature range from 97.0 to 5°K the intensity accurately follows a power law with $\beta = 0.15$. This should be compared with $\beta = \frac{1}{3}$ for a three-dimensional Heisenberg model and $\frac{1}{8}$ for the two-dimensional Ising model. At the present time no theoretical estimates of β are available for the two-dimensional Heisenberg model with small anisotropy. However, the closeness to the two-dimensional Ising value is consistent with an interpretation of the results in which the system is regarded as a genuine two-dimensional antiferromagnet even in the ordered phase, the ordering between the layers simply following as a necessary consequence of the establishment of LRO within the planes.

These experiments immediately suggest a number of topics for further investigation. Firstly, it seems likely that the magnons within the planes will be well-defined over most of the Brillouin zone at temperatures well above $T_N = 97.1$°K. Secondly, proper energy analysis and resolution corrections should enable us to obtain rather detailed information about the staggered susceptibility, correlation length, and fluctuations in the two dimensional phase above T_N. Thirdly, a proper measurement of the zero-point deviation should prove possible. Finally, experiments on isostructural systems such as Rb_2MnF_4 are also of considerable importance. Work is presently under way on each of the above.

We thank M. E. Lines for his constant interest in this work and for a number of helpful comments. Much of the interpretation of the results arose out of discussions with M. Blume, B. I. Halperin, P. C. Hohenberg, S. J. Pickard, and H. E. Stanley. We thank each of these for their valuable contributions.

---

*Work performed in part under the auspices of the U. S. Atomic Energy Commission.

[1]H. E. Stanley and T. A. Kaplan, Phys. Rev. Letters 17, 913 (1966).

[2]N. D. Mermin and H. Wagner, Phys. Rev. Letters 17, 1133 (1966).

[3]M. E. Lines, to be published.

[4]M. E. Fisher, Rept. Progr. Phys. 30, 615 (1967), Pt. I.

[5]D. Balz and K. Pleith, Z. Elektrochem. 59, 545 (1955).

[6]R. Plumier, J. Appl. Phys. 35, 950 (1964), and J. Phys. Radium 24, 741 (1963).

[7]M. E. Lines, Phys. Rev. 164, 736 (1967).

[8]E. P. Maarschall, A. C. Botterman, S. Vega, and A. R. Miedema, Physica 44, 224 (1969).

[9]We have also observed scattering from K_2NiF_4 powder which shows the same sharp transition at 97.1°K.

[10]M. J. Cooper and R. Nathans, J. Appl. Phys. 37, 1041 (1966).

[11]K. V. Srivastava, Phys. Letters 4, 55 (1963).

Brillouin Spectrum of Xenon Near Its Critical Point*

David S. Cannell† and George B. Benedek
Department of Physics and Center for Materials Science and Engineering, Massachusetts Institute of Technology, Cambridge, Massachusetts 02139
(Received 17 July 1970)

We have accurately measured the Brillouin spectrum of pure xenon along the critical isochore using two high-resolution spherical Fabry-Perot interferometers in tandem. The spectrum, which contained an extra diffusive mode, is analyzed in terms of a hydrodynamic model employing a relaxing bulk viscosity. We obtain the temperature dependence of the relaxation time, the bulk viscosity, the specific heat ratio C_p/C_v at finite k and ω, the correlation range, and the $k=0$, $\omega=0$ values for the compressibility and C_p-C_v.

This Letter reports accurate measurements of the Brillouin portion of the spectrum of light scattered by a pure fluid, xenon, near its critical point. The measurements were made along the critical isochore at temperatures ranging from 20°C above the critical temperature T_c to within 0.10°C of T_c. The spectral measurements were made using two high-resolution spherical Fabry-Perot interferometers in tandem. This technique enabled us to resolve clearly the weak Brillouin portion of the spectrum despite the presence of the extremely intense Rayleigh component. In addition to the normal Rayleigh and Brillouin components the spectrum contained an additional diffusive mode centered at the frequency of the incident light. The intensity of this extra mode increased as the critical point was approached, and for the lowest temperature studied, $T_c+0.10$°C, its integrated intensity was at least twice the integrated intensity of one Brillouin component. The general appearance of the spectrum as well as its dependence upon temperature is shown in Fig. 1.

The experimental setup consisted of a single-mode, frequency-stabilized, helium-neon laser; a high-pressure cell having two optical-quality glass windows; an axiconical collecting lens; a spectrometer consisting of two high-resolution spherical Fabry-Perot interferometers which were pressure swept in tandem; a photomultiplier tube; and a strip-chart recorder. The cell was carefully cleaned and filled to within 0.1% of the critical density with xenon containing less than 18 ppm of impurities. The cell temperature was controlled to within ±0.001°C, and was measured using a platinum resistance thermometer. The meniscus was observed to disappear at a temperature of (16.597 ± 0.01)°C, which was taken as the critical temperature, in good agreement with the accepted value of 16.590°C. Light scattered at an angle of 170°, corresponding to a scattering wave vector $k=2.25\times10^5$ cm^{-1}, was collected by an axiconical lens and spectrally analyzed using the tandem interferometer. The extremely high contrast of the interferometer, and its narrow instrumental width of 20

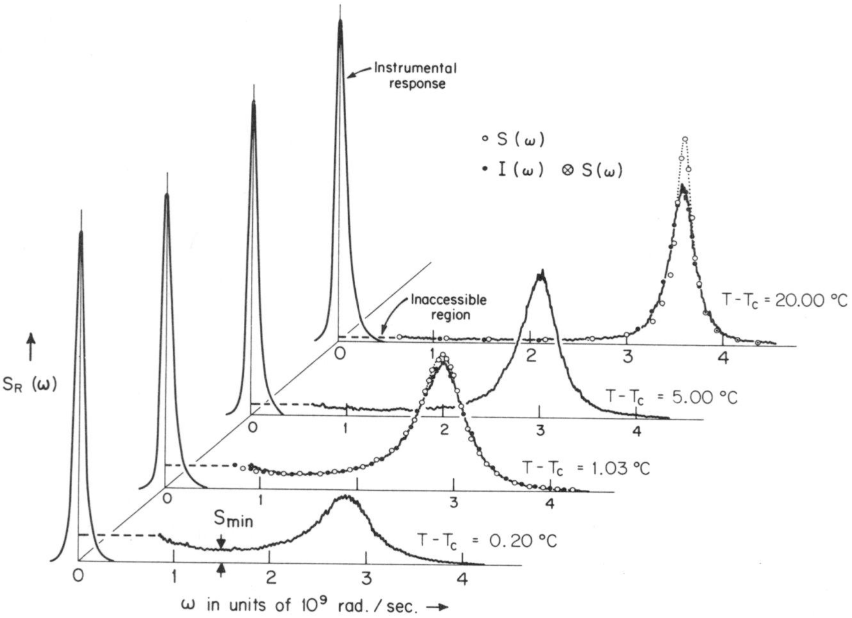

FIG. 1. The Brillouin spectrum of xenon for four temperatures along the critical isochore. The experimental traces are shown as solid curves. The open circles represent the spectrum $S_T(\omega)$ calculated from our hydrodynamic model without including the effects of the instrument. The closed circles are the convolution of $S_T(\omega)$ with the instrumental response function $I(\omega)$, and these agree accurately with the experimental spectra. The increase of spectral power between $\omega = 0$ and the Brillouin component clearly shows the growth of the diffusive mode as $T - T_c$ goes to zero.

MHz, permitted us to measure accurately the weak Brillouin portion of the spectrum to within 125 MHz of the extremely intense Rayleight line, and to discover a previously undetected[1-3] diffusive mode whose amplitude midway between the Rayleigh and Brillouin components is denoted as $S_{\min}$ in Fig. 1. The instrument response function $I(\omega)$ was determined independently for each spectrum recorded by measuring the folding of $I(\omega)$ with the very narrow Rayleigh line.[4] This also determined the relative intensities of the Rayleigh and Brillouin components.

The recorded spectrum $S_R(\omega)$ is the convolution of $I(\omega)$, measured as described above, with the true spectrum $S(\omega)$:

$$S_R(\omega) = \int_{-\infty}^{\infty} I(\omega - \omega')S(\omega')d\omega' \equiv I(\omega) \otimes S(\omega).$$

For each spectrum, an analytic function $S_T(\omega)$ was generated, which, when numerically convolved with the instrumental response function $I(\omega)$, yielded a good fit to the recorded spectrum $S_R(\omega)$. The physical model used for $S_T(\omega)$ was that of the spectrum of a fluid whose bulk viscosity relaxes with a single relaxation time. Such a spectrum has the form of a normal Rayleigh-Brillouin spectrum with an additional Lorentzian mode centered at the origin,[5] as is needed to explain our spectra as shown in Fig. 1. This physical model provides the simplest mathematical description which includes the possibility of a critical relaxation in the transport coefficients, as is expected from the mode-mode coupling schemes. The entire spectrum is completely determined by specifying the values of six parameters appearing in the expression for the spectrum. The six parameters involved are the low-frequency sound speed C_0, the infinite-frequency sound speed C_∞, the relaxation time for the bulk viscosity τ, the specific-heat ratio C_p/C_v, the ratio $\Lambda/\rho_0 C_v$, where Λ is the thermal conductivity and ρ_0 the average density, and the nonrelaxing viscosity term $\zeta_0 = (\frac{4}{3}\eta_s + \eta_{v0})/\rho_0$, where η_s is the shear viscosity and η_{v0} the nonrelaxing part of the bulk viscosity. The exact form for $S_T(\omega)$ which we employed is

$$S_T(\omega) = \mathrm{Re}[(F(s)/G(s))_{s=i\omega}], \qquad (1)$$

where

$$F(s) = s^2 + [a + b(s)]k^2 s + ab(s)k^4 + C_0^2 k^2(1 - 1/\gamma) \qquad (2)$$

and

$$G(s)=s^3+[a+b(s)]k^2s^2+[C_0^2k^2+ab(s)k^4]s+aC_0^2k^4/\gamma. \quad (3)$$

Following Mountain we have used the notation $a=\Lambda/\rho_0 C_v$ and $b(s)=\zeta_0+\eta_v(s)/\rho_0$, where

$$\eta_v(s)=\rho_0(C_\infty^2-C_0^2)\tau/(1+s\tau).$$

The actual deconvolution of a recorded spectrum was accomplished in the following manner: Values for the six parameters were estimated and the resulting $S_\tau(\omega)$ was numerically convolved with the instrumental response function $I(\omega)$, using the discrete fast Fourier transform.[6] The convolved spectrum was then compared with the recorded spectrum and a better estimate of the parameters made. This procedure was repeated until satisfactory agreement between $S_\tau(\omega)\otimes I(\omega)$ and $S_R(\omega)$ was obtained, thereby providing, at all temperatures, an accurate analytic representation of the true deconvolved spectrum $S(\omega)$. It is important to recognize that the values of the six parameters were not uniquely established by this procedure because it was not possible to measure the Brillouin spectrum within 125 MHz of the Rayleigh component. In fact, the experimental data effectively permitted the determination of five of the six parameters. In order to select physically meaningful values of all six parameters which enter into this single relaxation hydrodynamic model, one of the parameters must be independently specified. The two possible choices for specification are the low-frequency sound speed C_0 and the nonrelaxing viscosity term $\zeta_0=(\frac{4}{3}\eta_s+\eta_{v0})/\rho_0$. The ultrasonic sound speed has been measured[7,8] and the nonrelaxing viscosity term can be reasonably estimated by assuming that $\eta_{v0}\approx\eta_s$ [9] and using measured values for η_s.[10] It was assumed that η_s is independent of the temperature, as has been experimentally observed for the case of CO_2.[11] The estimated value of ζ_0 was 1.2×10^{-7} m^2/sec independent of temperature

In our analysis we considered each of the following possibilities. First, following Mountain,[12] we set ζ_0 equal to 1.2×10^{-7} m^2/sec and used the ultrasonic values of the sound speed for C_0. This proved to be an overspecification of the parameters as no possible choice of the remaining parameters resulted in agreement with the deconvolved spectra. The second possibility consisted of using the ultrasonic sound speed for C_0 and allowing ζ_0 to vary freely. In this case, although excellent agreement with the deconvolved spectra could be obtained, the resulting values for ζ_0 were ~4 times larger than the value estimated above. Furthermore, the value obtained for τ, the relaxation time, was $\sim2\times10^{-9}$ sec and was independent of the temperature. This corresponds to a relaxation frequency of ~80 MHz. While this choice of parameters accounts for all of the dispersion between the ultrasonic values for C_0 and the hypersonic (~500 MHz) sound speed[13] as determined by this experiment, it fails to account for the fact that dispersion in the sound speed is observed at frequencies as low as 250kHz.[7,8] The third approach which was tried, and which we propose to adopt for our analysis, was that of estimating ζ_0 as mentioned above and allowing C_0 to vary freely. For this purpose ζ_0 was set equal to 1.2×10^{-7} m^2/sec and assumed to be temperature independent. Again good fits were obtained, but in this case, at each temperature, C_0 had to be assigned a value well above the ultrasonic sound speed as shown in Table I. This table also shows the temperature dependence of C_∞, the infinite-frequency sound speed predicted by this fit, as well as the measured ultrasonic (0.55 MHz) sound velocity,[8] and the hypersonic (~500 MHz) velocity measured in this experiment. The difference between the ultrasonic sound speed and the values of C_0 necessary to fit the Brillouin data could be accounted for by a second relaxation in the bulk viscosity, one having a relaxation frequency well below 500 MHz. For our spectra obtained in the frequency regime of 500 MHz, the lower-frequency relaxation would have the effect of raising the values of C_0 necessary to describe the spectra. The existence of two relaxations would explain both our observed spectra and the observations of dispersion in the ultrasonic speed at low frequencies. The possibility of two relaxations in the bulk viscosity is supported by the theoretical work of Kadanoff and Swift.[14] Of course it is also possible that a continuous distribution of relaxation times may be necessary to account for the full frequency dependence of the sound speed and attenuation. In a separate Letter Garland, Eden, and Mistura analyze the existing Brillouin and ultrasonic absorption data using theories of Fixman[15] and Kawasaki[16] which contain such a distribution of relaxation times.

Table I also shows for each temperature the values of the three remaining parameters, τ, C_p/C_v, and $\Lambda/\rho_0 C_v$, needed to fit the deconvolved

Table I. Values of the parameters C_0, C_∞, τ, C_p/C_v, and $\Lambda/\rho_0 C_v$ used in the hydrodynamic model, for each temperature studied. The measured ultrasonic sound velocity V_{ult} (see Ref. 9) and the hypersonic sound velocity V_{hyp} measured by this experiment are also given

$T - T_c$ °C	C_o m/sec	V_{ult} m/sec	V_{hyp} m/sec	C_∞ m/sec	τ 10^{-9} sec	C_p/C_v	$\Lambda/\rho_o C_v$ 10^{-7} m²/sec
0.10	111.0	94.8	123.7	127.5	.54	1983	3.4
0.20	114.9	97.0	124.3	130.8	.38	951	2.7
0.30	116.6	98.8	127.3	133.1	.41	656	2.5
0.40	115.4	100.6	124.5	131.0	.37	497	2.4
0.50	116.5	102.2	126.8	132.4	.41	391	2.2
0.70	116.2	104.9	124.3	131.6	.34	279	2.1
1.03	118.2	108.4	126.8	134.0	.35	183	1.8
2.24	122.7	117.0	128.9	136.5	.29	82	1.5
3.00	127.9	121.8	132.7	141.6	.23	59	1.3
5.00	130.8	129.6	134.3	143.0	.20	35	1.1
7.50	138.6	136.7	140.5	150.4	.13	23	1.0
10.00	143.8	142.4	146.7	150.7	.24	18	1.0
15.00	150.0	151.8	152.4	155.4	.25	12	0.9
20.00	160.4	159.4	162.4	164.6	.25	10	0.9

spectra. As a final check the spectrum predicted by this model was numerically convolved with the instrumental response function and compared with the observed spectrum for all temperatures studied. As can be seen from Fig. 1, excellent agreement was obtained. Aside from a possible Botch-Fixman[17] correction to the Rayleigh linewidth, the spectrum predicted by this model reproduces for each temperature all of the known features of the Rayleigh-Brillouin spectrum.

The values of C_p/C_v listed above correspond to finite k and ω. It is possible to connect this quantity with the $k=0$, $\omega=0$ susceptibilities as follows. We apply the Ornstein-Zernike model to account for the k dependence and the hydrodynamic relaxation model to account for the frequency dependence. This gives

$$\frac{C_p}{C_v}=\frac{C_p(k,\omega)}{C_v(k,\omega)}(1+k^2\xi^2)\frac{C_{00}^2}{C_0^2}, \tag{4}$$

where C_{00} is the true zero-frequency sound speed, and ξ is the Ornstein-Zernike correlation range. Since C_{00} has not been measured and ξ is not known along the critical isochore, we eliminate the quantity $C_p/C_v C_{00}^2$, using the thermodynamic identity $\rho_0 K_T = C_p/C_v C_{00}^2$ where K_T is the isothermal compressibility. Since $K_T/K_I=\xi^2/R^2$, where $K_I = 1/nk_B T$ is the isothermal compressibility of an ideal gas of number density n and R is the direct correlation range, we find

$$\frac{\xi^2}{R^2}=\frac{C_p(k,\omega)}{C_v(k,\omega)}\frac{(1+k^2\xi^2)}{\rho_0 C_0^2}nk_B T. \tag{5}$$

This equation can be solved, using our experimental data, to give ξ^2 along the critical isochore. We used a value for R of 5.6 Å.[18] Using these values of ξ we then calculate $(\partial\rho/\partial\mu)_T = \rho_c^2 K_T$, where μ is the chemical potential. Finally, from K_T we obtain

$$C_p - C_v = \frac{T}{\rho_0}\left(\frac{\partial p}{\partial T}\right)_\rho^2 KT,$$

using a value[19] of 1.176 atm/°K for $(\partial p/\partial T)_\rho$ independent of the temperature. We find the following results for ξ, $(\partial\rho/\partial\mu)_T$, and $C_p - C_v$ along the critical isochore:

$\xi = 3.02\epsilon^{-0.60}$ Å,

$(\partial\rho/\partial\mu)_T = 1.63\epsilon^{-1.21}\times 10^{-9}$ g²/erg cm³,

$C_p - C_v = 1.70\epsilon^{-1.20}$ cal/mole °K.

Here $\epsilon = (T-T_c)/T_c$ is the reduced temperature. Our values for $C_p - C_v$ are in excellent agreement with those obtained from analysis of PVT data.[20] The numerical values of the relaxation time τ are comparable with the time necessary for a sound wave to travel one correlation length, as would be expected from the mode-mode coupling theory.[14] However, the temperature dependence of τ ($\tau \sim 10^{-10}\epsilon^{-0.2}$ sec) is very weak compared with the temperature dependence of the correlation range.

---

*Research supported with funds provided by the Advanced Research Projects Agency under Contract No. SD-90 with the Massachusetts Institute of Technology.

†Fannie and John Hertz Foundation Fellow. Now at Department of Physics, University of California at Santa Barbara, Santa Barbara, Calif. 93106.

[1]G. B. Benedek and D. S. Cannell, Bull. Amer. Phys. Soc. 13, 182 (1968).

[2]R. W. Gammon, H. L. Swinney, and H. Z. Cummins, Phys. Rev. Lett. 19, 1467 (1967).

[3]N. C. Ford, Jr., K. H. Langley, and V. G. Puglielli, Phys. Rev. Lett. 21, 9 (1968).

[4]D. Henry, H. Z. Cummins, and H. L. Swinney, Bull. Amer. Phys. Soc. 14, 73 (1969), and to be published. The values of $\Lambda/\rho_0 C_p$ determined by the experiment were used in our model.

[5]R. D. Mountain, J. Res. Nat. Bur. Stand., Sect. A 70, 207 (1966).

[6]W. T. Cochran, J. W. Cooley, D. L. Favin, H. D. Helms, R. A. Kaenel, W. W. Lang, G. C. Maling, D. E. Nelson, C. M. Rader, and P. D. Welch, Proc. IEEE 55, 1664 (1967).

[7]A. G. Chynoweth and W. G. Schneider, J. Chem. Phys. 20, 1777 (1952).

[8]C. W. Garland, D. Eden, and L. Mistura, following Letter [Phys. Rev. Lett. 25, 1161 (1970)].

[9]J. O. Hirschfelder, C. F. Curtiss, and R. B. Bird, *Molecular Theory of Gases and Liquids* (Wiley, New York, 1964), p. 643 ff.

[10]E. G. Reynes and G. Thodos, Physica (Utrecht) 30, 1529 (1964).

[11]J. Kestin, J. H. Whitelaw, and T. F. Zien, Physica (Utrecht) 30, 16 (1964).

[12]R. D. Mountain J. Res. Nat. Bur. Stand., Sect. A 72, 593 (1969).

[13]The values of the hypersonic sound speed listed in Table I were obtained by using the parameters given in Table I to evaluate the coefficients of the dispersion equation $G(s)=0$, whose roots were then obtained numerically. The imaginary parts of the complex roots are $\pm[V_{\rm hyp}(k)]k$.

[14]L. P. Kadanoff and J. Swift, Phys. Rev. 166, 89 (1968).

[15]M. Fixman, J. Chem. Phys. 36, 1961 (1962).

[16]K. Kawasaki, Phys. Rev. A 1, 1750 (1970).

[17]M. Fixman, J. Chem. Phys. 33, 1357 (1960).

[18]M. Giglio and G. B. Benedek, Phys. Rev. Lett. 23, 1145 (1969).

[19]H. W. Habgood and W. G. Schneider, Can. J. Chem. 32, 98 (1954).

[20]H. W. Habgood and W. G. Schneider, Can. J. Chem. 32, 164 (1954).

Dynamics of Concentration Fluctuations Near the Critical Mixing Point of a Binary Fluid

R. F. Chang, P. H. Keyes, J. V. Sengers, and C. O. Alley
Department of Physics and Astronomy and Institute for Molecular Physics, University of Maryland, College Park, Maryland 20742
(Received 10 November 1971)

The decay rate Γ of the concentration fluctuations near the critical mixing point of 3-methylpentane-nitroethane has been measured as a function of the correlation length ξ and the wave number k in the range $0.1 \leq k\xi \leq 20$. We have confirmed the theoretical prediction that Γ varies as k^3 in the critical regime. An analysis of Γ as a function of $k\xi$ reveals deviations from the theory of Kawasaki, which are discussed.

A system near a critical point exhibits large fluctuations in the order parameter, which decay very slowly in time. A central assumption in the current theoretical descriptions of this phenomenon is that the anomalous behavior of both static and dynamic properties is governed by a single correlation length ξ. It is, therefore, desirable to measure the relaxation time of the fluctuations and the equilibrium correlation length simultaneously. For this purpose we have measured the total intensity and the spectral width of light scattered by concentration fluctuations near the critical mixing point of 3-methylpentane-nitroethane. Details of the experimental arrangement will be published elsewhere.

The scattering angle θ is related to the wave number k of the fluctuations by the Bragg relation $k = 2k_0 \sin\frac{1}{2}\theta$, where k_0 is the wave number of the incident light in the medium. In the hydrodynamic regime ($k\xi \ll 1$), the decay rate Γ of the concentration fluctuations is given by $\Gamma = Dk^2$, where D is the binary diffusion coefficient.[1] The modern theories of dynamical scaling postulate that for all $k\xi$ the decay rate Γ should be a homogeneous function of k and ξ^{-1},[2]

$$\Gamma = \varphi(k, \xi^{-1}) = k^z \varphi(1, 1/k\xi) = k^z \Phi(k\xi), \tag{1}$$

where $z = 3$ according to the mode-mode coupling theory.[3,4] A specific form for the function $\Phi(k\xi)$ has been derived first by Kawasaki[4] and subsequently by Ferrell[5]:

$$\Gamma/k^3 = A(2/\pi)\{(k\xi)^{-3} + (k\xi)^{-1} + [1 - (k\xi)^{-4}]\arctan k\xi\}, \tag{2}$$

with $A = k_B T/16\bar{\eta}$, k_B being Boltzmann's constant and $\bar{\eta}$ a shear viscosity which is assumed to be independent of k and ξ in the theory.

We have measured the decay rate Γ from the linewidth of the central component in the spectrum of scattered light by the method of self-beat spectrometry.[6] A total of 88 data points were obtained at the critical concentration in the temperature interval $0.001 \leq T - T_c \leq 0.3$°C for seven different scattering angles. The scaling hypothesis (1) implies that in the critical regime ($k\xi \gg 1$), Γ should vary as k^z. To test this prediction we have plotted the linewidth data obtained at the seven scattering angles at $\Delta T = 0.001$ and 0.003°C as a function of k in Fig. 1. For the exponent z we find 2.99 ± 0.05 at $\Delta T = 0.001$°C

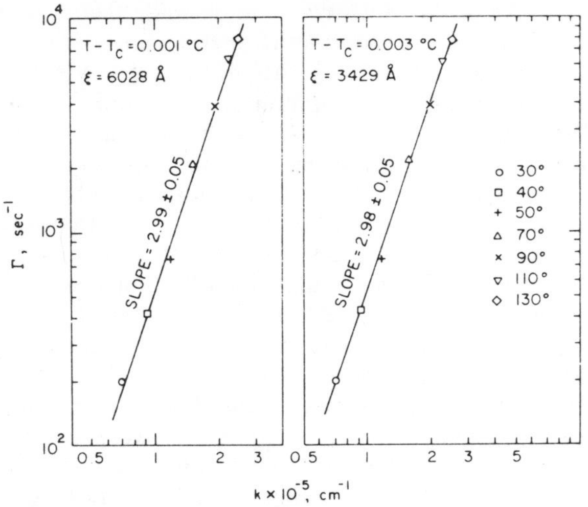

FIG. 1. Log-log plot of the decay rate Γ as a function of k at $\Delta T = 0.001$ and $0.003°C$.

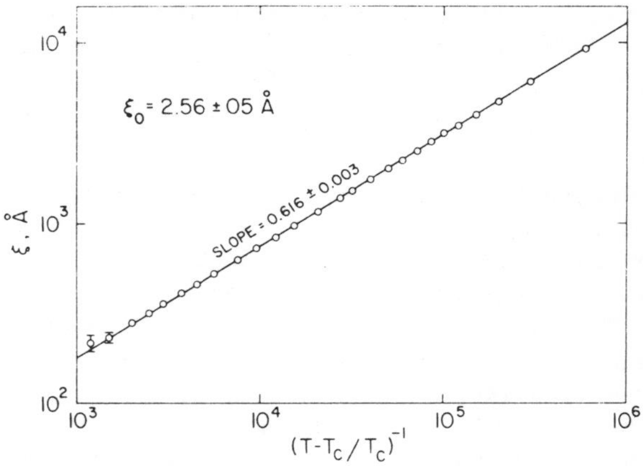

FIG. 2. Log-log plot of the correlation length ξ as a function of inverse reduced temperature.

and 2.98 ± 0.05 at $\Delta T = 0.003°C$, in excellent agreement with the value 3 predicted by the mode-mode coupling theory.

To test the validity of the Kawasaki function (2) we need to know the correlation length ξ. For that purpose we have measured the intensity of the scattered light $I(k, \xi)$ relative to the intensity of the incident light. The data were obtained as a function of temperature at three scattering angles ($\theta = 40°$, $60°$, and $90°$). In the analysis of the intensity measurements care was taken to correct for background scattering due to density fluctuations, residual dust, etc., as well as a small attenuation of light traveling through the medium.

The correlation length ξ diverges according to a power law $\xi = \xi_0 \epsilon^{-\nu}$, where $\epsilon = (T - T_c)/T_c$. The experimental values of ξ as determined by the slope of $I^{-1}(k, \xi)/I^{-1}(0, \xi)$ as a function of k^2, are shown in Fig. 2. We did not observe any statistically significant deviations from the Ornstein-Zernike theory, and all 82 intensity data points could be represented within their experimental precision by

$$I(k, \xi)^{-1} = I_0^{-1} \epsilon^{\gamma} [1 + k^2 \xi_0^2 \epsilon^{-2\nu}] \quad (3)$$

with

$$\gamma = 1.231 \pm 0.008, \quad \nu = 0.616 \pm 0.003,$$
$$\xi_0 = 2.56 \pm 0.05 \text{ Å}, \quad (4)$$

corresponding to a standard deviation $\sigma_I = 1.6\%$ for the intensity data. The errors quoted represent standard deviations; they are, of course, highly correlated. The standard deviation σ_I stays below 2% when γ is varied by ± 0.04, ν by ± 0.013, and ξ_0 by ± 0.14 Å. The latter errors may be more realistic for an assessment of the absolute accuracy of the parameters. The important point for the present paper is that at each temperature where the decay rate Γ was determined, we also have a reliable knowledge of ξ. Our value for γ is in a good agreement with the value 1.22 ± 0.04 obtained by Wims for the same system.[7]

The experimental data for Γ/k^3 are plotted as a function of $k\xi$ in Fig. 3. The solid curve represents the behavior predicted by Eq. (2). There exists some ambiguity in the meaning of the parameter $\bar{\eta}$ and consequently of A, since the ac-

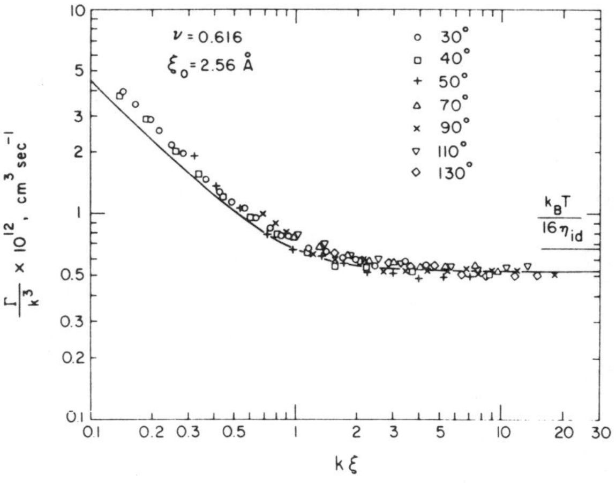

FIG. 3. Γ/k^3 as a function of $k\xi$ with ξ deduced from the intensity measurements. The solid line represents the behavior predicted by the theory of Kawasaki.

tual shear viscosity is not independent of k and ξ. Therefore, we have identified A with the value

$$\lim_{k\xi\to\infty} \Gamma/k^3 = (0.53 \pm 0.03)\times 10^{-12}\ \mathrm{cm^3\ sec^{-1}},$$

observed experimentally. With this identification, however, we note a different between theory and experiment in the hydrodynamic regime which amounts to as much as 20%.

This result is in apparent contradiction with the work of Bergé *et al.* who have concluded that the Kawasaki function describes the decay rate of the concentration fluctuations very well.[8] However, these authors did not calculate the Kawasaki function from an *a priori* knowledge of ξ, but instead fitted the decay rate with the theoretical Eq. (2) using ξ_0 and ν, in addition to A, as adjustable parameters. If we carry out the same procedure for our experimental data, we obtain the results shown in Fig. 4 with the parameters $A = (0.537 \pm 0.001)\times 10^{-12}$ cm^3 sec^{-1}, $\xi_0 = 1.47 \pm 0.06$ Å, and $\nu = 0.667 \pm 0.005$. With these values the Kawasaki function represents the linewidth data with a standard deviation of 4.6%. We conclude that this agreement is an apparent one; it is not substantiated when ξ is determined independently from the intensity measurements.

We present some arguments that the disagreement between theory and experiment is due to (a) the presence of a nonsingular background contribution to the Onsager coefficient, and (b) the fact that the shear viscosity $\bar{\eta}$ cannot be treated as a constant.

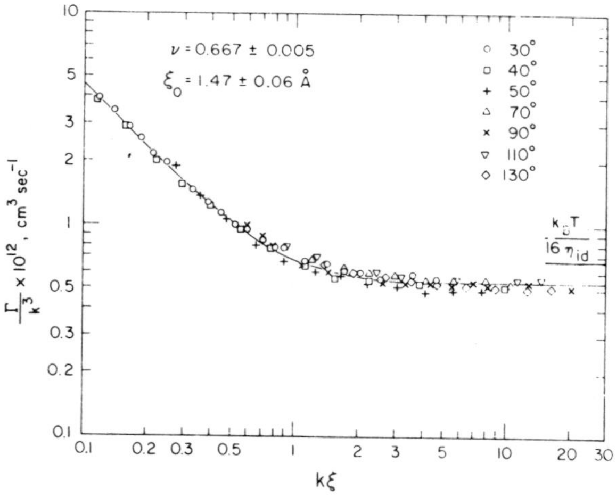

FIG. 4. Comparison of the linewidth data with the theory of Kawasaki using A, ξ_0 and ν as adjustable parameters.

(a) A similar disagreement between theory and experiment was noted by Henry, Swinney, and Cummins for the decay rate of the entropy fluctuations near the gas-liquid critical point.[9] The discrepancy could be resolved by accounting for a nonsingular background contribution to the thermal conductivity.[10] This observation has been confirmed by Swinney, Henry, and Cummins for xenon[11] and by Maccabee and White for carbon dioxide.[12] The diffusion coefficient D is the product of an Onsager kinetic coefficient L and $(\partial\mu/\partial c)_{T,P}$, where μ is the chemical potential and c the concentration of either component.[13] By analogy with the gas-liquid phase transition one would expect the presence of a background term L_0 in the Onsager coefficient. Thus the theory should not be compared with the full experimental decay rate Γ, but only with the asymptotic part,

$$\Gamma - \Gamma_0 = \Gamma - L_0 k^2 (\partial p/\partial c)_{T,P}(1 + k^2\xi^2)$$

$$= \Gamma - Bk^2\epsilon^\gamma(1 + k^2\xi^2), \quad (5)$$

where L_0, and consequently B, can be approximated by a constant in the small temperature range of our measurements. The effect of this background is to reduce the linewidth data at the higher temperatures, i.e., in the hydrodynamic regime ($k\xi \ll 1$) in Fig. 3. The coefficient B in (5) is 6.3×10^{-3} cm^2 sec^{-1} for CO_2 and 3.3×10^{-3} cm^2 sec^{-1} for Xe. For 3-methylpentane-nitroethane we do not have any independent experimental information for the Onsager coefficient. However, if we replace Γ in (2) with (5), evaluate the Kawasaki function on the basis of the equilibrium ξ, but use B as an adjustable parameter, we find $B = (4.1 \pm 0.2)\times 10^{-4}$ cm^2 sec^{-1}. With this value for B the agreement between theory and experiment is improved substantially and the Kawasaki function does represent the corrected linewidth data to within 4.8%.

(b) Even if the background term were known precisely, one still should not expect a complete agreement between theory and experiment. The viscosity of 3-methylpentane-nitroethane exhibits an anomaly which is close to logarithmic as shown by Stein, Allegra, and Allen[14] and Tsai and McIntyre.[15] In the temperature range of our measurements the viscosity varies from $\eta = 0.458 \times 10^{-2}$ g cm^{-1} sec^{-1} at $\Delta T = 0.3$°C to $\eta = 0.540 \times 10^{-2}$ g cm^{-1} sec^{-1} at $\Delta T = 0.002$°C. Our experimental value $A = k_B T/16\bar{\eta} = (0.53 \pm 0.03)\times 10^{-12}$ cm^3 sec^{-1} corresponds to $\bar{\eta} = (0.49 \pm 0.03)\times 10^{-2}$ g cm^{-1} sec^{-1}. This value is 29% larger than the ideal viscosity $\eta_{id} = 0.38 \times 10^{-2}$ g cm^{-1} sec^{-1}, but is of the same order as the experimental viscosity. The assumption that the viscosity is a constant is clearly not justified and one should expect that η should vary both with ξ and k. If we naively neglect the variation of η with k and identify the parameter $\bar{\eta}$ in (2) with the hydrodynamic viscosity $\eta(0, \xi)$ as measured by Stein, Allegra, and Allen,[14] we find a remarkably good agreement without introducing a background term. The Kawasaki function with the equilibrium ξ appears to represent the product $\Gamma\eta/k^3$ to within 4.7%; the coefficient $A\eta$ is only 12.5% larger than its theoretical value $k_B T/16$. A similar observation was recently made by Bergé and Dubois for the decay rate in the hydrodynamic regime.[16] Unfortunately, this procedure is not justified theoretically, since the viscosity in the theory is not the hydrodynamic viscosity $\eta(0, \xi)$ but a viscosity $\eta(k, \xi)$ at large values of k.[4] These results do illustrate, however, that at least part of the disagreement could be caused by the anomalous behavior of the viscosity.

We conclude that the difference between theory and experiment improves significantly (a) if we allow the possibility of a background term in the Onsager coefficient and (b) if the theoretical parameter $\bar{\eta}$ for the viscosity is allowed to vary with temperature. In the absence of independent experimental information concerning the background in the Onsager coefficient and a theoretical understanding of the viscosity anomaly and its effect on the decay rate, it is difficult to separate the two effects.

We express our appreciation to Mr. J. Maurey of the National Bureau of Standards for his aid in preparing the liquid mixture. We thank Mr. M. Wigdor for his assistance in the experiment, Professor G. F. Allen and Professor D. McIntyre for making their viscosity data available prior to publication, Professor R. A. Ferrell for his stimulating interest, and Dr. M. H. Ernst for several illuminating discussions. We also acknowledge the collaboration with Dr. J. S. Osmundson during the initial phase of this project.

*Work supported by the Office of Naval Research and the Advanced Research Projects Agency. The computer time for this project was supported by the Computer Science Center of the University of Maryland.

[1]P. Debye, Phys. Rev. Lett. 14, 783 (1965); R. D. Mountain, J. Res. Nat. Bur. Stand., Sect. A 69, 523 (1965).

[2]B. I. Halperin and P. C. Hohenberg, Phys. Rev. 177, 952 (1969).

[3]L. P. Kadanoff and J. Swift, Phys. Rev. 166, 89 (1968); J. Swift, Phys. Rev. 173, 257 (1968).

[4]K. Kawasaki, Ann. Phys. (New York) 61, 1 (1970).

[5]R. A. Ferrell, Phys. Rev. Lett. 24, 1169 (1970).

[6]N. C. Ford and G. B. Benedek, Phys. Rev. Lett. 15, 649 (1965); H. Z. Cummins and H. L. Swinney, in *Progress in Optics*, edited by E. Wolf (North-Holland, Amsterdam, 1970), Vol. 8, Chap. 3.

[7]A. M. Wims, thesis, Howard University, 1967 (unpublished).

[8]P. Bergé, P. Calmettes, C. Laj, M. Tournarie, and B. Volochine, Phys. Rev. Lett. 24, 1223 (1970).

[9]D. L. Henry, H. L. Swinney, and H. Z. Cummins, Phys. Rev. Lett. 25, 1170 (1970).

[10]J. V. Sengers and P. H. Keyes, Phys. Rev. Lett. 26, 70 (1971); J. V. Sengers, in "Varenna Lectures on Critical Phenomena," edited by M. S. Green (Academic, New York, to be published).

[11]H. L. Swinney, D. L. Henry and H. Z. Cummins, to be published.

[12]B. S. Maccabee and J. A. White, Phys. Rev. Lett. 27, 495 (1971).

[13]See, e.g., R. Haase, *Thermodynamics of Irreversible Processes* (Addison-Wesley, Reading, Mass., 1969).

[14]A. Stein, J. C. Allegra, and G. F. Allen, to be published.

[15]B. Tsai, Master's thesis, University of Akron, 1970 (unpublished).

[16]P. Bergé and M. Dubois, Phys. Rev. Lett. 27, 1124 (1971).

PHYSICAL REVIEW LETTERS

VOLUME 16 | 16 MAY 1966 | NUMBER 20

TEMPERATURE DEPENDENCE OF THE SUPERFLUID DENSITY IN He II NEAR T_λ *

James R. Clow and John D. Reppy†

Department of Physics, Yale University, New Haven, Connecticut

(Received 1 April 1966)

The phase transition in liquid He^4 at the λ point is of interest for the general theory of phase transitions as well as for an understanding of the quantum properties of the superfluid. Liquid helium below the λ transition is characterized by the macroscopic occupation of a single quantum state. This state determines a superfluid velocity field, $\vec{v}_s$, throughout the liquid. There is a mass flow $\vec{j}_s$ associated with the superfluid velocity. The relation $\vec{j}_s = \rho_s \vec{v}_s$ defines a superfluid "density," ρ_s. The superfluid density becomes equal to the total density, ρ, at absolute zero and decreases rapidly to zero at T_λ.

In this Letter we wish to report a direct determination of the temperature dependence of ρ_s in the neighborhood of the transition. The apparatus and experimental method is similar to that reported earlier.[1] A persistent superfluid current is formed in an annular container of He II. The angular momentum of the persistent current is measured by a gyroscopic technique.

The previous experiment demonstrated that the quantum state of macroscopic occupation can remain fixed even when the temperature is cycled to within 10^{-5} °K of the λ transition. As a consequence of this stability the angular momentum of a persistent current is a reversible function of temperature, proportional to the superfluid density. Thus the measurement of persistent-current angular momentum provides a direct means for determining the temperature dependence of ρ_s. An increase in sensitivity and a reduction in vibrational noise were required in the earlier gyroscopic apparatus before useful measurements could be made closer than 10 mdeg to T_λ.

The experimental procedure was as follows. First, the λ point was identified by a discontinuity in the resistance of a doped-germanium thermometer. Then a persistent current was formed by rotating the container while cooling through T_λ. A sequence of values for the angular momentum was obtained, approaching but not exceeding T_λ. The angular-momentum measurement was then repeated at a lower temperature to check on any possible attenuation of the current. Finally the resistance of the thermometer was remeasured at T_λ. The values of angular momentum were normalized to give ρ_s/ρ, using the asymptotic value at lower temperatures where ρ_s approaches the total density, ρ.

The values ρ_s/ρ obtained are shown in Fig. 1. The data from near T_λ to about 100 mdeg below T_λ are consistent with a power law (solid line), $\rho_s/\rho = A(T_\lambda - T)^\alpha$ where A is a constant and $\alpha = 0.67 \pm 0.03$. The data of Dash and Taylor,[2] obtained by the classic method of Andronikashvilli, are shown (open triangles) for comparison. It is gratifying, considering the difference in experimental method, that the measurements agree within 2% in ρ_s/ρ. The data

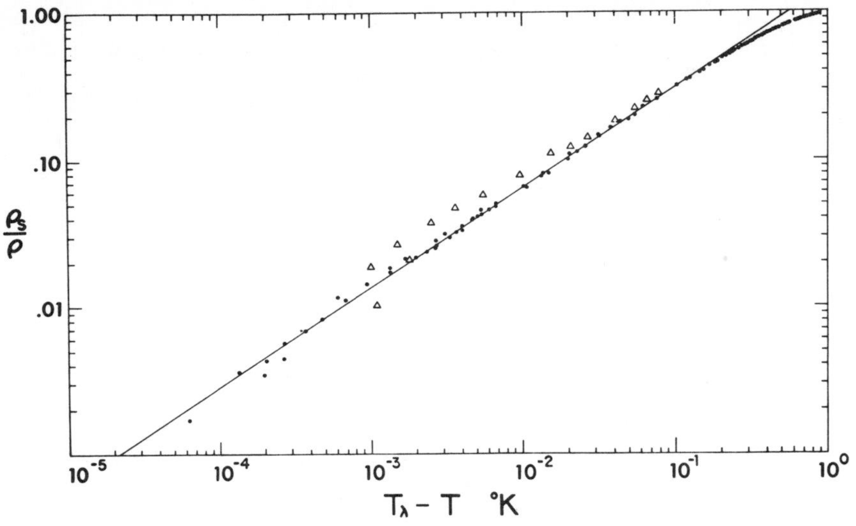

FIG. 1. Values of ρ_s/ρ obtained in the present experiment are plotted as solid circles as a function of $(T_\lambda - T)$. The data of Dash and Taylor[2] are plotted as open triangles.

near T_λ are shown in greater detail in the linear plots of Fig. 2.

In discussion of second-order phase transitions it is convenient to define an order parameter which goes to zero at the transition. In the case of liquid helium the order parameter ψ is taken to be complex,[3] $\psi = \eta e^{i\varphi}$ where $\eta^2 = \rho_s$. The superfluid velocity is reflected in the phase factor through the relation $\vec{v}_s = (\hbar/m)\nabla\varphi$. In terms of this definition, our data give a magnitude for the He II order parameter proportional to $(T_\lambda - T)^\beta$ where $\beta = 0.335 \pm 0.015$.

A similar dependence is observed in other systems with a λ singularity in the specific heat. For example, the sublattice magnetization in the antiferromagnetic material[4] MnF_2 is proportional to $(T_N - T)^\beta$ where $\beta = 0.335 \pm 0.010$ and T_N is the paramagnetic-antiferromagnetic critical temperature. Measurements on the liquid-gas density difference for a simple fluid such as xenon in the coexistence region[5] indicate a power-law dependence with $\beta = 0.345 \pm 0.015$. In addition, the series expansion calculations[6] for the three-dimensional lattice gas indicate a value for β of $\frac{5}{16} = 0.3125$. These similarities suggest that the square root of the superfluid density is the correct parameter to be used in making an identification between the λ transition in liquid helium and the λ transitions of other systems.

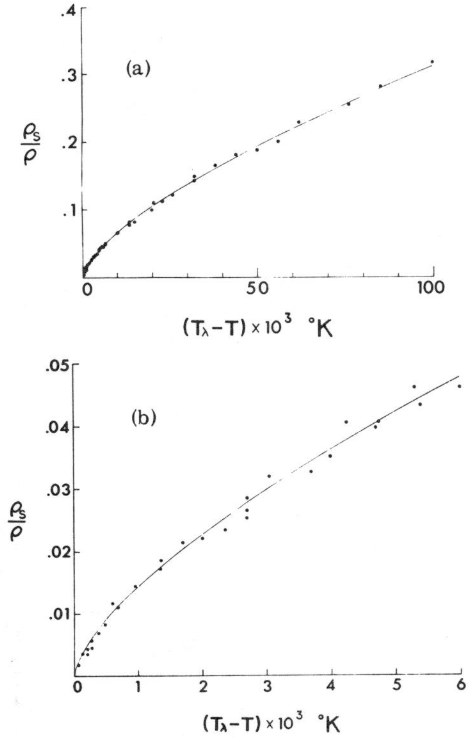

FIG. 2. (a) Values of ρ_s/ρ obtained for the first 100 mdeg below T_λ; (b) detail of the first 6 mdeg. The solid line is the function $\rho_s/\rho = 1.438(T_\lambda - T)^{2/3}$.

*Work assisted by grants from the National Science Foundation and the Army Research Office (Durham).

†Present address: Laboratory of Atomic and Solid State Physics, Cornell University, Ithaca, New York.

[1]J. D. Reppy, Phys. Rev. Letters 14, 733 (1965).
[2]J. G. Dash and R. Dean Taylor, Phys. Rev. 105, 7

(1957).
[3]I. M. Khalatnikov, Introduction to Theory of Superfluidity (W. A. Benjamin, Inc., New York, 1965), Chap. 17.
[4]P. Heller and G. B. Benedek, Phys. Rev. Letters 8, 428 (1962).
[5]M. A. Weinberger and W. G. Schneider, Can. J. Chem. 30, 422 (1952).
[6]J. W. Essum and M. E. Fisher, J. Chem. Phys. 38, 802 (1963).

CRITICAL BEHAVIOR OF IRON ABOVE ITS CURIE TEMPERATURE*

M. F. Collins,† R. Nathans, L. Passell, and G. Shirane
Brookhaven National Laboratory, Upton, New York 11973
(Received 20 May 1968)

High-resolution measurements of the energy distribution of neutrons critically scattered in Fe establish that the hydrodynamic region in which conventional theory is valid extends only to values of the scattering $q \approx K_1$, where K_1 is the inverse of the spin correlation range. Analysis of data restricted to this region indicates that the susceptibility varies as $(1-T_c/T)^{1.30 \pm 0.06}$. Contrary to expectations, the spin-diffusion constant Λ shows considerably less dependence on temperature, varying as $(1-T_c/T)^{0.14 \pm 0.04}$.

High-resolution neutron-scattering measurements have been made of the generalized magnetic susceptibility $X(q,\omega)$ of iron near its Curie temperature T_c for small wave vectors q and for small frequencies ω. A notable feature of the data is that the spin diffusion constant Λ is found to vary more slowly with temperature than has been predicted theoretically on the basis of molecular-field theories,[1,2] Kawasaki's treatment,[3] or the dynamic scaling laws.[4] The spatial correlations are confirmed as having much the expected form[5,1,6] and show that the susceptibility varies with temperature according to a simple power law whose exponent is, within experimental error, that calculated for a Heisenberg ferromagnet.

The data represent a significant improvement over previous measurements[7-12] since the intense slow-neutron fluxes available at the Brookhaven High Flux Beam Reactor have enabled us to measure $X(q,\omega)$ generally whereas measurements depended for their interpretation on assumptions about the frequency dependence of the susceptibility. The current data show clearly that these assumptions are only valid in restricted range of q.

Before describing the actual results of the measurements, we will say a few words about the experimental technique. The scattering was measured with a triple-axis spectrometer in the constant-Q mode of operation. The [111] reflections from squashed germanium monochromators and analyzers were used in conjunction with a pyrolytic graphite filter to give a very "clean" neutron beam of 3-Å wavelength. Resolution corrections were applied to the data using the approach of Cooper and Nathans.[13] This required numerical integration over the four-dimensional resolution function. The semimajor axes of this function were 0.047 (vertical), 0.029, 0.004 (along $\vec{q}$) and 0.002 Å^{-1}, where the frequency scale has been expressed in the reduced units defined by Eq. (32a) of Ref. 13. Measurements were taken for small q values near to the (1, 1, 0) reciprocal lattice point; this has the advantage over taking data near the forward direction that there is no practical restriction of the amount of energy that can be transferred for a given Q value. Temperatures were known to about 0.2°K and the spread in temperature over the specimen was not more than 0.5°K. The same iron single crystal was used as that previously employed by Shirane, Minkiewicz, and Nathans[14] for spin-wave investigations.

At small wave vectors and at low frequencies, conventional theory[1] shows that the scattering should be of the form

$$\sigma(q,\omega) \sim \frac{1}{r_1^2}\frac{1}{K_1^2+q^2}\frac{\Lambda q^2}{(\Lambda q^2)^2+(\hbar\omega)^2}. \tag{1}$$

The region of q and ω where this relation is valid has been called the hydrodynamic region.[4] It can be shown on quite general grounds that this small-q and -ω expansion must break down when either q or ω gets very large; Marshall[15] in particular pointed out that the diffusion form is only valid when $q \lesssim K_1$ and that this may be the origin of some of the difficulties over the interpretation of the linewidths indicated by the data of Refs. 9 and 10. The current measurements have confirmed that there is a range of q values over which the form (1) for the cross section gives an excellent fit to the experimental data. This is shown in Fig. 1 for data at $T_c + 14.1$°K; the scattering is well described by the use of just the two disposable parameters K_1 and Λ at all q values up to 0.0875 Å^{-1}. But for higher q values, the scattered distributions depart from the form (1) as can be seen in Fig. 2. In the upper part of Fig. 2 we have replotted the two largest q values in Fig. 1 on an expanded scale and have indicated the statistical accuracy of the individual data points. A small but significant broadening of the observed distributions is apparent. To show this more clearly we can express the adequacy of fit

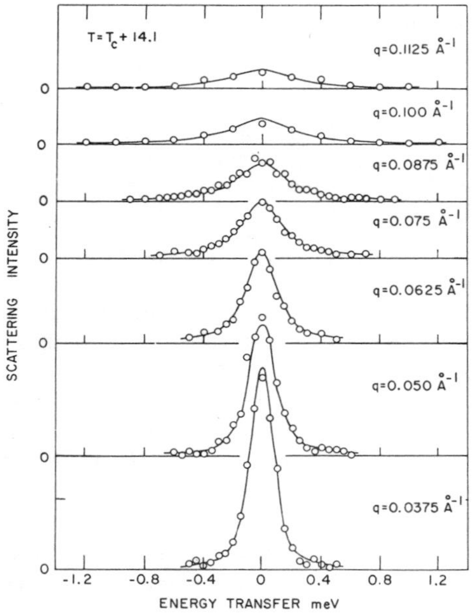

FIG. 1. Critical scattering from iron at T_c +14.1°K for a range of q values between 0.0375 and 0.1125 Å^{-1}. The open circles indicate the observed intensities but not their statistical accuracy. Up to tenfold longer counting times were employed for the larger q values. The solid line is the predicted scattering from conventional theory when convoluted with the instrumental resolution function.

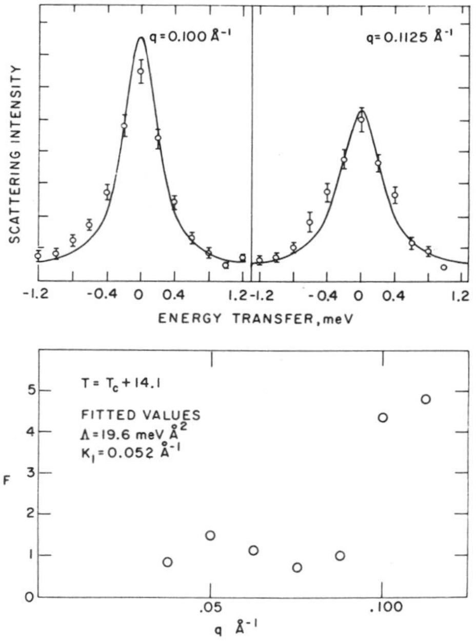

FIG. 2. The upper part shows the two largest q values of Fig. 1 plotted on an expanded scale with the statistical accuracy of the observed intensitites indicated. The lower part shows the F factor for goodness of fit for the data of Fig. 1 after a least-squares fitting with disposable parameters K_1, Λ, and r_1. The fit is good for $q \leq 0.0875$ Å^{-1} but significant departures occur at larger values of q.

in terms of the statistical criterion

$$F = \sum_{n=1}^{N} \frac{1}{(N-m)} \left[\frac{I_n(\text{obs}) - I_n(\text{calc})}{\sigma_n} \right]^2,$$

where I_n(obs) and I_n(calc) represent, respectively, the observed and calculated intensities for each pair of values of q and ω, σ_n is the standard deviation of the observed value, N is the total number of observations, and m is the number of fitted parameters. The lower part of Fig. 2 shows that for $q \leq 0.0875$ Å^{-1}, the F values are near unity indicating a good fit. At the largest values of q, however, F increases abruptly showing that the data are not adequately represented by the calculated curves. It is generally true of all of our data above the critical temperature that the breakdown value of q is in the range $q = 1.5K_1$ to $2.0K_1$. This breakdown of the cross-sectional form (1) appears to be quite sharp and cannot be explained solely in terms of a q^4 term in the cross section.

It is perhaps of interest to note that in the region of q outside the hydrodynamic region, where spin-wave-like modes are thought to exist,[16] the frequency distribution of the lines is always more square shaped than that of the Lorentzian form (1). Some of the earlier measurements on iron included data in this region and in the light of our observed line shapes, it is clear that their interpretation is open to question.

Least-squares-fit analyses have been performed on our data in the hydrodynamic region in order to obtain values of K_1, Λ, and r_1. Figure 3 shows Λ and $(K_1r_1)^2$ plotted against $1-T_C/T$ on a log-log scale. $(K_1r_1)^2$ should vary as the static zero-wave-vector susceptibility, a quantity which has been investigated theoretically in some detail. The theory indicates a divergence at T_C with an exponent of between 1.33 and 1.43.[17-19] Experimental measurements of the static susceptibility indicate an exponent of 1.33 $\pm$ 0.01.[20-22] Our data give a divergence with exponent 1.30 $\pm$ 0.06. The point nearest to T_C in Fig. 3 is significantly below the fitted line; however, in this region, the cross section is varying rapidly and the resolution analysis is more prone to systematic errors.

The value 1.30 for the exponent is in excellent agreement with values obtained from other experiments in the hydrodynamic region.[9,10,12] Unfortunately, there is less agreement on the values of K_1. On examination, we find that K_1 cannot be as accurately determined by neutron measurements as can the product $(K_1r_1)^2$; hence increased scatter in the values of K_1 is to be expected.

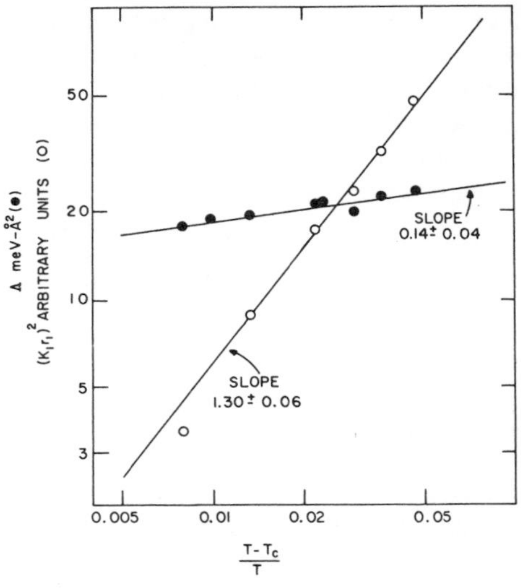

FIG. 3. $(K_1r_1)^2$, which is directly proportional to the static susceptibility, and Λ, the diffusion constant, plotted against $(1-T_c/T)$ on a log-log scale. Power laws with exponents 1.30 $\pm$ 0.06 and 0.14 $\pm$ 0.04, respectively, are obtained. (To obtain the value of the dimensionless parameter $2m\Lambda/\hbar$, multiply the corresponding value of Λ by 0.483.)

Turning now to the parameter Λ, molecular field theory predicts[1-3] that it should go to zero at T_C with a power law whose exponent is 1.0 while recently Kawasaki[3] and Halperin and Hohenberg,[4] using different approaches, have predicted an exponent of $\frac{1}{3}$. While earlier experiments did not indicate any temperature variation, our present data show that Λ does vary slowly with temperature. If this variation is interpreted in terms of a simple power law, we obtain an exponent of 0.14 $\pm$ 0.04. Although there is no experimental reason to disbelieve that Λ is following a power law, it must nevertheless be pointed out that several other analytic forms would fit our data equally well. The hydrodynamic region very close to T_C is so restricted as to preclude our investigation of the question of whether or not Λ actually goes to zero at T_C.

The observed very weak temperature dependence of Λ is not easy to explain, particularly in view if the fact that the Heisenberg antiferromagnet $RbMnF_3$ showed no such anomalous behavior.[23] There seems no obvious way of determining experimentally whether the difficulties lie with inadequacies of current theories of critical phenomena or with the metallic nature of iron but for the moment we are inclined to lean toward the first of these two possibilities.

We are grateful to V. J. Minkiewicz and J. Skalyo, Jr., for helpful discussions during the course of this work.

---

*Work performed under the auspices of the U. S. Atomic Energy Commission.

†On leave from Atomic Energy Research Establishment, Harwell, England.

[1]L. Van Hove, Phys. Rev. 93, 1374 (1954).

[2]H. Mori and K. Kawasaki, Progr. Theoret. Phys. (Kyoto) 27, 529 (1962).

[3]K. Kawasaki, J. Phys. Chem. Solids 28, 1277 (1967).

[4]B. I. Halperin and P. Hohenberg, Phys. Rev. Letters 19, 700 (1967).

[5]L. S. Ornstein and F. Zernike, Koninkl. Ned. Akad. Wetenschap., Proc. 17, 793 (1914).

[6]M. E. Fisher, Rept. Progr. Phys. 30, 615 (1967).

[7]H. A. Gersch, C. G. Shull, and M. K. Wilkinson, Phys. Rev. 103, 525 (1956).

[8]R. D. Lowde, Rev. Mod. Phys. 30, 69 (1958).

[9]B. Jacrot, J. Konstantinovic, G. Parette, and D. Cribier, in Proceedings of the Symposium on Inelastic Scattering of Neutrons in Solids and Liquids, Chalk River, Canada, September, 1962 (International Atomic Energy Agency, Vienna, Austria, 1963), p. 317.

[10]L. Passell, K. Blinowski, T. Brun, and P. Nielsen, Phys. Rev. 139, A1866 (1965).

[11]S. Spooner and B. L. Averbach, Phys. Rev. 142, 291 (1966).

[12]D. Bally, B. Grabcev, A. M. Lungu, P. Papovici, and M. Totia, J. Phys. Chem. Solids 28, 1947 (1967).

[13]M. J. Cooper and R. Nathans, Acta Cryst. 23, 357 (1967).

[14]G. Shirane, V. J. Minkiewicz, and R. Nathans, J. Appl. Phys. 39, 383 (1968).

[15]W. Marshall, in Critical Phenomena, Proceedings of a Conference, Washington D. C., 1965, edited by M. S. Green and J. V. Sengers, National Bureau of Standards Miscellaneous Publication No. 273 (U.S. Government Printing Office, Washington, D.C., 1966).

[16]J. L. Beeby and J. Hubbard, Phys. Letters 26A, 376 (1968).

[17]C. Domb and M. F. Sykes, Phys. Rev. 128, 168 (1962).

[18]J. L. Gammel, W. Marshall, and L. Morgan, Proc. Roy. Soc. (London), Ser. A 275, 257 (1963).

[19]G. A. Baker, H. E. Gilbert, J. Eve, and G. S. Rushbrooke, Phys. Rev. 164, 800 (1967).

[20]S. Arajs and R. V. Colvin, J. Appl. Phys. Suppl. 35, 2424 (1964).

[21]J. E. Noakes, N. E. Tornberg, and A. Arrott, J. Appl. Phys. 37, 1264 (1965).

[22]G. Develey, Compt. Rend. 260, 4951 (1965).

[23]R. Nathans, F. Menzinger, and S. J. Pickart, J. Appl. Phys. 39, 1237 (1968).

BRILLOUIN LINEWIDTHS IN CO_2 NEAR THE CRITICAL POINT*

N. C. Ford, Jr., K. H. Langley, and V. G. Puglielli
Department of Physics and Astronomy, University of Massachusetts, Amherst, Massachusetts
(Received 21 May 1968)

We have measured the linewidth of light Brillouin scattered from CO_2 for $0.015 \leq T-T_c \leq 15°C$. As T_c is approached from high temperature, the linewidth increases rapidly but then saturates at $T-T_c \approx 1°C$ and remains roughly constant to $T-T_c = 0.015°C$. These results are discussed in the light of the dynamic scaling laws.

We report here measurements of Brillouin scattering from CO_2 near its critical point over the temperature range $0.015 \leq T-T_c \leq 15°C$ along the critical isochore. From these measurements we obtain the sound velocity, Brillouin linewidth, and ratio of the integrated intensity of the Rayleigh line to that of the Brillouin line. The new features we have found are the following: (1) The linewidth increases rapidly from $T-T_c = 15°$ to $T-T_c = 1°$ but then remains constant to $T-T_c = 0.015°$. (2) The sound velocity decreases to a constant value of 190 ± 3.6 m/sec as T approaches T_c at scattering wave vector $q = 2.18 \times 10^5$ cm^{-1}, a value about 5% larger than that found by Gammon, Swinney, and Cummins[1] at $q = 1.54 \times 10^5$ cm^{-1}. (3) The intensity ratio diverges as $(T-T_c)^{1.02 \pm 0.03}$ over the entire temperature range. The data for $(T-T_c) < 1°$ were obtained using a technique which is described here for the first time.

Several authors have reported observations of Brillouin scattering near the critical point in both one-component[1,2] and binary-liquid systems.[3] Measurements in the one-component systems have been limited to the temperature range $T-T_c > 2°C$ because the instrumental wings of the very intense Rayleigh component of the scattered light obscure the Brillouin components as the critical point is approached. In order to overcome this difficulty we have developed apparatus utilizing a Michelson interferometer with unequal optical path lengths to filter out the unwanted light. Using a path difference of 23 cm, the Brillouin component is transmitted with neligible loss while the Rayleigh component is attenuated by a factor of 60 or more.

Light from a stabilized, single-frequency, 6328-Å, He-Ne laser (Spectra Physics model 119) was scattered from a sample of Matheson research-grade CO_2 containing impurities of 2.5 ppm nitrogen and 0.6 ppm oxygen. The sample was held in a copper-jacketed stainless steel cell fitted with glass windows sealed to the cell by gold-plated stainless steel V rings. The oil bath surrounding the cell was temperature stabilized to ±0.001°C. The critical temperature was determined by observing the vanishing of the meniscus; $T-T_c$ measurements were made to an accuracy of ±0.003°C using a Leeds and Northrup platinum resistance thermometer. The cell was filled to the critical density by first over-filling and then allowing some CO_2 to escape so that the meniscus vanished at the critical temperature within 3 mm of the cell center. We measured the width of the Rayleigh line at several temperatures obtaining good numerical agreement with the results of Swinney and Cummins,[4] indicating that our density was the same as theirs. Light scattered through 166° was collimated by a conical window, filtered by the Michelson interferometer (at $T-T_c < 1.0$°C only), and analyzed by a piezoelectrically scanned Fabry-Perot interferometer with a free spectral range of 1970 MHz and an average finesse of 40.

The Fabry-Perot was swept once each second through a range large enough to include one Brillouin line and one Rayleigh line. Pulses from a cooled RCA 7265 photomultiplier were counted into the memory of a computer of average transients (TMC model 4606), whose addresses were swept in synchronism with the Fabry-Perot. Readout of the data accumulated in the computer memory on each sweep displayed the Brillouin line with a signal-to-noise ratio that increased as the square root of the number of sweeps. Except at $T-T_c = 0.015$°C, where the Brillouin component was significantly reduced by multiple scattering, an adequate signal-to-noise ratio was reached in 20 to 50 min. The narrow Rayleigh line gave an average instrumental line shape that was used in analyzing the Brillouin lines; furthermore, the Rayleigh line was intense enough to be observed directly on an oscilloscope, permitting the continuous monitoring of both interferometers. The spectra for several values of $T-T_c$ are shown in Fig. 1.

The ratio of the integrated intensity of the Rayleigh component to that of the Brillouin components is plotted as a function of $T-T_c$ in Fig. 2. Far from T_c the Michelson interferometer was not required, and the intensities of both lines were obtained from data collected during a single run. When the Michelson interferometer was used, two separate runs were made—one of about 3000 sweeps with the Michelson transmitting the Brillouin line, and a second of 25 sweeps

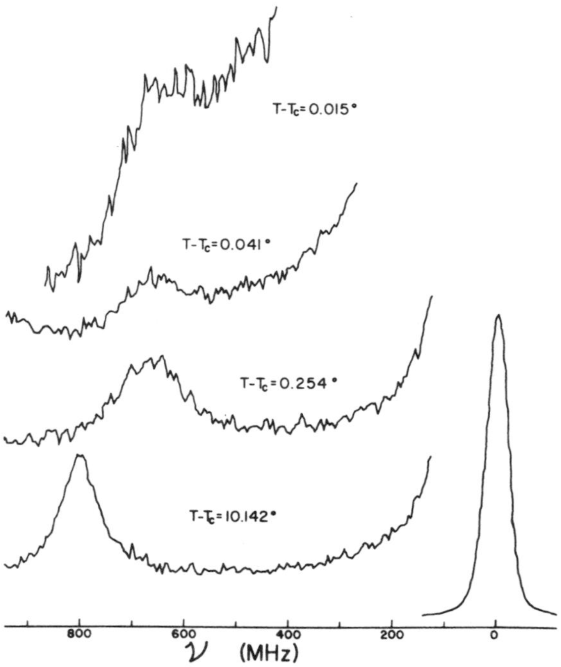

FIG. 1. Brillouin spectra of CO_2 near the critical point showing the Rayleigh line at reduced gain on the right. Data at the three lowest temperatures were obtained using a Michelson interferometer as a filter. Peak intensity of the Brillouin line is about 100 counts/sec.

with the Michelson transmitting the Rayleigh line and the laser intensity attenuated 10 to 100 times to avoid saturating the detection system. It can be shown[5] that the integrated intensity of a spectral line transmitted through the Michelson set with the maximum transmission at the center

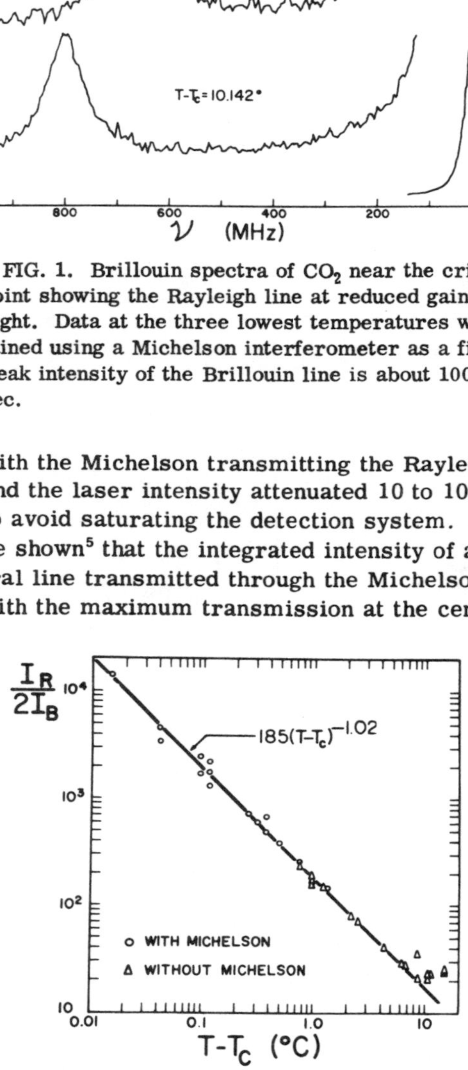

FIG. 2. Ratio of the integrated intensity of the Rayleigh component to that of the Brillouin components along the critical isochore.

frequency of the line is proportional to $\frac{1}{2}[1 + \exp(-\Delta\omega_{1/2}\tau)]$, where $\Delta\omega_{1/2}$ is the half-width at half-maximum of the line and τ is the time required for light to travel the difference in path length between the two arms. To correct for this effect the measured ratios for data taken using the Michelson were reduced by 9.6%. At temperatures near T_c the intense scattering of light greatly increases the probability of multiple scattering which might be expected to influence the intensity-ratio data. Geometrical considerations show, however, that the collection efficiency of the conical collimating lens is more than 10^3 times as great for light scattered within 0.013 cm of the lens axis than for light scattered further from the axis. Since the laser beam was focused in the scattering region, we expect multiply scattered light to contribute negligibly to the measured intensity ratios. At high temperatures we expect the data to be less reliable because stray light may make up as much as 50% of the light scattered at the incident frequency, causing the Rayleigh peak to appear larger than it really is.

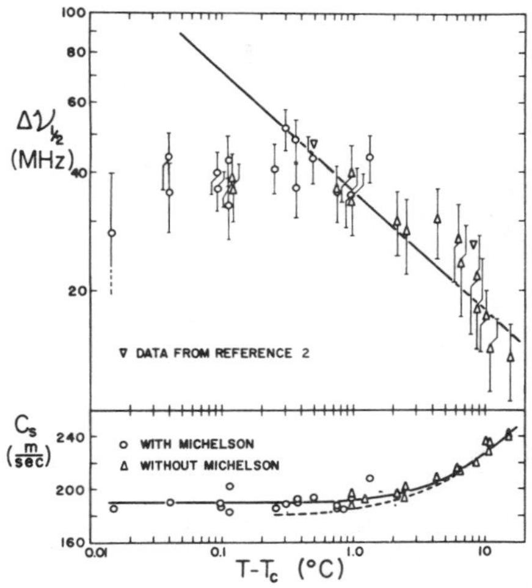

FIG. 3. Upper: Brillouin-line half-width at half-maximum. The solid line of slope −0.29 represents the contribution of λ/C_v in Eq. (2). Lower: Sound velocity calculated from Brillouin scattering at 166° using index of refraction data from Ref. 6. The dashed curve is taken from Brillouin scattering data at 88.6° from Ref. 1.

The natural width of the Brillouin component and the sound velocity are shown as functions of $T-T_c$ in Fig. 3. Linewidths were obtained by subtracting the instrumental width from that of the Brillouin line, an accurate procedure for lines fitted well by a Lorentzian shape. Linewidths obtained in this way should contain a negligible contribution due to either a finite acceptance angle or multiple scattering. The half acceptance angle of the detector is limited by the Fabry-Perot to 2.5×10^{-4} rad; the definition of the scattering angle was limited by alignment of the optical axes to 4×10^{-3} rad. These effects contribute an estimated 200 kHz to the measured Brillouin linewidth, far below the limits of resolution. Multiple scattering is not expected to influence the measured linewidths because of the low collection efficiency for light scattered off axis. The sound velocities were calculated from the frequency shift of the Brillouin line using $n = 1.1077$ obtained from Straub's[6] results for the index of refraction of CO_2 at the critical density. The sound velocity for small $T-T_c$ shows very little temperature dependence; a linear least-squares fit to these data gives, for $T-T_c < 1$°C,

$$c = 190 - 1.6(T-T_c) \pm 3.6 \text{ m/sec}. \qquad (1)$$

Recently, Kadanoff and Swift[7] have made predictions of the behavior of the transport coefficients near T_c using the scaling-law idea. We consider our results in the light of their predictions, assuming the divergences in C_p, C_v, and the order parameter ξ to be proportional to $\epsilon^{-\gamma}$, $\epsilon^{-\alpha}$, and $\epsilon^{-\nu}$, respectively, with $\epsilon = |(T-T_c)/T_c|$. Before discussing the transport coefficients we estimate these exponents from our intensity data and earlier experiments.

Standard treatment of the light-scattering problem[8] predicts that the ratio of the Rayleigh to Brillouin intensities is given by $I_R/2I_B = (C_p/C_v) - 1$; this is strictly correct only if the sound velocity measured at the Brillouin frequency is equal to the sound velocity at zero frequency,[9,1] a condition that is not satisfied in our experiment. We can conclude only that the slope of the intensity ratio must lie between γ and $\gamma-\alpha$, $1.02 \pm 0.03 \leq \gamma \leq 1.02 \pm 0.03 + \alpha$, in agreement with the value 0.95 ± 0.15 in Ref. 1. Combining this result with Swinney and Cummins's measurement of the thermal diffusivity, $\lambda/(\rho C_p) \propto \epsilon^{0.73\pm0.02}$, we find for the low-frequency divergence of the thermal conductivity λ $0.29 \pm 0.04 \leq \theta \leq 0.29 \pm 0.04 + \alpha$ where $\lambda(0) \propto \epsilon^{-\theta}$. The prediction of Kadanoff and Swift for $\lambda(0)$ depends on whether the shear viscosity, η, is weakly divergent or strongly cusped at high frequency. They expect $\lambda(0) \sim \epsilon^{-\nu-\frac{1}{2}\alpha}$ if η diverges, and $\lambda(0) \sim \epsilon^{-\gamma+\nu}$ if η is cusped. If we assume that η does diverge, this yields 0.29

$\pm 0.04-\frac{1}{2}\alpha \leq \nu \leq 0.29 \pm 0.04+\frac{1}{2}\alpha$, which is quite low in view of the currently accepted value $\nu \approx \frac{2}{3}$. On the other hand, if η is cusped we have $\lambda \sim \epsilon^{-\gamma+\nu}$, and the thermal diffusivity, $\lambda/(\rho C_p)$, is proportional to ϵ^{ν}; measurements of thermal diffusivity in Ref. 5 give directly that $\nu = 0.73 \pm 0.02$, closer to the accepted value. We take this as evidence that η is cusped and not divergent at high frequency.

The standard hydrodynamic result for the Brillouin half-width at half-maximum is[7,8]

$$\Delta\omega_{1/2} = \frac{q^2}{2\rho}\left[\frac{4}{3}\eta(\omega) + \zeta(\omega) + \left(\frac{\lambda(\omega)}{C_v} - \frac{\lambda(\omega)}{C_p}\right)\right], \quad (2)$$

where ρ is the density and η and ζ are the shear and bulk viscosities evaluated at the frequency of the Brillouin line. Kadanoff and Swift define low-, intermediate-, and high-frequency regions separated by temperature-dependent boundaries. As they have summarized in Sec. IV of Ref. 7, $\lambda(\omega)$ is expected to diverge at low and intermediate frequencies as discussed above and to be at most weakly divergent at high frequencies; $\eta(\omega)$ is weakly divergent or strongly cusped at all frequencies, although the behavior may be different in the low- and intermediate- or high-frequency regions; and $\zeta(\omega)$ is expected to diverge strongly as $\epsilon^{\alpha-3\nu}$ at low frequency and $\epsilon^{-\nu+\frac{1}{2}\alpha}$ at intermediate and high frequencies. Using rough estimates of ζ and assuming that $\eta(\omega)$ is not too different from its low-frequency value, we expect to cross from the low- to intermediate-frequency regions at $T-T_c \approx 3°C$ and from the intermediate- to high-frequency regions at $T-T_c \approx 0.5°C$. Applying these results to Eq. (2) and neglecting the effect of η, at temperatures above $T-T_c \approx 3°$ we would expect to observe a divergence of the linewidth roughly proportional to $\epsilon^{-0.29}$ if the dominant contribution is from the $\lambda(\omega)/C_v$ term; if the $\zeta(\omega)$ term dominates, the divergence is much stronger, roughly ϵ^{-2}. Below $T-T_c = 0.5°$, the only divergence arises from $\zeta(\omega) \sim \epsilon^{-\nu+\frac{1}{2}\alpha}$. We do not observe the predicted high-frequency (low-temperature) divergence in the linewidth, possibly because the contribution from ζ, although divergent, is not large at $T-T_c \leq 0.015°C$. Taking this as an indication that the larger contribution is from λ/C_v, we would expect a slope of -0.29 at the higher temperatures as shown by the solid line in Fig. 3. A steeper slope is not ruled out, indicating that a strongly divergent low-frequency ζ contribution may be present.

From Eq. (1) at $T-T_c = 0.2°$, corresponding to a sound-wave frequency $\nu = 650$ MHz, we find the sound velocity $c(650 \text{ MHz}) = 189 \pm 3.6$ m/sec. This value is to be compared with the Brillouin scattering results of Ref. 1 at the same temperature but at a different angle and hence a different frequency: $c(440 \text{ MHz}) = 180 \pm 1.5$ m/sec. The difference between the two results, $c(650 \text{ MHz}) - c(440 \text{ MHz}) = 9 \pm 3.9$ m/sec, may be attributed to an internal relaxation frequency ν_R according to the relation

$$c(\nu) = c_\infty - \frac{(c_\infty - c_0)}{1+(\nu/\nu_R)^2}.$$

We find $\nu_R \approx 200$ MHz, much too large for molecular relaxation processes but lying approximately on the boundary separating the intermediate- and high-frequency regions of Ref. 7. The fact that this velocity dispersion apparently vanishes far from T_c provides additional evidence that it is not molecular relaxation but rather is characteristic of the critical region as has been suggested by Gammon, Swinney, and Cummins[1] for the larger dispersion between ultrasonic data and their hypersonic data.

The authors wish to thank Spectra Physics for the use of a single-frequency laser and Linn Mollenauer for the loan of a Fabry-Perot interferometer. We are indebted to Michael Stephen, Stanley Engelsberg, and Leo Kadanoff for valuable discussions and suggestions.

---

*Research supported in part by the National Science Foundation under Grants Nos. GP5158 and GP7912.

[1]R. W. Gammon, H. L. Swinney, and H. Z. Cummins, Phys. Rev. Letters 19, 1467 (1967).

[2]G. B. Bendek and D. S. Cannell, Bull. Am. Phys. Soc. 13, 182 (1968).

[3]S. H. Chen and N. Polonsky, Phys. Rev. Letters 20, 909 (1968).

[4]H. L. Swinney and H. Z. Cummins, Phys. Rev. 171, 152 (1968).

[5]N. C. Ford, Jr., and K. H. Langley, to be published.

[6]J. Straub, thesis, Technische Hochschule, München, Germany, 1965 (unpublished); J. Straub, J. M. H. Levelt Sengers, and M. Vicentini-Missoni, to be published.

[7]L. P. Kadanoff and J. Swift, Phys. Rev. 166, 89 (1968).

[8]See, for example, R. D. Mountain, Rev. Mod. Phys. 38, 205 (1966).

[9]H. Z. Cummins and R. W. Gammon, J. Chem. Phys. 44, 2785 (1966).

Critical Sound Absorption in Xenon*

Carl W. Garland, Don Eden, and Luciano Mistura†
Department of Chemistry and Center for Materials Science and Engineering, Massachusetts Institute of Technology, Cambridge, Massachusetts 02139
(Received 17 July 1970)

Ultrasonic attenuation in Xe has been measured along a near-critical isochore at frequencies in the range 0.4-5 MHz and at temperatures above T_c. Hypersonic attenuation values obtained from Brillouin linewidths are also cited. It is shown that the critical attenuation per wavelength depends on temperature and frequency through a single reduced variable $\omega^* = \omega/\omega_D$, where the characteristic frequency $\omega_D = (2\Lambda/\rho C_p)\xi^{-2}$. The experimental results are compared with numerical calculations based on a recent theoretical formulation by Kawasaki.

In this Letter we wish to report and interpret recent measurements of the sound absorption in Xe near its critical point. Data obtained as a function of frequency and temperature for $\rho \simeq \rho_c$ and $T > T_c$ will be discussed. Following a brief description of the experimental procedures, a modified version of the pertinent theory will be outlined and the results will be discussed in terms of this theory. The essential result of both theory and experiment is that the critical attenuation per wavelength depends only on a single reduced variable $\omega^* = \omega/\omega_D$.

Previous ultrasonic investigations have clearly indicated that α_λ, the attenuation per wavelength, shows an anomalous behavior near the critical point.[1] However, none of these investigations presented sufficient data to allow a quantitative comparison with the predictions of recent theoretical studies.[2-6] With this in mind, a modification of the traditional pulse interferometer has been developed[7] in which the received signal is compared with a continuous coherent reference signal. Phase-sensitive detection and signal averaging provided an output which is sensitive to small delay changes and has a very high signal-to-noise ratio. Signals attenuated by more than 80 dB can be easily detected, and very accurate α_λ values are directly obtained from differential path measurements (the path can be continuously varied between 0.1 and 2 cm). Thus, this method is ideally suited for the measurement of ultrasonic velocity and attenuation in fluids near their critical point.

Both sound velocity and attenuation measurements have been carried out in high-purity xenon using X-cut 1-MHz quartz transducers. Data were obtained in the critical region along several isotherms and isochores at frequencies of 0.4, 0.55, 1, 3, and 5 MHz. A complete account of this experimental work will be presented elsewhere[8] and will include a comparison with the previous ultrasonic results of Chynoweth and Schneider.[9] In this Letter, we report only the attenuation data obtained along a single near-critical isochore at temperatures from 0.08 to 20°C above T_c.[10,11] The 0.55-MHz velocity data obtained along this isochore are cited in a separate Letter on critical light scattering in xenon.[12] Additional ultrasonic velocity data will be cited in an independent analysis of velocity dispersion.[13]

Our ultrasonic attenuation results are shown in Fig. 1 where the critical absorption per wavelength is plotted versus the reduced frequency $\omega^* = \omega/\omega_D$. Also shown are hypersonic attenuation values obtained from Brillouin linewidths.[12] The characteristic frequency ω_D is defined by

$$\omega_D \equiv (2\Lambda/\rho C_p)\xi^{-2}, \tag{1}$$

where Λ is the coefficient of thermal conductivity, C_p is the specific heat (i.e., the heat capacity per gram), and ξ is the correlation length. The critical attenuation is defined as the difference between the observed total attenuation and the so-called classical contribution $\alpha_\lambda(\text{class}) = \pi\omega u^{-2}[(\frac{4}{3}\eta + \zeta)/\rho + \Lambda(C_v^{-1} - C_p^{-1})/\rho]$, where ζ denotes the normal ("nonrelaxing") bulk viscosity. The value of $\Lambda(C_v^{-1} - C_p^{-1})/\rho$ is known as a function of temperature along the critical isochore,[12] and $(\frac{4}{3}\eta + \zeta)/\rho$ is assumed to have a temperature-independent value of 1.2×10^{-3} cm^2 sec^{-1} (see Ref. 12). At ultrasonic frequencies, $\alpha_\lambda(\text{class})$ is quite small and makes a significant contribution only when the total attenuation is very low (i.e., far from the critical point). At hypersonic frequencies, $\alpha_\lambda(\text{class})$ makes a sizable contribution throughout the critical region.

The most striking feature of Fig. 1 is that all the points, taken over a wide range of temperature and frequency, fall on a single curve. This is a new result, and the rest of this Letter is devoted to explaining to what extent the observed behavior is to be expected on theoretical grounds.

Some time ago, Fixman[2] and Botch and Fixman[3] proposed that energy transfer between

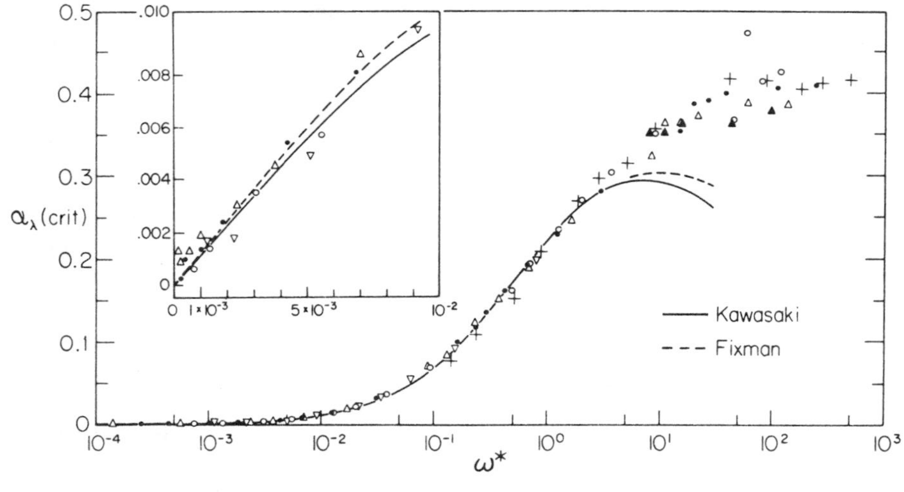

FIG. 1. Critical sound attenuation per wavelength in Xe as a function of the reduced frequency $\omega^* = \omega/\omega_D$. Ultrasonic data are shown at 0.4 (solid triangle), 0.55 (open triangle), 1 (solid dot), 3 (open dot), and 5 (inverted triangle) MHz. Hypersonic values at ~500 MHz (pulses) were obtained from Brillouin linewidths. All data were obtained along the critical isochore at temperatures above T_c. The theoretical curves represent Eq. (6) with the value of B chosen empirically so as to obtain the best fit for $\omega^* < 1$.

sound waves and density fluctuations might explain the anomalous acoustic behavior in the critical region. We shall show that a modified version of Fixman's theory will give results identical to those obtained recently by Kawasaki[6] on the basis of an analysis of anomalous contributions to the bulk viscosity. Let $\sigma(\vec{q}, t)$ be an order-parameter fluctuation with wave vector $\vec{q}$ at time t. In our case $\sigma(\vec{q}, t)$ will be the Fourier transform of $\rho(\vec{r}, t)-\rho_c$. Following Mistura and Sette,[14] we shall assume that the energy of the system associated with the density fluctuations which interact with the sound wave has the form

$$\delta U(t) \equiv U(t)-\langle U\rangle = \frac{1}{2V}\sum_{\vec{q}} b(q)|\sigma(\vec{q}, t)|^2 + \cdots, \tag{2}$$

where $b(q) = -V\langle|\sigma(q)|^2\rangle^{-1}\partial \ln\langle|\sigma(q)|^2\rangle/\partial\beta$ and $\beta = 1/k_B T$. This expression is based on the assumption that fluctuations with different wave numbers q are statistically independent and Gaussian variables.[14] Let us now define a complex, frequency-dependent excess specific heat $\Delta(\omega)$ in terms of a time-dependent correlation function:

$$\Delta(\omega) = \frac{i\omega}{k_B T^2 \rho V}\int_0^\infty dt\, e^{i\omega t}\langle\delta U(0)\delta U(t)\rangle. \tag{3}$$

Furthermore, we shall assume that the order-parameter fluctuations decay according to the diffusion equation $\dot{\sigma}(\vec{q}, t) = -\tau^{-1}(q)\sigma(\vec{q}, t)$.[15] This assumption, together with Eq. (2), allows us to write Eq. (3) in the form

$$\Delta(\omega) = \frac{i\omega}{k_B T^2 \rho}\frac{1}{(2\pi)^3}\int d^3q[\partial \ln\langle|\sigma(q)|^2\rangle/\partial\beta]^2 \frac{1}{2\tau^{-1}(q)-i\omega}. \tag{4}$$

For the static behavior of the fluctuations at small values of q, we assume the Ornstein-Zernike form, as modified by the small Fisher correction: $\langle|\sigma(q)|^2\rangle = A(q^2+\kappa^2)^{-1+\eta/2}$, where $\kappa \equiv \xi^{-1}$ is the inverse correlation length. A recent investigation of the Ising model[16] indicates that this form should be valid up to $x \equiv q\xi \lesssim 2$. The situation is less clear with respect to the decay rate. As $q \to 0$ hydrodynamics must become valid and we therefore have $\tau^{-1}(q) \to (\Lambda/\rho C_p)q^2$. If one accepts dynamic scaling ideas,[17] a more general expression can be written in the form $\tau^{-1}(q) = (\omega_D/2)K(x)$, where ω_D is the characteristic frequency defined in Eq. (1), and the function $K(x)$ must have the following properties:

$$\lim_{x\to 0} K(x) \propto x^2, \quad \lim_{x\to\infty} K(x) \propto x^3.$$

Since the critical absorption per wavelength is directly related to the imaginary part of $\Delta(\omega)$,[2] we obtain

$$\alpha_\lambda(\text{crit}) = \frac{k_B T^3}{\pi\rho^3}\left(1-\frac{\eta}{2}\right)^2 \frac{1}{u^2 C_V^2}\left(\frac{\partial P}{\partial T}\right)_V^2 \kappa\left(\frac{\partial\kappa}{\partial T}\right)^2 \int_0^\infty \frac{x^2 dx}{(1+x^2)^2}\frac{\omega^* K(x)}{K^2(x)+\omega^{*2}}. \tag{5}$$

This result is essentially identical to an expression obtained by Kawasaki[6] from a very similar analysis of the complex bulk viscosity. However, our use of the reduced variable ω^* simplifies the result in an important way. On the basis of scaling-law predictions of u, C_V, and κ, the quantity which appears in front of the integral in Eq. (5) should be, at most, a very slowing varying function of temperature. Indeed, our ultrasonic data indicate that this quantity (call it B) is a constant. Thus,

$$\alpha_\lambda(\text{crit}) = BI(\omega^*), \tag{6}$$

where the integral $I(\omega^*)$ defined in Eq. (5) depends on temperature and frequency through the single variable ω^*. The conclusion that B is independent of temperature is greatly strengthened by the excellent agreement between our data and the high-frequency Brillouin results. Since the hypersonic and ultrasonic frequencies differ by several orders of magnitude, a given ω^* value corresponds to quite different temperatures in each case.

Before discussing Fig. 1 in more detail, it must be stressed that the feasibility of fitting the data with a unique curve is rather sensitive to the temperature dependence chosen for ω_D. Assuming power-law divergences for ξ, C_p, and Λ which are characterized by the critical exponents ν, γ, and ψ, respectively, we find from Eq. (1) that $\omega_D = a\epsilon^{2\nu+\gamma-\psi}$, where $\epsilon \equiv (T-T_c)/T_c$. On the basis of mode-mode coupling results,[4] we have taken $2\nu+\gamma-\psi = 3\nu = 2.0$. From the available thermal diffusivity data[18] and correlation lengths[19] we can then determine the value 5.38×10^{12} sec^{-1} for the coefficient a.

Let us first comment on the hydrodynamic

region which must occur at sufficiently low values of ω^*. According to the mode-mode coupling prediction of Kadanoff and Swift,[4] α_λ(crit) $\sim \omega\epsilon^{-2} \sim \omega^*$ in this region. The inset in Fig. 1 shows that our data conform to this limiting theoretical behavior for $\omega^* \lesssim 5\times 10^{-3}$.

In order to compare theory and experiment at higher values of ω^*, the function $K(x)$ must be specified and the integral $I(\omega^*)$ evaluated numerically. Kawasaki[5,6] has recently proposed an explicit form for $K(x)$ which has the correct limiting behavior at low and high x values: $K(x) = \frac{3}{4}[1+x^2+(x^3+1/x)\tan^{-1}x]$. It should be noted that for small values of x this Kawasaki form reduces to $K(x)=x^2(1+\frac{3}{5}x^2)$, which is a modification of the form originally proposed by Fixman[3]: $K(x) = x^2(1+x^2)$. For comparison, we have calculated $I(\omega^*)$ for ω^* values up to 30 using both the Kawasaki form and Fixman form of $K(x)$. As mentioned in connection with our choice of $\langle|\sigma(q)|^2\rangle$, the theoretical expression given in Eq. (5) should be correct as long as the major contributions to the integral come from the interval $x<2$. Table I shows that this condition is satisfied when $\omega^* \lesssim 2$. In order to obtain the best visual fit of $BI(\omega^*)$ to the experimental α_λ(crit) points in the range $\omega^* \lesssim 2$, we have adopted the empirical value of 2.3 for B in the case of the Fixman calculation and 1.9 for B in the case of the Kawasaki calculation. Such empirical B values can be compared with the "experimental" value of 1.6 ± 0.2 obtained in the range $\epsilon = (2-6)\times 10^{-4}$ from our present sound velocity data and available data for C_V,[20] $(\partial P/\partial T)_V$,[10] and κ.[19] This comparison suggests that the Kawasaki form of $K(x)$ is to be preferred over the Fixman form. However, it must be pointed out that although the value used for a in $\omega_D = a\epsilon^2$ does not influence the shape of the semilog plot shown in Fig. 1 it will influence the B values. In summary, the experimental α_λ(crit) values for $\omega^* \lesssim 2$ can be well represented by Eq. (5) but they do not provide a very sensitive test of the form of $K(x)$.

Table I. Values of $I(\omega^*)$ and the contributions to this integral from the regions $x=0$ to 1 and $x=0$ to 2 for the choice of the Fixman (F) and the Kawasaki (K) form of $K(x)$.

ω^*		$I(\omega^*)=\int_0^\infty$	$\int_0^1$ (%)	$\int_0^2$ (%)
0.02	F	0.0088	90	99
	K	0.0100	86	98
0.2	F	0.0472	81.4	98.5
	K	0.0562	74.8	96.6
2	F	0.1135	38.3	93.3
	K	0.1384	28.4	86.6
20	F	0.1282	4.6	53.9
	K	0.1448	3.3	33.6

At still higher reduced frequencies ($\omega^* \gtrsim 2$), Eq. (5) is no longer in agreement with the experimental data. Table I shows that x values greater than 2 begin to contribute significantly to $I(\omega^*)$, and it is reasonable to expect a breakdown in the theory due to the inadequacy of the Ornstein-Zernike form for $\langle|\sigma(q)|^2\rangle$. On the basis of rather general assumptions about the form of $\langle|\sigma(q)|^2\rangle$ for large values of q, Kawasaki[6] has recently estimated that the absorption per wavelength in a sufficiently high-frequency region must be practically independent of both temperature and frequency. This prediction is confirmed by our experimental data for $\omega^* > 30$. Although an analysis of the possible contributions from higher-order processes has not yet been carried out, this qualitative agreement between theory and experiment supports the idea that the relaxation process first proposed by Fixman may be the only process responsible for the excess absorption in the critical region.

In conclusion, we have demonstrated experimentally that the critical attenuation per wavelength depends essentially on a single variable —the reduced frequency $\omega^* = \omega/\omega_D \sim \omega\epsilon^{-2}$. In the range $0 \le \omega^* < 2$, experiment is in good agreement with existing theory. At very low frequencies ($\omega^* \le 5\times 10^{-3}$) we find hydrodynamic behavior with a temperature dependence in agreement with the Kadanoff and Swift prediction.[4] At intermediate frequencies ($\omega^* \le 2$), we have shown that Eq. (5) can provide a good representation of the experimental data. Although these data are not extremely sensitive to the choice of $K(x)$, it would appear that the Kawasaki form is somewhat preferable to the Fixman form. For higher frequencies ($\omega^* > 2$) we have observed deviations of the experimental points from the values predicted by Eq. (5) and we have given a possible explanation for these deviations within the framework of a single relaxation mechanism with a continuous distribution of relaxation times. Finally, we have shown that the existing experimental data support the recent estimate by Kawasaki[6] of the limiting attenuation behavior at very high frequencies ($\omega^* \gtrsim 30$).

---

*Research supported in part by the Advanced Research Projects Agency and in part by the National Science Foundation.

†On leave from the Istituto di Fisica, Facoltà id Ingegneria, Università di Roma, and Gruppo Nazionale di Struttura della Materia del Consiglio Nazionale delle Ricerche, Roma, Italy.

[1]See, for example, the review article by C. W. Garland, in *Physical Acoustics*, edited by W. P. Mason and R. N. Thurston (Academic, New York, 1970), Vol. 7, Chap. 2.

[2]M. Fixman, J. Chem. Phys. 36, 1961 (1962).

[3]W. Botch and M. Fixman, J. Chem. Phys. 42, 199 (1965).

[4]L. P. Kadanoff and J. Swift, Phys. Rev. 166, 89 (1968).

[5]K. Kawasaki, to be published.

[6]K. Kawasaki, Phys. Rev. A 1, 1750 (1970).

[7]R. C. Williamson and D. Eden, J. Acoust. Soc. Amer. 47, 1278 (1970).

[8]P. E. Mueller, D. Eden, C. W. Garland, and R. C. Williamson, to be published.

[9]A. G. Chynoweth and W. G. Schneider, J. Chem. Phys. 20, 1777 (1952).

[10]H. W. Habgood and W. G. Schneider, Can. J. Chem. 32, 98 (1954).

[11]In analyzing these data, we have used $T_c = 16.955°C$, which was the observed temperature of the velocity minimum. The difference between this value and the accepted literature value (see Ref. 10) of 16.59°C may be due to gravitational effects since the transducers are located below the center of the cell, or to impurities which might possibly have been introduced during the filling procedure.

[12]D. S. Cannell and G. B. Benedek, preceding Letter [Phys. Rev. Lett. 25, 1157 (1970)].

[13]H. Z. Cummins and H. L. Swinney, following Letter [Phys. Rev. Lett. 25, 1165 (1970)].

[14]L. Mistura and D. Sette, J. Chem. Phys. 49, 1419 (1968).

[15]This approximation neglects the fact that density fluctuations can also decay as sound waves even near the critical point. See Refs. 3, 4, and 6 for a discussion of this point.

[16]M. Ferer, M. A. Moore, and M. Wortis, Phys. Rev. Lett. 22, 1382 (1969).

[17]B. J. Halperin and P. C. Hohenberg, Phys. Rev. 177, 952 (1969).

[18]D. L. Henry, H. L. Swinney, and H. Z. Cummins, second following Letter [Phys. Rev. Lett, 25, 1170 (1970)].

[19]M. Giglio and G. B. Benedek, Phys. Rev. Lett. 23, 1145 (1969).

[20]C. Edwards, J. A. Lipa, and M. J. Buckingham, Phys. Rev. Lett. 20, 496 (1968).

IDEAL RESISTIVE TRANSITION OF A SUPERCONDUCTOR

R. E. GLOVER
Department of Physics and Astronomy. University of Maryland. College Park. Maryland. USA

Received 11 September 1967

Precise measurements of the change of conductivity with temperature have been made for stabilized amorphous bismuth films. The experiments indicate that the initial rapid change of conductivity at the superconducting transition depends on temperature as $(T-T_c)^{-\frac{2}{3}}$. This extends over a temperature range of about 3 millidegrees. At higher temperatures the dependence is found to be $(T-T_c)^{-1}$.

The work was initiated at the suggestion of R. Ferrell and H. Schmidt and was intended to check on predictions they have recently made about the observability of critical phenomena associated with the phase change from the normal to the superconducting state.

A short electron mean free path results in a short coherence length and would be expected to expand the temperature internal over which fluctuation effects might be observable. The best substances from this point of view are the "amorphous" superconducting materials discovered by Buckel and Hilsch [1, 2]. As a result of a liquid-like atomic disorder [2, 3], electron mean free paths are of the order of a few atomic spacings. Amorphous bismuth was chosen for the present study in large part because of the work of Shier and Ginsberg [4] showing that films could be prepared with relatively sharp resistive transitions and with transition temperatures which varied only slightly from sample to sample. This suggests that the effect of any macros-

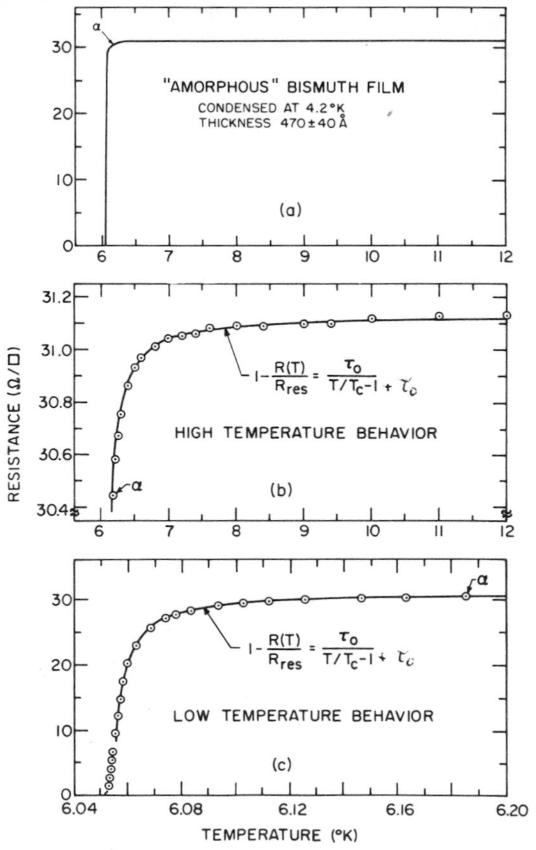

Fig. 1. Measured resistance per square as a function of temperature for an amorphous bismuth film. Absolute temperatures are good to ± 0.1°K. Relative temperatures in the vicinity of the 6° are measured to $\pm 2 \times 10^{-4}$°K. Points are measured values. The solid line in b) and c) corresponds to a $(T/T_c-1)^{-1}$ temperature dependence of the decrease in resistance. The point at T = 6.185°K is marked "α" in (a), (b) and (c) as an aid to orientation. In (b) the resistance scale has been expanded by a factor of 40 over that of (a) in order to better show the first decrease of resistance with falling temperature. Section (c) shows details of the final disappearance of resistance using a temperature scale expanded by a factor of 40 over that of (a) and (b).

copic inhomogeneities on the superconducting transition is small. Amorphous bismuth was prepared by the usual technique [1] of vacuum evaporation onto substrates held at 4.2°K. The vapor source material was 97.1 atomic % Bi - 2.9 atomic % Tl. Presumably, the composition of the films was similar. The purpose of the Tl is to help stabilize the amorphous structure.

The solid curve in fig. 1a) shows the measured change of resistance with temperature. The rounding of the curve near the transition indicates that the loss of resistance takes place over an extended temperature interval. As is seen, a change in resistance with temperature was detected up to almost twice the transition temperature.

H. Schmidt has recently suggested that the temperature dependence of the conductivity of a metal above its superconductive transition temperature might be expected to approximate the Curie-Weiss (molecular field) dependence found for the magnetic susceptibility above a ferromagnetic transition. According to this idea, the conductivity in the limit of zero current, σ, would be given by $\sigma-\sigma_{res} = A/(T-T_c)$, where A is the material constant, σ_{res} is the residual conductivity and the equation holds for temperatures T higher than the superconducting transition temperature T_c. In reduced temperature $\tau \equiv (T-T_c)/T_c$ this can be written as $R/R_{res} = \sigma_{res}/\sigma = (1+\tau_0/\tau)^{-1} \approx 1-\tau_0/\tau$. The material constant τ_0 is expressed in reduced temperature units. It is numerically equal to the displacement between the critical temperature and that at which the material reaches twice its residual conductivity. The solid curve shown in fig. 1b is that of the equation with the values $\tau_0 = 0.467 \times 10^{-3}$, T_c = 6.0548°K, and R_{res} = 31.13Ω/□.

Fig. 1c shows the transition curve in a small temperature region near T_c with the temperature expanded by a factor of forty. As a reference, the point at T = 6.185°K is indicated by the letter α and an arrow in all three graphs. The curve is the same as before. The fit to the $(T/T_c-1)^{-1}$ curve is good from twice the transition temperature all the way down to the region where the samples has lost half its resistance. A theoretical discussion of the $(T/T_c-1)^{-1}$ dependence is given by H. Schmidt in a paper being prepared for publication.

The fit to the equation is poor at temperatures so close to T_c that the film has lost more than about half its resistance. The data in this region have been analyzed by Ferrell and Schmidt [5] and are found to agree with their prediction of a $(T/T_c-1)^{-\frac{2}{3}}$ temperature dependence of the conductivity in the immediate vicinity of the superconducting transition. The region in question is found to be only about 3 millidegrees wide. Another estimate of the width of the critical region is given by the value of τ_0 in the above equation. For the data shown in fig. 1 this is found to correspond to a temperature interval of 2.8 millidegrees.

References
1. W.Buckel and R.Hilsch. Z. Physik. 138 (1954) 109.
2. W.Buckel. Z. Physik. 138 (1954) 136.
3. W.Rühl. Z. Physik. 138 (1954) 121.
4. J.S.Shier and D.M.Ginsberg. Phys. Rev. 147 (1966) 384.
5. R.A.Ferrell and H.Schmidt. Phys. Letters 25A (1967) 544.

* * * * *

Chemical Potential of He^3-He^4 Solutions Near the Tricritical Point*

Gregory Goellner and Horst Meyer
Department of Physics, Duke University, Durham, North Carolina 27706
(Received 5 April 1971)

New vapor-pressure measurements of He^3-He^4 solutions in the region between 0.5 and 1.2°K have been made and are used to calculate changes in the chemical potential. Special emphasis is directed towards the region of the tricritical point. The variation of the chemical potential and its derivative along certain paths leading to the critical point yields critical indices which satisfy the scaling relation proposed by Griffiths.

In recent years, interest in the liquid He^3-He^4 phase-separation critical point[1] at a mole fraction $X^*=0.67$ of He^3 in the liquid and a temperature $T^*=0.87$°K was further heightened by the experiments of Graf, Lee, and Reppy.[2] These authors observed that at this point, the two branches of the phase-separation curve for $X<X^*$ and $X>X^*$, respectively, form an angular top in the X-T plane. The liquid mixture then has a superfluid, a normal, and a stratified region. In the X-T plane, the first two regions are separated by the λ line that ends at the critical point. Griffiths,[3] who explored the theoretical limiting behavior of certain thermodynamic quantities in the neighborhood of this point, suggested the name tricritical point.

In analogy with the usual binary fluid mixing point and the liquid-gas critical point in pure fluids, we assume that the asymptotic behavior for He^3-He^4 solutions near (T^*, X^*) can be described by power laws. For a usual binary mixture with a phase separation point (X_c, T_c), the asymptotic behavior of X, T, and Δ (the difference between the chemical potentials of the two components) can be assumed to be given by

$$|X-X_c| \propto (T_c-T)^{\beta} \quad \text{for the coexistence curve,} \qquad \text{(1a)}$$

$$\partial X/\partial\Delta \propto (T_c-T)^{-\gamma'} \quad \text{along the coexistence curve,} \qquad \text{(1b)}$$

$$\partial X/\partial\Delta \propto (T-T_c)^{-\gamma} \text{ for } X=X_c \text{ and } T>T_c, \qquad \text{(1c)}$$

$$[\Delta(X)-\Delta(X_c)] \propto |X-X_c|^{\delta} \text{ for } T=T_c. \qquad \text{(1d)}$$

For the He^3-He^4 solutions we will characterize the exponents just defined by the subscripts "+" and "−" referring to the normal and superfluid phases, respectively. It is not obvious that corresponding plus and minus exponents are necessarily equal, but from experimental evidence[2] $\beta_+=\beta_-=1$. Griffiths[3] predicts the divergence

$$(\partial X/\partial\Delta)_T = A(T-T^*)^{-\gamma_+} \qquad (2)$$

at $X=X^*$. He made a "first guess" that γ_+ would be of the order of unity and suggested a scaling relation for both normal and superfluid phases, which implies that γ and γ' are identical and

$$\gamma_\pm = \delta_\pm - 1. \qquad (3)$$

Similarly one might guess the relations

$$[\Delta(X)-\Delta(X^*)]_{T^*} = B_\pm|X-X^*|^{\delta_\pm}[1+O(X-X^*)+\cdots] \qquad (4)$$

and

$$(\partial\Delta/\partial X) = C_\pm(T^*-T)^{-\gamma_\pm'}[1+O(T^*-T)+\cdots] \qquad (5)$$

along the phase-separation line.

These considerations suggested measurements of the saturated vapor pressure $P_{sat}(X)$ of the solutions since from these the chemical potential can be obtained quite readily. This paper describes such measurements and their analysis in terms of the critical exponents. Actually, the careful specific-heat measurements by Alvesalo *et al.*[4] give $\partial\Delta/\partial X$ along the phase-separation curve and when analyzed according to Eq. (5) give $\gamma_+'=0.95$ and $\gamma_-'=1.05$ with an uncertainty of ~10%.

A careful and extensive series of measurements of $P_{sat}(X)$ was already reported by Sydoriak and Roberts,[5] who used a conventional manometer system at room temperature and made corrections for the effect of the thermomolecular pressure at sufficiently low temperatures. Their data were taken at mole-fraction intervals of 0.1 and hence their measurements did not cover the region very close to T^* and X^*. It was accordingly difficult to derive from these measurements the variations of P_{sat} in this region and to calculate the change of Δ with X and T.

To eliminate the problem of thermomolecular pressures and to make data taking more convenient, we have designed a pressure-measuring strain gauge in thermal equilibrium with the vapor-pressure cell capable of resolving about 3 μm pressure. The solution fills to about 85% a flat cell of 2 mm height and 2.5 cm diam filled with copper wires to reduce thermal gradients. The pressure is transmitted via a hole at the top of the cell to a cylindrical space with a 4-mil stainless-steel membrane. This membrane in turn forms the electrically grounded plate of a capacitor, the second plate being a concentric steel ring with a clearance of 1 mil. The capacitor is part of an LC circuit powered by a tunnel diode and oscillating at a frequency of about 15 MHz measured by an electronic counter. The circuit has a frequency stability of 0.1 Hz over periods up to about 20 min, and a systematic drift only rarely exceeding 0.3 Hz over the time of an experiment, about 4 h. A germanium resistance thermometer is attached to this unit.

The strain gauge was calibrated against P_{sat} of pure He^3 at temperatures between 0.85 and 1.2°K, using a liquid-He^3 bath connected to a conventional mercury and oil manometer system as a thermometer. The vapor pressure as measured by the strain gauge was then used to calibrate the germanium thermometer below 0.85°K. Various solutions, prepared by mixing known amounts of He^3 and He^4, were introduced into the cell and condensed at 0.5°K. The pressure P_{sat} was then measured at increasing and decreasing T, the intervals varying between about 2 and 20 mdeg K. About 3 min were needed to achieve equilibrium after heating, as indicated by the constancy of pressure and temperature. Reproducibility of P_{sat} upon successive thermal cyclings was usually better than about 8 μm. Some of this uncertainty may have been caused by drift, hysteresis of the strain gauge, or even by some irreproducibility of the conditions in the liquid; and this needs further study.

As expected from thermodynamic considerations and also from previous work,[5] $P_{sat}(X)$ below the temperature of stratification is independent of X. However, the vapor-pressure measurements were not found as sensitive in determining the coexistence curve as have been sound-velocity measurements.[6] This already was evident in the measurements of Sydoriak and Roberts.[5] Hence we select as the coordinates for the tricritical point the values $X^* = 0.673$ and $T^* = 0.871$°K, most recently determined with precision by Alvesalo *et al.*[4] from specific-heat measurements.

Once pressure measurements at mole-fraction intervals of 0.1 and a few at 0.05 intervals were taken, the pressure of a solution of known volume with $X = 0.76$ was measured. This solution

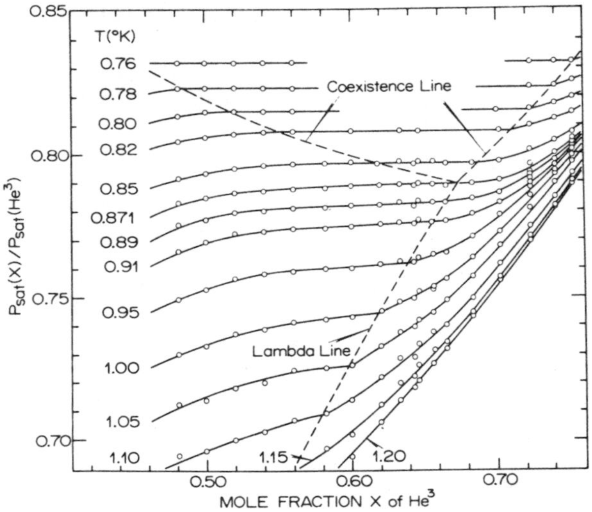

FIG. 1. The ratio $\mathcal{R} = P_{sat}(X)/P_{sat}(He^3)$ as a function of X (the mole fraction of He^3 in the liquid) for several representative isotherms, in the neighborhood of the tricritical point. The region where the isotherms are horizontal is the stratification region.

was then successively diluted in steps of 0.02 in the mole fraction, by condensing a metered amount of He^4 into the cell, until a solution of $X = 0.64$ was obtained. A similar series of dilutions was conducted for $0.48 \leq X \leq 0.66$ to explore carefully the region around the critical point. A total of about one thousand vapor-pressure readings were taken. Our values for $P_{sat}(X)$ were found to be systematically lower than those of Sydoriak and Roberts by an average of about 0.7% over $0.7 \lesssim T \lesssim 1.2$°K and $0.2 \lesssim X \lesssim 0.9$.

The data were analyzed by first plotting the ratio $\mathcal{R} = P_{sat}(X)/P_{sat}(He^3)$ vs T for the 28 solutions investigated, and smooth curves were drawn. This ratio was then replotted at given temperatures versus X. The portion of greatest interest in the diagram is shown in Fig. 1 for some assorted isotherms. In this representation, the phase separation curve appears "upside down" because for a given X, $\mathcal{R}$ increases with decreasing T, finally tending to unity.

Using the relations between the vapor pressure and the chemical potential, the Gibbs-Duhem relations, and making appropriate approximations,[1] one can show that

$$\left(\frac{\partial \Delta}{\partial X}\right)_T \approx \frac{RT}{1-X}\left(\frac{\partial \ln \mathcal{R}}{\partial X}\right)_T. \tag{5}$$

Using the plot given in Fig. 1, we obtained the quantity $(\partial \Delta/\partial X)_T$ as a function of X for a large number of isotherms. At the coexistence curve,

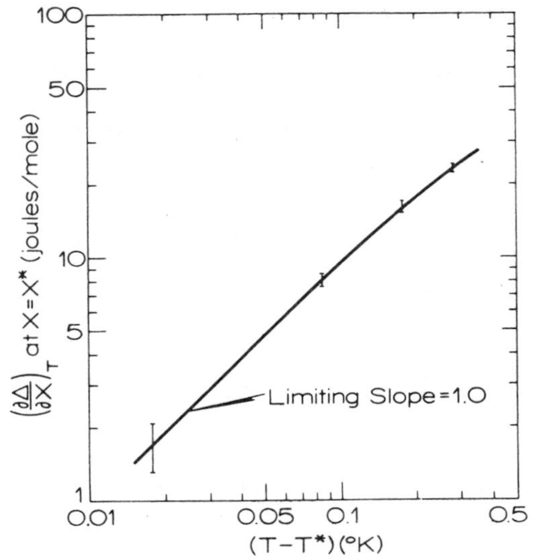

FIG. 2. Plot of the chemical potential gradient $(\partial\Delta/\partial X)_T$ at $X=X^*$ as a function of $T-T^*$. The error bars have been estimated as explained in the text.

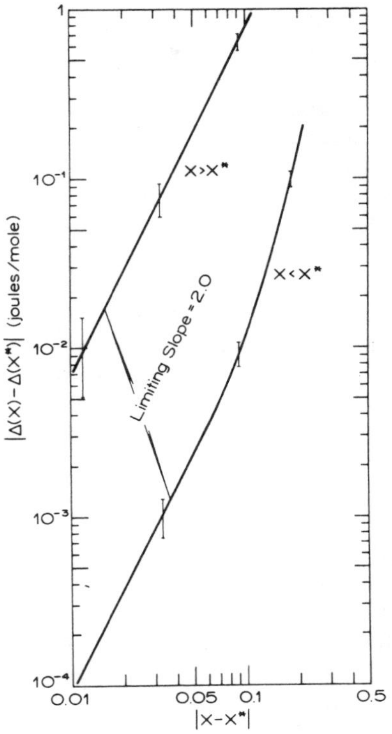

FIG. 3. Plot of $|\Delta(X)-\Delta(X^*)|$ vs $|X-X^*|$ along the critical isotherm, as explained in the text.

the values of $(\partial\Delta/\partial X)_T$ so obtained extrapolated smoothly into those calculated by Alvesalo *et al.*[4] from their specific-heat measurements for $0.51 < X < 0.72$. Evaluating $(\partial\Delta/\partial X)_T$ at $X = X^*$ and plotting this quantity versus $T-T^*$ (Fig. 2) gives a limiting slope of $\gamma_+ = 1.0 \pm 0.1$ as $T \to T^*$. In Fig. 2, no points have been plotted since $\partial\Delta/\partial X$ is the product of two smoothing processes—the first one in generating the isotherms in Fig. 2, the second one in differentiating the $\mathcal{R}$ vs X curve and in plotting the isotherms $(\partial\Delta/\partial X)_T$. The uncertainty bars at arbitrarily chosen $T-T^*$ values have been estimated from the possible systematic uncertainties traced back to the original measurements.

Similarly, with the use of these isotherms, the difference $\Delta(X)-\Delta(X^*)$ was formed by numerical integration along the critical isotherm. At $X = X^*$, there is a point of inflection and the initial slope $(\partial\Delta/\partial X)_T$ near X^* is considerably smaller for $X < X^*$ than for $X > X^*$, as can be seen from Fig. 3. This is already apparent from the calorimetric measurements of Alvesalo *et al.*[4] In fact for $X^*-X < 0.08$, where the change of $\mathcal{R}$ with X is smaller than its uncertainty due to the small irreproducibilities, the calorimetric results give the more reliable values of $(\partial\Delta/\partial X)_T$. Using an estimate of $\partial^2\Delta/\partial X\partial T$ near $T = T^*$ from the saturated-vapor-pressure measurements, we obtained $(\partial\Delta/\partial X)_T$ at $T = T^*$ in this mole-fraction range by an extrapolation of the calorimetric results. Within experimental uncertainty, the exponents δ_+ and δ_- were found to be equal to 2.0 $\pm$ 0.25. The error in our analysis leading to γ_+, $\gamma_\pm'$, and $\delta_\pm$ takes into account possible uncertainties in X^* and T^* of ± 0.005 and ± 0.003°K, respectively.

In summary, the limiting behavior of Δ (expressed in joules/mole) appears to be given by

$$\partial\Delta/\partial X \cong 0.7\times 10^2 (T-T^*)^{1.0} \quad (\text{at } X = X^*),$$

$$[\Delta(X)-\Delta(X^*)]_{T=T^*} \cong 75(X-X^*)^{2.0} \quad (X > X^*),$$

$$[\Delta(X^*)-\Delta(X)]_{T=T^*} \cong 9.7\times 10^{-1}(X^*-X)^{2.0} \quad (X < X^*).$$

The scaling relation, Eqs. (1), is therefore satisfied within the admittedly large experimental uncertainty. It is rather intriguing that for this tricritical point, the critical exponents appear to be positive integers within experimental error.

The λ line is located by the inflection point of the $\mathcal{R}$ vs X isotherms. If one accepts the Griffiths and Wheeler[7] theory of weak divergences, one would expect $(\partial\mathcal{R}/\partial X)_T = 0$ along this line.[8] However, the expected (weak) divergence in $\partial X/\partial\mathcal{R}$, even if it is present, may take place over such a small interval in X that it is, for practical purposes, unobservable.

In conclusion, we have established the limiting behavior of thermodynamic potentials at the critical point and have given preliminary values for the critical exponents. These satisfy the scaling relation [Eq. (2)], and one has in addition $\gamma_+ = \gamma_+'$ within experimental error. Hence the superfluid and normal regions appear to have at least closely the same critical indices. Experiments with a still more sensitive strain gauge are being attempted with the purpose of obtaining more accurate values of these exponents. Furthermore, a study of small instabilities and the hysteresis effects of the pressure readings will be made.

The authors are grateful to Dr. R. C. Richardson, Dr. B. Widom, Dr. W. Goldburg, Dr. R. B. Griffiths, Dr. E. K. Riedel, and Dr. W. Zimmermann for stimulating discussions, correspondence, and advice on the manuscript. They acknowledge the valuable help of Mr. R. Behringer in some of the analysis and the data taking. They are grateful to Mr. B. Wallace for the preparation of the solutions.

*Research supported by a grant from the National Science Foundation. The results were presented in a preliminary fashion in G. Goellner and H. Meyer, Bull. Amer. Phys. Soc. 14, 1308 (1970).

[1]For a review see the article by K. W. Taconis and R. DeBruyn Ouboter, in *Progress in Low Temperature Physics*, edited by C. J. Gorter (North-Holland, Amsterdam, 1964), Vol. 4, p. 38.

[2]E. H. Graf, D. M. Lee, and J. D. Reppy, Phys. Rev. Lett. 19, 417 (1967).

[3]R. B. Griffiths, Phys. Rev. Lett. 24, 715 (1970). In his paper, Griffiths used for the critical exponents the symbols μ $[=\delta^{-1}]$ and ϵ $[=\gamma]$.

[4]T. Alvesalo, P. Berglund, S. Islander, G. R. Pickett, and W. Zimmermann, Jr., in Proceedings of the Twelfth International Conference on Low Temperature Physics, Kyoto, Japan, September 1970, edited by E. Kanda (to be published).

[5]S. G. Sydoriak and T. R. Roberts, Phys. Rev. 118, 901 (1960).

[6]T. R. Roberts and S. G. Sydoriak, Phys. Fluids 3, 895 (1960).

[7]R. B. Griffiths and J. C. Wheeler, Phys. Rev. A 2, 1047 (1970).

[8]R. B. Griffiths, private communication.

Critical Magnetic Thermal Expansivity of $RbMnF_3$

Brage Golding
Bell Laboratories, Murray Hill, New Jersey 07974
(Received 8 April 1971)

Precision measurements of the magnetic thermal expansivity of $RbMnF_3$ near its critical (Néel) temperature are reported. The critical exponents are $\alpha = +0.007 \pm 0.02$ and $\alpha' = -0.100 \pm 0.03$, in disagreement with scaling predictions. The rounding about T_c due to sample inhomogeneity is observed to be asymmetric.

The experimental determination of the asymptotic temperature dependence of a magnetic solid's specific heat or thermal expansivity as its critical point is approached has proven extremely difficult. In contrast to the situation at the λ point of liquid helium,[1] or the gas-liquid critical point,[2,3] the rounding near T_c caused by the solid's inhomogeneity is large and poorly understood.[4] The resultant ambiguity in locating T_c and in determining the extent of the rounded region usually leads to large uncertainties in critical exponents. We report here measurements of the magnetic thermal expansivity of the isotropic cubic antiferromagnet $RbMnF_3$ in the vicinity of its critical (Néel) temperature, $T_c \simeq 83$ K. As a result of the sensitivity and accuracy obtainable with capacitive dilatometry, as well as the perfection of our $RbMnF_3$ crystal,[5] we have been able to establish with a high degree of certainty the location of T_c and the extent of the rounded region about T_c. The region over which the magnetic thermal expansivity obeys a simple power law shows a pronounced asymmetry about T_c, with substantially more rounding occurring below T_c. A simple power law with critical exponents $\alpha = +0.007$ and $\alpha' = -0.100$ gives the best fit to the expansivity data.

A brief discussion of the experimental techniques has been presented previously.[6] The magnetic part of the volume thermal expansivity[7] (at constant pressure and zero magnetic field), β_m, is shown in Fig. 1 as a function of $\log_{10}\epsilon$, where $\epsilon \equiv |T - T_c|/T_c$. We shall be concerned with determining the values for the exponents α and α' defined by the asymptotic forms[8,9]

$$\beta_m = (A/\alpha)(\epsilon^{-\alpha} - 1) + B, \quad T > T_c; \tag{1a}$$

$$\beta_m = (A'/\alpha')(\epsilon^{-\alpha'} - 1) + B', \quad T < T_c. \tag{1b}$$

In practice, experimental data can be expected to

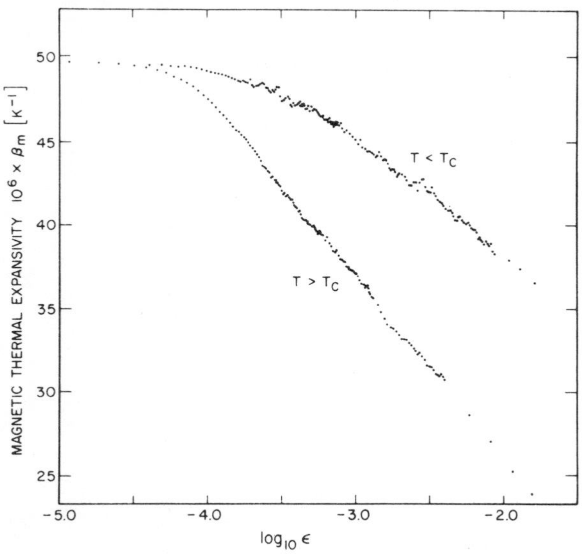

FIG. 1. The magnetic part of the volume thermal expansivity of $RbMnF_3$ as a function of $\log_{10}\epsilon$ ($\epsilon = |T - T_c|/T_c$).

obey Eqs. (1) only within bounds ϵ_{max} to ϵ_{min} for $T > T_c$ and ϵ_{max}' to ϵ_{min}' for $T < T_c$. The bounds ϵ_{max} and ϵ_{max}' are the largest values of ϵ within which corrections to Eqs. (1) are insignificant. The temperature dependence of β_m in the region ϵ_{min} to ϵ_{min}' containing T_c is dominated by phenomena related to the solid's inhomogeneity. Since the magnitude of β_m need not be symmetric about T_c, locating the thermal-expansivity maximum of a rounded transition does not, in general, constitute the establishment of T_c. In order to determine reliable values for α and α' we have utilized a self-consistent procedure for obtaining estimates of the ϵ ranges and the location of T_c which we now discuss in detail.

We fit Eq. (1a) [Eq. 1(b)] to the equally weighted experimental data using a linear least-squares procedure, regarding α, T_c^+, and ΔT^+ (α', T_c^-, and ΔT^-) as externally variable parameters. The last quantity in each case is the temperature interval between the lowest (highest) temperature point included in the fit and the expansivity maximum, and are introduced instead of ϵ_{min} (ϵ_{min}') since T_c is initially undetermined. The choice of ϵ_{max} (ϵ_{max}') is not critical for the data in the range shown in Fig. 1. Above T_c, for example, we fix ΔT^+ and vary α and T_c^+ so as to minimize the rms deviation σ^+ of the data points from the fitted curve. Figure 2(a) [Fig. 2(b)] shows α, T_c^+, and σ^+ (α', T_c^-, and σ^-) as functions of ΔT^+ (ΔT^-). The optimum range of the fit is chosen by the value of ΔT^+ (ΔT^-) which minimizes σ^+ (σ^-). T_c is determined by choosing that temperature which minimizes the total rms deviation $\sigma^2 = (\sigma^+)^2 + (\sigma^-)^2$.[10] This criterion is equivalent to minimizing σ subject to the constraint $T_c^+ = T_c^-$. We obtain in this manner the following parameters[11] for $RbMnF_3$:

$$\alpha = +0.007 \pm 0.02, \quad 2.5 \times 10^{-4} \leq \epsilon \leq 8.2 \times 10^{-3};$$

$$\alpha' = -0.100 \pm 0.03, \quad 4.4 \times 10^{-4} \leq \epsilon' \leq 1.65 \times 10^{-2};$$

$$T_c = 83.0425 \pm 0.002.$$

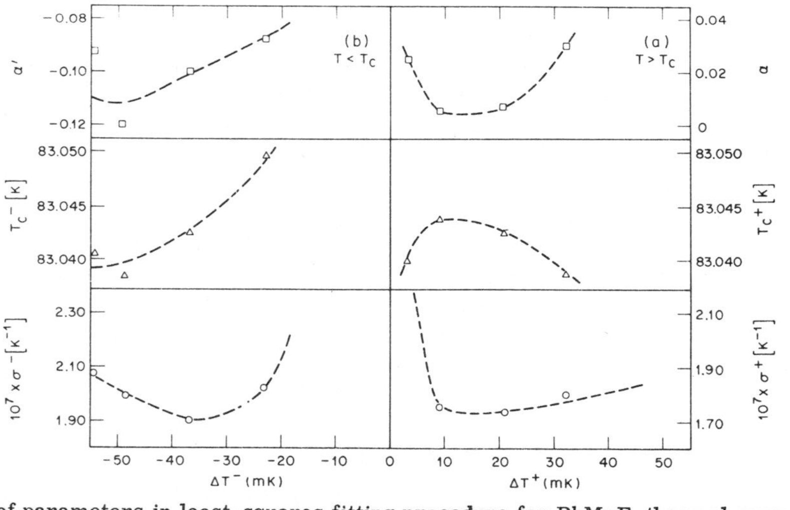

FIG. 2. Summary of parameters in least-squares fitting procedure for $RbMnF_3$ thermal expansivity as a function of range of fit. The ΔT's are the temperature differences between the expansivity maximum and the closest point allowed in the fit. For a given ΔT, the α and T_c shown are those values which minimize the estimated standard deviation σ.

The errors quoted for α and α' are approximately *two* standard deviations, calculated on the assumption that T_c and the ϵ ranges are known without error and that Eqs. (1a) and (1b) apply. These uncertainties, however, encompass the spread in α, α', and T_c values shown in Fig. 2 and are therefore a reasonable estimate of 90% confidence limits, including the uncertainties in T_c and the ϵ ranges. The location of T_c thus obtained is very close to the expansivity maximum T_m with $T_c - T_m = 0.0 \pm 0.004$ K.

Since there is as yet no detailed description of the deviations from a magnetic solid's ideal behavior near T_c due to its inhomogeneity, we shall define as the onset of rounding that temperature at which the measured thermal expansivity deviates systematically by 2σ from the best fit to Eq. (1). Above T_c, we observe an unusually small degree of rounding for a solid, with systematic deviations beginning only for $\Delta T \lesssim 6$ mK. Therefore, transition rounding does not become significant until $\epsilon \simeq 8 \times 10^{-5}$ above T_c, whereas below T_c, $\epsilon' \simeq 5 \times 10^{-4}$. This striking asymmetry in the rounded region about T_c has not been observed previously at a continuous magnetic transition. The asymmetry suggests that even in a solid with high crystalline homogeneity, there exists some inhomogeneous characteristic of the ordered phase. A source of inhomogeneity in a magnetically isotropic solid such as $RbMnF_3$ is the antiferromagnetic domain structure whose existence can be inferred below T_c. Since, however, domain boundaries in an antiferromagnet are stabilized at lattice imperfections (which are presumably not affected by magnetic ordering), it is not clear that the boundaries alone can give rise to the observed asymmetrical behavior.

The conclusion is reached that for $RbMnF_3$, $\alpha \neq \alpha'$, a result in disagreement with the equality of critical exponents about T_c predicted by static scaling theories.[12,13] The present analysis is entirely consistent with $\alpha = 0$, implying a logarithmic divergence of the specific heat above T_c for this three-dimensional isotropic Heisenberg antiferromagnet. Direct specific heat measurements have previously suggested a logarithmic divergence in $RbMnF_3$.[14] Our analysis cannot rule out the possibility that α is slightly negative, an occurrence which leads to an expansivity finite at T_c, but with a large jump at T_c, since $(A/\alpha - B) \times (A'/\alpha' - B')^{-1} \simeq 7$ (for $\alpha' = -0.01$).

We have considered the possibility that higher-order terms may contribute significantly in the region of fitting, i.e., that we are not able to penetrate the asymptotic region before rounding becomes significant, particularly below T_c. Such a possibility with the postulate $\alpha' \simeq 0$ would necessarily lead to $A > A'$, also in violation of scaling.[13,15,16] One might further speculate that the entire region $\epsilon' < 10^{-2}$ is rounded—a possibility that is suggested by our measurements and, for example, the specific heat of EuS,[17] which for $-2 < \log_{10}\epsilon < -1$ could be loosely interpreted as $\alpha = \alpha' = 0$, $A = A'$. Although this possibility could exist, it would seem highly fortuitous that one should arrive at $T_c^+ \cong T_c^-$ on the basis of independent fitting trials on either side of T_c unless the procedure and the analytic function had some physical content.

It is of interest to examine the scaling relations $d\nu = 2 - \alpha$ $(d\nu' = 2 - \alpha')$, where d is the dimensionality and ν (ν') is the critical exponent of the correlation length.[12,13] Recent neutron-scattering measurements have yielded $\nu = 0.72$ and $\nu' = 0.59$ for $RbMnF_3$.[18] The ratios $3\nu/(2-\alpha) = 1.08$ and $3\nu'/(2-\alpha') = 0.84$ differ considerably from the scaling prediction of unity. Since, however, the neutron data were obtained in a region rather far from T_c, e.g., $10^{-2} \lesssim \epsilon \lesssim 10^{-1}$ below T_c, it is our opinion that these discrepancies need not be significant. It would be necessary to have more extensive neutron measurements in a region much closer to T_c to afford a reliable test of these scaling relations.

The measurements and analysis presented here have allowed us to make, for an extremely simple magnetic system, what we believe to be highly reliable determinations of the critical exponents α and α'. The precision of the measurements and the high crystalline perfection of the $RbMnF_3$ sample have permitted the observation of a marked asymmetry about T_c in the influence of the solid's inhomogeneity on its critical thermal expansivity. Our observations strongly suggest that the dominant inhomogeneity below T_c is not structural, but some inhomogeneous characteristic of the ordered phase, possibly related to antiferromagnetic domains. The result $\alpha \neq \alpha'$ is in disagreement with scaling predictions.

The author thanks G. Ahlers for discussions and for his valuable suggestions. The experimental assistance of D. L. Simon is gratefully acknowledged.

[1]G. Ahlers, Phys. Rev. Lett. 23, 464 (1969), and Phys. Rev. A 3, 696 (1971).

[2]M. Barmatz and P. C. Hohenberg, Phys. Rev. Lett

24, 1255 (1970).

[3]J. A. Lipa, C. Edwards, and M. J. Buckingham, Phys. Rev. Lett. 25, 1086 (1970).

[4]By rounding we shall mean the systematic deviation of a measured quantity from its ideal behavior as $T \to T_c$.

[5]Grown by A. Linz, Center for Materials Science and Engineering, Massachusetts Institute of Technology.

[6]B. Golding, to be published.

[7]It is assumed that $RbMnF_3$ exhibits the macroscopic properties of a cubic crystal in the transition region. We therefore have $\beta = 3\alpha_L$, where $\alpha_L = (1/l)(dl/dT)$ is the measured linear thermal expansivity and l is a linear dimension of the crystal.

[8]See, for instance, M. E. Fisher, Rep. Progr. Phys. 30, 615 (1967).

[9]β and C_p are linearly related by the *exact* thermodynamic relation $\beta = aC_p + b$, where $a = (VT)^{-1}(\partial T/\partial P)_t$ and $b = -V^{-1}(\partial S/\partial P)_t$. The partial derivatives are evaluated along a path of constant $t = T_c - T$ and are only weakly temperature dependent [see M. J. Buckingham and W. M. Fairbank, in *Progress in Low Temperature Physics*, edited by C. J. Gorter (North-Holland, Amsterdam, 1961)]. Therefore, β and C_p will have the same asymptotic temperature dependence, i.e., both may be described by the critical exponents α and α'. The relative magnitudes of higher-order terms will not be the same generally, and the asymptotic regions for β and C_p may be considerably different.

[10]We emphasize that the minimum in σ has no fundamental or statistical significance. Significance is attached, however, to the rise in σ as ΔT is decreased beyond this point. The minimum in σ is convenient, therefore, as an approximate delimiter of the "rounded" region.

[11]We also have $A = 4.56$, $B = 4.68$, $A' = 5.49$, and $B' = 17.82$ in $10^{-6}\ K^{-1}$. The error in T_c is stated relative to the expansivity measurements. The possible error in the absolute temperature scale is ± 0.06 K.

[12]B. Widom, J. Chem. Phys. 43, 3892, 3898 (1965).

[13]R. B. Griffiths, Phys. Rev. 158, 176 (1967).

[14]D. T. Teaney, V. L. Moruzzi, and B. E. Argyle, J. Appl. Phys. 37, 1122 (1967).

[15]Because of the uncertainties in T_c and the extent of the rounding near T_c, a detailed investigation of the form of possible higher-order contributions has not proved fruitful. In liquid helium near the λ point, where T_λ can be measured very precisely and where the effect of the gravitational inhomogeneity on the specific heat can be easily calculated (Ref. 1), estimates of the magnitude of higher-order terms have been made and contribute significantly to C_p below T_λ in the region $\epsilon \gtrsim 3 \times 10^{-4}$.

[16]M. E. Fisher, to be published; see also the "Note added in proof" in Ref. 1.

[17]B. J. C. van der Hoeven, Jr., D. T. Teaney, and V. L. Moruzzi, Phys. Rev. Lett. 20, 719 (1968).

[18]H. Y. Lau, L. M. Corliss, A. Delapalme, J. M. Hastings, R. Nathans, and A. Tucciarone, J. Appl. Phys. 41, 1384 (1970).

ULTRASONIC PROPAGATION NEAR THE MAGNETIC CRITICAL POINT OF NICKEL

B. Golding and M. Barmatz
Bell Telephone Laboratories, Murray Hill, New Jersey 07974
(Received 17 June 1969)

Measurements of the attenuation and velocity of megahertz longitudinal-acoustic waves have been made near the magnetic critical point of nickel. The temperatures of the attenuation maxima and velocity minima occur below T_c and are frequency dependent. These phenomena, which have not been observed previously in a magnetic system, show striking similarities to the behavior of first sound near the λ point of liquid helium.

Anomalous sound propagation near magnetic critical points has been predicted theoretically and observed experimentally in many substances.[1,2] Previous ultrasonic investigations have associated the transition temperature with the measured attenuation maximum.[3] We have studied megahertz sound propagation near the critical point of nickel ($T_c \simeq 630°$K) and have observed that the temperatures of the attenuation maxima and velocity minima occur below T_c and are frequency dependent. Although these phenomena have not been seen previously in a magnetic system, our results show striking similarities to the behavior of first sound near the λ point of liquid helium.[4,5]

In our experiments we utilized high-purity ⟨001⟩ oriented Ni cylinders with resistivity ratios $R_{300}/R_{4.2} \simeq 1300$. The furnace-sample-holder design allowed temperature stability of ±1 mdeg (10^{-3} deg) for periods of several hours. Temperature was measured with a platinum resistance surface sensor in an ac bridge circuit. Changes in velocity and attenuation could be measured continuously (and simultaneously) with sensitivities $\Delta V/V \sim 10^{-7}$ in velocity and $\sim 5\times 10^{-3}$ dB/cm in attenuation. Critical-point sound-attenuation measurements were made in the temperature interval $10^{-6} < \epsilon < 3\times 10^{-3}$, where $\epsilon \equiv |1-T/T_c|$.

The attenuation changes for ⟨001⟩ longitudinal sound near T_c at 20 and 60 MHz are shown in Fig. 1(a). All data were taken after the establishment of equilibrium (~0.5 h/point). The temperature origin T_0 is arbitrarily referenced to the 20-MHz attenuation peak. At these frequencies the critical attenuation is quite small and appreciable only at temperatures within a few degrees of T_c. Well below T_c we attribute the increased attenuation to arise from the presence of ferromagnetic domains. Above T_c the attenuation approached a constant background level. The corresponding velocity changes are shown in Fig. 1(b) referred to the same temperature scale as the attenuation. The temperature dependence of the sound velocity well above the transition is linear with negative slope, but on approaching T_c the velocity undergoes a sharp decrease and passes through a minimum. The total velocity change near the transition is small, $(\Delta V/V)_{max} \sim 3\times 10^{-4}$.

Close examination of the temperature region near the 20- and 60-MHz attenuation maxima and velocity minima reveals that these extrema do not occur at the same temperature. This situation is shown in Fig. 2 and may be summarized as follows: (1) The 60-MHz attenuation peak is displaced 80 ± 20 mdeg below the 20-MHz peak, (2) at 20 MHz, the velocity minimum occurs 80 ± 15 mdeg below the attenuation peak, and (3) the separation between 20- and 60-MHz velocity minima is 100 ± 20 mdeg. These data represent the results of a number of experiments in which the velocity and attenuation were monitored continuously as the temperature was swept through the extrema at drift rates ~±0.5 mdeg/min. In some instances the velocity, attenuation, and temperature were recorded simultaneously.

These phenomena have not been seen in other magnetic systems for several reasons. First, the presence of impurities tends to obscure intrinsic critical behavior near T_c.[6] Also, these effects, which occur very close to T_c, necessitate a higher degree of temperature resolution and stability than achieved previously. In our nickel specimens, the temperature displacement of the attenuation and velocity extrema as well as the sharpness of the 20-MHz attenuation peak indicate that impurities or imperfections do not obscure the region near T_c. In liquid helium near the λ point, where problems due to impurities are minimal, similar intrinsic acoustic phenomena have been observed by several investigators.[4,5] We suggest that the shape and amplitude of the attenuation in the region of the maximum are characteristics of the finite frequency used in the measurement, as will be discussed below in greater detail.

Our interpretation of the attenuation measurements near the peaks is based on the basic idea that, at a given frequency, the temperature of the

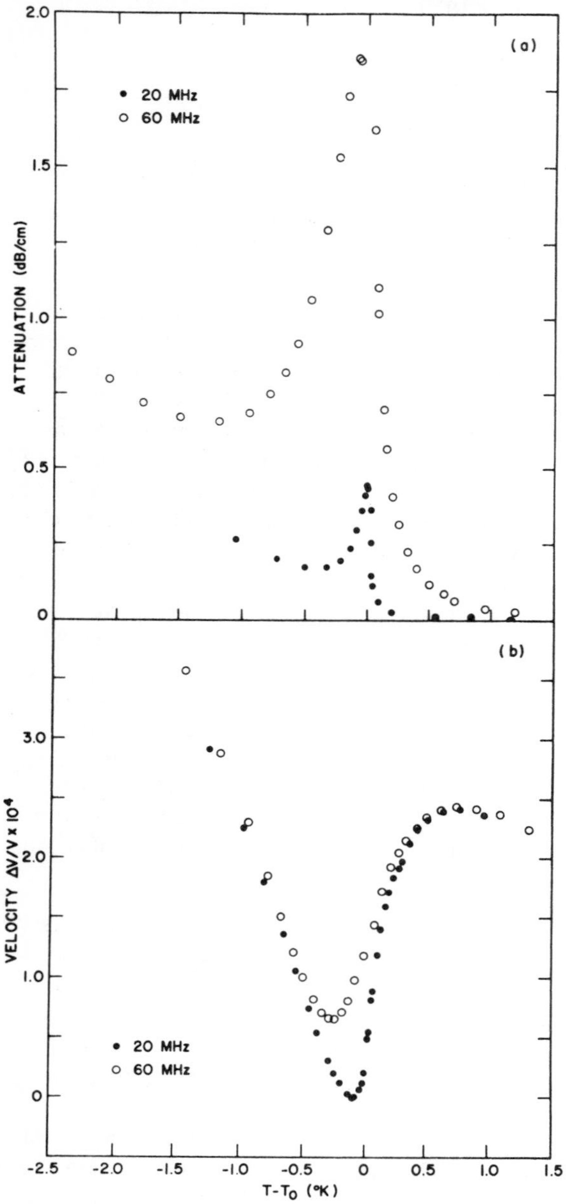

FIG. 1. Propagation of 20- and 60-MHz longitudinal sound along ⟨001⟩ in nickel near T_c. The origin of the temperature scale T_0 is arbitrarily referred to the 20-MHz attenuation peak. (a) The attenuation changes measured relative to the constant background attenuation well above T_c. (b) The dispersive effects in the velocity very near T_c can be clearly observed.

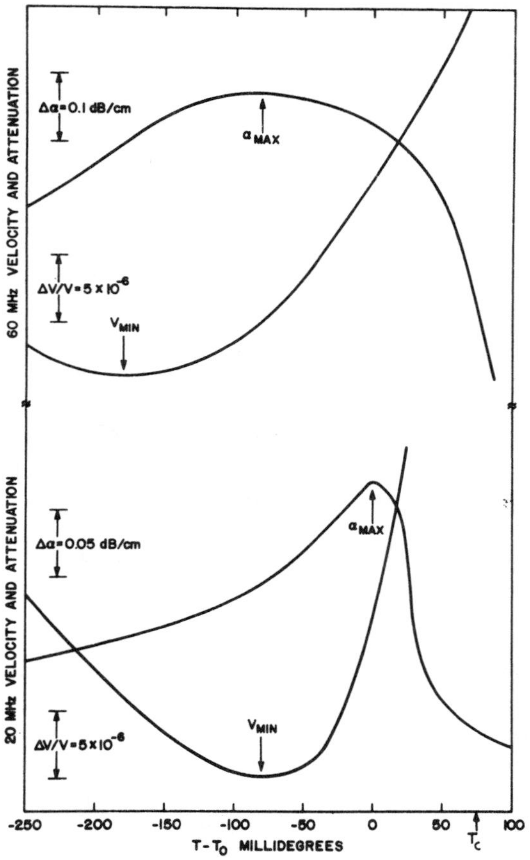

FIG. 2. Attenuation and velocity of 20- and 60-MHz sound on an expanded temperature scale near T_c. The zero of the temperature scale T_0 is arbitrarily referenced to the 20-MHz attenuation peak. The attenuation and velocity extrema shift to lower temperatures as the acoustic frequency increases. The arrow on the temperature axis indicates the position of T_c used in the data analysis.

maximum attenuation divides regions of different characteristic behavior. Since the attenuation maxima shift to lower temperatures with increasing frequency, we infer that T_c lies above these extrema. In liquid helium, because of the precision with which T_λ can be measured, it is known that similar finite-frequency acoustic extrema occur below T_λ.

To understand our results we make use of recently proposed arguments for the application of dynamic scaling ideas to sound propagation.[5,7,8] We begin by recognizing the importance of the temperature-dependent critical frequency ω_c of the critical mode,[9] i.e., that mode which dominates the frequency spectrum of the magnetization fluctuations at long wavelengths near the critical point. As $T-T_c \to 0^\pm$, $\omega_c^\pm \to 0$, which implies the existence of a temperature $T^\pm(\omega)$ at which $\omega_c^\pm = \omega$, where ω is the angular sound frequency. The temperatures $T^\pm$ divide regions with different acoustic properties.[10] In nickel, the critical frequency ω_c can be determined for $\epsilon > 4 \times 10^{-3}$ from recent inelastic-neutron-scattering experiments.[11] One obtains $\omega^+ = 1.9 \times 10^{14} \epsilon^{1.84}$ $\sec^{-1}$ and $\omega^- = 3.1 \times 10^{14} \epsilon^{1.72}$ $\sec^{-1}$ which can be compared with the dynamic scaling prediction for the temperature dependence of an isotropic Heisenberg ferromagnet, $\omega_c^{\pm} \sim \epsilon^{3\nu-\beta} \sim \epsilon^{5/3}$, where $\nu = \frac{2}{3}$ and $\beta = \frac{1}{3}$.

To indicate the validity of this interpretation we have made estimates of the attenuation-peak separation using two slightly different approaches. Using the ω_c^- derived from the neutron-scattering measurements, we calculate the difference between T^- at 20 and 60 MHz to be 160 mdeg, a result in rather good agreement with our experimental result of ~80 mdeg. It should be noted, however, that our measurements are confined to a region much nearer T_c than the neutron-scattering experiments. There is evidence from magnetization measurements[12] in Ni that the critical index β changes from a value near $\frac{1}{3}$ for $\epsilon > 5 \times 10^{-3}$ to a value near $\frac{1}{2}$ closer to T_c. These results raise the question of the correct temperature dependence of the critical frequency for $\epsilon < 5 \times 10^{-3}$. We have accordingly computed the 20- and 60-MHz peak separation consistent with $\beta = \frac{1}{2}$ and obtain a separation of ~40 mdeg. Either calculation yields a result sufficiently consistent with experiment to give confidence in this interpretation. Using the latter temperature dependence of ω_c^- and the experimental attenuation-peak separation, we estimate T_c to be ~75 mdeg above the 20-MHz attenuation maximum. This choice of T_c is indicated by the arrow on the temperature axis in Fig. 2.

We assume the critical attenuation α far from T_c to have the low-frequency form $\alpha = A\omega^2\tau$, where τ is a relaxation time which diverges at T_c as $\epsilon^{-\zeta}$. A reliable measure of the temperature dependence of α below T_c is complicated not only by the presence of a peak near T_c but by a noncritical attenuation probably associated with domains. For the 60-MHz measurements this noncritical contribution is approximately linear in $T_c - T$. An estimate of the critical exponent yields $0 \lesssim \zeta < 0.2$. Below T_λ in liquid helium the temperature dependence of the attenuation is the same as that of the appropriate inverse critical frequency, viz., $\tau = \omega_c^{-1}$. It is clear that this relation is not valid for Ni below T_c as it predicts $\zeta_- \simeq 1.8$. The possibility exists, therefore, that more than one relaxation time is necessary to explain the attenuation in this region.

Above T_c, we find a region in which the attenuation fits a power law. In order to estimate the critical exponent ζ_+ we have analyzed the 60-MHz data in the decade $3 \times 10^{-3} > \epsilon > 3 \times 10^{-4}$. Using a weighted least-squares fit we obtain $\zeta_+ = 1.4 \pm 0.1$, where the error represents the statistical accuracy of the fit (standard error) and does not reflect the uncertainty resulting from the choice of T_c. An uncertainty of T_c of ± 15 mdeg produces a variation of ζ_+ of ± 0.1. This measured ζ is less than the value 1.7 obtained from the relation $\tau = \omega_c^{-1}$. A recent calculation of ζ for a Heisenberg ferromagnet using the mode-mode coupling formalism[13] predicts $\zeta = 5/3 - 11\alpha/6 - \frac{1}{3}\eta + \frac{1}{8}\alpha\eta \simeq 1.45$, using critical exponents appropriate for Ni. We have also made the observation that the temperature at which the 20-MHz attenuation data depart from a power law is closer to T_c than for the 60-MHz data. This result is consistent with the idea that deviations are to be anticipated when $\omega \simeq \omega_c^+$.

The behavior of the velocity in the vicinity of T_c can be understood using an approach which has proven successful in liquid helium near the λ point.[5] The temperature dependence of the velocity near T_c arises from a thermodynamic ($\omega = 0$) term and a dispersive ($\omega \neq 0$) contribution. The velocity changes of the ⟨001⟩ longitudinal mode in our experiments can be regarded, to a good approximation, as representing the changes in the adiabatic bulk modulus. It can therefore be shown that the thermodynamic velocity must (1) decrease near T_c, reflecting the increase in specific heat and (2) reach a finite minimum at T_c. For finite frequency, dispersion will increase as the transition is approached and a velocity minimum may occur at $T \neq T_c$, if dispersion can overcome the decrease in thermodynamic velocity. As ω increases the minimum moves away from T_c since dispersive effects become more pronounced when $\omega \simeq \omega_c$ as can be observed in Fig. 2. A velocity minimum resulting from dispersion is not observed above T_c. This observation is due, in part, to the asymmetry in specific heat which leads to a stronger temperature dependence of the thermodynamic velocity above T_c.

In conclusion, our measurements are the first at a magnetic critical point in which intrinsic acoustic behavior very near T_c has been observed. We have shown that the critical extrema do not occur at T_c for finite frequency and that the temperatures of the attenuation maxima can be quantitatively related to the spin-wave mode. An acoustic relaxation time τ can be related to the critical frequency ω_c in the region well above T_c.

Below T_c, however, no similar association is possible, in contrast to the acoustic behavior of liquid helium in the region away from T_λ. In the near vicinity of T_c in nickel, both the attenuation and velocity of sound show features similar to those seen near T_λ in liquid helium.

We are indebted to P. C. Hohenberg for frequent discussions of dynamic scaling ideas. We also wish to acknowledge the experimental assistance of V. G. Chirba.

---

[1]L. P. Kadanoff, to be published.

[2]P. P. Craig and W. I. Goldburg [J. Appl. Phys. 40, 964 (1969)] give an excellent review of transport properties near a magnetic transition.

[3]See, for example, B. Golding, Phys. Rev. Letters 20, 5 (1968).

[4]C. E. Chase, Phys. Fluids 1, 193 (1958).

[5]M. Barmatz and I. Rudnick, Phys. Rev. 170, 224 (1968).

[6]R. J. Pollina and B. Lüthi, Phys. Rev. 177, 841 (1969).

[7]B. I. Halperin and P. C. Hohenberg, Phys. Rev. 177, 952 (1969).

[8]K. Kawasaki, Solid State Commun. 6, 57 (1968), and Progr. Theoret. Phys. (Kyoto) 39, 285 (1968).

[9]The critical frequency is defined as $\omega_c = \omega_0|_{k=\xi^{-1}}$, where ξ_0 is the frequency of the critical mode and ξ is the correlation length. For an isotropic ferromagnet ω_0 corresponds to spin waves for $T < T_c$ and spin diffusion for $T > T_c$.

[10]There is another dividing line defined by $k\xi = 1$, where k is the sound wave vector and ξ is the correlation length. For our frequency range this boundary occurs $\sim 10^{-5}$°K from T_c which is beyond our experimental resolution.

[11]M. F. Collins, V. J. Minkiewicz, R. Nathans, L. Passel and G. Shirane, Phys. Rev. 179, 417 (1969).

[12]D. G. Howard, B. D. Dunlap, and J. G. Dash, Phys. Rev. Letters 15, 628 (1965).

[13]G. E. Laramore, thesis, University of Illinois, 1969 (unpublished).

NUCLEAR MAGNETIC RESONANCE IN MnF_2 NEAR THE CRITICAL POINT*

P. Heller†
Gordon McKay Laboratory, Harvard University, Cambridge, Massachusetts
and
G. B. Benedek
Department of Physics, Massachusetts Institute of Technology, Cambridge, Massachusetts
(Received May 8, 1962)

The F^{19} nuclear resonance in MnF_2, first observed by Shulman and Jaccarino,[1,2] has been studied in detail in the very interesting temperature region around the paramagnetic-antiferromagnetic critical temperature $T_N = 67.4°K$. Our experimental system permits temperature measurement and control to within one millidegree for as long as six hours, any temperature variation across the sample being much less than a millidegree. Temperatures were measured with a platinum resistance thermometer calibrated very carefully using the vapor pressure-temperature curve for nitrogen.[3] An inductive coupling arrangement used with a very high frequency spectrometer,[4] plus a standard Pound-Knight-Watkins spectrometer, made possible nuclear magnetic resonance observations at frequencies from 23 to 110 Mc/sec without making any changes inside the low-temperature apparatus. Nuclear resonances of width up to 3 Mc/sec could be detected with this system.

We have determined the temperature dependence of the F^{19} resonance frequency ν_{19} in zero applied field in the antiferromagnetic state between T_N and 52°K. Our data start at $T = T_N - 0.005K°$ at which temperature ν_{19} is about 8 Mc/sec. This frequency corresponds to a reduction of the sublattice magnetization to 5% of the magnetization at absolute zero. For the first few degrees below the Néel point, the data very accurately fit the expression

$$\frac{M(T)}{M(0)} = \frac{\nu_{19}(T)}{\nu_{19}(0)} = A(T_N - T)^R, \quad (1)$$

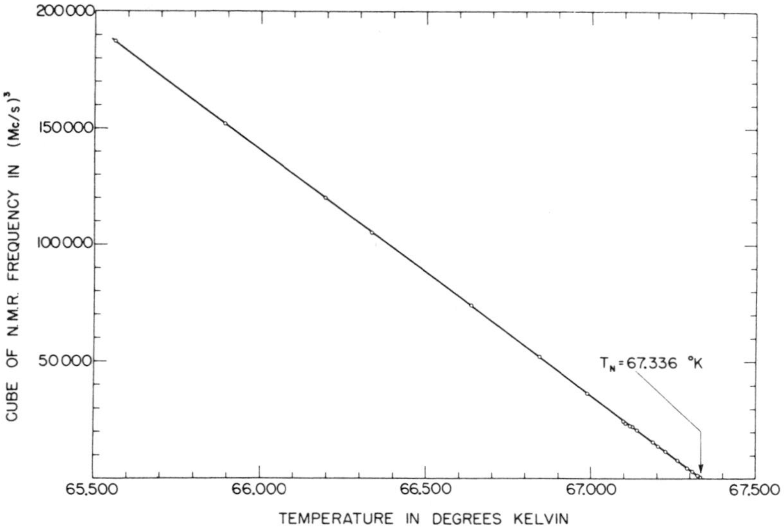

FIG. 1. Temperature dependence of the cube of the F^{19} nuclear resonance frequency for the first 1.8 degrees below T_N. The points lie on the straight line shown to within the experimental uncertainty of about 5 millidegrees.

where $R = 0.335 \pm 0.01$, $A = (0.295 \pm 0.001)(\mathrm{K}^\circ)^{-R}$, and $\nu_{19}(0) = 159.978$ Mc/sec. The accuracy with which Eq. (1) fits the data can be seen in Fig. 1 which plots $\nu_{19}{}^3(T)$ versus the temperature for the first 1.8 degrees below T_N. Extrapolation over the last 5 millidegrees enables us to determine that

$$T_N = (67.336 \pm 0.003)^\circ\mathrm{K}. \qquad (2)$$

Since Eq. (1) does not have the correct asymptotic behavior as $T \to 0$, it cannot be expected to hold far below T_N. The experimental results depart from the values given by Eq. (1) by 0.02K°, 0.10K°, and 0.5K° at 61°K, 58.5°K, and 53°K, respectively. The experimental data for the range $52^\circ\mathrm{K} < T < T_N$ are plotted in Fig. 2. The cube root law that we observe shows that the sublattice magnetization rises much faster than the law $M \propto (T_N - T)^{1/2}$ predicted by the molecular field theory just below T_N. The molecular field theory prediction is shown together with our data for $\nu_{19}(T)$ in Fig. 2. The precision of the nuclear resonance measurements, as illustrated in Figs. 1 and 2 and Eq. (1), demonstrates the power of this method[5,6] in determining the sublattice magnetization in the theoretically difficult, but important, critical region.

By observing the shift of the nuclear resonances on applying a magnetic field H_0 along the C (antiferromagnetic) axis, we were able to determine the susceptibility $\chi_\parallel$ for each of the two Mn^{++} sublattices separately near T_N. The susceptibilities of the up and down sublattices were found to be equal and their sum is in good agreement with the ordinary macroscopic susceptibility measurements.[7]

In the antiferromagnetic state the F^{19} linewidths are strongly anisotropic, depending on the direction of the vector sum $\vec{H}_{\mathrm{nucl}}$ of the applied field and the field produced by the Mn^{++} spins at the F^{19} nucleus. The lines are broadest for $\vec{H}_{\mathrm{nucl}}$ along C and narrowest for $\vec{H}_{\mathrm{nucl}}$ along A. (See Fig. 3.) For $\vec{H}_{\mathrm{nucl}}$ along C, the linewidth is given to within 20% by

$$\delta\nu_C = 490\ [\mathrm{kc/sec}\ (\mathrm{K}^\circ)^{1/2}](T_N - T)^{-1/2};$$

$$0.02\mathrm{K}^\circ < T_N - T < 15\mathrm{K}^\circ. \qquad (3)$$

In this 15-degree temperature range the linewidth increases from 100 kc/sec to 3000 kc/sec. In the paramagnetic state the line shows an extraordinarily rapid broadening as T_N is approached. The broadening is anisotropic, being more marked at a given temperature with the applied field along C than along A. (See Fig. 4.) For $0.04\mathrm{K}^\circ < T - T_N < 10\mathrm{K}^\circ$ the linewidths are given to within 15% by

$$H_0 \text{ along } C\text{:}\quad \delta\nu_C = 95\ \mathrm{kc/sec}\ [1 + 0.33\mathrm{K}^\circ/(T - T_N)],$$

$$H_0 \text{ along } A\text{:}\quad \delta\nu_A = 85\ \mathrm{kc/sec}\ [1 + 0.12\mathrm{K}^\circ/(T - T_N)]. \qquad (4)$$

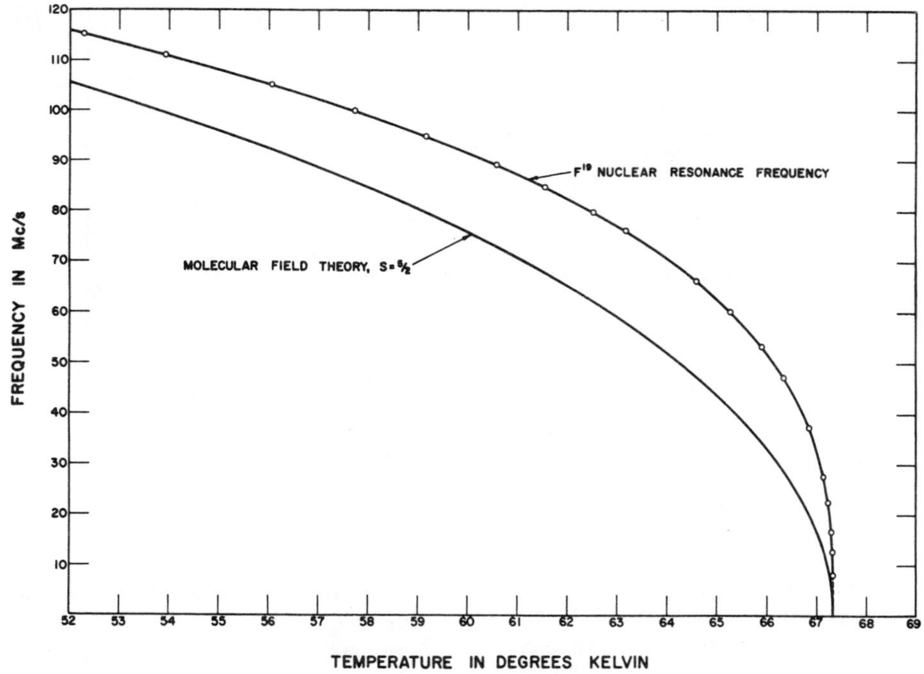

FIG. 2. Temperature dependence of the F^{19} resonance frequency in the range 52°K $< T < T_N$. The circles greatly exaggerate the size of the experimental uncertainties. The lower curve is computed using the molecular field theory.

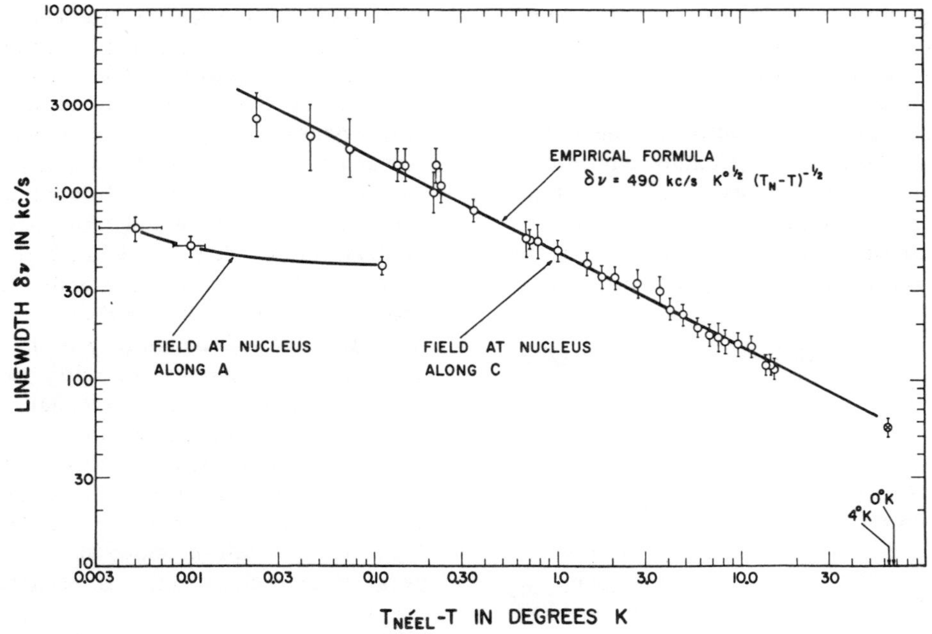

FIG. 3. Temperature dependence of the F^{19} linewidths in the antiferromagnetic state. The single point at 4°K is from reference 2.

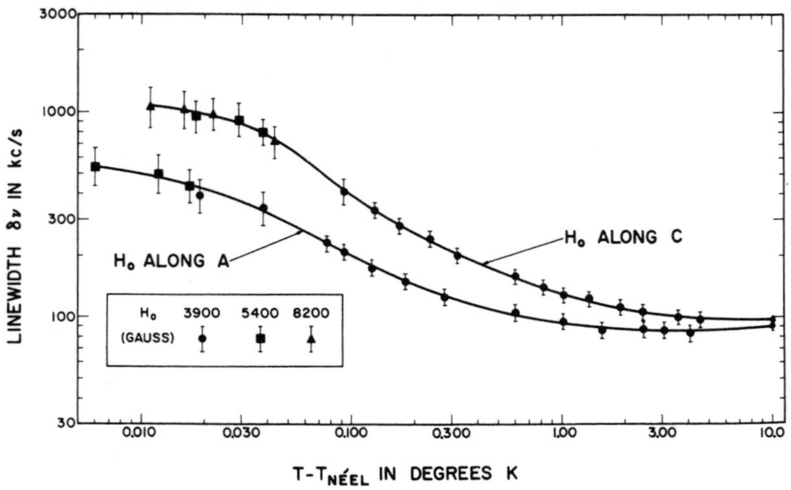

FIG. 4. Temperature dependence of the F^{19} linewidths in the paramagnetic state just above the Néel point.

Note that in the last degree above T_N the line broadens by about a factor of ten. It has been proposed[1,8] that the linewidth of the F^{19} resonance is determined by exchange narrowing of the hyperfine interaction between the F^{19} nucleus and the nearest neighbor Mn^{++} spins. If this is the case, the present measurements provide rather direct experimental information on the temperature dependence of the time-space correlation function for the manganese spins in the critical region.

The rapid fade-out of the line which accompanies this broadening provides an extremely sensitive method for measuring changes in the Néel temperature. In particular we have measured the depression of the Néel temperature due to an applied field. In Table I we list the results obtained, together with the predicted Néel point shifts on the molecular field model. We have also measured the effect of hydrostatic pressure on T_N. Our result,

$$dT_N/dP = 303 \pm 3 \text{ millidegrees per 1000 kg/cm}^2, \tag{5}$$

Table I. Effect on T_N of raising applied field H_0 from 5.4 kG to 8.25 kG.

	ΔT_N in millidegrees	
Field direction	Expt.	Mol. field theory
A axis	-2.8 ± 1	-0.87
C axis	-7.5 ± 1	-2.6

agrees well with the estimate of Benedek and Kushida[4] obtained by a quite different method. The pressure dependence of T_N shows that the Néel point is itself a function of the temperature because of the effect of thermal expansion. Using our pressure data and the thermal expansion measurements of Gibbons,[9] we have estimated the importance of this effect in fitting the $\nu_{19}(T)$ data to Eq. (1) with $T_N = T_N(T)$. Our estimate shows that, with this correction, the coefficient A in (1) should be about one percent less than stated above. The effect on the exponent R is negligible.

Just below T_N we observe, to our surprise, lines corresponding to nuclei in both the antiferromagnetic and paramagnetic phases, the line at the paramagnetic location fading rapidly with decreasing temperature and disappearing entirely at about 25 millidegrees below T_N. At 5 millidegrees below T_N the areas under the two lines are roughly equal. We do not understand this effect. One obvious possibility is that it is due to an inhomogeneity in the sample's Néel temperature caused, perhaps, by internal strains. Such a mechanism would broaden the line below T_N. However, the actual linewidths for $\vec{H}_{nucl}$ along A are several times larger than possible for such a mechanism. Furthermore, the linewidths for $\vec{H}_{nucl}$ along C fall off more slowly as the temperature is reduced than can be predicted on this basis.

We are greatly indebted to Dr. V. Jaccarino and Dr. H. Guggenheim of the Bell Telephone Laboratories for providing us with oriented single crystals of MnF_2. It is a pleasure to thank Dr. D. Gill, Dr. J. Jeener, and Dr. G. Seidel for many helpful discussions.

---

*Research supported by the U. S. Joint Services and Advanced Research Projects Agency.

†Raytheon Predoctoral Fellow 1959-60; Texaco Predoctoral Fellow 1960-61; now at Massachusetts Institute of Technology, Cambridge, Massachusetts.

[1]R. G. Shulman and V. Jaccarino, Phys. Rev. 108, 1219 (1957).

[2]V. Jaccarino and R. G. Shulman, Phys. Rev. 107, 1196 (1957).

[3]G. T. Armstrong, J. Research Natl. Bur. Standards 53, 263 (1954).

[4]G. B. Benedek and T. Kushida, Phys. Rev. 118, 46 (1960).

[5]N. J. Poulis and G. E. G. Hardeman, Physica 18, 391 (1952).

[6]V. Jaccarino and L. R. Walker, J. phys. radium 20, 341 (1959).

[7]J. W. Stout and M. Griffel, J. Chem. Phys. 18, 1455 (1950).

[8]T. Moriya, Progr. Theoret. Phys. (Kyoto) 16, 641 (1956).

[9]D. F. Gibbons, Phys. Rev. 115, 1194 (1959).

MAGNETIC EQUATION OF STATE OF $CrBr_3$ NEAR THE CRITICAL POINT*

John T. Ho and J. D. Litster
Physics Department and Center for Materials Science and Engineering, Massachusetts Institute of Technology, Cambridge, Massachusetts 02139
(Received 2 January 1969)

We present the results of analysis of our measurements of the magnetic properties of the insulating ferromagnet $CrBr_3$ near its critical temperature. From this analysis we obtain a mathematical representation of an equation of state valid over the entire critical region.

We have used the Faraday effect to measure the magnetization M of the insulating ferromagnet $CrBr_3$ as a function of field along 30 isotherms in the temperature range $T_C-0.9°K < T < T_C+6.7°K$, where the critical temperature is $T_C = 32.844°K$. In this Letter we present the results of an analysis of our measurements and report the first determination of a mathematical representation for the equation of state of a ferromagnet which is valid over the entire critical region. Our measurements[1] provided the first verification of the scaling hypothesis[2-4] for an insulating or localized-spin ferromagnet. Experiments to test the scaling hypothesis in metallic ferromagnets have been done by more conventional methods.[5-7]

We define the reduced magnetization $\sigma = M/M(0°K)$ and the scaled magnetization $m = \sigma|1-T/T_C|^{-\beta}$, where β gives the shape of the coexistence curve. Our data cover the range $0.01 \leq m \leq 6.5$ plus the critical isotherm where m is infinite. Scaling has been verified[5] in CrO_2 for $0.12 \lesssim m \lesssim 2.3$ but only for $T > T_C$. The experimental results of Kouvel and Comly[6] ($0.42 \lesssim m \lesssim 2.8$) and of Arrott and Noakes[7] ($0.23 \lesssim m \lesssim 4.6$) for nickel are consistent with scaling laws for both $T > T_C$ and $T < T_C$. A scaling law equation of state has been shown to hold for a number of fluids in the critical region.[8]

We first analyzed data near the critical isochore, the coexistence curve, and along the critical isotherm to determine the values $\gamma = 1.215 \pm 0.015$, $\beta = 0.368 \pm 0.005$, and $\delta = 4.28 \pm 0.1$. These exponents were used to compute the scaled magnetic field $h = H|t|^{-\beta\delta}$ and the scaled magnetization $m = \sigma|t|^{-\beta}$, where $t = (T-T_C)/T_C$.

The scaling laws predict that h is a function of m only, viz.,

$$h = h(m).$$

This is equivalent to the homogeneity argument of Widom[2] or the equation of state proposed by Griffiths.[4] Our measurements enable an experimental determination of the scaling function $h(m)$. The accuracy of our magnetization measurements is comparable to that of nuclear magnetic resonance and our data are sufficiently precise that we may use the experimental scaling function to establish the mathematical form of $h(m)$. This mathematical representation constitutes an equation of state valid over the entire critical region. (We define the critical region to be that over which the scaling laws hold. This is the region $-0.03 \lesssim t \lesssim 0.20$ with $\sigma \lesssim 0.45$.)

To obtain a mathematical representation for the scaling function we have followed three approaches. First, we have assumed[2,4] that the free energy is an analytic function of σ, H, and t everywhere in the one-phase region. We have expanded the magnetic field in a power series about the critical isochore ($\sigma = 0$, $t > 0$) and also in a power series about the critical isotherm ($t = 0$, $\sigma \neq 0$). This approach was very successful and one or the other of two simple series represents $h(m)$ well within experimental error everywhere within the one-phase region.

Our second approach was to attempt to find a single equation of simple form that would serve approximately over the entire critical region. Here we were not so successful. We found an equation superior to any that had been suggested previously, but it was clearly incorrect near the critical isochore. Our third approach was to seek a parametric equation of state following a suggestion of Schofield. This approach was more successful in that we obtained simple functions that fit almost within the scatter of experimental points throughout the critical region.

We first discuss the power series representations for the equation of state. To be consistent with the scaling laws,[4] the power series near the critical isochore must have the form

$$H = a_1 \sigma t^{\gamma} + a_3 \sigma^3 t^{\gamma - 2\beta} + a_5 \sigma^5 t^{\gamma - 4\beta} + \cdots, \tag{1a}$$

or the equivalent

$$h = a_1 m + a_3 m^3 + a_5 m^5 + \cdots. \tag{1b}$$

We found that this series with just three terms fits well within the experimental error for $0 \leqslant m \leqslant 1.8$ and $t > 0$. The coefficients are $a_1 = 0.89$ ($\pm 1\%$), $a_3 = 0.85$ ($\pm 6\%$), and $a_5 = 0.23$ ($\pm 20\%$) when the magnetic field is expressed in dimensionless units $\hat{H} = (g\mu_B / SkT_c)H$.

The power series about the critical isotherm has the form

$$H = b_0 \sigma^{\delta} + b_1 t \sigma^{\delta - 1/\beta} + b_2 t^2 \sigma^{\delta - 2/\beta} + \cdots, \tag{2a}$$

or its equivalent

$$h = b_0 m^{\delta} \pm b_1 m^{\delta - 1/\beta} + b_2 m^{\delta - 2/\beta} \pm \cdots. \tag{2b}$$

[The minus sign in (2b) is used for $t < 0$.] We found that three terms of this series fit our data well within experimental error from the coexistence curve ($t < 0$) to $m \geqslant 1.0$ (for $t > 0$). The coefficients are $b_0 = 0.63$ ($\pm 3\%$), $b_1 = 1.14$ ($\pm 5\%$), and $b_2 = 0.20$ ($\pm 25\%$). Widom[2] suggested that Eqs. (1) and (2) would be asymptotically valid for small and large values of m, respectively. The striking result we find is that with only three terms in each series, Eq. (1) represents our data for $0 \leqslant m \leqslant 1.8$ and Eq. (2) serves for $m \geqslant 1.0$. Therefore at any point in the critical region, one of the two equations is valid and there is considerable region of overlap (for $1.0 \leqslant m \leqslant 1.8$ and $t > 0$) where both series will suffice. In Fig. 1 we show the experimentally determined scaling function $h(m)$ and the fit provided by Eq. (1) (dashed line) and by Eq. (2) (solid line).

The two power series are the most accurate representation of the scaling function; however, a single function which is approximately correct over the entire critical region can be found rath-

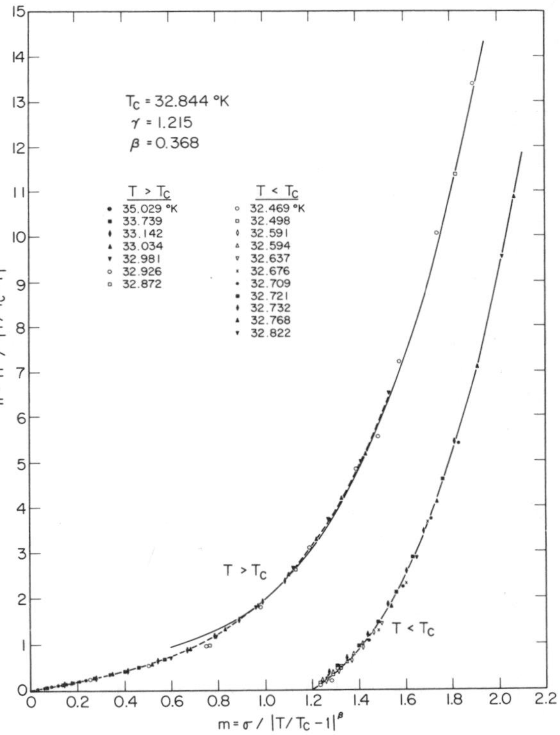

FIG. 1. A linear plot of scaled magnetic field $h(m)$ as a function of scaled magnetization m. The dashed line represents Eq. (1), and fits experimental points from $0 < m < 1.8$ when $T > T_c$. The solid lines are Eq. (2), and the dotted lines show Eq. (4) when it departs significantly from the experimental points. On this plot the origin is the critical isochore, the critical isotherm is at $h = m = \infty$, and the coexistence curve is represented by the point $h = 0$, $m \simeq 1.2$.

er easily. In the molecular field approximation the equation of state is

$$\hat{H} = A\sigma(t + B\sigma^{1/\beta}) \tag{3}$$

with $\beta = \frac{1}{2}$. Widom[2] has pointed out that (3) will yield the scaling laws if multiplied by a homogeneous function of degree $\gamma - 1$ of the variables t and $\sigma^{1/\beta}$. Choosing the simplest function we can think of, we obtain

$$\hat{H} = A\sigma(t + B\sigma^{1/\beta})(t + C\sigma^{1/\beta})^{\gamma-1}. \tag{4}$$

Although Eq. (4) is not analytic in σ along the critical isochore, it is a better approximation than other simple functions that have been tried.[7] If (4) is expanded about $\sigma = 0$, the second term is proportional to σ to the power $1 + 1/\beta = 3.7$. A careful examination of our experimental $h(m)$ shows the second term in an expansion of the correct equation of state varies as σ to the power 3.0 ± 0.3. Therefore we expect Eq. (4) must fail near the critical isochore. Our best fit was obtained with $A = 0.89$, $B = 0.61$, and $C = 2.12$; this is shown as the dotted line in Fig. 1. Agreement with the experimental results is good except for the range $0.2 \leq m \leq 1.2$; here the difference is about 10%, clearly outside the range of experimental error.

A single representation of the scaling function which is a very good approximation over the entire critical region can be obtained using the parametric form suggested by Schofield.[9] We make the following transformation of our measured thermodynamic quantities:

$$\hat{H} = ar^{\beta\delta}\theta(1-\theta^2), \quad t = r(1-2\theta^2), \quad \sigma = r^{\beta}g(\theta). \tag{5}$$

The singular behavior at the critical point is determined by the behavior as $r \to 0$ and we expect $g(\theta)$ to be a well-behaved function. With this particular transform the critical isochore is represented by $\theta = 0$, the critical isotherm by $\theta = 1/\sqrt{2}$, and the coexistence curve by $\theta = 1$. Eliminating r, one finds the scaled magnetic field to be

$$h(m) = a\theta(1-\theta^2)/|1-2\theta^2|^{\beta\delta}, \tag{6}$$

and the scaled magnetization is

$$m = g(\theta)/|1-2\theta^2|^{\beta}. \tag{7}$$

This transformation is similar to one suggested to us by Josephson[10] in which one represents h by a function similar to (6), but replaces (7) by

$$m = U/|1-2U^2|^{\beta}. \tag{8}$$

Using either transform one finds that $U(\theta)$ and $g(\theta)$ have similar mathematical forms.

The function obtained for $g(\theta)$ depends sensitively on the details of the transform chosen for $\hat{H}$ and t. Using the particular transform of Eq. (5) (with $a = 1.1$), we obtain the function $g(\theta)$ shown in Fig. 2. As can be seen from the figure, this can be very closely approximated by the equation $g(\theta) = k\theta$ with $k = 1.24 \pm 0.04$. One advantage of this parametric equation of state lies in the fact that it is readily integrated to obtain the free energy, and data for the entire critical region may be represented on one finite graph.

In conclusion, we remark that we have obtained several mathematical representations of the equation of state for $CrBr_3$ in the critical region. When only an approximate form is desired, Eq. (4) is suitable and preferable to other approximations that have been suggested.[7] A better approximation is obtained in the parametric form of Eq. (5), and this form is most amenable to

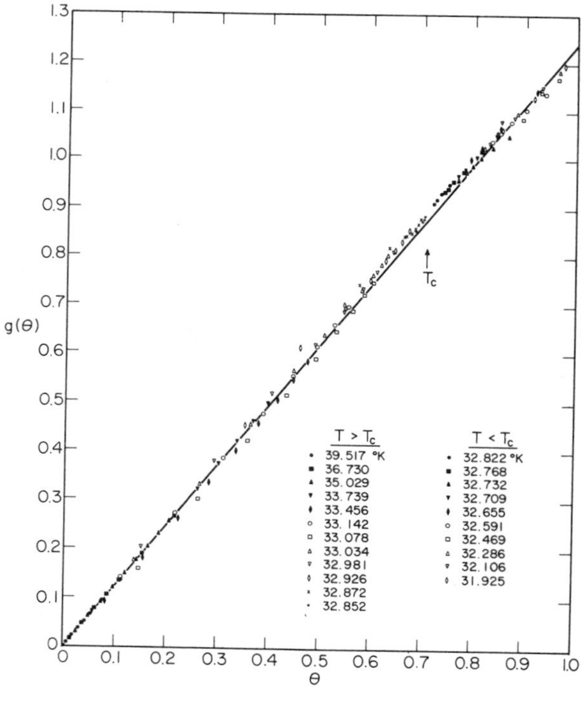

FIG. 2. A plot of the function $g(\theta)$ in the parametric equation of state, Eq. (5), obtained from an analysis of our $CrBr_3$ data with $a = 1.1$. The line has the equation $g(\theta) = 1.24\theta$.

mathematical analysis. The most accurate representation is provided by the two power series (1) and (2). These appear to be the best equations to check theoretical calculations when they become available and to compare various magnetic systems, pure fluids, and critical mixtures.

It is a pleasure to acknowledge a helpful collaboration with Dr. Peter Schofield and stimulating discussions with Professor George Benedek.

---

*Work supported by the Advanced Research Projects Agency (Contract No. SD-90) and by the National Aeronautics and Space Administration (Grant No. NGR-22-009-182).

[1]John T. Ho and J. D. Litster, in Proceedings of the 1968 International Conference on Magnetism and Magnetic Materials, Boston, Mass. (to be published).

[2]B. Widom, J. Chem. Phys. 43, 3898 (1965).

[3]L. P. Kadanoff, Physics 2, 263 (1966).

[4]R. B. Griffiths, Phys. Rev. 158, 176 (1967).

[5]J. S. Kouvel and D. S. Rodbell, Phys. Rev. Letters 18, 215 (1967).

[6]J. S. Kouvel and J. B. Comly, Phys. Rev. Letters 20, 1237 (1968).

[7]A. Arrott and J. E. Noakes, Phys. Rev. Letters 19, 786 (1967).

[8]M. S. Green, M. Vicentini-Missoni, and J. M. H. Levelt Sengers, Phys. Rev. Letters 18, 1113 (1967).

[9]P. Schofield, following Letter [Phys. Rev. Letters 22, 606 (1969)].

[10]B. Josephson, private communication.

Evidence of Mode-Mode Coupling and Nonlocal Shear Viscosity in a Binary Mixture near the Consolute Point*

C. C. Lai† and S. H. Chen
Nuclear Engineering Department, Massachusetts Institute of Technology, Cambridge, Massachusetts 02139
(Received 24 May 1972)

Linewidths of Rayleigh scattering from a binary liquid system of n-hexane plus nitrobenzene have been measured at two fixed scattering angles over a temperature range such that $\epsilon \equiv (T - T_c)/T_c$ extends from 10^{-2} to 10^{-6}. The ratio of the linewidths at two angles as a function of the correlation length ξ have been compared with recent self-consistent mode-mode coupling calculations of Kawasaki and Lo. The theory takes into account the nonlocality of shear viscosity near the critical point, and the data clearly establish the predicted effect in the critical region.

It has been well established both theoretically[1] and experimentally[2] that the decay rate Γ_k of the concentration fluctuation C_k deviates appreciably from the hydrodynamic expression K^2D as one approaches the critical region, as defined by $K\xi \geqslant 1$. In fact, an explicit $K\xi$ dependence of the diffusion coefficient $D(K\xi)$ has been given by Kawasaki[1] as

$$\Gamma_k \equiv K^2D(K\xi) = (k_B T/6\pi\eta^*)\xi^{-3}K_0(K\xi), \qquad (1)$$

where

$$K_0(x) = \tfrac{3}{4}[1 + x^2 + (x^3 - x^{-1})\tan^{-1}x] \qquad (2)$$

and η^* is loosely called the "high-frequency" shear viscosity. Experimental work of Berge *et al.*[3] indicated that with reasonable choices of ξ_0 and ν in the defining equation $\xi = \xi_0\epsilon^{-\nu}$, η^* could be taken as a constant in fitting the Rayleigh linewidth data over the entire temperature range covered. Since then there have been new measurements[4] indicating that far away from the critical point η^* approaches $\eta(T)$, the hydrodynamic shear viscosity, and near the critical point it deviates from $\eta(T)$ but not entirely temperature independently. Based on these observations, Kawasaki and Lo[5] recently made an attempt to clarify the meaning of η^* by solving self-consistently the coupled equations of C_k and V_q, the transverse local velocity fluctuation.

Far away from the critical point, the time dependence of these two fluctuating modes assumes the hydrodynamic expression $C_k(t) \sim \exp(-K^2Dt)$

and $V_q(t) \sim \exp(-q^2\eta t/\rho)$ and both transport coefficients D and η are local quantities. As one approaches the critical region the increased order-parameter fluctuations cause appreciable nonlinear coupling between C_k and V_q. Since the fluctuation of V_q is always much faster than C_k because of the finiteness of the shear viscosity and the critical slowing down of the order-parameter fluctuation, the stochastic variable $C_k(t)$ can be taken to be Markovian in the time scale of observation by light scattering. Thus the exponential decay of $C_k(t)$ is valid even when the coupling is taken into account except now $D \rightarrow D(K\xi)$, i.e., the diffusion coefficient becomes nonlocal. In the self-consistent treatment, this latter statement is also true for the shear viscosity.

The Kawasaki-Lo[5] treatment gave the following essential results:

$$\eta_q = \eta(t)[1 - F(q\xi)], \tag{3}$$

$$\eta^* = \eta(T) K_0(K\xi)/K(K\xi), \tag{4}$$

$$\Gamma_k = [k_B T/6\pi\eta(T)]\xi^{-3}K(K\xi), \tag{5}$$

where functions $F(x)$ and $K(x)$ are related through Eq. (8) of Ref. 5 and numerical values of both are given there. Two essential points are to be re-

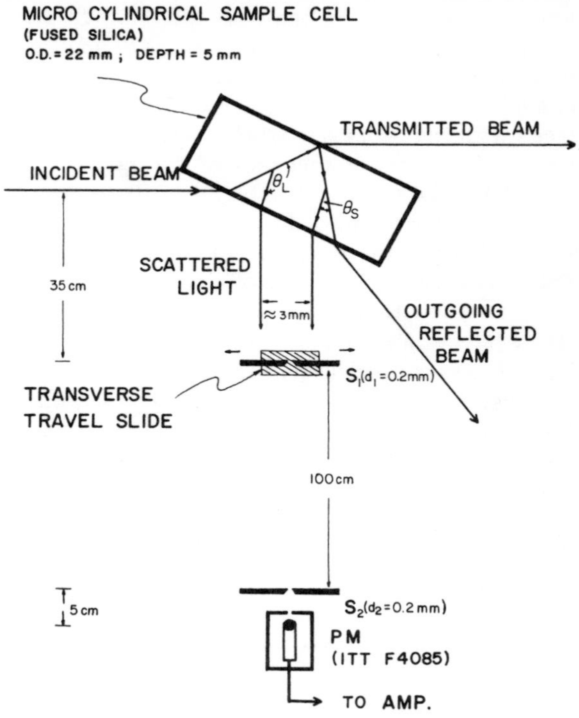

FIG. 1. Scattering geometry for measurement at two fixed scattering angles.

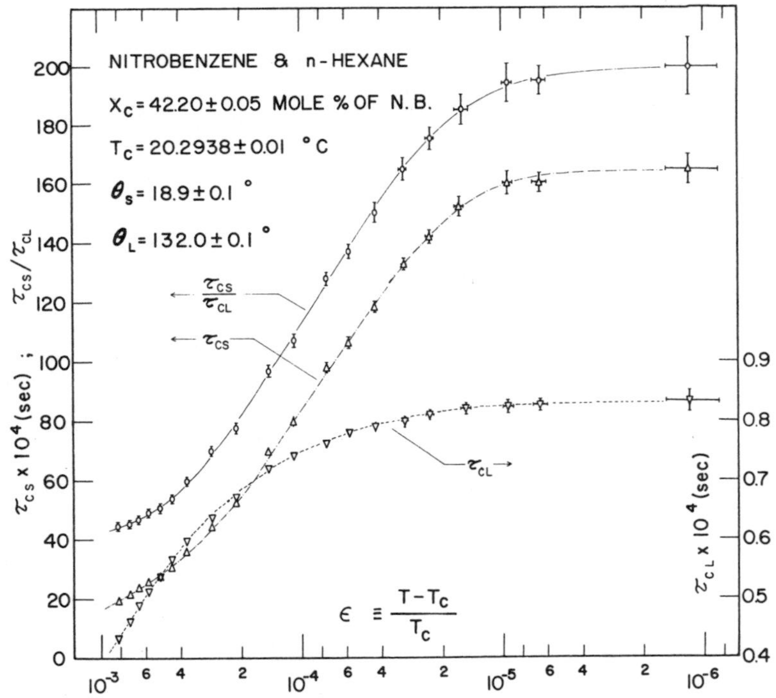

FIG. 2. Fluctuation decay constants and their ratio at small and large scattering angles as a function of reduced temperature.

marked here. Both the nonlocal and the high-frequency viscosities are smaller than $\eta(T)$ and the nonlocal shear viscosity reduces to $\eta(T)$ in the limit $K\xi \to 0$. In this limit $K(0)=1.055K_0(0)$ which agrees with the observation[4] that $\eta^* \to \eta(T)$. However, the real test of accuracy of the theory comes from the critical region $K\xi \geqslant 10$, where $K(x)$ can be 20% or more larger than $K_0(x)$. The obvious check of the theory is to use the experimental values of $\eta(T)$ and to plot the product $\Gamma_k \eta(T)$ as a function of $\xi(T)$ at a fixed K. Since in the system of n-hexane plus nitrobenzene an accurate measurement of $\eta(T)$ is not yet available,[6] we choose to test the theory in the following way.

We perform the linewidth measurement at two fixed K values as a function of temperature, taking advantage of the fact that the correlation length depends only on temperature along the critical isoconcentration line. Thus according to Eq. (5)

$$\frac{\Gamma_{kL}}{\Gamma_{kS}} \equiv \frac{\tau_{cS}}{\tau_{cL}} = \frac{K(K_L\xi)}{K(K_S\xi)}. \tag{6}$$

Note that if η_q were local, i.e., $F(x)=0$, $\eta^* \simeq \eta(T)$, then Eq. (1) would have predicted

$$\frac{\Gamma_{kL}}{\Gamma_{kS}} = \frac{\tau_{cS}}{\tau_{cL}} = \frac{K_0(K_L\xi)}{K_0(K_S\xi)}. \tag{7}$$

The purpose of this experiment is therefore to distinguish between two possibilities, (6) and (7).

The linewidth Γ_k is measured by the technique of intensity correlation spectroscopy.[7] The scattered photons are detected by a photomultiplier and the resulting pulses fed into a 128-channel digital autocorrelator.[8] The output of the correlator, except for a known dc background, is proportional to $|\langle C_{-k}(0)C_k(t)\rangle|^2 \sim \exp(-2\Gamma_k t)$. The line-

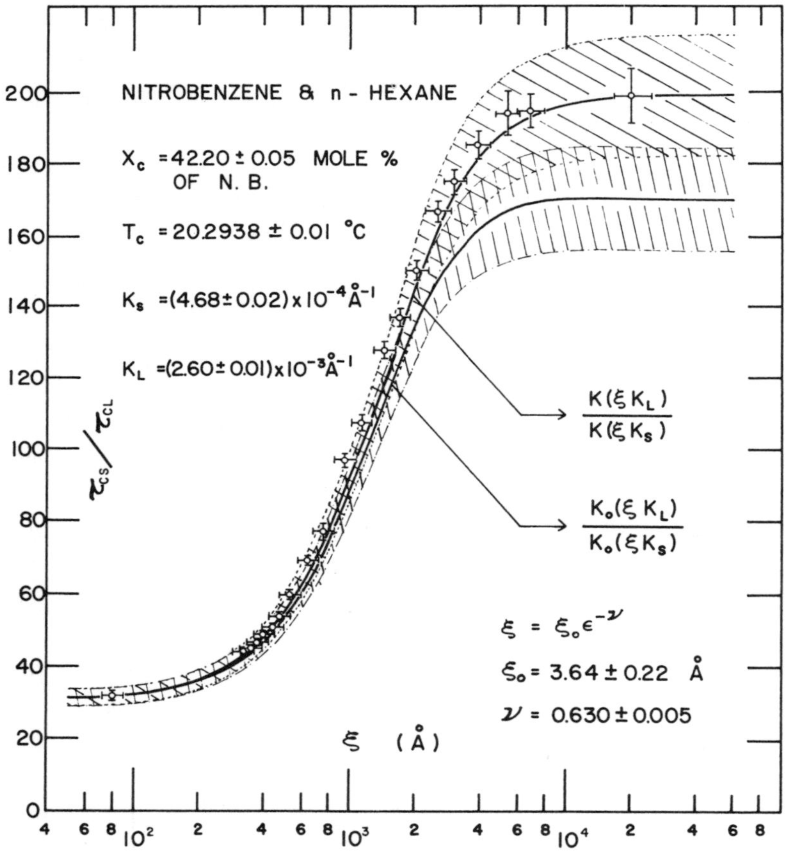

FIG. 3. Comparison of measured decay time constant ratio as a function of correlation length with theories of Kawasaki and Lo (Ref. 5) and Kawasaki (Ref. 1) The shaded areas around the theoretical curves take into account the indicated uncertainties in K values.

width thus determined is accurate to better than a percent. The schematic arrangement of the scattering geometry is given in Fig. 1. This particular geometry is chosen because it is extremely important to maintain a constant temperature for the measurement at both K_L and K_S. Two consecutive measurements at each temperature were made with only slight displacement of slit S_1. This allows us to measure scatterings from two angles with a minimal disturbance to the system. For example, we do not have to shift the laser beam during the whole measurement. The temperatures of the cell are constantly monitored by a thermistor and are maintained to about 0.1 mdeg stability. The part of the linewidth data which is relevant to the present article is given in Fig. 2. Figure 3 gives the comparison between the ratio of the linewidths measured and the two theoretical expressions (6) and (7).

In plotting Fig. 3 the crucial points are the conversion of measured temperature T to the correlation length ξ and the accurate determination of K_L and K_S. First, the T_c is located by observation of transmitted light and also by the linewidth data above and below the phase separation. The error of T_c given in Fig. 3 reflects only the calibration accuracy of the thermister, but the relative value of $\Delta T \equiv T - T_c$ is believed to be accurate to ± 0.1 mdeg. K_L and K_S are determined by both an optical method and the linewidth measurement in the hydrodynamic limit where one knows $\Gamma_k = K^2 D$.

The values of ξ_0 and ν are determined by a separate intensity measurement[9] which gives $\xi_0 = 3.64 \pm 0.22$, $\nu = 0.63 \pm 0.005$. These values are consistent with values reported in literature and also agree with the earlier measurement of Chen and Polonsky[7] on the same system.

In conclusion we have shown that accurate linewidth data decisively prefer expression (6) over (7) in the critical region. The change of values of ξ_0 and ν within a reasonable range would not change substantially this conclusion. This method of comparison with the theory has the advantage that one can avoid the uncertainty in values of $\eta(T)$ extremely close to the critical point due to the fact that the viscosity measurement is seldom reliable when ΔT is less than 1 mdeg.

The authors are grateful to Dr. K. Kawasaki for providing the numerical value of $K(x)$ prior to publication and for many informative discussions. We also acknowledge a conversation with Dr. B. Chu in which he informed us that his recent linewidth measurement in the system isobutyric acid plus water gives the expected variation of η^* with temperature.

*Research supported by the National Institute of Health under Contract No. NIH-70-2013 and by the Sloan Fund for basic research.

†Based in part on work submitted to the Nuclear Engineering Department of the Massachusetts Institute of Technology for the partial fulfillment of the Ph. D. degree requirement.

[1]K. Kawasaki, Ann. Phys. (Paris) 61, 1 (1970); R. A. Ferrell, Phys. Rev. Lett. 24, 1169 (1970), and references contained in these two papers.

[2]B. Chu, F. J. Schoenes, and W. P. Kao, J. Amer. Chem. Soc. 90, 3042 (1968); S. H. Chen and N. Polonsky, Opt. Commun. 1, 64 (1969); P. Berge *et al.*, Phys. Rev. Lett. 23, 693 (1969).

[3]P. Berge *et al.*, Phys. Rev. Lett. 24, 1223 (1970).

[4]P. Berge and M. Dubois, Phys. Rev. Lett. 27, 1706 (1971).

[5]K. Kawasaki and S. M. Lo, Phys. Rev. Lett. 29, 48 (1972). See also R. Perl and R. A. Ferrel, Phys. Rev. Lett. 29, 51 (1972).

[6]We have sent our sample to Dr. G. F. Allen for $\eta(T)$ determination and therefore this way of comparison should be possible in the near future.

[7]S. H. Chen and N. Polonsky, Opt. Commun. 1, 64 (1969); E. Jakeman and E. R. Pike, J. Phys. A: Proc. Phys. Soc., London 2, 411 (1969).

[8]The construction of the digital autocorrelator was described in a paper by R. Nossal and S. H. Chen.

[9]C. C. Lai and S. H. Chen, "Light Scattering Intensity and Correlation Length of a Binary Critical Mixture" (to be published).

TEST OF DYNAMIC SCALING BY NEUTRON SCATTERING FROM $RbMnF_3$ †

H. Y. Lau, L. M. Corliss, A. Delapalme,* J. M. Hastings, R. Nathans, and A. Tucciarone
Brookhaven National Laboratory, Upton, New York 11973
(Received 6 October 1969)

Measurements are presented of the inelastic magnetic scattering of neutrons from $RbMnF_3$ in the vicinity of the critical point. The data are analyzed quantitatively in terms of the dynamic scaling proposal of Halperin and Hohenberg and provide strong confirmation of the theory.

In a recent paper[1] Halperin and Hohenberg proposed a generalization of the static scaling laws to dynamic phenomena by making assumptions about the behavior of time-dependent correlation functions in the vicinity of the critical point of "second-order" phase transitions. They write the Fourier transform of the symmetrized space-time correlation function for the operator A, in terms of wave vector $\vec{q}$, frequency ω, and inverse range parameter κ, in the general form

$$C_\kappa^A(\vec{q}, \omega) = [2\pi/\omega_\kappa^A(\vec{q})]C_\kappa^A(\vec{q})f_{\vec{q},\kappa}^A[\omega/\omega_\kappa^A(\vec{q})],$$

where the characteristic frequency $\omega_\kappa^A(\vec{q})$ is defined by the condition

$$[\omega_\kappa^A(\vec{q})]^{-1}\int_{-\omega_\kappa^A(\vec{q})}^{\omega_\kappa^A(\vec{q})} f_{\vec{q},\kappa}^A[\omega/\omega_\kappa^A(\vec{q})]d\omega = \tfrac{1}{2},$$

and where the spatial transform $C_\kappa^A(\vec{q})$ is assumed to obey static scaling.[2] The dynamic scaling assumptions are (i) that the characteristic frequency $\omega_\kappa(\vec{q})$ is a homogeneous function of q and κ,

$$\omega_\kappa(\vec{q}) = q^E\Omega(q/\kappa),$$

and (ii) that the form of the frequency-dependent function $f_{q,\kappa}$ depends only on the ratio q/κ and not on $\vec{q}$ and κ separately. These scaling assumptions are used to relate the behavior of the correlation function in the "critical" region ($q/\kappa \gg 1$) to that calculated for the two "hydrodynamic" cases ($q/\kappa \ll 1$, with $T < T_N$ or $T > T_N$). Using this matching procedure, and the theory of spin waves for $T < T_N$, Halperin and Hohenberg predict that $\omega_\kappa(\vec{q}) \sim q^{1.5}$ at T_N and $\omega_\kappa(0) \sim \kappa^{1.5}$ for $T > T_N$.

We have measured the inelastic neutron scattering from $RbMnF_3$ in the "critical" region and have analyzed our data quantitatively, including resolution and instrumental effects. Our results strongly support the concept of dynamic scaling and agree well with the specific predictions of the theory.

$RbMnF_3$ has the simple cubic perovskite structure in the paramagnetic state and becomes antiferromagnetic below 83°K. In the ordered state it exhibits negligible magnetic anisotropy, no measurable distortion from cubic symmetry, and appears to be an ideal Heisenberg antiferromagnet.[3] Measurements were made near the $(\frac{1}{2}, \frac{1}{2}, \frac{1}{2})$ point of reciprocal space using longitudinal constant-$\vec{q}$ scans ($\vec{q} \parallel 2\pi\vec{\tau}$) and incident neutron energies of 6.6, 13.0, and 47.0 meV. The zone boundary in this direction is 0.644 Å^{-1}. The sample temperature was regulated to better than 10 mdeg using an ac resistance bridge[4] and a calibrated platinum resistance thermometer. The inverse correlation range κ was obtained as a function of temperature from the analysis of our two-axis data.[5]

The calculated scattered intensity at the instrumental setting $(\vec{q}_0, \omega_0)$ is obtained by convolution of the cross section with the resolution function:

$$I(\vec{q}_0, \omega_0) = P(\omega_0) \int \frac{d^2\sigma}{d\Omega d\omega}(q, \omega) R(\vec{q}-\vec{q}_0, \omega-\omega_0) d\vec{q} d\omega,$$

where P contains an instrumental constant and a slowly varying correction for the energy dependence of the counter sensitivity and the reflectivity of the analyzing crystal. $\vec{q}$ is the momentum transfer measured from the $(\frac{1}{2}, \frac{1}{2}, \frac{1}{2})$ spot to a general point in reciprocal space, and ω is the energy transfer. The resolution function was calculated analytically[6] using experimentally determined values of mosaic spread and collimation parameters. The cross section for $T \geqslant T_N$ can be written[7]

$$\frac{d^2\sigma}{d\Omega d\omega}(\vec{q}, \omega) \propto \frac{k_f}{k_i} |f(\vec{Q})|^2 \frac{1}{1+e^{-(\omega/kT)}} C_\kappa(\vec{q}, \omega),$$

where $\vec{k}_i$ and $\vec{k}_f$ are the initial and final neutron wave vectors, and $f(\vec{Q})$ is the magnetic form factor.

At $T = T_N$ and for values of q greater than about 0.1 Å^{-1}, where resolution corrections are small, the uncorrected energy spectra of scattered neutrons exhibit three unresolved peaks. We have chosen, therefore, to analyze the data at T_N for all q in terms of the dynamic correlation function

$$C_{\kappa=0}(\vec{q}, \omega) \propto \frac{A}{q^{2-\eta}} \left[\frac{\Gamma_1}{\Gamma_1^2+\omega^2} + \frac{B\Gamma_2}{\Gamma_2^2+(\omega+\omega_s)^2} + \frac{B\Gamma_2}{\Gamma_2^2+(\omega-\omega_s)^2}\right],$$

where $A/q^{2-\eta}$ is the static part, B is an arbitrary constant, and the energy widths Γ_1, Γ_2, and ω_s (all in meV) are taken to be proportional to q^E with proportionality constants C, G, and D, respectively. This choice for the energy widths automatically ensures that $C_\kappa(\vec{q}, \omega)$ satisfies the assumptions of dynamic scaling along the line $\kappa = 0$. Inserting this form of the correlation function into the expression for the cross section using $\eta = 0.067$,[5] we have performed least-squares fittings of the data for the range 0.05 Å$^{-1} \leqslant q \leqslant$ 0.25 Å^{-1}, which is experimentally accesible with an incident energy of 13 meV. A very good fit to the data is obtained with an exponent $E = 1.4 \pm 0.2$, in agreement with the predicted value of 1.5. Figure 1 shows the observed data taken at 13 meV and calculated curves based on the least-squares parameters $B = 1.82$, $C = 5.00$, $G = 9.66$, $D = 16.0$, and $E = 1.4$. As can be seen from the scale in Fig. 1, there is a factor of 100 between the peak intensities for $q = 0.05$ Å^{-1} and $q = 0.25$ Å^{-1}. This fit has been obtained using a single normalization constant to place all the intensity data on an absolute basis.

A second test of dynamic scaling was made in the hydrodynamic region ($q \ll \kappa$) above T_N, where $C_\kappa(\vec{q}, \omega)$ is assumed to have the "diffusion" form

$$C_\kappa(\vec{q}, \omega) \propto \frac{1}{(\kappa^2+q^2)^{1-\eta/2}} \frac{\Gamma(q, \kappa)}{[\Gamma(q, \kappa)^2+\omega^2]}.$$

As noted earlier, dynamic scaling theory predicts that the width $\Gamma(q, \kappa)$ (or characteristic frequency) should vary as $\kappa^{1.5}$ for $q = 0$. In making a measurement at a nominal setting of $q = 0$, nonzero values of q are simultaneously sampled because of finite instrumental resolution and hence the dependence of Γ on q is required in order to perform the convolution of the cross section with the resolution function. To accord with the theory of dynamic scaling and the expected analyticity for $q = 0$, $\Gamma(q, \kappa)$ has been taken to have the form

$$\Gamma(q, \kappa) = \Gamma_0(\kappa)[1 + c(q/\kappa)^2],$$

where $\Gamma_0(\kappa)$ is the characteristic frequency for $q = 0$. The insert in Fig. 2 shows a typical least-squares fit from which $\Gamma_0(\kappa)$ is obtained at a given value of κ. Figure 2 gives the observed $\Gamma_0(\kappa)$ in meV as a function of $\kappa^{1.4}$ together with the line $\Gamma_0(\kappa) = d\kappa^E$ drawn for the best values of d and E which are 10.8 and 1.4. The exponent E agrees well, within the estimated error of ±0.2, with the

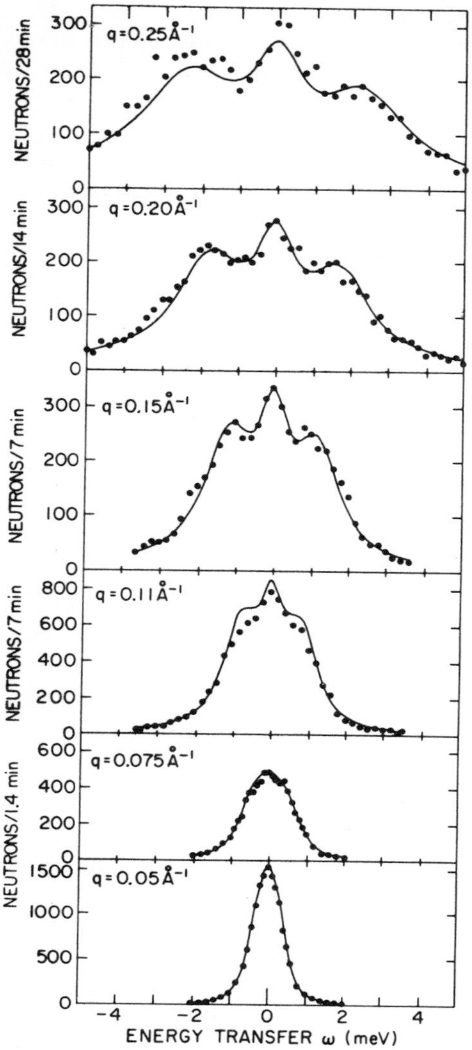

FIG. 1. Calculated and observed intensities as a function of energy transfer at T_N for different momentum transfers q. Observed data taken with incoming neutron energy of 13 MeV.

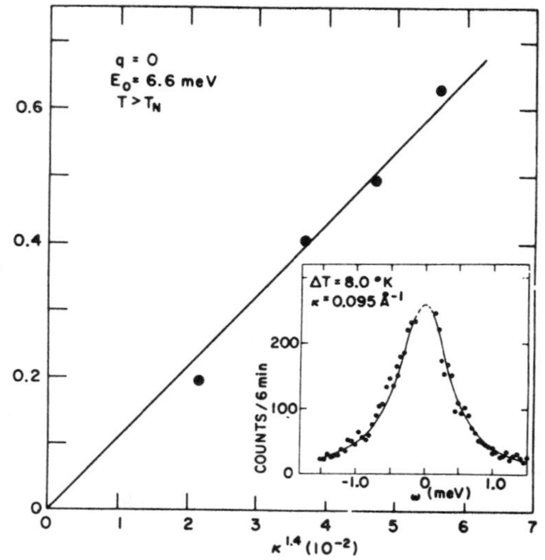

FIG. 2. Plot of $\Gamma_0(\kappa)$, the characteristic frequency at $q=0$, as a function of $\kappa^{1.4}$. The line corresponds to $\Gamma_0(\kappa)=d\kappa^E$ for the best values of d and E, which are 10.8 and 1.4. The insert shows a typical fit of calculated and observed intensities, for $q=0$ and $T_N = 8°K$. The central points have been deleted because of the interference of a weak nuclear reflection coming from a higher-order contamination.

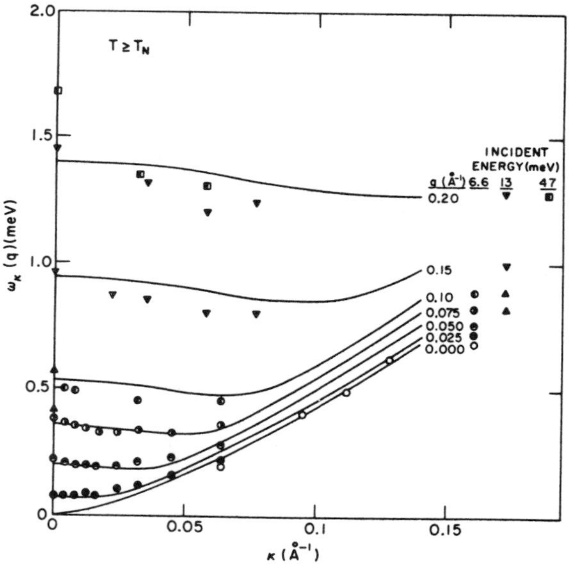

FIG. 3. Comparison of observed and calculated characteristic frequencies $\omega_\kappa(q)$ for $T \geq T_N$. Observed frequencies are half-area values, uncorrected for resolution, and are labeled by different symbols, according to momentum transfer and incident neutron energy. Calculated curves are labeled according to the value of q.

predicted value of 1.5.

The basic homogeneity assumption of dynamic scaling was further investigated by measurements at and above T_N for general points in the (q, κ) plane where resolution corrections to the characteristic frequency are negligible. Characteristic frequencies were determined directly from the uncorrected data by integration. Following Halperin and Hohenberg,[1] we assumed a simple homogeneous form for the characteristic frequency:

$$\omega_\kappa(q) = (aq^6 + bq^4\kappa^2 + c'q^2\kappa^4 + d'\kappa^6)^{E/6},$$

where a, b, c', d', and E are constants. This expression was used to fit 51 observed characteristic frequencies in the range $0<q \leq 0.2$ Å^{-1} and $0 \leq \kappa \leq 0.128$ Å^{-1}. The observed frequencies in meV are shown in Fig. 3 together with calculated curves based on the best values of the constants: $a=6.7\times10^4$, $b=-1.14\times10^5$, $c'=1.27\times10^5$, $d'=2.69\times10^4$, and $E=1.4$. In addition to providing a satisfactory fit to the observed characteristic frequencies, these constants are in good agreement with those previously obtained in fitting the data for $q=0$ and for $\kappa=0$.

In the present experiment three separate tests of dynamic scaling have been performed. In all cases quantitative agreement was obtained with the predictions of the theory. In addition, the three sets of results show a mutual consistency required by the theory, thus providing further confirmation of the concept of dynamic scaling.

We are grateful to P. C. Martin and P. C. Hohenberg for extensive discussions in the course of these experiments.

---

†Research performed under the auspices of the U. S. Atomic Energy Commission.

*Present address: Centre d'Etudes Nucléares, Grenoble, France.

[1]B. I. Halperin and P. C. Hohenberg, Phys. Rev. 177, 952 (1969). Related work has been done by R. A. Ferrell et al. and K. Kawasaki. See Halperin and Hohenberg for references.

[2]Our notation differs slightly from that of Halperin and Hohenberg. We use $\vec{q}$ and κ in place of $\vec{k}$ and ξ^{-1} for wave vector and inverse range parameter in order to conform to usage in neutron scattering. The superscript A, indicating the operator to which the correlation function refers, will henceforth be omitted for simplicity; in the present paper this operator is always the staggered magnetization. It is to be noted, also, that $\hbar$ has been set equal to unity throughout.

[3]D. T. Teaney, M. J. Freiser, and R. W. H. Stevenson, Phys. Rev. Letters 9, 212 (1962); D. T. Teaney, V. L. Moruzzi, and B. E. Argyle, J. Appl. Phys. 37, 1122 (1966).

[4]Designed by R. L. Chase, Brookhaven National Laboratory.

[5]L. M. Corliss, A. Delapalme, J. M. Hastings, H. Y. Lau, and R. Nathans, J. Appl. Phys. 40, 1278 (1969).

[6]M. J. Cooper and R. Nathans, Acta Cryst. 23, 357 (1967).

[7]The use of a symmetrized space-time correlation function introduces the thermal factor $(1+e^{-\omega/kT})^{-1}$ in the equation for the cross section. $C_\kappa(\vec{q},\omega)$ can be expressed in terms of the imaginary part of the generalized susceptibility $\chi''(\vec{q},\omega)$ by means of $C_\kappa(\vec{q},\omega)=\coth(\omega/kT)\chi''(\vec{q},\omega)$ [L. P. Kadanoff and P. C. Martin, Ann. Phys. (N.Y.) 24, 419 (1963)].

Kawasaki-Einstein-Stokes Formula and Dynamical Scaling in the Critical Region of a Binary Liquid Mixture: Isobutyric Acid in Water*

S. P. Lee, W. Tscharnuter, and B. Chu

Chemistry Department, State University of New York at Stony Brook, Stony Brook, New York 11790

(Received 3 April 1972)

The Rayleigh linewidth Γ of isobutyric acid in water has been measured by means of signal correlation at the critical solution concentration between 0.003 and 30°C above the critical mixing temperature T_c over a range of scattering angles θ varying from 20° to 140°. The results are analyzed according to the mode-mode coupling theory of Kawasaki in the $K\to 0$ limit, and at $K\xi \gg 1$.

Halperin and Hohenberg[1] have proposed that for all values of $K\xi$ the decay rate Γ should be a homogeneous function of K and ξ^{-1}, where the same correlation length ξ governs the critical behaviors of both static and dynamic properties of the system,

$$\Gamma = K^Z H(K\xi), \tag{1}$$

with Z being the degree of homogeneity and $K = (4\pi/\lambda)\sin(\frac{1}{2}\theta)$. According to the mode-mode coupling theory of Kadanoff and Swift,[2] Eq. (1) becomes $\Gamma = BK^Z$ in the critical limit ($\xi K \gg 1$), where B is a temperature-independent constant and $Z = 3$; and reduces to $\Gamma = DK^2$ in the hydrodynamic limit ($\xi K \ll 1$), where D is the binary diffusion coefficient. A specific form of $H(K\xi)$ has been derived first by Kawasaki[3] and subsequently by Ferrell,[4] using the approximate Ornstein-Zernike form of the correlation function to include all values of X [$=(K\xi)^{-1}$]:

$$\Gamma = AK^3(2/\pi)[X + X^3 + (1 - X^4)\arctan(X^{-1})], \tag{2}$$

where $A = k_BT/16\eta_{hf}^*$ with η_{hf}^* being the high-frequency shear viscosity which is assumed to be independent of $K\xi$ in the theory. According to Kawasaki, the value of η_{hf}^* should lie in the interval $\eta_r^* < \eta_{hf}^* < \eta_r^* + \eta_s^* = \eta^*$. η_r^* is the value of the shear viscosity if there is no critical anomaly, while η_s^* is the singular part of the shear viscosity exhibiting the asymptotic critical behavior. Furthermore, Kawasaki has obtained a formula for the mutual diffusion coefficient of a binary mixture,

$$D = k_BT/6\pi\eta_{hf}^*\xi, \tag{3}$$

which is analogous in form to the Einstein-Stokes equation for the mass diffusion coefficient of spheres of radius r. The correlation length ξ diverges as $\xi = \xi_0\epsilon^{-\nu}$ with $\epsilon = (T - T_c)/T_c$ and ν being another critical exponent.

During the past two years there have been many comparisons between experiment and theory. Unfortunately, the first verifications by Berge *et al.*[5] and by Henry, Swinney, and Cummins[6] were invalid. Firstly, a best fit of Eq. (2) using three adjustable parameters of A, ξ_0, and ν invariably gives a good fit but incorrect magnitudes for the three parameters, especially for ξ_0 and ν. In fact, the Kawasaki equation (2) is quite insensitive to multiparameter fits. Even fairly poor data produce impressive log-log plots of Γ/K^3 versus $K\xi$. Secondly, Eq. (2) represents the theoretical Kawasaki linewidth due to critical contribution. Scaling of the thermal conductivity of carbon dioxide near the critical point by Sengers and Keyes[7] shows the presence of an appreciable background for one-component fluid systems. Since then, deviations from Eq. (2) have been observed for binary fluid systems of perfluoromethyl cyclohexane in carbon tetrachloride[8] and 3-methylpentane in nitroethane[9] without using A, ξ_0, and ν as adjustable parameters. Thus, η_{hf}^* should not be assumed to be independent of temperature even if we neglect background contributions. The emphasis has been in comparing experimental Rayleigh linewidth data of one-component fluid systems, such as xenon[10,11] and sulfur hexachloride,[12] with the Kawasaki theory. In the $K\to 0$ limit, the thermal diffusivity χ has been separated into two parts: $\chi = \Lambda_r/\rho C_p + k_BT/6\pi\eta_{hf}^*\xi$, where Λ_r, ρ, and C_p are the regular nondivergent or background thermal conductivity, the density, and the specific heat at constant pressure, respectively. The second term is the Kawasaki critical contribution. η_{hf}^* is assumed to be constant in Eq. (3). On the other hand, the hydrodynamic shear viscosity may be expressed as[13]

$$\eta^* = (E/\alpha)(\epsilon^{-\alpha} - 1) + F\epsilon + G, \tag{4}$$

where E, F, and G are constants and α is another exponent. Thus, the separation of χ (or D) into regular and singular parts may not be straightforward since η_{hf}^* appears in the denominator.

A slightly different approach for comparing experiment with the Kawasaki theory is to utilize Eq. (3). Berge and Dubois[14] have shown the validity of Eq. (3) over large temperature ranges by taking $\eta_{hf}^* = \eta^*$. Their results were preliminary and agreements to within 10% could be considered as good. We want to determine whether (a) $Z = 3$, (b) η_{hf}^* is dependent upon temperature, (c) $\eta_{hf}^* < \eta^*$, (d) Eq. (3) is valid, (e) a background contribution, if any, exists in critical binary mixtures, and finally whether deviations between experiment and theory can be attributed to the approximate Ornstein-Zernike form of the correlation function in the Kawasaki equation (2) and to the vertex correction.[15] For these purposes we have measured the Rayleigh linewidth of concentration fluctuations of isobutyric acid in water as a function of the correlation length ξ and the momentum transfer vector $\vec{K}$ in the range $11 \geq \xi K \geq 0.0067$. Details of our experiments will be published elsewhere.

A total of 505 linewidths were measured at the critical solution concentration in the temperature interval $0.003°C \leq T - T_c \leq 30°C$ for scattering angles varying from 20° to 140°. Each linewidth was obtained from a 95- to 100-point least-squares fit of the exponential current correlation function to $\pm(0.5-1)\%$. Thus, our data represent over 50 000 measurements. In this Letter, we shall limit our discussions to results in the $K \to 0$ limit and in the critical region ($\xi K \gg 1$). Our conclusions concerning (a)-(e) are as follows:

(a) According to dynamical scaling, Γ varies as K^3 in the critical region. Figure 1 shows a typical log-log plot of Γ versus K at $\Delta T = T - T_c = 0.003°C$. For the exponent Z we find $2.976 \pm 1.5\%$ at $\Delta T = 0.003°C$ and $3.046 \pm 1.5\%$ at $\Delta T = 0.006°C$. Thus the mean value of Z from two separate independent determinations is 3.01 ± 0.03, which is in excellent agreement with $Z = 3$ as predicted by the mode-mode coupling theory. The fact that $Z = 3$ also shows that background contributions must be negligible in the critical region.

(b) If we take $Z = 3$, we find $A = (1.054 \pm 0.3\%) \times 10^{-13}$ cm^3/sec at $\Delta T = 0.003°C$ and $(1.065 \pm 0.3\%) \times 10^{-13}$ cm^3/sec at $\Delta T = 0.006°C$. The errors quoted are standard deviations. We further obtained $A = 1.115 \times 10^{-13}$ cm^3/sec at $\Delta T = 0.025°C$. The slight variations in A are outside of our experimental error limits and show that η_{hf}^*, like the hydrodynamic shear viscosity η^*, depends upon temperature in the critical region.

(c) A comparison of η_{hf}^* $(= k_B T/16A)$ with the

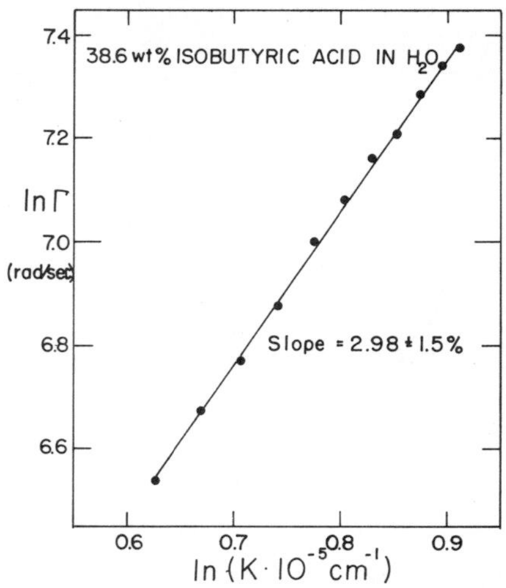

FIG. 1. Log-log plot of the decay rate Γ as a function of K at $T - T_c = 0.003°C$.

hydrodynamic shear viscosity[16] shows that $\eta_{hf}^* < \eta^*$. However, this conclusion is less certain since the η^* measured by the capillary method is susceptible to error in the critical region because of gravitational effects. Furthermore, the simplest vertex correction to the decay rate of concentration fluctuations contributes to 0.4%,[15] while with the modified Ornstein-Zernike correlation function η_{hf}^* (e.g., for xenon) is increased by about 6%.[11] Thus, the experimental data showing $\eta^* \approx 1.25\eta_{hf}^*$ are approximate in the critical region. Furthermore, with $\eta^* \approx 1.06\eta_{hf}^*$ in the hydrodynamic region, it appears necessary to account for this change by introducing a correction term f which depends upon $K\xi$. Thus, $\eta_{hf}^* = \eta^* f(K\xi)$.

(d), (e) To test the validity of Eq. (3), we need to know η_{hf}^*, D, and ξ from independent measurements. The mutual diffusion coefficient was obtained from $D = (\lim K \to 0)\Gamma/K^2$ in the nonlocal hydrodynamic ($\xi K \leq 1$) and the hydrodynamic ($\xi K \ll 1$) regions by means of optical-mixing spectroscopy. The correlation length ξ was obtained from independent measurements of the angular distribution of scattered intensity. In the analysis of our intensity data, we have corrected for volume, attenuation, density fluctuations, stray light, and residual dust scattering. A least-squares fit of 84 selected intensity data points[17] gives $\xi_0 = (3.57 \pm 0.07) \times 10^{-8}$ cm and $\nu = 0.613 \pm 0.001$. The errors quoted again represent standard deviations. Finally, we take the viscosity data of Woermann and Sarholz[18] and those of Allegra, Stein, and Allen,[16] and assume that η_{hf}^*

TABLE I. A typical comparison of D (in 10^{-8} cm^2/sec) with $k_BT/6\pi\eta^*\xi$.

$T-T_c$	D	D_{WS}[a]	D_{ASA}[b]	D/D_{ASA}
0.250	3.21_2	3.007	3.017	1.06
0.500	5.06_4	4.767	4.810	1.05
5.010	26.3_4	24.64	24.74	1.06
10.95	52.6_8	48.24	48.58	1.08
30.00	186_4	⋯	(185.1)[c]	1.01

[a]WS denotes viscosity data from Woermann and Sarholz (Ref. 18).
[b]ASA denotes viscosity data from Allegra, Stein, and Allen (Ref. 16).
[c]Extrapolated value.

$=\eta^*$. Table I shows a typical comparison of the measured diffusion coefficients and those computed by means of Eq. (3) with $\eta_{hf}^*=\eta^*$. The computed D differs from the measured D by about 6% over the entire temperature range (0.75 to 30°C). This signifies that η^* and η_{hf}^* must have the same temperature dependence.

If we take $\eta_{hf}^*=k_BT/6\pi D\xi$, a least-squares fit of our data according to Eq. (4) gives $E=1.40\pm1.31$, $G=-0.894\pm2.11$, $F=-3.33\pm4.53$, and $\alpha=-0.388\pm0.152$. The corresponding values from a least-squares fit using the viscosity data of Woermann and Sarholz[18] are $E=1.33\pm0.40$, $G=-0.68\pm0.04$, $F=-3.3\pm1.4$, and $\alpha=-0.37\pm0.04$. The parameters in Eq. (4) are very sensitive to minor variations in experimental data and we should not take their magnitudes seriously. Nevertheless, we show evidence that $-1<\alpha<0$, signifying the presence of a cusp in the critical viscosity anomaly. This viscosity anomaly must be a weak one, and the exact mathematical character is not known since present-day data are not sufficiently precise to make the very fine but definitive distinctions among a logarithmic divergence, a very weak power-law divergence, and a cusp. We do feel that the possibility of a cusp is very high in view of such good agreement with the modified mode-mode coupling theory of Kawasaki.

Recently, Kawasaki and Lo[19] obtained a relation $f(K\xi)$ between the so-called high-frequency viscosity η_{hf}^* and the hydrodynamic shear viscosity η^*. Their results show that for $K\xi\ll1$, $f(K\xi)\simeq0.948$, which agrees to within 1% with our measured $\eta_{hf}^*/\eta^*\simeq0.944$. In the critical region for $7.74<K\xi<10.3$, as shown in Fig. 1, $f(K\xi)$ varies from 0.833 to 0.813, which agrees to within a few percent with our measured $\bar{\eta}_{hf}^*$(averaged over $7.74<K\xi<10.3)/\eta^*\simeq0.80$. There η^* was computed from Eq. (4) using the viscosity data of Allegra, Stein, and Allen.[16] If we take their measured viscosity, then $\bar{\eta}_{hf}^*/\eta^*\simeq0.84$. The agreement between theory and experiment in the hydrodynamic as well as the critical region is indeed amazing. By neglecting the weak vertex corrections, it appears that we need not even invoke a breakdown of the approximate Ornstein-Zernike correlation function. It should be noted that, with $D=k_BT/6\pi\eta^*\xi f(K\xi)$, where $f(K\xi)$ corrects for the nonlocal shear viscosity, we have obtained agreement to within the error limits of our experiments for $K\xi\ll1$ and $\gg1$ from independent measurements of D, ξ, and η^*.

We thank Professor G. F. Allen for making his viscosity data available prior to publication.

*Work supported by the National Science Foundation.

[1]B. I. Halperin and P. C. Hohenberg, Phys. Rev. 177, 952 (1969).
[2]L. P. Kadanoff and J. Swift, Phys. Rev. 166, 89 (1968); J. Swift, Phys. Rev. 173, 257 (1968).
[3]K. Kawasaki, Phys. Lett. 30A, 325 (1969), and Ann. Phys. (New York) 61, 1 (1970), and Phys. Rev. A 1, 1750 (1970), and in Dynamical Theory of Fluctuations near Critical Points, Proceedings of the International School of Physics "Enrico Fermi," Course LI, Varenna, Italy, 1970 (Academic, New York, to be published).
[4]R. A. Ferrell, Phys. Rev. Lett. 24, 1169 (1970).
[5]P. Berge, P. Calmettes, C. Laj, M. Tournarie, and B. Volochine, Phys. Rev. Lett. 24, 1223 (1970).
[6]D. L. Henry, H. L. Swinney, and H. Z. Cummins, Phys. Rev. Lett. 25, 1170 (1970).
[7]J. V. Sengers and P. H. Keyes, Phys. Rev. Lett. 26, 70 (1971).
[8]B. Chu, D. Thiel, W. Tscharnuter, and D. V. Fenby, in Proceedings of the International Conference on Light Scattering in Liquids, Paris, France, July 1971 (unpublished), and to be published.
[9]R. F. Chang, P. H. Keyes, J. V. Sengers, and C. O. Alley, Phys. Rev. Lett. 27, 1706 (1971).
[10]I. W. Smith, M. Giglio, and G. B. Benedek, Phys. Rev. Lett. 27, 1556 (1971).
[11]H. L. Swinney, D. L. Henry, and H. Z. Cummins, to be published.
[12]T. K. Lim, H. L. Swinney, K. H. Langley, and T. A. Kachnowski, Phys. Rev. Lett. 27, 1776 (1971); G. T. Feke, G. A. Hawkins, J. B. Lastovka, and G. B. Benedek, Phys. Rev. Lett. 27, 1780 (1971).
[13]M. E. Fisher, Rep. Progr. Phys. 30, 615 (1967).
[14]P. Berge and M. Dubois, Phys. Rev. Lett. 27, 1125 (1971).
[15]S. M. Lo and K. Kawasaki, Phys. Rev. A 5, 421 (1972).
[16]J. C. Allegra, A. Stein, and G. F. Allen, to be pub-

lished.
[17]B. Chu, F. J. Schoenes, and W. P. Kao, J. Amer. Chem. Soc. 90, 3042 (1968).
[18]D. Woermann and W. Sarholz, Ber. Bunsenges. Phys. Chem. 69, 319 (1965).
[19]K. Kawasaki and S. M. Lo, "Nonlocal Shear Viscosity and Order Parameter Dynamics near the Critical Point of Fluid" (to be published).

PRECISION MEASUREMENT OF THE SPECIFIC HEAT OF CO_2 NEAR THE CRITICAL POINT*

J. A. Lipa,† C. Edwards, and M. J. Buckingham
Physics Department, University of Western Australia, Western Australia 6009, Australia
(Received 16 June 1970)

An experiment is reported in which the specific heat of a 1 mm high sample of CO_2 has been measured to within 12 mdeg of its critical temperature. The results show that the exponent $\alpha = \frac{1}{8}$ both above an below T_c, a value in agreement with the Ising model but inconsistent with scaling law analyses of existing *PVT* data.

The reliable measurement of the specific heat of a fluid near its critical point presents particular difficulties because of the effects of gravity and the problem of achieving equilibrium in the face of "critical slowing down." Distortion produced by the earth's gravity has been the main limitation on the validity and resolution of attempts reported so far to establish experimentally the form of the specific-heat singularity and in particular the value of α. At the critical density, C_v is assumed to diverge at the critical temperature T_c like $|t|^{-\alpha}$, where α is possibly different for the branches of the expression above and below T_c and where $t = 1 - T/T_c$. The effects of gravity[1] distort the specific heat if the relative temperature difference t is much less than a characteristic value t_h dependent on the height h and the atomic mass m of the fluid sample, where $t_h = (mgh/2kT_c)^{1/\beta\delta}$; $\beta\delta \simeq \frac{5}{3}$ (e.g., for Xe with $h = 1$ cm, t_h corresponds to a temperature difference of 0.13 K).

We report here measurements of the specific heat of CO_2 of nominal purity 99.996%, permanently sealed at a density of 0.4660 ± 0.0008 gm cm^{-3} inside a calorimeter of only 1 mm height, for which t_h corresponds to 17 mdeg. The disk-shaped calorimeter was constructed of high-strength stainless steel with internal reinforcing members to maintain its volume constant while keeping its contribution (about 40% at the specific heat maximum) to the total heat capacity as small as possible. Critical slowing down was evident in the rapid increase (approximately as the inverse $\frac{2}{3}$ power of the temperature interval) in the thermal relaxation time, τ, observed as the temperature was raised towards T_c, the value of τ becoming as large as 1000 sec 2 mdeg below T_c. Precise temperature control of the surroundings of the calorimeter was achieved by the use of a multistage thermal environment combined with a constant–ramp-rate method of measurement, similar to that previously described,[1] but refined by the addition of a further stage of thermal isolation which greatly improved the performance.

At a rate of change of temperature, or ramp rate $\dot{T}$, of 10^{-6} deg sec^{-1}, for which the signal-to-noise ratio permits a precision of about 1% in the specific heat of the CO_2 specimen, the resolution is limited to about 10^{-2} deg each side of the critical temperature. Within this interval the increasing thermal time constant τ causes a significant distortion of the measurements. Also the calculated distortion due to gravity for the 1 mm high sample becomes about 1% at this temperature and, while this could be allowed for as in Ref. 1, the necessary information concerning the equation of state so near the critical point is in any case too uncertain to permit a useful extension of the range of measurement. Our reported[2] results therefore are limited to magnitudes of the relative temperature interval $t > 4\times10^{-5}$, corresponding to 12 mdeg. In all, 1100 data points were collected in 35 runs at various ramp rates ranging from 10^{-4} to 3×10^{-7} deg sec^{-1}, each point representing an average of 900 sec of record. Cooling runs were used only for checking purposes. The scatter in the observations ranges from $\frac{1}{4}$% far from T_c to 1% at the inner limit. This increase is due to the necessary use close to T_c of smaller ramp rates and the consequent lower signal-to-noise ratio. A possible constant error estimated as ± 10 J/mole deg arises from uncertainty in the heat capacity of the empty calorimeter.

The present report is concerned with the asymptotic temperature dependence of the specific heat near the critical temperature. For this purpose we have compared our results with functions of the class

$$C_v/R = A|t|^{-\alpha} + B, \qquad (1)$$

where A, B, T_c, and α are separate parameters for the branches of the function above and below T_c, and R is the gas constant.

For any particular test function, the value can be computed of the weighted sum of the squares of the deviations of the data points from that function. The minimum value of this sum as the parameters are varied subject to any constraint imposed on the test function is defined as Σ^2, which is thus a function of the particular constraints imposed as well as the range of data points included.

The class of functions (1) proves inadequate to describe the data if points too far from T_c are retained. However if the range of $|t|$ is restricted to $|t| < 5\times10^{-3}$ no significant reduction in Σ^2 can be obtained by extending the class of functions with the inclusion of further parameters; an additional term, for example one linear in t, can cause a significant improvement if data beyond that limit are included but not otherwise. About 440 of the data points lie within the range $4\times10^{-5} < |t| < 5\times10^{-3}$ and further discussion is restricted to this asymptotic region.

For a specified pair of values of α^+ and α^- the value of Σ^2 is by definition minimized with respect to all the other six parameters $A^\pm$, $B^\pm$, $T_c^\pm$. (The superscripts refer to the two branches $T > T_c$ and $T < T_c$, respectively.) The dependence of this Σ^2 on α^+ and α^- is shown in Fig. 1 by the broken line which is the locus of pairs α^+, α^- for which the value of Σ^2 is one standard deviation greater than the minimum possible value, Σ_0^2. This contour line thus represents the 68% confidence level, if the data are normally distributed. If any further constraint is imposed on the test function the value of Σ^2 can only increase, but if the constraint is consistent with the data there will remain a significant fraction of the area enclosed by the contour still within one standard deviation of Σ_0^2. The effect of constraining the value of $T_c^+ - T_c^- = \Delta T_c$ to be zero is to make the one standard deviation contour that shown by the full line in Fig. 1. The area of this contour would be a maximum if the constrained value of ΔT_c were -2 mdeg and would shrink to zero at the limits

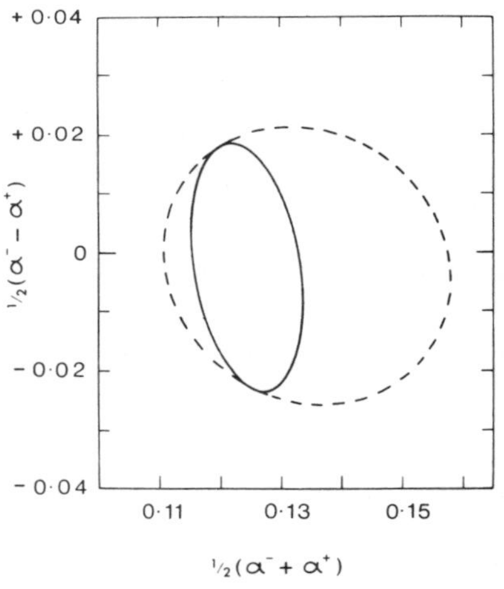

FIG. 1. Showing the goodness of fit of the observed specific heat of CO_2 in the asymptotic region $|T-T_c| < 5\times10^{-3}T_c$ by test functions with various values of α^- and α^+, the critical exponents for the branches below and above T_c, respectively. The curves are contour lines at which the fit is one standard deviation worse than the best possible. The broken curve results when the two branches are considered independently, with possibly different values for T_c; the full curve when T_c is constrained to be the same for each branch. The figure shows that there is no evidence that it should not be the same. The figure also shows no evidence that the values of α^- and α^+ should differ.

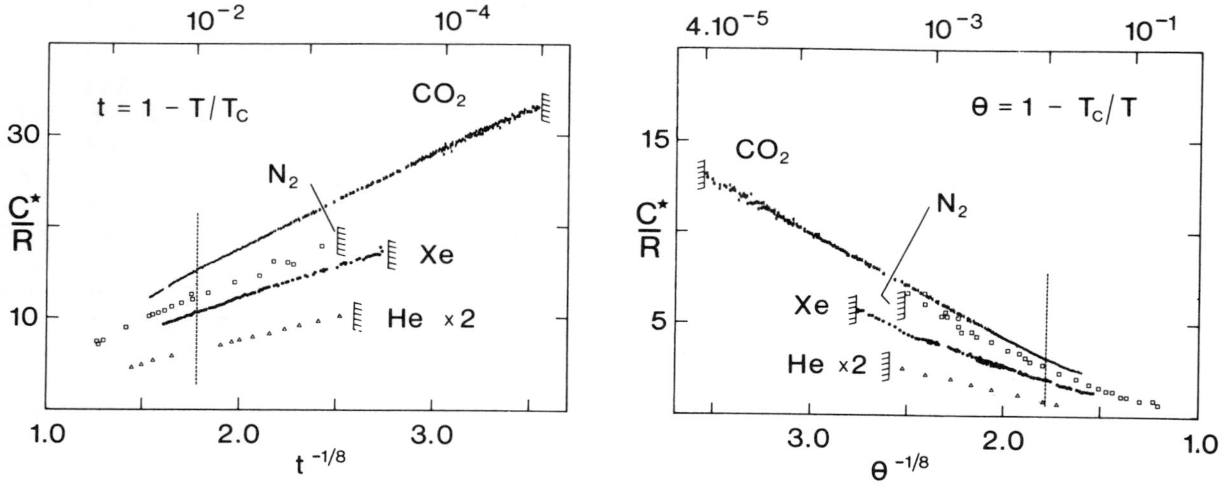

FIG. 2. The upper sets of points show the values observed for the specific heat of CO_2 less its ideal gas value at T_c; $C^* = C_v - C_{ideal}$. (For CO_2, $C_{ideal} \simeq 7R/2$.) They are plotted against the inverse one-eighth power of the interval from the critical temperature and show the straight line dependence in the asymptotic region (inside the vertical dotted lines) indicating $\alpha = \frac{1}{8}$. Nearer to T_c than the cross-hatched markings, measured values are seriously affected by gravity and are not displayed. The other sets of points displays in the same fashion results that have been published for some other fluids. (Note that there is a factor 2 different in the vertical scale of the two parts of the figure; also that C^* for He has been multiplied by 2 for clarity.)

of the range

$$T_c^+ - T_c^- = \Delta T_c = -2 \pm 5 \text{ mdeg}.$$

The results are thus fully consistent with the expectation that $T_c^+ = T_c^-$.

Figure 1 also demonstrates that the data require α^+ and α^- to be essentially equal. In fact, taking $T_c^+ = T_c^-$,

$$\tfrac{1}{2}(\alpha^- - \alpha^+) = 0 \pm 0.02; \quad \tfrac{1}{2}(\alpha^- + \alpha^+) = 0.12_5 \pm 0.01.$$

If both T_c and α are constrained to have the same values for each branch we find the best fit function is given by Eq. (1) with $\alpha = \frac{1}{8}$ and

$$A = 10.473, \quad B = -0.024, \quad T < T_c;$$

$$A = 5.583, \quad B = -3.457, \quad T > T_c.$$

The observed values for the specific heat of CO_2 are presented in Fig. 2, plotted against the inverse one-eighth power of the interval from the critical temperature. Also shown for comparison are results that have been published for some other fluids: Xe,[3] N_2,[4] and ^{4}He.[5] In each case however the data have been restricted by the hatched markings to the region unaffected by gravity for the appropriate experimental arrangement. Other fluids[6] show similar features but, for clarity in the figure, are not displayed.

It is apparent that a linear dependence in this diagram, corresponding to $\alpha = \frac{1}{8}$, is consistent will all the measurements within the asymptotic temperature interval up to 1 % of T_c, indicated in the figure by the vertical broken line. Thus there is no evidence from direct measurements unaffected by gravity that for any gas the value of α is significantly different from $\frac{1}{8}$.

The effect of gravity shows itself clearly in Fig. 3 which demonstrates that the specific heat

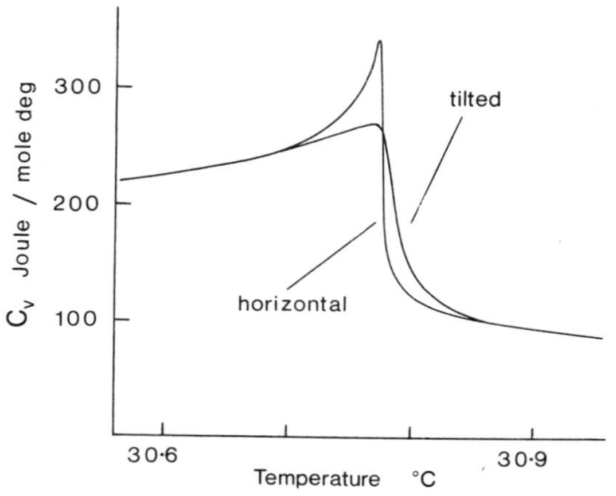

FIG. 3. The measured specific heat of a 1 mm high sample of CO_2 near the critical temperature. The effect of gravity is demonstrated by the curve labeled "tilted" which shows the results measured with the same disk-shaped sample tilted out of the horizontal plane so that its total height is 14 mm.

is grossly distorted in the qualitative way theoretically expected.[7] From the features shown by this figure it is not hard to see that retention in an asymptotic analysis of results affected by gravity would tend to reduce the apparent value of α for the low-temperature branch while increasing it on the high-temperature side.

It has been noted that in order to reduce the effect of gravity the vertical dimension h must be made small. If it is too small, however, finite-size effects would distort the results. It is reasonable to suppose that so long as the correlation length remains sufficiently small compared with the linear dimensions of the fluid such finite-size effects (and explicit gravity-dependent effects[1]) will be negligible. As h is reduced, resolution can be achieved closer to the critical point. However, at that point the correlation length diverges, so that its value at the resolution limit correspondingly increases. There is therefore an optimum height which is a compromise between gravity and finite-size effects. If we adopt the arbitrary criterion that at its largest the correlation length must not be more than $10^{-3}h$ we find that for CO_2 the optimum value for h in the earth's gravity is about 0.3 mm. This value would allow some extension of the range of our C_v measurements—in fact, to the inner edges of the diagrams in Fig. 2, which can in this sense be regarded as the limit of possible measurement in an earth-bound laboratory. It might be thought that an experiment in an orbiting satellite and free of gravity would permit a substantial improvement but this is not so, unless a new and more sensitive type of thermometry becomes available. It is only possible to measure C_v with precision at a resolution about 10^3 times the noise level of temperature measurement. Using thermistors and phase-sensitive detection the latter can be made as low as $10^{-8}T_c$ but much improvement beyond this figure is not yet in sight.

The value $\frac{1}{8}$ found for α is in striking agreement with the value obtained from numerical studies[8] of the three-dimensional Ising model. On the other hand the value $A^-/A^+ = 1.88$ found for the ratio of the coefficients below and above T_c is much larger than the Ising-model values, e.g., 1.33 found by Gaunt and Domb[9] for the tetrahedral lattice. This ratio may be expected to depend on more features of a statistical system than does an exponent such as α, however. The exponents are only known for certain to depend on two features of a system with short-range forces. These are the dimensionality d of space and D of the order parameter. Of three-dimensional fluid systems, $D = 2$ for the λ transition of liquid helium but $D = 1$ for the gas-liquid critical transition, as for the Ising model. Thus if the critical exponents depend on no other property of a system, we could expect a value the same for Ising and gas-liquid systems but different for the helium λ transition. This indeed appears to be the case as far as the specific-heat exponent α is concerned. On this basis one could also expect agreement with the Ising model for the exponents describing the equation of state. For the Ising model[10] $\gamma \simeq 5/4$, $\delta \simeq 5$, and $\beta \simeq 5/16$, where γ, δ, and β are the exponents characterizing the compressibility, critical isotherm, and coexistence curve, respectively. A recent analysis by Vicentini-Missoni, Levelt Sengers, and Green[11] of PVT measurements on several fluids including CO_2, in terms of a scaled equation of state, led to the conclusion that the results were consistent with the scaling laws, with values $\gamma \simeq 5/4$, $\delta \simeq 4.6$, and $\beta \simeq 0.35$. Consistency furthermore required that α was about 0.04, and in any case less than 0.1. The value $\frac{1}{8}$ now found for α suggests a serious inconsistency, if scaling is to be maintained, between the calorimetric and equation-of-state measurements.

The strict requirements of scaling when $\alpha^+ = \alpha^- \neq 0$ imply values of the coefficient A different for the two branches of the function (1) but the same value for B. This is not consistent with the present results: Imposition of the condition $B^+ = B^-$ would lead to an increase in Σ^2 of no less than 9 standard deviations. Extension of the class of function (1) to include nonasymptotic terms in the scaling theory may resolve this problem but such extended analysis of our results is left for another occasion.

The value of T_c for CO_2 found from the present analysis is 303.925 ± 0.005 K (30.775°C) and is significantly lower than the generally accepted value. For example, Michels, Blaisse, and Michels[12] and Lorentzen[13] find $T_c = (31.04 \pm 0.01)$°C, although a more recent analysis by Vicentini-Missoni, Levelt Sengers, and Green[11] of the former worker's equation of state measurements leads to a value (30.94 ± 0.04)°C. It is by no means exceptional—indeed it is usual[6,14]—for calorimetric measurements to yield smaller values for critical temperatures than PVT or optical measurements. In spite of the great care taken it is difficult to rule out the possible existence of some contamination, perhaps by a small percentage of air which could be expected to shift the critical temperature (but by less than a part in a thousand) without significantly altering the form of the specific heat singularity.

We are indebted to F. J. van Kann for his assistance with the data analysis and one of us (J.A.L.) wishes to thank the Commonwealth Scientific and Industrial Research Organization for the award of a Senior Postgraduate Studentship.

---

*Research supported by a grant from the Australian Research Grants Committee.

†Now at the Physics Department, Stanford University, Calif. 94305.

[1]C. Edwards, J. A. Lipa, and M. J. Buckingham, Phys. Rev. Lett. 20, 496 (1968).

[2]A listing of the data can be made available on request.

[3]See Ref. 1. The data plotted are slightly different from those previously reported, adjustment having been made for certain small but systematic corrections. The result originally given as $\alpha = 0.08 \pm 0.08$ now becomes 0.14 ± 0.07.

[4]A. V. Voronel', V. G. Gorbunova, Yu. R. Chashkin, and V. V. Shchekochikhina, Zh. Eksp. Teor. Fiz. 50, 897 (1966) [Sov. Phys. JETP 23, 597 (1966)].

[5]M. R. Moldover, Phys. Rev. 182, 342 (1969).

[6]See P. Heller, Rep. Progr. Phys. 30, 731 (1967), and references given there.

[7]See Ref. 1. See also M. Barmatz and P. C. Hohenberg, Phys. Rev. Lett. 24, 1225 (1970).

[8]D. S. Gaunt, Proc. Phys. Soc., London 92, 151 (1967).

[9]D. S. Gaunt and C. Domb, J. Phys. C: Proc. Phys. Soc., London 1, 1038 (1968).

[10]M. E. Fisher, Rep. Progr. Phys. 30, 615 (1968).

[11]M. Vicentini-Missoni, J. M. H. Levelt Sengers, and M. S. Green, J. Res. Nat. Bur. Stand., Sect. A 73, 563 (1969).

[12]A. Michels, B. Blaisse, and C. Michels, Proc. Roy. Soc., Ser. A 160, 358 (1937).

[13]H. L. Lorentzen, Acta Chem. Scand. 7, 1335 (1953).

[14]J. A. Lipa, thesis, University of Western Australia, 1970 (unpublished).

ACTA CHEMICA SCANDINAVICA 7 (1953) 1335—1346

Studies of Critical Phenomena in Carbon Dioxide contained in Vertical Tubes

HANS LUDVIG LORENTZEN

Universitetets Kjemiske Institutt, Blindern-Oslo, Norway

An accurate knowledge of the isotherms in the critical region is of great importance for problems of equations of state, for studies of liquid-gas phase interaction and many others. For the computation of the isotherms close to the critical point, the determination of the mass distribution of the substance in vertical tubes as a function of distance perpendicular to the original liquid-vapour interface might be useful. Previously, techniques such as the measurement of the position of small glass floats of known density or of a single spring-actuated float [1], and the measurement of radiation intensity from radioisotope loaded samples [2] have been used for the determination of mass distributions in vertical tubes.

It has appeared to me that a technique of greater simplicity and of potentially far greater resolving power could be developed by taking advantage of the relationship between density and refractivity. In this paper the apparatus developed for this work will be briefly described, and preliminary results for carbon dioxide presented.

The assembled equipment is shown in Fig. 1. It consists essentially of a cylindrical glass-jacketed thermostat containing the sample tube made of ordinary Pyrex tubing, and an optical system designed to focus an image of a pair of parallel vertical lines (illuminated by a sodium lamp) as seen through the sample tube and thermostat on a photographic plate. The thermostat and sample tube together may be regarded as a part of the optical system since they form a cylindrical lens. The telescope is focussed a few millimeters beyond the sample tube.

The photographic plate shows the image of two vertically running lines, the horizontal distances between which are determined by the extent of the refraction of the light which passes through the tube at whatever the height of the sample tube, and hence by the density of the substance in that region. One must, of course, establish the relationship between density and the distances between the lines, that is, calibrate the apparatus.

The line distance is found to increase nearly linearly with increased density. Therefore, each of the two lines on the photographic plate represents a mass

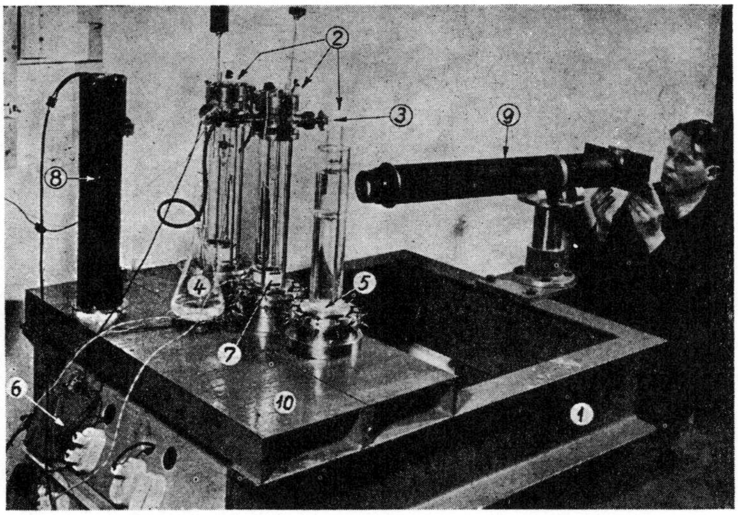

Fig. 1. Apparatus.

1) Steel frame (300 kg) mounted on 3 rubber blocks.
2) 3 thermostats (the one in front not fully mounted).
3) Screw for selecting the temperature of the thermostat. One turn of the screw changes the temperature about 50 millidegrees.
4) Reservoir of thermostat-water. Is used in the case of rough temperature selection.
5) Circulation pump, magnetically driven. The motor is inside the hollow foundation.
6) Electronic relays for switching off and on heating current on one of the thermostats.
7) Coil for additional heating of thermostat.
8) Lamp carrying vertical, illuminated lines (Na-light). The lamp is movable and can be placed in fixed position behind each thermostat.
9) Telescope can face any of the thermostats and can be raised and lowered for measuring at any level of the thermostats. (Screw and wheel not visible.)
10) Planed iron table which allows exact vertical placing of lamp and thermostats.

distribution curve of the carbon dioxide in the sample tube at the temperature of the thermostat. The same principle has been employed previously, but only by making direct visual observations and therefore without significant results [3].

THE THERMOSTAT

The thermostat consists basically of a glass cylinder (390 mm long, 60 mm inner diameter) hermetically sealed at both ends and totally filled with de-aired water. A magnetically-driven pump wheel spinning like a top on the bottom of the thermostat forces the water up through the 5 mm intermediate space created by the wall of the thermostat and a second glass cylinder placed inside and coaxial with it. (About 50 recirculations per minute.)

The spinning of the pump wheel is maintained by a motor below the thermostat bottom, the motor being supported by a rubber ring and placed coaxially inside the hollow foundation of the thermostat. The top of the vertical shaft of the motor carries a permanent horseshoe magnet. The pump wheel inside the thermostat also consists basically of a permanent magnet formed as a cylindrical disc. Naturally, the magnetic connection between the motor and the pum p wheel does not transfer the vibrations of the motor to the thermostat construction. In fact, in spite of the relatively high speed of the motor and the pump wheel — 1 400 rpm — the thermostat is practically free from vibrations — this being of great importance for the measurements close to the critical point.

Heat is supplied by means of a low-voltage alternating current (about 4 watts) conducted by naked resistant wires directly immersed in the water. The heating current is controlled by a relay actuated by a mercury switch, which is constructed in the form of a U-tube with a tungsten wire contact inside one arm. The contact end of the tube is sealed off, and the space above the mercury is filled with hydrogen. The other arm is open and the whole U-tube immersed in the thermostat water, so that the thermostat construction operates as the bulb in a great water contact thermometer.

The intense (not visible) stirring of the water provides for homogeneous temperature in the thermostat. The mercury switch reacts almost immediately by changes in the average temperature of the water. The resistant wires, spun of fine nichrome filaments, allows for an almost immediate heat transfer to the water when the heating current is switched on. Thus the inertia factors of an off-on thermostat are greatly reduced and, in fact, the calculated "hunting" (energy × time/heat capacity) of this thermostat can be as low as less than 0.001° in periods of one second. In cases where the calculated "hunting" even exceeded the last-mentioned value, the actual "hunting" was not registered by the resistant thermometer employed.

A rough temperature selection is obtained by adding or withdrawing water through a valve: withdrawing water causes, for example, the mercury contact to be broken and heat to be supplied to the system. The specific volume of the water will increase and the reduced mass of water will, due to the increased temperature, re-establish the balance of the mercury switch. An accurate temperature selection (0.001° intervals over about 1°) is obtained by screwing a plunger in or out of the thermostat, thus changing the volume. The easy regulation and the small size of the thermostat allow a convenient and rapid selection of desired temperatures. For precise work the thermostat has to be shielded by reflecting aluminium sheets.

The main difficulty which I met when designing this new type of thermostat was to secure a constant interior volume in the thermostat construction, a volume that was to be independent of exterior temperatures and air pressure changes, and to avoid leaks which tend to cause a constant rise or fall in temperature.

Through the top enclosure and almost all the way down the entire length of the thermostat a coaxial glass tubing (closed at the bottom and open at the top) form a "well", inside of which the test-tubes can be conveniently placed. The test-tubes are fully immersed in an oil with almost the same refractive index as the pyrex glass, thus reducing optical imperfections and securing good thermal conduction between the thermostat water and test tube. In the same oil and above the test tube, the thermometer is placed. Above the oil, but still inside the thermostat, a bobbin of wool yarn provides for adequate thermal insulation. FS Precision-Bore Tubing of Pyrex glass is used for the construction of the thermostat. The outside of these tubes is ground cylindrically and polished for optical perfection.

PREPARATION OF CARBON DIOXIDE

Carbon dioxide is prepared by dripping sulfuric acid into a solution of potassium-hydrocarbonate in an apparature described by Reihler [4]. This author described the carbon dioxide to be airfree. The carbon dioxide is dried by passing through 4 U-tubes containing concentrated sulfuric acid and glass beads, while the last of the 4 tubes is cooled by solid carbon dioxide. At this temperature the drying power of the solid sulfuric acid distributed on the surface of the beads is calculated to be about as high as that of phosphorus pentoxide at ordinary temperature.

From the fourth drying tube the carbon dioxide is led through a valve into an all-fused glass tubing system, consisting of the sample tubes to be filled, a trap cooled with solid carbon dioxide and an open mercury manometer. Finally, a glass valve leads to a diffusion pump and a mechanical vacuum pump. The glass tubing system is repeatedly filled with carbon dioxide heated and evacuated below 10^{-4} mm Hg, and, filled with carbon dioxide, allowed to rest for several days. Then it is evacuated and refilled before the carbon dioxide is finally precipitated by immersing the bottom of the sample tubes in liquid air. The sample tubes are then individually fused off and thereby sealed. The purity of the enclosed carbon dioxide was probably better than 99.999 %. I am indebted to Cand.real. Arne Almenningen, who has done the preparation work of the carbon dioxide.

When using heavy-walled capillary tubing as sample tube, the tubes are deliberately under-filled when fused off. By warming the tubes to a temperature close below the critical, inaccuracies in the filling proportions can be determined. When recooling the bottom of the sample tube in liquid air the carbon dioxide will again precipitate and vacuum be reestablished in the tube. If now the top of the tube is fused, the interspace of the tube will be reduced. By repeating the test and thus shortening the tube the interior volume, corresponding to the mass of the enclosed carbon dioxide and its critical density, can be attained with adequate exactness.

GENERAL DESCRIPTION OF THE MASS DISTRIBUTION PHENOMENA

The mass distribution in tubes filled with CO_2 and heated to temperatures near the critical was studied in 3 series, the first as early as in the autumn of 1951. The purity of the CO_2 enclosed in the tubes has been successively improved and only the tube of the last series was prepared with the accuracy described above. In the first two series the tubes were about 20 cm long and with an inner diameter of 6 mm. In the last series the tube used for the mass distribution measurements was 5 cm long and of 1.8 mm inner diameter.

Qualitatively, the mass distribution found in the tubes in all 3 series was such as is to be expected when determined by isotherms of classical type and if the pressure increase downward in the tube is computed as being due to the weight of the substance itself.

At temperatures below the critical the carbon dioxide is separated into vapour (upper part) and liquid (lower). On approaching the critical temperature pronounced density gradients are built up in the two formerly homogeneous phases, the vapour being denser near the surface than higher up, and the liquid less dense near the surface than lower down. At a certain temperature the brilliant reflecting surface is replaced by a region with a sharp density gradient between the still fairly homogeneous fillings in the extremities of the tube. The density difference between both ends of the tube is then 10—15 % depending on the length of the tube investigated. When the temperature is further increased the densities at the extremities approach each other. More marked, however, is the increase in height of the region with density gradient. In the case of proper filling proportions the density gradient will soon reach both ends of the tube, and the fairly homogeneous filling, formerly present in the extremities of the tube, will have disappeared.

At temperatures just below the critical, the surface in tubes containing too much carbon dioxide will, with increasing temperatures, move upwards, in the opposite case downwards. Above the critical temperature the region with density gradient which has replaced the surface continues the movement of the surface and will, on further heating, in the case of improper filling, soon "move out of" one of the ends of the tube. If the two phases were homogeneous at all temperatures up to the critical point, then the surface would always "move out of" one of the ends of the tube already before the critical temperature had been reached. As stated above, this movement of the surface will always exist but, due to the fact that the density difference between the fillings in the extremities of the tube is greater than the density difference at the surface, the surface will not reach one of the ends of the tube before a region with density gradient has been established. This happens when the filling proportion does not deviate more than about $\pm$ 10 % from that corresponding to the critical density. These limits must depend on the height of the tube employed.

In my experiments the method of raising or lowering the temperature in steps was employed, thus allowing the tubes to remain for a considerable time at the selected temperatures before the photographs were made.

In practice the mass transportation which takes place when changing the temperature is slow. When selecting the time in which the tube was kept at the chosen temperature, regard was paid to the changes in mass distribution to be expected. In some cases, the time during which the tube was kept at constant temperature was as long as 8 days. Working carefully enough it must be expected that the distribution actually measured will not be far from an equilibrium distribution.

The new material seems to indicate very strongly that the mass distribution in the tubes in equilibrium is determined by the isotherm of the substance at the temperature selected. If this is so, the mass distribution curves will show parts of the isotherms, and the determination of the mass distribution in the vertical tube will make possible a determination of the isotherms substantially closer to the critical point than any previously known method.

If a simultaneous determination of the pressure inside the tube had been carried out, the position of the isotherm in the *pv*-diagram would have been known. However, in the region in question which is very close to the critical point, the $\frac{dp}{dt}$ of substances at critical density can be regarded as constant. At the critical point $\frac{dp}{dt}$ is known for a series of substances.

The relation between line distances and actual densities

For the calibration of the apparatus the sample tube was heated to the temperatures 29.01°, 29.98° and 30.76° (2.03°, 1.06° and 0.28° below the critical). The horizontal distances between the vertical lines on the photographic plate which recorded the densities of the liquid and vapour phases were at these temperatures 1.745, 1.382; 1.698, 1.414; 1.639 and 1.460 (mm).

Fig. 2 shows the top of the coexistence curve. The densities of the liquid and vapour phase at the temperatures mentioned according to the curve are as follows: 0.623, 0.308; 0.593, 0.338; 0.545 and 0.380 (g/cm^3). Using these density data and the corresponding line distances the calibration curve was constructed. It is a slightly curved line which is almost straight at the densities found in the tube near and above the critical point.

For future work a more satisfactory way of calibration will be used.

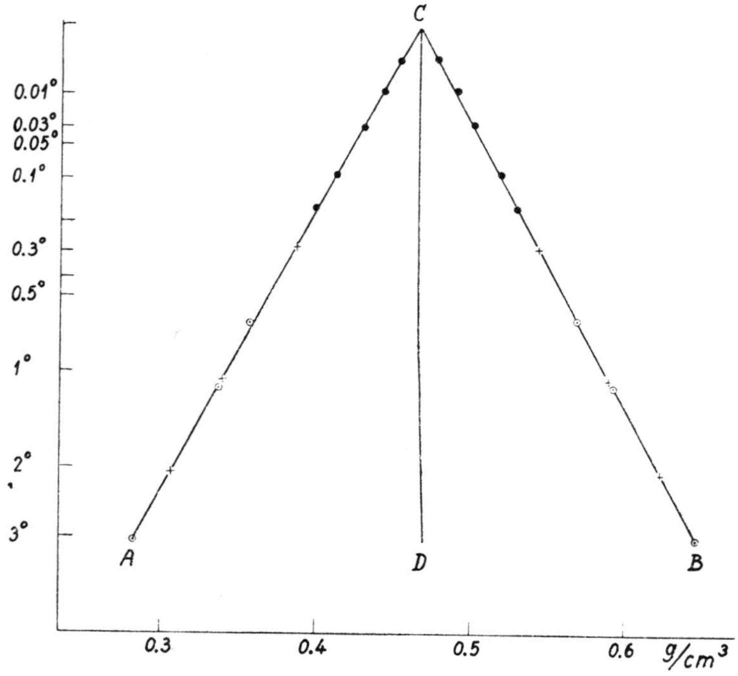

Fig. 2. Top of Coexistence Curve.

C is the critical point. The ordinate is in the scale of $\triangle T^{0.357}$.

AC and BC are the vapour and liquid parts of the coexistence curve, CD is the rectilinear diameter.

Down to $\triangle T = 1°$ *CD is a straight line. Below it has slightly positive curvature. The lines AC and BC should have corresponding curvature. For simplification, however, the curves are drawn as straight lines and the points below* $\triangle T = 1°$ *herefore slightly corrected. The 6 ⊙ marked points correspond to the values of MBM, the 6 × marked points to those used for calibration of the measurements and the ● marked points correspond to the determined densities of vapour and liquid at the surface at the temperatures 1.4, 9.6, 18, 90, 158 millidegrees below the critical.*

The Top of the Coexistence Curve

A. Michels, B. Blaisse, and C. Michels [5] (in the following abbreviated to MBM) have found that the coexistence curve for carbon dioxide at temperatures up to about 1° below the critical can be represented by the equation $d_l - d_v = C \cdot \triangle T^{0.357}$ where d_l and d_v are the densities of liquid and vapour and $\triangle T$ the distance from the critical temperature. Furthermore, the Matthias rule of rectilinear diameter was found to be correct up to the temperature mentioned. Above this temperature the coexistence curve was found to be less regular. If, however, the coexistence curve is supposed to be regular all the way up to the critical point, it will, when drawn with the temperature scale $\triangle T^{0.357}$ as ordinate, be represented by two symmetrical straight lines

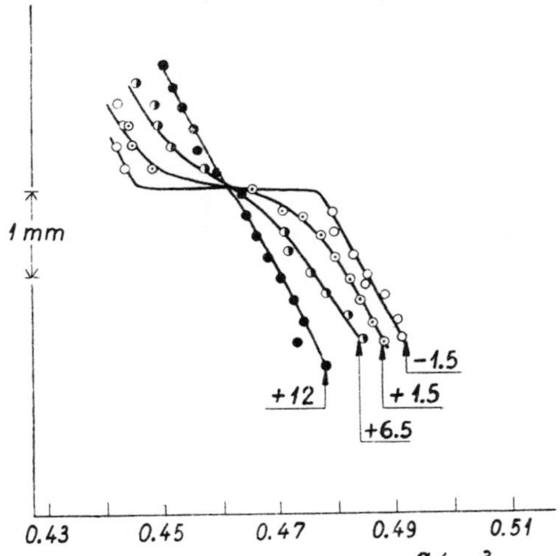

Fig. 3. Mass Distributions Close To the Critical Point.

The temperature of the thermostat was raised in steps to the temperatures of 1.5 millidegrees below and 1.5, 6.5 and 12 millidegrees above the critical. The curves show the mass distribution in the middle section of the tube. The line marked 1 mm corresponds to 1 mm height difference in the tube. The horizontal part of the curve —1.5 corresponds to the density difference at the surface.

meeting at the critical point. Down to 1° below the critical temperature the inclination of the rectilinear diameter is of no practical importance. In Fig. 2 the coexistence curve is drawn in the way just described. The 6 centered circles correspond to the values of MBM for the temperatures 2.99°, 1.11° and 0.63° below the critical with a minor correction for the lower two. The 6 points marked by crosses correspond to those used for the calibration as described above. The filled circles thus correspond to my measurements of the coexistence curve up to 1.4 millidegrees below the critical temperature.

Mass distribution close to the critical point

In the second series carried out in the autumn of 1952 the tube was heated in steps of 3 millidegrees per hour past the critical temperature. Some of the mass distribution curves obtained are shown in Fig. 3. In the third series the tube was cooled from a temperature some millidegrees above the critical, past the critical temperature in steps of about 1 millidegree per 20 minutes. In Fig. 4 the mass distribution curves for the temperatures + 1.2 and —1.4 millidegrees are shown. A mass distribution curve made at a temperature between these two is not shown on the figure, but runs, as is to be expected, in between the curves + 1.2 and — 1.4.

The agreement between the curves + 1.5 and — 1.5 in Fig. 3, and + 1.2 and — 1.4 in Fig. 4, is as good as can be expected. In the second series (Fig. 3)

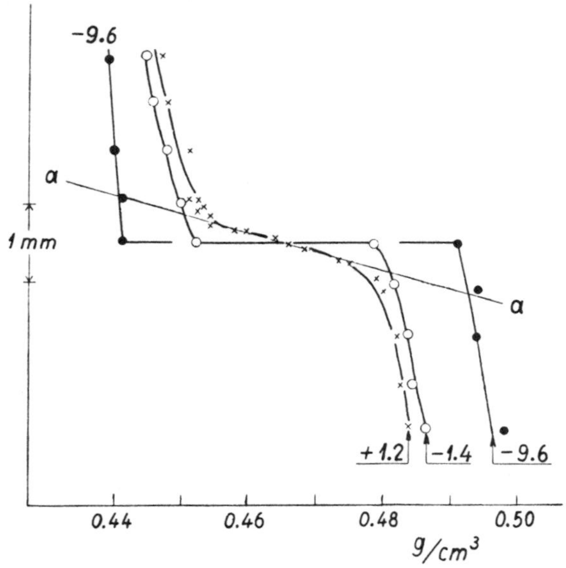

Fig. 4. Mass Distributions Close to the Critical Point.

The temperature of the thermostat was lowered in steps to 1.2 millidegrees above and 1.4 and 9.8 millidegrees below the critical. Ordinate and abscissa as in Fig. 3.
The straight line (a—a) represents a calculated tangent of the mass distribution curve +1.2 at the inflexion point.

the inner diameter of the tube was 6 mm. For optical reasons this diameter is too large for really good measurements of a very sharp density gradient. Moreover, as the calibration of the apparatus was not carefully done, the density scale of Fig. 3 is uncertain by about ± 10 %.

The line *a*—*a* in Fig. 4 is the calculated tangent of the isotherm + 1,2 millidegrees of carbon dioxide at the inflexion point. The calculation is based on the data given by MBM in their Fig. 9 and the average pressure increase downward in the tube of 0.000046 atm. per mm. In addition, the tangents for the isotherm + 7 millidegrees and + 90 millidegrees have been calculated and found to be in adequate agreement with the mass distribution curves + 7 and + 90 in Figs. 5 and 6.

Fig. 4 also shows the liquid and vapour curves at — 9.8. The inhomogeneity of the two phases can be seen.

The critical temperature

In the second series (Fig. 3) the critical temperature was determined as the temperature at which the surface was no longer seen as a horizontal, brilliantly reflecting disc. In the third series (Fig. 4) the density gradient was visible at the temperature + 1.2 millidegrees. Moreover, the photograph shows clearly the smooth transition between the densities in the upper and lowei part of the tube. In the next step it was not possible to determine visually whether a very large density gradient or a real surface existed. Nor did the photograph allow this determination. The temperature of this step was chosen as the critical in this series. At the temperature — 1.4 millidegrees, however, the brilliantly reflecting disc of the surface was clearly visible and, moreover, the photograph showed definitely the density step at the surface.

The critical temperature was found to be 31.04°, the same as found by MBM.

Mass distributions at equilibrium

Already in the first series the mass distributions were determined in the tube at a temperature (+ 25 millidegrees) which was reached by cooling as well as by heating. The mass distributions were found to be similar, but definitely different from the distributions which were found at higher and lower temperatures. Corresponding experiments which, however, showed poorer agreement between the final mass distributions at + 20, are reported in the next paragraph.

Identical mass distributions by heating and by cooling have never been obtained, most probably because insufficient time has been allowed for equilibration at the temperatures selected.

More precise experiments will be carried out when the accuracy of the system is improved. Any serious hysteresis effect, however, would have been detected in the great number of mass distribution measurements which have been carried out.

Mass transportation phenomena

When the mass distribution in the tube, at temperatures below the critical, is in equilibrium, the surface will appear at certain heights in the tube, dependent upon the temperature and the ratio between the interior volume of the tube and the mass of the enclosed substance.

On rapid heating, however, the surface will start an unexpected movement upwards in the tube. The reason for this may be that the thermal expansion of the liquid is faster than the mass transfer from liquid to vapour. In the case of heating rapidly past the critical temperature the surface will degenerate at a position too high in the tube.

Fig. 6 demonstrates that a similar phenomenon takes place when the tube is rapidly heated from + 7 up to + 20 millidegrees above the critical. Curve + 7 shows the mass distribution in a tube which had remained at this temperature over night. Curve *a* shows the mass distribution 10 minutes after the temperature was raised up to + 20. The density gradient has moved upwards in the tube. Curve + 20 shows the mass distribution in the tube after having been kept for 3 hours at this temperature.

On the other hand, by rapid cooling of the tube past the critical temperature, the reappearence of the surface will take place in a lower position in the tube than expected. By cooling the tube to a temperature above the critical

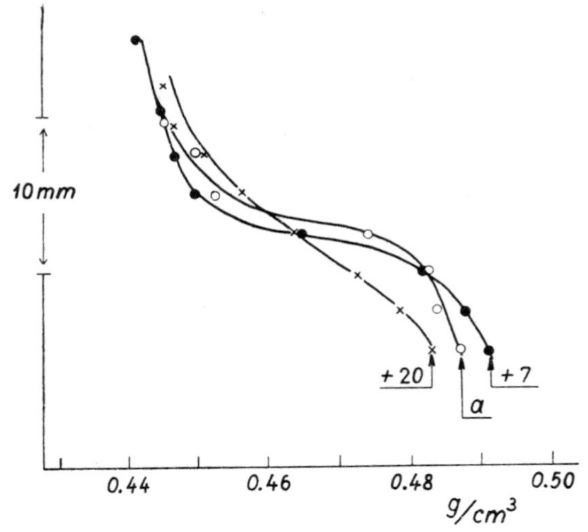

Fig. 5. Mass Transportation.

Ordinate and abscissa as in Fig. 3. The temperature of the thermostat was raised from 7 millidegrees above the critical (curve +7) to 20 millidegrees above the critical. Curve a shows the mass distribution when tube had remained 10 minutes at this temperature, curve +20 shows the mass distribution after 3 hours.

a similar phenomenon seems to take place. In Fig. 6, curve + 90 shows the mass distribution in the tube after the tube had stayed at + 90 millidegrees over night. By lowering the temperature the density of the lower part of the tube increases almost immediately and a density gradient is built up which moves very slowly upwards in the tube — *cf.* curves *a* and *b*. Curve + 20 shows the mass distribution after 48 hours at 20 millidegrees above the critical.

Fog

Fog formation, commonly found by cooling the tubes past the critical region, may be connected with the mass transportation phenomena. In the tubes of the two first series heavy fog was formed by cooling and gave the impression of being a subcooling phenomenon. Fog was generally not produced by heating. However, when the tube was slowly heated past the critical temperature (second series Fig. 3), a light fog was observed in the region of a sharp density gradient.

In the tubes prepared for the last series, and with the care described above, no fog formation at all was produced by either heating or slow cooling. On cooling rapidly past the critical temperature, however, a 5 mm high white belt introduced the reappearance of the surface.

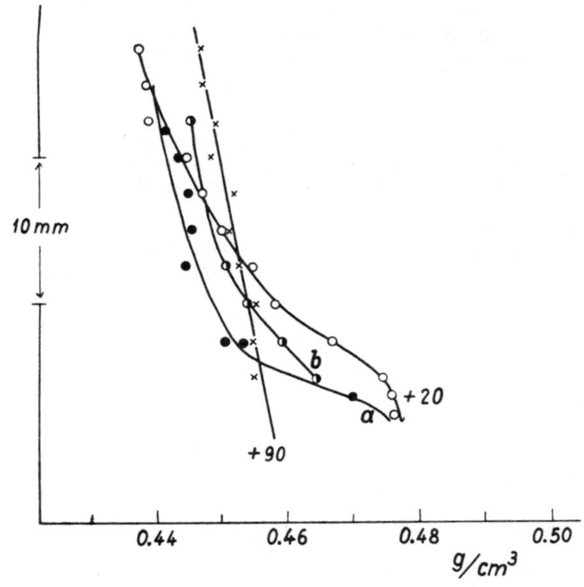

Fig. 6. Mass Transportation.

Ordinate and abscissa as in Fig. 3. The temperature of the thermostat was lowered from 90 millidegrees above the critical (curve +90) to 20 millidegrees above the critical. Curve a shows the mass distribution immediately after the thermometer had reached the new temperature, curve b the mass distribution 3 hours later. Curve +20 shows the mass distribution when the tube had remained for 48 hours at 20 millidegrees above the critical.

SUMMARY

The initial purpose of this work was to study the supposed deviation from classical behaviour of a substance (carbon dioxide) enclosed in a vertical tube of adequate filling and heated within the critical region. However, no anomalies could be observed. The mass distributions were as expected if determined by isotherms of the classical type. Furthermore, the coexistence curve has a rounded top. The measurements of mass distribution in vertical tubes can very likely provide a method for the determination of the isotherm in the very neighbourhood of the critical point.

Although the equipment has not yet reached the desired perfection for precise quantitative determinations, it appears most likely that the isotherms have been determined and further that the top of the coexistence curve conforms to the "cubic rule".

The carbon dioxide, which was prepared with the greatest caution, did not show the commonly observed fog-formation when heated or cooled in the critical region.

Measurements in the close neighbourhood of the critical point require a very good thermostat. A new thermostat with the temperature control based on the thermal expansion of the thermostat water itself has been designed and built. For the measurements of the mass distributions in the vertical tubes the refractivity of the enclosed carbon dioxide could be measured at any desired level. For this purpose a cylindrical lens method was developed.

First of all I wish to thank Professor Dr. Odd Hassel for his interest in this work and for his permission to let me work in his laboratory. Thanks are also due to several members of the staff for valuable discussions, and to Dr. K. W. Hedberg, Pasadena, temporarily in Oslo.

I am indebted to my father for the economical support which allowed the accomplishment of this work. I also thank "*Almenvidenskapelige Forskningsråd*", which has paid the wages of an instrument maker put at my disposal. Mr. Amund Helgesen I thank for his loyal interest and a series of ideas in the design of the equipment.

REFERENCES

1. Maass, O., and co-workers *e. g.* Tapp, J. S., Steacie, E. W. R., and Maass, O. *Can. J. Research* **9** (1933) 217.
2. Weinberger, M. A., and Sneider, W. G. *Can. J. Chem.* **30** (1952) 847.
3. Young, F. B. *Phil. Mag.* [6] **20** (1910) 793.
4. Reihlen, H. *Ber.* **72** (1939) 112.
5. Michels, A., Blaisse, B., and Michels, C. *Proc. Roy. Soc.* **160 A** (1937) 358.

For a general survey concerning the critical phenomena and the supposed anomalies reference can be made to Partington, J. R. *An Advanced Treatise on Physical Chemistry I*, Longhams, Green and Co., London, 1949, p. 630.

Received September 7, 1953.

Long-Range Correlation Length and Isothermal Compressibility of Carbon Dioxide Near the Critical Point

Joseph H. Lunacek and David S. Cannell
Department of Physics, University of California, Santa Barbara, California 93106
(Received 6 July 1971)

We have measured the angular dependence and total intensity of light scattered by carbon dioxide near its critical point, shown that the Ornstein-Zernike theory is inadequate, and determined the temperature dependence of the isothermal compressibility κ_T and the long-range correlation length ξ. The exponents describing the divergences of κ_T and ξ above the critical temperature are $\gamma = 1.219 \pm 0.01$ and $\nu = 0.633 \pm 0.01$, respectively. The exponent describing the departure from the Ornstein-Zernike theory is $\eta = 0.074 \pm 0.035$.

It has been predicted by Ornstein and Zernike (OZ),[1] and verified experimentally,[2] that the angular distribution of the intensity of light scattered by a pure fluid near its critical point is accurately given by

$$I(\vec{K}) = A\kappa_T \sin^2\varphi/(1 + K^2\xi^2), \qquad (1)$$

where κ_T is the isothermal compressibility, A is a constant which can be evaluated at any temperature, φ is the angle between the electric field of the incident light and the scattered wave vector; and K is the scattering wave vector $2k_0 \sin\frac{1}{2}\theta$, where k_0 is the wave vector of the incident light in the fluid and θ is the scattering angle. The above expression defines the OZ correlation length ξ. As the critical point is approached, ξ and κ_T diverge as $(T - T_c)^{-\nu}$ and $(T - T_c)^{-\gamma}$, respectively. The fluctuation theorem for the isothermal compressibility together with the OZ form of the correlation function yields the result $2\nu = \gamma$. Both ξ and κ_T can be determined independently by measuring the asymmetry and total intensity of light scattered by the fluid, thus offering an ideal method of checking the theory. In addition, measurements of ξ can be used to check predictions[3] concerning the dynamic behavior of critical fluids.

We have made accurate measurements of the differences in and magnitudes of the intensity of light scattered through the angles $\theta_F = 17.5°$ and $\theta_B = 168.5°$ in carbon dioxide. The measurements were made on the critical isochore over the temperature range $0.023 \leq T - T_c \leq 10.00°C$. We also measured the total scattered intensity over essentially the same range. Since the anisotropy becomes very small at higher temperatures,[2] the measurements were made using a differential technique with the apparatus shown in Fig. 1.

The scattering cell was constructed of beryllium copper, optical glass, and indium. Its temperature was controlled to $\pm 0.001°C$, and measured using a platinum resistance thermometer. It was filled to within 0.1% of the critical density with carbon dioxide containing less than 50 ppm of impurities. The density was determined by measuring the meniscus height as a function of temperature using the data of Michels, Blaisse, and Michels[4] for the shape of the coexistence curve. This also determined the critical temperature, and yielded the value $31.000 \pm 0.001°C$, which coincided with the temperature of meniscus disappearance. A differential scattered-intensity measurement was made by removing the beam splitter and blocking the transmitted beam. Light scattered through the angle θ_F was collected using the optics L_1, L_2, and L_3. The variable aperture A determined the collection solid angle. Lenses L_1 and L_2 imaged the scattering region on the slit S, thus determining that section of the beam from which scattering was accepted. The optics for θ_B functioned identically. Lenses L_3 and L_3' focused the light from each channel

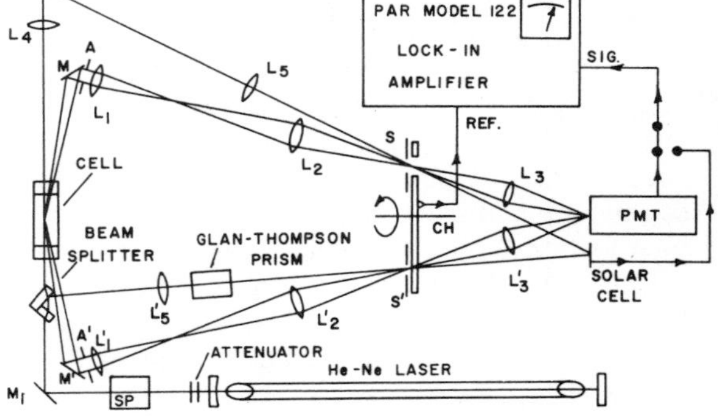

FIG. 1. Schematic diagram of the apparatus.

onto the same spot on the photocathode of an RCA 7265 photomultiplier. The chopper *CH* alternately accepted light from each channel. The photocurrent was passed to a lock-in amplifier whose reference was generated by the chopper. Its output was proportional to the difference in the intensity of the light collected by each channel. The intensity collected by either channel alone was measured by blocking the other channel.

To measure the total scattered intensity, the attenuation of the beam by the fluid was determined by comparing the reference and transmitted beams.[5] The beams passed through the chopper and fell on a solar cell whose output went to the lock-in. The Glan-Thompson prism served as a variable attenuator.

In making either differential or total scattered-intensity measurements, the lock-in output can be written as $V = G_F I_F - G_B I_B$. For a differential scattered-intensity measurement, I_F and I_B are the scattered powers per unit solid angle for the angles θ_F and θ_B, and the gains G_F and G_B are determined by the collection solid angles; lens, window, and mirror losses; photomultiplier gain; and electronic gains. For a total scattered-intensity measurement, I_B and I_F are the respective intensities of the incident and transmitted beams *in the cell*. The gains are determined by the beam splitter, Glan-Thompson setting, etc.

We now consider the differential scattered intensity. For each temperature T_i the ratio of the lock-in output with both channels open to that with only the forward channel open is

$$S_\xi(T_i) = 1 - \frac{G_B}{G_F}\frac{I_B}{I_F} \equiv 1 - \frac{G_B}{G_F} f_\xi(T_i), \qquad (2)$$

where $f_\xi(T_i) = [1 + K_F^2 \xi^2(T_i)]/[1 + K_B^2 \xi^2(T_i)]$ with K_F and K_B the scattering wave vectors for the angles θ_F and θ_B. For each pair of temperatures the ratio

$$R_{ij}^{\ \xi} \equiv \frac{1 - S(T_i)}{1 - S(T_j)} = \frac{f_\xi(T_i)}{f_\xi(T_j)} \qquad (3)$$

was computed. Thus the measurements determine the ratios $f_\xi(T_i)/f_\xi(T_j)$. If ξ is assumed to have the form $\xi_0[(T - T_c)/T_c]^{-\nu}$, then ξ_0 and ν can be determined. The best fit for ξ was used to calculate G_B/G_F by using Eq. (2) with data points at $T - T_c = 4.420°\mathrm{C}$. The correlation range for all other points was calculated with this value for G_B/G_F. The points and the best fit to ξ are shown in Fig. 2.

A similar procedure was utilized in interpreting the total scattered-intensity measurements. In this case the $S_\tau(T_i)$ to be used is

$$S_\tau(T_i) = 1 - \frac{G_F I_F}{G_B I_B} \equiv 1 - \frac{G_F}{G_B}\exp[-\tau(T_i)L], \qquad (4)$$

where $\tau(T_i)$ is the attenuation per unit length in the fluid, L is the path length in the cell, and I_B and I_F are the intensities of the incident and transmitted beam *in the cell*. The transmission and reflectivity of the beam splitter and the cell-window losses are included in the gains. The experimentally determined ratios thus take the form

$$R_{ij}^{\ \tau} \equiv \frac{1 - S_\tau(T_i)}{1 - S_\tau(T_j)} = \exp\{[\tau(T_j) - \tau(T_i)]L\}, \qquad (5)$$

providing the difference in the attenuation for any two temperatures. The attenuation τ may be related to κ_T by integrating Eq. (1) over angles, yielding[5]

$$\tau = \pi A \kappa_T \times \left[\frac{(2\alpha^2 + 2\alpha + 1)}{\alpha^3}\ln(1 + 2\alpha) - \frac{2(1+\alpha)}{\alpha^2}\right], \qquad (6)$$

where $\alpha = 2k_0^2\xi^2$; $A = (\pi^2 k_B T/\lambda_0^4)(\rho\partial\epsilon/\partial\rho)_T^2$, where λ_0 is the vacuum wavelength of the light, k_B is Boltzmann's constant, T is the absolute temperature, ρ is the density, and ϵ is the dielectric constant. The quantity $(\rho\partial\epsilon/\partial\rho)_T^2 = [(n^2 - 1)(n^2 + 2)]^2/9$ was calculated by using the Lorentz-Lorenz relation and measured values of the index of refraction[7] n. If κ_T is assumed to have the form $\kappa_0[(T$

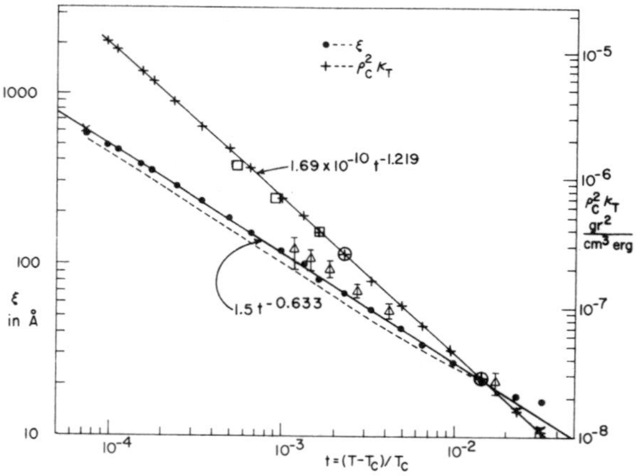

FIG. 2. The long-range correlation length and susceptibility $\rho_c^2\kappa_T$ of carbon dioxide along the critical isochore as a function of the reduced temperature $t = (T - T_c)/T_c$. The triangles are x-ray scattering results from Ref. 6. The squares are from the static *PVT* data of Ref. 4. The circles are our own determination of the absolute compressibility.

$-T_c)/T_c]^{-\gamma}$ then κ_0 and γ can be determined. Figure 2 shows the fit and the experimental values for the isothermal compressibility, presented as $\rho_c^2\kappa_T$. The points were obtained in the same manner as those for ξ. We also determined the compressibility absolutely by measuring the ratio of the forward scattered intensity at two temperatures. As seen from Eq. (1) this determines the *ratio* of the compressibilities at those temperatures. A different relationship involving the compressibility is obtained from R_{ij}^{τ} which gives the *difference* in the attenuation τ for the two temperatures. In computing the ratio we were able to eliminate the effect of attenuation by the fluid, using our value for the difference in τ at those two temperatures. By using Eq. (6) and the measured values for ξ, this difference may be combined with the ratio of the compressibilities to yield the compressibility at those two temperatures. These values are shown as open circles in Fig. 2. The analysis yields the following results for ξ and κ_T:

$$\xi = (1.50 \pm 0.09)[(T - T_c)/T_c]^{-0.633 \pm 0.01}\ \text{Å},$$

$$\kappa_T = (7.75 \pm 0.46) \times 10^{-10} \times [(T - T_c)/T_c]^{-1.219 \pm 0.01}\ \text{cm}^2/\text{dyn}.$$

The errors were obtained by varying one parameter while adjusting the other in order to maintain a best fit, until the rms deviation of the fit from the points was twice that of the best fit.

The OZ expression for the angular dependence of the scattered intensity accurately represents our data with the above power-law dependences for ξ and κ_T. However, the theory also predicts that $2\nu = \gamma$, which disagrees with our results. We conclude that the OZ theory is inadequate to describe completely the long-range correlation near the critical point. This failure has been anticipated theoretically and modifications have been worked out.[8] The principal modification is the introduction of an exponent η describing the departure of the angular dependence from that of Eq. (1). The exponent relationship becomes $(2-\eta)\nu = \gamma$, which with our values for ν and γ yields $\eta = 0.074 \pm 0.035$. We also analyzed our data using the modified theory and this value of η, but no significant change occurred either in the fits or in ν or γ.

Recently, Kawasaki[3] has derived an expression for the critical part Γ_c of the Rayleigh linewidth $\Gamma_R = \Lambda K^2/\rho_0 C_p$ of a pure fluid near its critical point, namely,

$$\Gamma_c = \frac{k_B T}{8\pi\eta_s\xi^3}\left\{1 + (K\xi)^2 + \left[(K\xi)^3 - \frac{1}{K\xi}\right]\tan^{-1}K\xi\right\}, \quad (7)$$

where η_s is the shear viscosity. The derivation used the OZ correlation function, but for our range of $K\xi$ it is very insensitive to $\eta \leq 0.1$. By using measured values for the thermal conductivity[9] Λ, it is possible to obtain values for Γ_c from the measured values[10] Γ_R. Since η_s is known,[11] our measurements of ξ provide the means of checking Eq. (7). We have done so by using the measured values of Γ_c and the $K=0$ limit of Eq. (7) to compute ξ in the temperature range $0.023 \leq T - T_c \leq 4.0$°C. The values of ξ obtained in this manner are shown as the dashed line in Fig. 2, where they may be compared with our measured values. The agreement is within the error for the linewidths and corrections.

The stability was enhanced by mounting the system on a massive table and by temperature controlling the photomultiplier. The value of the difference in the scattered intensities at any temperature was reproducible to within $\pm 0.05\%$ of the scattered intensity. The error caused by stray light increases as $T - T_c$ is increased because of the reduction in the scattered intensity. We estimated its effect by extending our measurements to $T - T_c = 17.8$°C, and attributing all of the difference between the measured asymmetry and the value obtained by extrapolating our fit to stray light. The one-sided error bars on the two highest temperature points in Fig. 2 represent the possible effect of stray light. At $T - T_c = 4.42$°C the stray light contributed less than 0.03% to the asymmetry of 0.23%. The extremely low stray-light level was achieved by orienting the cell so that the background viewed by either set of optics was a cell window viewed in a plane containing no specular reflections. For temperatures near T_c, the primary difficulty arises from multiple scattering. At $T - T_c = 0.023$°C we estimated its effect by observing the asymmetry and intensity of the multiple scattering occurring immediately above the beam. Correcting for this effect moves the point at $T - T_c = 0.023$°C in Fig. 2 to the point indicated by the X. To eliminate heating by the beam we attenuated it until the results were independent of beam power. An error present at all temperatures arises from the attenuation of light by the fluid. It arises if the path length in the cell differs for the two channels. For our geometry this effect changes all the values of ξ by essentially the same percentage, the maximum difference in the changes being less than 0.5% and the changes being less than 4%.

In conclusion, our measurements of the long-range correlation length and isothermal compressibility of CO_2 indicate that the Ornstein-Zernike theory is inadequate to describe completely the form of the long-range correlation function in a pure fluid near its critical point. Our measurements together with the exponent relation $(2-\eta)\nu=\gamma$ lead to the result that $\eta=0.074 \pm 0.035$. In addition, our measurements, together with measurements of the thermal conductivity,[9] Rayleigh linewidth,[10] and shear viscosity,[11] show that the critical part of the Rayleigh linewidth is adequately described by the dynamical theory due to Kawasaki.[3] We should mention that White and Maccabee[12] have recently measured κ_T for CO_2, obtaining $\gamma=1.17\pm 0.02$ which disagrees with our result.

The authors wish to thank Professor Richard A. Ferrell for many valuable suggestions. We also acknowledge with thanks stimulating conversations with Dr. Marzio Giglio, and wish to extend special thanks to Professor Stuart B. Dubin who aided us immeasurably in the original design and testing of the apparatus. We also wish to thank Professor George B. Benedek for providing us with the temperature controller used in the experiment. We were aided immeasurably in our efforts by the skill and meticulous craftsmanship of Mr. Hans R. Stuber.

[1]L. S. Ornstein and F. Zernike, Proc. Acad. Sci. Amsterdam 17, 793 (1914), and Phys. Z. 19, 134 (1918).

[2]M. Giglio and G. B. Benedek, Phys. Rev. Lett. 23, 1145 (1969).

[3]K. Kawasaki, Phys. Rev. A 1, 1750 (1970).

[4]A. Michels, B. Blaisse, and C. Michels, Proc. Roy. Soc., Ser. A 160, 358 (1937).

[5]V. G. Puglielli and N. C. Ford, Jr., Phys. Rev. Lett. 25, 143 (1970).

[6]B. Chu, J. S. Lin, and J. A. Duisman, Phys. Lett. 32A, 95 (1970).

[7]J. Straub, Ph. D. thesis, Technische Hochschule, München, 1965 (unpublished).

[8]M. E. Fisher, J. Math. Phys. 5, 944 (1964).

[9]J. V. Sengers and P. H. Keyes, Phys. Rev. Lett. 26, 70 (1971).

[10]H. L. Swinney and H. Z. Cummins, Phys. Rev. 171, 152 (1968).

[11]J. Kestin, J. H. Whitelaw, and T. F. Zien, Physica (Utrecht) 30, 16 (1964).

[12]J. A. White and B. S. Maccabee, Phys. Rev. Lett. 26, 1468 (1971).

The Isotherms of CO_2 in the Neighbourhood of the Critical Point and Round the Coexistence Line*

By A. Michels, B. Blaisse and C. Michels

(*Communicated by F. A. Freeth, F.R.S.—Received* 18 *May, Revised* 11 *August* 1936)

The isotherms of CO_2 between 0 and 150° C. and up to 3000 atm. have been previously published by two of the authors (Michels, A. and C. 1935). The method used for these measurements was not suitable, however, for determinations in the neighbourhood of the critical point and the co-existence line. A second method has therefore been developed by which both the critical data and the coexistence line can be determined. This method and the results obtained are described in the present paper.

The Method and Apparatus

The method was based on the one developed by Michels and Nederbragt (1934) for the determination of the condensation points of a binary mixture. While, however, for the measurements of condensation points, it was not necessary to know the quantity of gas in the apparatus, this knowledge is

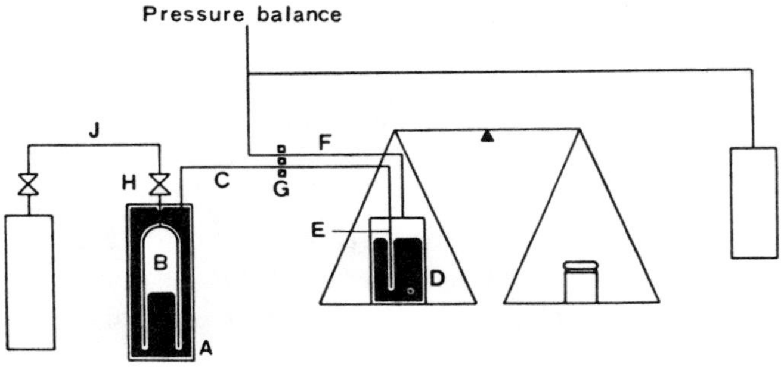

Fig. 1

essential for the determination of isotherms. A new apparatus was therefore constructed in which this quantity could be determined. A diagrammatic sketch showing the principle employed is given in fig. 1. In a steel vessel A, a glass bell B is suspended which is connected through the steel valve H and the capillary J to a cylinder containing a supply of the gas to be

* 50th publication of the van der Waals Fund.

examined. A steel capillary *C* connects *A* with a second steel vessel *D*, placed on one scale pan of a balance. Inside *D* a steel tube *E*, which is coupled to *C*, reaches to the bottom. The capillary *F* is connected to the top of *D* and leads to a cylinder of pure nitrogen and to an apparatus for measuring the gas pressure. The capillaries *C* and *F* are flexible, and are supported at *G* at such a distance from the scale pan that the variations in the forces acting on the latter during the swinging can be neglected. Before starting the measurements, the vessel *A*, the glass bell *B* and the tube *C* are completely, and the vessel *D* is partly filled with mercury. The valve *H* is then opened and CO_2 gas admitted to the glass bell, driving mercury out of *A* into *D*. The pressure in *D* is balanced by nitrogen introduced through *F*. When sufficient CO_2 has entered the glass bell, the valve *H* is shut. As the filling operation is carried out at a temperature and pressure at which the isotherms of CO_2 are known, the amount of gas in *B* can be calculated from a knowledge of the volume.

The pressure of the gas can be derived from the pressure of the nitrogen by applying a correction for the difference of the mercury levels in the glass bell *B* and the steel vessel *D*. The volume is equal to the volume of the mercury displaced and can be calculated from the increase of the weight of *D*. Here, however, three corrections have to be applied: one for the weight of the nitrogen in *D*, one for the expansion of the two cylinders by the pressure and one for the compression of the mercury.

As it is also necessary to know the temperature of the gas, the bomb *A* is placed in an oil thermostat. Having determined the quantity of gas in *B*, the isotherms can be measured by changing the pressure of the nitrogen and determining the new volume and pressure of the gas in *B* in the same way as has been already described. For measurements at other temperatures the bath in which bomb *A* is placed can be heated or cooled and the procedure repeated. In this case another correction must be applied for the thermal expansion of the steel and the mercury in *A*.

The thermostat was regulated at 25° C. and higher temperatures in the way described by two of the authors (Michels, A. and C. 1933, p. 416) and below 25° C. as described by Michels and Nederbragt (1934).

Details of the Apparatus

A more detailed drawing of the apparatus is given in fig. 2. The glass bell *B*, which had a volume of about 300 c.c., was sealed at *K* to a chromium steel tube, which was screwed, vacuum and pressure tight, into the top of the steel bomb *A*. The tube *C*, leading from the bomb in the thermostat to the one on the balance, D, was divided into two parts by a small valve L. The lower part, C_1, had an internal diameter of 2 mm., while the rest, C_2, was a flexible capillary of 1·2 mm. outer and 0·8 mm. inner diameter. Preliminary experiments had shown that tubes of such diameters allowed pressure equilibrium to be reached within 1/1000 of an atmosphere in a very short time. The length of C_2 from G, where it was supported, to the balance bomb was about 40 cm. The variation on the forces exerted by this tube on the scale pan of the balance during its swing was less than 50 mg.

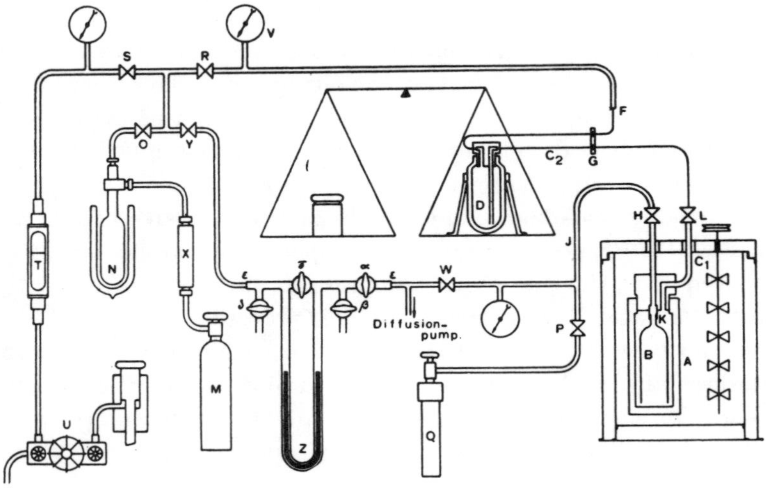

FIG. 2

The volume of the balance bomb was about 850 c.c. The nitrogen used for transmitting the pressure to the top of the mercury in it was supplied from the storage cylinder M. This nitrogen contained impurities amounting to 1 %, consisting of inert gases. As an extra precaution it was passed through a filter X and a stainless steel catchpot N, cooled with solid CO_2. The nitrogen was admitted through the valves O and R and the capillary F to the balance bomb; this capillary F had the same dimensions as C_2 and was also supported at G. The pressure of the nitrogen could be measured to within 0·1 of an atmosphere by a spring manometer, V. For accurate measurements of the pressure, the valve S was opened, thus applying pressure on the oil surface in the glass levelling gauge T. This levelling gauge was coupled through the hydraulic press U to a pressure balance, with which the pressure could be determined to 1/1000 of an atmosphere. The steel valves W and Y led to a mercury diffusion pump and a glass apparatus for the determination of some of the corrections.

Determination of the Corrections and Experimental Procedure

After the apparatus had been assembled, the balance bomb was filled with about 10·5 kg. of mercury. The valves O, S, L, and the glass taps β, and δ were shut and all the other valves and taps opened while the whole apparatus was evacuated with the diffusion pump. When a vacuum of about 1 cm. had been obtained, L was opened and the pumping continued for several days. Then the valves H and Y were shut and nitrogen admitted to the balance bomb by opening O, thus forcing mercury from this bomb into the thermostat bomb B. The pressure was then raised to 1 atm. L was shut and nitrogen again let into the balance bomb at about 30 atm. From the increase of weight, the temperature of the nitrogen and the known isotherms of this gas, the free volume of the balance bomb could be calculated. The procedure was repeated to 70 and 100 atm. Since the figures for the free volume agreed within the experimental accuracy (about 100 mg.) it was possible to neglect the expansion of the balance bomb with pressure. With a pressure in D of 100 atm. the valve L was opened. From the small decrease in the weight of D, the sum of the compression of the mercury and the expansion of the bomb A was calculated. This process was repeated at other pressures and it was found that the change in weight was proportional to the pressure within the experimental accuracy. From the results, the correction to be applied to the final measurements for the compression and expansion could be deduced. Pressure was released to 1 atm., the bath regulated at 50° C., and from the increase of the weight of the balance bomb and the density of mercury at 25 and 50° C., the thermal expansion of the bomb A was calculated.

The temperature was then again reduced to 25° C. and the difference between the mercury levels in A and D measured at different heights of the mercury in A, or at different weights of the balance bomb D, using the mercury differential manometer Z. This manometer was made of glass and was joined to the apparatus at E by soldered joints. The valve γ was shut and the balance bomb, together with the left-hand side of the manometer, opened to the atmosphere through δ (by opening R and Y). The other side of the differential manometer was connected with the capillary J (by opening α and W). A small amount of CO_2 was admitted from the supply bomb until the differential manometer showed a pressure of about 0·5 atm. The valve H was then opened and some CO_2 entered the glass bell till the pressure of the CO_2 was balanced by the difference in the mercury heights in B and D. This difference was then read directly from the levels of the mercury in Z. At the same time the weight of D was determined. By repeating this procedure at different positions of Z, the difference between the levels in B and D as a function of the amount of mercury in the balance bomb was determined.

Filling of the Apparatus with CO_2

The valve L was shut and, as a precautionary measure, the CO_2 was pumped out of B with the diffusion pump till a high vacuum was obtained. W was then shut and L, P and Q opened. CO_2, under the full pressure of the storage bomb, entered A, forcing the mercury into the balance bomb. When sufficient CO_2 had been let in, H was shut and the pressure counterbalanced in D by allowing nitrogen to enter the balance bomb. The nitrogen pressure was then measured with a pressure balance. It may be noted that a correction must be applied for the position of the oil level in the levelling gauge, which could be read to within 1 mm. The difference in the height of the mercury level in D and of the oil level in T caused a small hydrostatic head of compressed nitrogen which could be neglected. Applying the different corrections mentioned, the pressure of the CO_2 gas could be calculated from the load on the pressure balance, and the volume from the weight of the balance bomb.

The temperature of the thermostat was measured with mercury thermometers, divided in 0·01° C. and calibrated by the P.T.R. at Berlin. P, V and T being known, the amount of CO_2 and thus the normal volume, could be calculated from the isotherm data, published previously by two of the authors (Michels, A. and C. 1935). The determination of the normal volume was carried out at three pressures and the results agreed with each other to within 1/5000. Two sets of measurements were carried out, the first with a normal volume of 9951 l. and the second with a normal volume of 20,612 l.

The volumes at other pressures and temperatures, including points in the liquid and coexistence region, were determined in exactly the same way.

The gas used was prepared from commercial CO_2 of "medical" quality* by a triple distillation in the way previously described by two of the authors (Michels, A. and C. 1933, p. 421). It was found that the vapour pressure during the condensation did not vary more than 1/20,000, showing that the gas contained only minute quantities of impurities.

The Results

The results obtained are given in Table I, the isotherms measured with the normal value 20,612 l. being marked with an asterisk. The critical region

* The authors wish to thank the Ammoniakfabriek in Weesp who kindly supplied the gas.

is given separately in fig. 3. Four isotherms, 0·027° C. below and 0·145, 0·280 and 0·483° C. above the critical temperature respectively, are plotted on a larger scale in fig. 4.

TABLE I

T (° C.)	d	p	v	pv
$2{\cdot}85_3$	52·288	36·318	0·0191248	0·69458
	53·159	36·663	0·0188115	0·68968
	53·852	36·932	0·0185694	0·68580
	54·751	36·992	0·0182645	—
	114·681	37·000	0·0087198	—
	433·67	37·000	0·0023059	—
	461·94	37·146	0·0021648	0·08041
	462·50	37·836	0·0021622	0·08181
	462·93	38·577	0·0021602	0·08333
$10{\cdot}82_2$	66·622	44·151	0·0150101	0·66271
	68·019	44·610	0·0147018	0·65585
	69·483	45·053	0·0143920	0·64840
	74·705	45·256	0·0133860	—
	144·06	45·264	0·0069416	—
	370·30	45·262	0·0027005	—
	433·77	45·414	0·0023054	0·10470
	434·89	46·686	0·0022994	0·10735
	435·13	46·944	0·0022982	0·10788
$19{\cdot}87_4$	92·825	55·360	0·0107730	0·59640
	94·939	55·785	0·0105331	0·58759
	97·346	56·224	0·0102726	0·57757
	99·042	56·261	0·0100967	—
	224·30	56·273	0·0044583	—
	342·56	56·270	0·0029192	—
	393·17	56·401	0·0025434	0·14345
	394·13	57·037	0·0025372	0·14472
	395·43	57·775	0·0025289	0·14611
$25{\cdot}07_0$	119·54	63·114	0·0083654	0·52797
	120·66	63·247	0·0082878	0·52417
	122·13	63·403	0·0081880	0·51915
	124·42	63·451	0·0080373	—
	127·92	63·446	0·0078174	—
	137·91	63·450	0·0072511	—
	188·89	63·468	0·0052941	—
	290·88	63·467	0·0034378	—
	292·65	63·456	0·0034171	—
	359·82	63·474	0·0027792	0·17641
	360·00	63·513	0·0027778	0·17642
	360·31	63·593	0·0027754	0·17650
	360·99	63·760	0·0027702	0·17663
	362·45	64·151	0·0027590	0·17699
	366·24	65·259	0·0027305	0·17819
	366·72	65·370	0·0027269	0·17826
	372·83	67·514	0·0026822	0·18108

TABLE I (*continued*)

T (° C.)	d	p	v	pv
$25{\cdot}07_0$	118·08	62·946	0·0084688	0·53308
	120·58	63·231	0·0082933	0·52439
	121·95	63·377	0·0082001	0·51969
	126·04	63·450	0·0079340	—
	196·98	63·462	0·0050767	—
	329·42	63·459	0·0030356	—
	360·42	63·550	0·0027745	0·17632
	362·44	64·073	0·0027591	0·17678
$25{\cdot}29_8$	119·42	63·284	0·0083738	0·52993
	123·95	63·750	0·0080678	0·51432
*	124·34	63·780	0·0080425	0·51295
	124·65	63·781	0·0080225	—
	124·68	63·785	0·0080205	—
	125·11	63·793	0·0079930	—
	126·11	63·791	0·0079296	—
	324·45	63·798	0·0030821	—
	353·69	63·796	0·0028273	—
	357·82	63·812	0·0027947	0·17834
	358·53	63·975	0·0027892	0·17844
	359·13	64·124	0·0027845	0·17856
$28{\cdot}05_2$	129·76	66·767	0·0077065	0·51454
	135·50	67·297	0·0073801	0·49666
	143·45	67·856	0·0069711	0·47303
	151·24	68·019	0·0066120	—
	194·85	68·015	0·0051322	—
	296·84	68·018	0·0033688	—
	332·66	68·128	0·0030061	0·20480
	336·72	68·597	0·0029698	0·20372
	341·73	69·350	0·0029263	0·20294
$29{\cdot}92_9$	160·37	70·678	0·0062356	0·44072
	164·20	70·819	0·0060901	0·43129
	168·26	70·916	0·0059432	0·42147
	170·87	70·974	0·0058524	0·41537
	178·77	70·993	0·0055938	—
	205·87	71·000	0·0048574	—
	213·24	71·007	0·0046896	—
	232·93	71·003	0·0042931	—
	268·04	71·002	0·0037308	—
	296·13	71·002	0·0033769	—
	303·78	71·063	0·0032919	0·23393
	307·15	71·148	0·0032557	0·23164
	311·03	71·326	0·0032151	0·22932
	313·21	71·443	0·0031928	0·22810
$30{\cdot}40_9$	168·62	71·513	0·0059305	0·42411
	174·58	71·653	0·0057280	0·41043
	179·37	71·737	0·0055751	0·39994
	184·25	71·775	0·0054274	—
	221·34	71·780	0·0045179	—
	275·97	71·787	0·0036236	—

TABLE I (*continued*)

T (° C.)	d	p	v	pv
$30{\cdot}40_9$	293·09	71·834	0·0034119	0·24509
	298·91	71·953	0·0033455	0·24072
	306·06	72·232	0·0032673	0·23600
$31{\cdot}01_3$	85·415	59·237	0·0117076	0·69353
	185·12	72·632	0·0054019	0·39235
*	192·69	72·710	0·0051897	0·37734
	200·02	72·757	0·0049995	0·36375
	206·22	72·778	0·0048492	0·35292
	209·21	72·787	0·0047799	0·34791
	214·24	72·794	0·0046677	0·33978
	227·82	72·796	0·0043894	—
	258·35	72·797	0·0038707	—
	267·51	72·821	0·0037382	0·27222
	272·69	72·846	0·0036672	0·26714
	286·91	73·002	0·0034854	0·25444
	300·94	73·419	0·0033229	0·24396
	85·418	59·239	0·0117071	0·69351
$31{\cdot}18_5$	85·423	59·322	0·0117065	0·69445
	187·00	72·894	0·0053476	0·38981
*	200·77	73·016	0·0049808	0·36368
	211·68	73·055	0·0047241	0·34512
	220·16	73·067	0·0045422	0·33188
	227·88	73·078	0·0043883	0·32069
	237·23	73·085	0·0042153	0·30808
	247·96	73·094	0·0040329	0·29478
	259·13	73·111	0·0038591	0·28214
	272·27	73·171	0·0036728	0·26874
	286·25	73·345	0·0034935	0·25623
	85·437	59·327	0·0117045	0·69439
$31{\cdot}32_0$	85·422	59·387	0·0117066	0·69522
	187·06	73·072	0·0053459	0·39063
*	200·66	73·207	0·0049836	0·36483
	211·65	73·260	0·0047248	0·34614
	220·30	73·281	0·0045393	0·33263
	227·89	73·294	0·0043881	0·32162
	237·24	73·307	0·0042151	0·30900
	247·88	73·319	0·0040342	0·29578
	259·45	73·347	0·0038543	0·28270
	272·51	73·426	0·0036696	0·26944
	286·31	73·622	0·0034927	0·25714
	85·426	59·389	0·0117060	0·69521
$31{\cdot}52_3$	174·09	73·044	0·0057442	0·41958
	186·58	73·332	0·0053596	0·39303
	201·15	73·505	0·0049714	0·36542
	216·34	73·585	0·0046224	0·34014
	235·70	73·636	0·0042427	0·31241
	259·19	73·701	0·0038582	0·28435
	271·43	73·789	0·0036842	0·27185
	284·69	73·992	0·0035126	0·25990
	307·75	74·949	0·0032494	0·24354

TABLE I (*continued*)

T (° C.)	d	p	v	pv
$32{\cdot}05_4$	85·381	59·731	0·0117122	0·69958
	157·99	72·954	0·0063295	0·46177
*	167·57	73·457	0·0059677	0·43837
	178·36	73·844	0·0056066	0·41402
	191·60	74·146	0·0052192	0·38699
	208·45	74·361	0·0047973	0·35674
	224·61	74·478	0·0044522	0·33159
	245·84	74·586	0·0040680	0·30339
	269·80	74·807	0·0037065	0·27727
	289·54	75·269	0·0034538	0·25996
	307·08	76·216	0·0032565	0·24820
	85·397	59·736	0·0117100	0·69951
$34{\cdot}72_1$	85·341	61·008	0·0117177	0·71487
	115·14	69·709	0·0086851	0·60544
*	123·59	71·408	0·0080913	0·57778
	132·87	72·969	0·0075262	0·54917
	132·91	72·969	0·0075239	0·54901
	144·07	74·459	0·0069411	0·51683
	156·02	75·690	0·0064094	0·48513
	171·91	76·858	0·0058170	0·44708
	189·49	77·726	0·0052773	0·41018
	210·80	78·420	0·0047438	0·37201
	237·00	79·066	0·0042194	0·33361
	259·80	79·691	0·0038492	0·30674
	287·37	80·982	0·0034798	0·28181
	318·82	83·484	0·0031366	0·26185
	85·352	61·011	0·0117162	0·71481
$40{\cdot}08_7$	162·14	82·101	0·0061675	0·50636
	173·93	83·407	0·0057494	0·47954
	188·00	84·713	0·0053192	0·45060
	203·00	85·891	0·0049261	0·42311
	220·18	87·094	0·0045417	0·39556
	238·95	88·380	0·0041850	0·36987
	263·53	90·318	0·0037946	0·34272
	292·72	93·730	0·0034162	0·32020
	316·70	98·447	0·0031576	0·31085

To find the correlation between the isotherms obtained by the present method and those mentioned above, measurements were carried out at 40° C. and at densities within the density-range of the previous measurements.

The values of pv at the required densities were calculated from the previous measurements by interpolation. The results after correction for a small temperature difference are shown in columns 3 and 4 of Table II, while column 5 shows the differences. The mean difference is 1/5500, and the greatest is only 1 : 3000. This agreement can be considered good, as the method

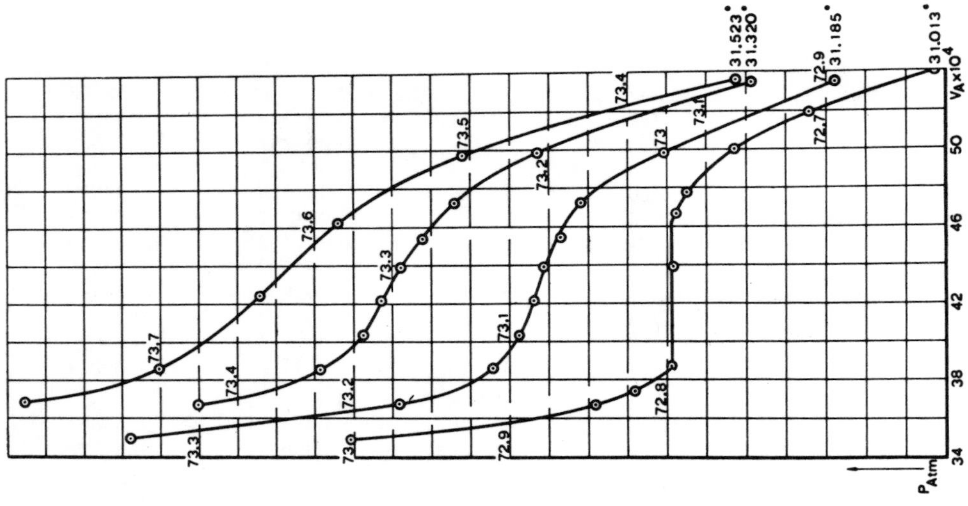
31.523°
31.320°
31.185°
31.013°
$V_A \times 10^4$
P_{Atm}
34
38
42
46
50
73.7
73.6
73.5
73.4
73.3
73.2
73.1
73
72.9
72.8
72.7
73.4
73.3
73.2
73.1
73.0
72.9
73.3

Fig. 4

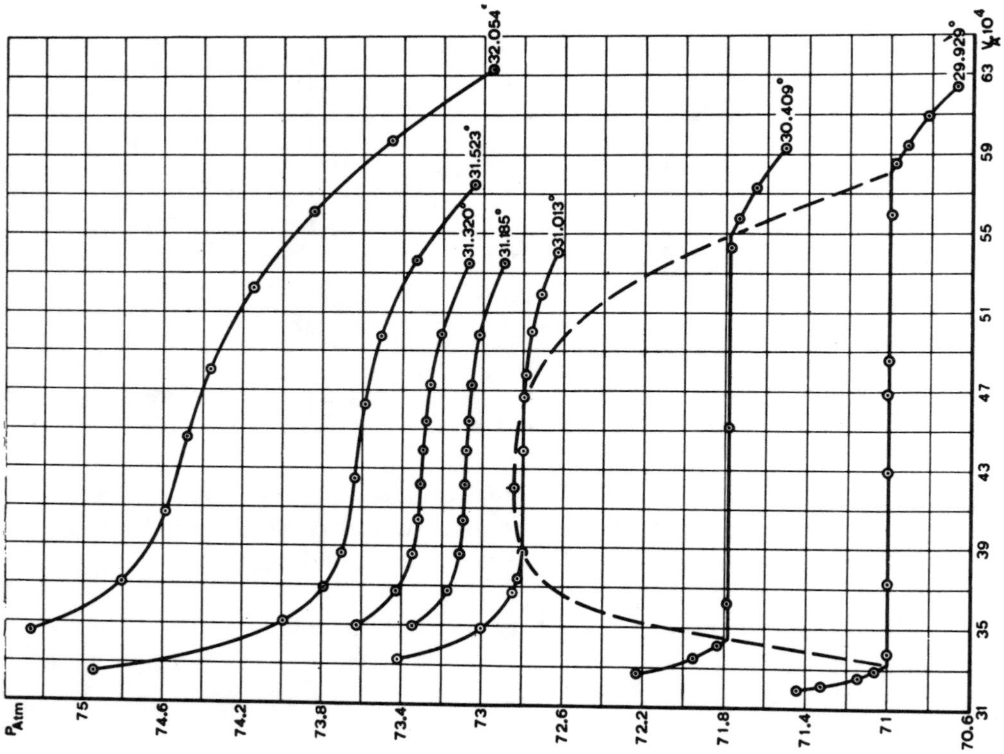
P_{Atm}
75
74.6
74.2
73.8
73.4
73
72.6
72.2
71.8
71.4
71
70.6
31
35
39
43
47
51
55
59
63
$V_A \times 10^4$
32.054°
31.523°
31.320°
31.185°
31.013°
30.409°
29.929°

Fig. 3

described here, particularly for the higher densities, is not as accurate as the one used formerly. As pointed out above, however, the present method was developed for measurements in a region in which the previous method is inapplicable.

TABLE II

d	pv (40°·087)	pv (40°·105)	$pv_{\text{ser. ev.}}$	$\Delta \times 10^5$
162·14	0·50636	0·50647	0·50638	9
173·93	0·47954	0·47966	0·47958	8
188·00	0·45060	0·45072	0·45066	6
203·00	0·42311	0·42324	0·42314	10
220·18	0·39556	0·39569	0·39575	−6
238·95	0·36987	0·37000	0·36708	−8
263·53	0·34272	0·34286	0·34293	−7
292·72	0·32020	0·32035	0·32043	−8
316·70	0·31085	0·31101	0·31091	10·5

DISCUSSION OF THE RESULTS

The Vapour Pressure

The vapour pressure of liquid CO_2 at the different temperatures can be obtained from the horizontal part of the isotherms. The results are given in Table III.

TABLE III—VAPOUR PRESSURE OF CO_2

t (° C.)	p (int. amt.)	p_{calc}	$\Delta = p_{\text{exp}} - p_{\text{calc}}$
2·853	36·997	36·997	0·000
10·822	45·261	45·409	−0·148
19·874	56·268	56·536	−0·268
25·070	63·456	63·736	−0·280
25·298	63·791	64·065	−0·274
28·052	68·017	68·146	−0·129
29·929	71·002	71·028	−0·026
30·409	71·780	71·778	+0·002
31·013	72·797	72·732	+0·065

The relationship $$^{10}\log p = \frac{-875{\cdot}186}{t+273{\cdot}15} + 4{\cdot}73909$$

can be used for interpolation. The deviation between the experimental values and those calculated from this formula are given in column 4 of Table III and are plotted in fig. 5. Meijer and van Dusen (1933) have published values for the vapour pressure of CO_2; to compare these values with ours, they have also been substituted in the formula and the deviations are plotted in fig. 5.

The Coexistence Line

The volumes of saturated vapour and liquid were determined from the intersection of the vapour and liquid curves with the horizontal part of the

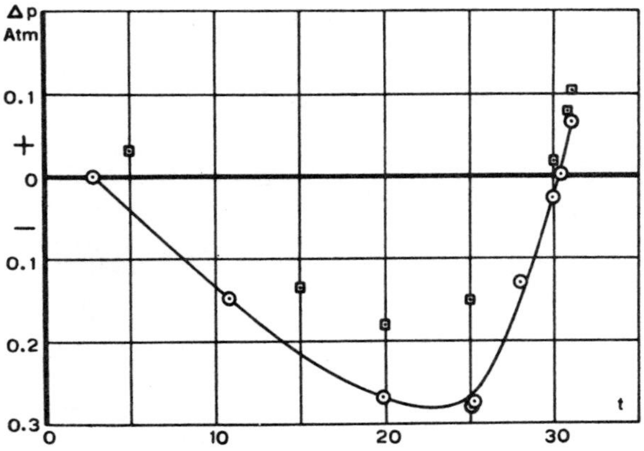

FIG. 5

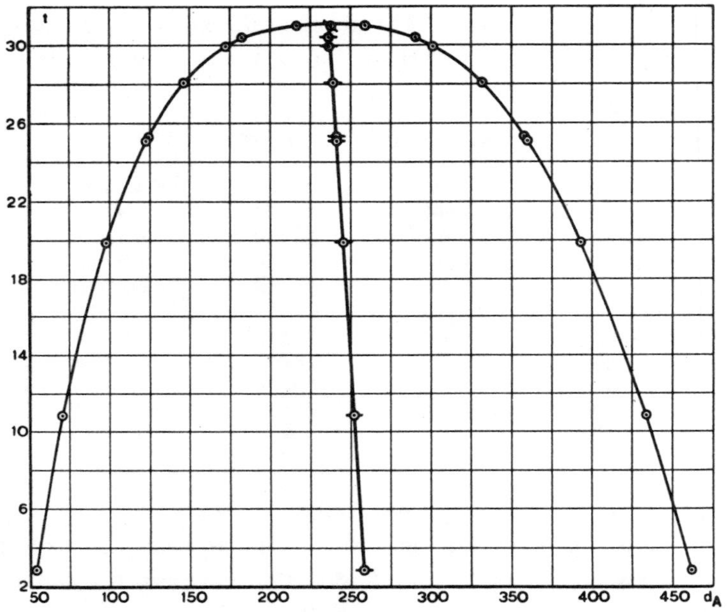

FIG. 6

isotherm. These volumes, and the corresponding densities, are given in Table IV. The densities are plotted as a function of temperatures in fig. 6.

TABLE IV—COEXISTENCE LINE OF CO_2

t (° C.)	v_{liq}	v_{gas}	d_{liq}	d_{gas}
2·853	0·002165	0·018506	461·8	54·04
10·822	0·002306	0·014244	433·6	70·21
19·874	0·002545	0·010244	393·0	97·62
25·070	0·002780	0·008152	359·8	122·67
25·298	0·002796	0·008034	357·7	124·47
28·052	0·003015	0·006828	331·7	146·46
29·929	0·003322	0·005804	301·0	172·30
30·409	0·003445	0·00549	290·3	182·22
31·013	0·00386	0·00462	259·0	216·4

The values of $\frac{d_{vap}+d_{liq}}{2}$ are also plotted in fig. 6. A straight line has been drawn in the diagram and it can be seen that, within the experimental accuracy, all the points lie on it showing that the empirical law of Matthias, for the rectilinear diameter, holds even in the neighbourhood of the critical point. The deviations of the experimental points from this line are given in Table V.

TABLE V—DEVIATIONS FROM RECTILINEAR DIAMETER

t (° C.)	$\frac{d_{liq}+d_{gas}}{2}$ (exp)	$\frac{d_{liq}+d_{gas}}{2}$ (calc)	Δ
2·85	257·93	257·93	0·0
10·82	251·92	251·97	−0·05
19·87	245·30	245·20	+0·10
25·07	241·21	241·31	−0·10
25·30	241·09	241·14	−0·05
28·05	239·07	239·07	0·00
29·93	236·66	237·67	−1·01
30·41	236·25	237·31	−1·06
31·01	237·7	236·9	+0·8

The values of $d_{liq}-d_{vap}$ could be represented by the well-known expression

$$d_{liq}-d_{vap} = C(T_k-T)^n,$$

except in the immediate neighbourhood of the critical point. Fig. 7 shows the difference between the experimental values and those calculated from the formula, with $C = 127{\cdot}60$, $n = 0{\cdot}357$. The formula, together with the deviation curve and the equation for the rectilinear diameter

$$\frac{d_{liq}+d_{vap}}{2} = 260{\cdot}07-0{\cdot}7484t$$

give an easy method for the calculation of d_{liq} and d_{vap} at intermediate temperatures.

Lowry and Erickson (1927) have summarized all the data published on the liquid and gaseous densities of CO_2. In the temperature range above 23° C., the only published data are those of Amagat (1892) and two points on the gas side published by Cailletet and Matthias (1891). The agreement between these and the values calculated from the present figures is of the order of 3 %.

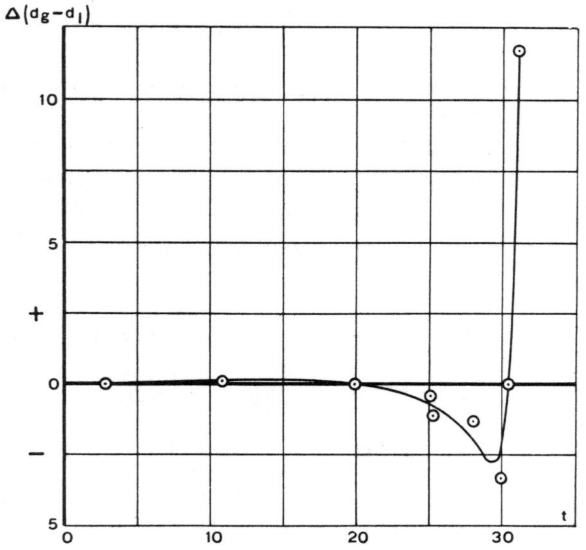

Fig. 7

The Latent Heat

The latent heat can be derived in the usual way from the slope of the vapour-pressure line and the volumes of saturated vapour and liquid using Clapeyron's equation

$$T\frac{dp}{dT} = \frac{\lambda}{v_2 - v_1}.$$

This latent heat can be divided into the internal latent heat and the work for the change in volume $p(v_2 - v_1)$. The results are given in Table VI and fig. 8.

It can be seen from the drawing that, in the immediate neighbourhood of the critical temperature, the values of λ fall rather abruptly to zero.

TABLE VI—HEAT OF VAPORIZATION

t (° C.)	(cal./mol.)	$p(v_{gas} - v_{liq})$ (cal./mol.)
2·85	2343·3	327·5
10·82	2047·5	292·5
19·87	1616·3	234·5
25·07	1269·3	185·6
25·30	1244·8	180·9
28·05	968·9	140·3
29·93	659·0	95·4
30·41	551·1	79·5
31·01	209·7	29·9

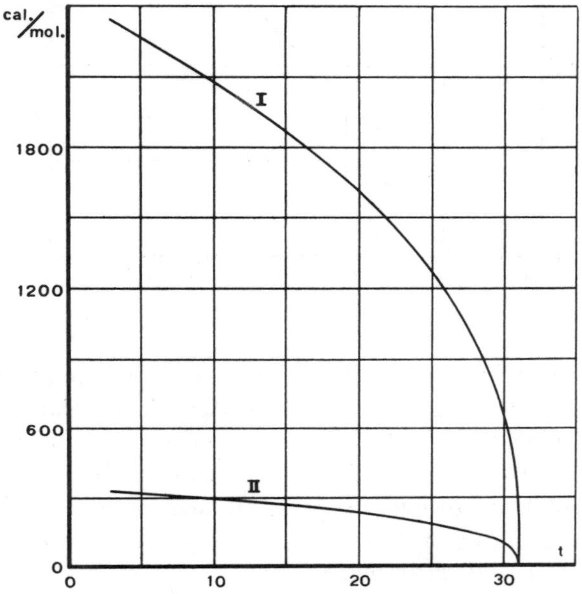

FIG. 8

Position of the Critical Point

The isotherms in the immediate neighbourhood of the critical temperature were measured with sufficient accuracy to allow the critical temperature to be determined from the inflexion points of the isotherms. The minimum values of $\left(\frac{dp}{dv}\right)_T$, found graphically from fig. 4, were plotted as a function of T; within the experimental accuracy the graphical representation appeared to be a straight line. The intersection of this line with the T axis gives the temperature where the tangent in the inflexion point is horizontal, i.e. the critical temperature.

Fig. 9 shows T_k to be 31·03° C. ± 0·01.

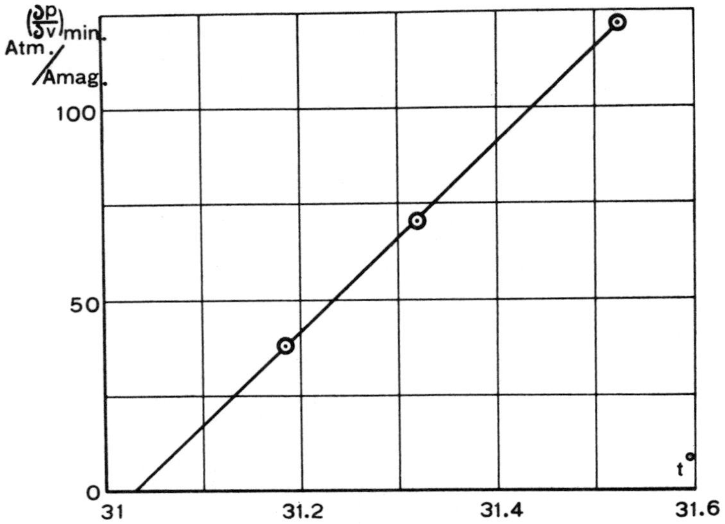

FIG. 9

In a similar way, by plotting the pressure of these inflexion points against p, the critical pressure is found to be $72{\cdot}83^5$ atm. Extrapolation of the vapour pressure line gives $72{\cdot}82^5$ atm. By extrapolation of the volumes of the inflexion points, the critical volume was found to be 0·00424 in Amagat units. Extrapolation however of the rectilinear diameter of Matthias gives the more reliable value of $d_k = 236{\cdot}7$ Amagat units corresponding with a critical volume of 0·004224 Amagat units.

A check on the position of the critical point can be obtained from the formula mentioned above:

$$d_{\text{liq}} - d_{\text{vap}} = C(T_k - T)^n.$$

If this formula is extrapolated for $T = T_k$, both sides of the equation must become 0. The experimental values for $d_{\text{liq}} - d_{\text{vap}}$ lead to $T_k = 31°{\cdot}05 \pm 0{\cdot}01$.

The most probable value of T_k can be taken as 31·04° C.

The corresponding value of p_k then becomes 72·85 atm.

These figures are compared with those of other authors in Table VII.*

* The figures of the other authors have been taken from van Dusen.

TABLE VII—CRITICAL DATA OF CO_2

	t_k	p_k	V_k
Andrews, Th. (1869)	30·92	73	—
Amagat, E. H. (1892)	31·35	72·9	0·464
Kuenen, J. P. (1897)	31·1	73·26	—
Keesom, W. H. (1903)	30·98	72·93	0·440
Dorsman, C. (1908)	31·10	73·00	0·451
Cordoso, E. and Bell, T. M. (1912)	31·00	72·85	—
Meijer, C. H. and v. Dusen, M. S. (1933)	31·0	72·80	—
A. and C. Michels and Blaisse, B.	31·04	72·85	0·467

The Critical State

Several authors recently have expressed a doubt as to the ability of the continuity theory of van der Waals-Andrews to describe the phenomena in the neighbourhood of the critical point.

Jacyna (1935) developed a theory leading to an equation, called by him "the thermodynamical equation of state" from which he deduced that a critical region should exist rather than a critical point. On the boundary of this region the value $\left(\frac{\partial p}{\partial v}\right)_T$ should be zero. In the case of CO_2 this region should extend over a temperature range of about 0·43° C. The results as plotted in fig. 4 make it doubtful whether this theory is reliable.

Winkler and Maass (1933) published some papers on density discontinuities in the critical region; they concluded that density differences should persist above the critical temperature, which can be compared with a continuation of the two phases.

In some cases this leads to a picture of a coexistence line with a discontinuity above the critical point. The results given above do not support this suggestion, but it may be mentioned that similar phenomena are predicted by Kuenen (1907) and others if minute quantities of impurities are present.*

Ostwald (1933) attempted to give a qualitative description of the critical phenomena in the terminology of colloid-chemistry. From the results described above it may be concluded that for a qualitative description the concise theory of van der Waals suffices.

SUMMARY

In previous papers isotherm data of CO_2 have been published between 0 and 150° C. up to 3000 atm. The method used could not be applied in the

* Verschaffelt calculated that the presence of 0·1 % H_2 in CO_2 would lead to density differences of 24 % at a temperature of 0·5° C. above the critical point.

direct neighbourhood of the critical point and round the coexistence line. The present paper gives a description of a new method to measure isotherms in this region with an average accuracy of 1 : 5500, and the data obtained for CO_2 by this method.

The principle of the method is to compress gas with mercury and to determine the change of volume of the gas from the weight of the mercury displaced. This weight is measured without the mercury being taken out of the apparatus.

Besides the isotherm results, the vapour pressure curve, the critical data and the latent heat are given. The results do not support the doubts, which have been expressed lately in literature as to the real existence of a critical point.

References

Andrews, Th. 1869 *Philos. Trans.* **159**, 11, 575.
Amagat, E. H. 1892 *C.R. Acad. Sci., Paris*, **114**, 1093.
Cailletet, L. and Matthias, E. 1891 *C.R. Acad. Sci., Paris*, **112**, 1170.
Cordoso, E. and Bell, T. M. 1912 *J. Chim. Phys.* **10**, 500.
Dorsman, C. 1908 Diss. Amsterdam.
Jacyna, W. 1935 *Z. Phys.* **95**, 253.
Keesom, W. H. 1903 *Comm. Phys. Lab. Leiden*, **88**.
Kuenen, J. P. 1897 *Phil. Mag.* **44**, 179.
— — 1907 "Die Zustandsgleichung." Vieweg.
Lowry, H. H. and Erickson, W. R. 1927 *J. Amer. Chem. Soc.* **49**, 2729.
Meijer, C. H. and v. Dusen, M. S. 1933 *Bur. Stand. J. Res.* **10**, 381.
Michels, A. and Michels, C. 1933 *Philos. Trans.* A, **231**, 416.
— — 1935 *Proc. Roy. Soc.* A, **153**, 201.
Michels, A. and Nederbragt, G. W. 1934 *Industr. Engng Chem.* (Anal. ed.), **6**, 135.
Ostwald, W. 1933 *Kolloidzschr.* **64**, 50.
Winkler, C. A. and Maass, O. 1933 *Canad. J. Res.* **9**, 613.

Static Critical Exponents at Structural Phase Transitions

K. A. Müller and W. Berlinger
IBM Zurich Research Laboratory, 8803 Rüschlikon, Switzerland
(Received 27 October 1970)

The temperature dependence of the rotational displacement parameters below the second-order phase transitions in $SrTiO_3$ and $LaAlO_3$ at $T_a = 105.5$ and 797°K is described by an exponent $\beta = 0.33 \pm 0.02$ down to $t = T/T_a = 0.95$. For smaller t's there occurs a change to Landau behavior approximately followed between $t = 0.9$ and 0.7. The observation of static critical exponents near displacive phase transitions confirms now the notion of universality in this field.

The static scaling theories introduced for fluids and for the Ising model have been quite successful in describing phase transitions[1-4] in the critical region. The critical exponents deduced for magnetic systems and fluids are close to one another[4,5] and the notion of universality has been advocated[6]; its essence being that, in the critical region, physically measurable variables become, to a substantial extent, independent of interatomic interactions. Due to the cooperative phenomena the correlation lengths are much larger than the range of forces between particles making the phenomena nearly independent of them.

Although the universality notion is quite suggestive and successful in most of the order-disorder transformations, an important void existed to fully confirm its validity. This concerns the large group of order-order transformations such as displacive purely structural or ferroelectric ones. Indeed up to recently these have all been analyzed within the framework of the Landau approach or the mean-field approximation, which yields critical exponents $\alpha = 0$, $\beta = \frac{1}{2}$, $\gamma = 1$, etc. (for a definition see Ref. 4). However, very recently it was found that the uniaxial stress dependence of the order parameters in $SrTiO_3$ within one degree of the transition T_a of the unstressed crystal deviated from the classically computed value.[7] In $LaAlO_3$ the temperature dependence of the EPR parameter $D(T)$ of Fe^{3+} substitutional for Al^{3+} deviated for $T < T_a$ near the structural phase transition from the straight line it should have followed if the system had behaved Landau-like[8] (see below). Hereafter it will be shown quantitatively that in both crystals the exponent β of the displacement parameter differs indeed from the mean field result ($\beta \neq \frac{1}{2}$).

The phase transitions in $SrTiO_3$ and $LaAlO_3$ set in when, in the cubic crystals, alternate static rotations of nearly rigid TiO_6 and AlO_6 octahedra occur around the [100] and [111] axes, respectively.[7] The rotation angle φ is a measure of the oxygen displacement, and corresponds to the order parameter of an order-disorder transition.[9] Its temperature dependence has been determined with limited accuracy by paramagnetic resonance of Fe^{3+} ions substitutional for Al^{3+} in $LaAlO_3$ and Ti^{4+} in $SrTiO_3$.[8] In the latter case there is a charge misfit, and in addition to the Fe^{3+} spectrum due to nonlocal charge compensation, one with a nearest-neighbor oxygen vacancy (Fe^{3+}-V_O) is observed.[8] Very recently proper shaping of $SrTiO_3$ crystals led to samples which became nearly monodomain below the phase transitions.[10] These samples allowed us a heretofore unattained accuracy in the determination of φ. The c axis of the monodomain was aligned parallel to the rotation axis of the magnet. Under this geometry there are essentially two Fe^{3+} spectra rotated relatively to each other by 2φ, the difference in TiO_6 sublattice rotations. Due to local distortion the corresponding Fe^{3+}-V_O spectra are rotated by the smaller amount $2\bar{\varphi}$; $\varphi = (1.59 \pm 0.05)\bar{\varphi}$.[7] For temperatures very close to T_a, where φ is small, the field splitting of the lines proportional to φ or $\bar{\varphi}$ was scanned on a recorder and analyzed. Maximum sensitivity occurs for maximum slope of the resonance magnetic fields versus field angle, i.e., ~30° away from [100] for the Fe^{3+} and along [110] for the Fe^{3+}-V_O spectrum (H_c in Fig. 1 of Ref. 7). For the latter an ultimate accuracy of 1×10^{-2} angular degrees was achieved corresponding to a determination of relative oxygen position by $\Delta x = 4\times10^{-12}$ cm. The temperature was determined with a gold-2.1% cobalt-copper thermocouple, a constant-voltage source, and a Keithley model No. 149 millimicrovolt null detector to a relative accuracy of 4/100 of a degree.

To determine β in the exponential relation

$$\varphi \propto \epsilon^{\beta}, \quad \epsilon = (T_a - T)/T_a, \qquad (1)$$

we plotted $\varphi^{1/\beta}$ as a function of T with assumed exponents β for two data sets; one obtained for the "cubic" Fe^{3+} and the other for the Fe^{3+}-V_O center. A near straight line was obtained for $\beta = 0.33$ rather than 0.32 or 0.34. Extrapolating this line back to $\varphi^{1/\beta} = 0$ yields the transition temperature $T_a = 105.50 \pm 0.04$. With this we obtain $t = T/T_a$ and $\epsilon = 1-t$ to an accuracy of 4×10^{-4}. From Fig. 1 one sees that down to $t = 0.94$ the straight line is well followed, but afterwards it deviates. To estimate the range of ϵ in which (1) is valid a least-squares fit (LSF) computed for the first eleven points with the more precise data of the Fe^{3+}-V_O center gives $1/\beta = 3.004$, and for the first thirteen points (down to $t = 0.92$) $1/\beta = 2.991$. With a range in ϵ of 5×10^{-2} to 1.5×10^{-3} we estimate the absolute limit $\beta = 0.33 \pm 0.02$. The determined T_a is of value only for the particular sample. Due to differences in the oxidation and impurity content, deviations as large as 3°K have been observed for certain crystals.[7]

In $LaAlO_3$, determination of φ necessitates a rotation of the monodomain sample around the domain axis (inclined by ~60° to the magnetic field). It is far too imprecise for present needs, as is also a measurement of appropriate line splittings, due to large EPR line widths. These are narrowest for H parallel to the domain axes. There the lines are insensitive to φ but sensitive to D, the accurately determined crystalline field parameter,[8] which reflects the local distortion of the octahedra along the trigonal axis. D must be proportional to an expansion in even powers of φ because the trigonal $R\bar{3}c$ structure has a center of inversion. Thus for small φ, $D \propto \varphi^2$. Assuming the critical exponent of φ is close to $1/\beta = 3$ as in $SrTiO_3$, $D^{3/2}$ should be proportional to t. In Fig. 2 this is seen to hold for the first seven points down to $t = 0.94$ with $T_a = 797$°K for our sample. An LSF for these gave $1/\beta = 3.03$.

In our opinion it is remarkable how well the critical exponents β coincide in magnitude as well as in range of validity for $SrTiO_3$ and $LaAlO_3$, despite one crystal being tetragonal and the other trigonal. In addition, the transition temperature for $LaAlO_3$ is eight times higher than that of $SrTiO_3$. However it should be noted that in both cases the transformations occur from a cubic structure. In this context is it interesting that the value of β is, within the experimental error, the same as the one determined by Heller and Benedek[11] in ferromagnetic EuS although this is an order-disorder system. The high-temperature phase of EuS is also O_h^1 as in $LaAlO_3$ and

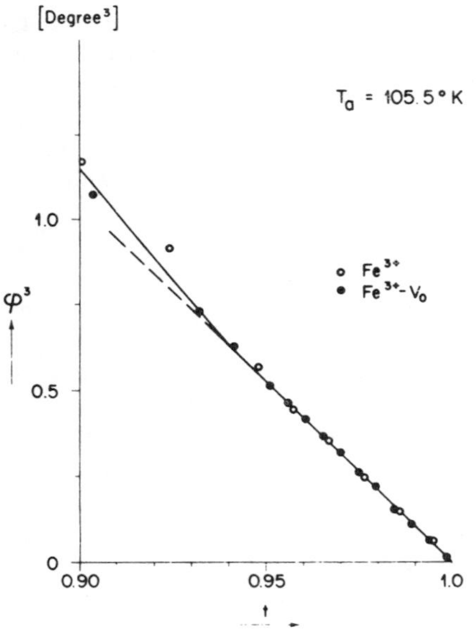

FIG. 1. The cube of the rotational parameter φ versus reduced temperature $t = T_a$ in $SrTiO_3$.

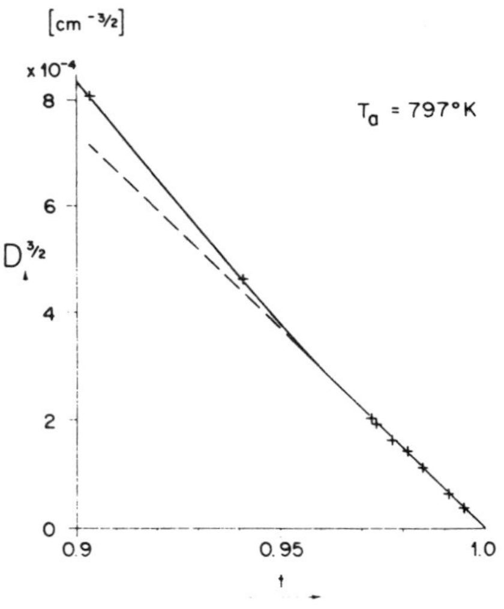

FIG. 2. EPR parameter $D^{3/2}$ versus reduced temperature t in $LaAlO_3$.

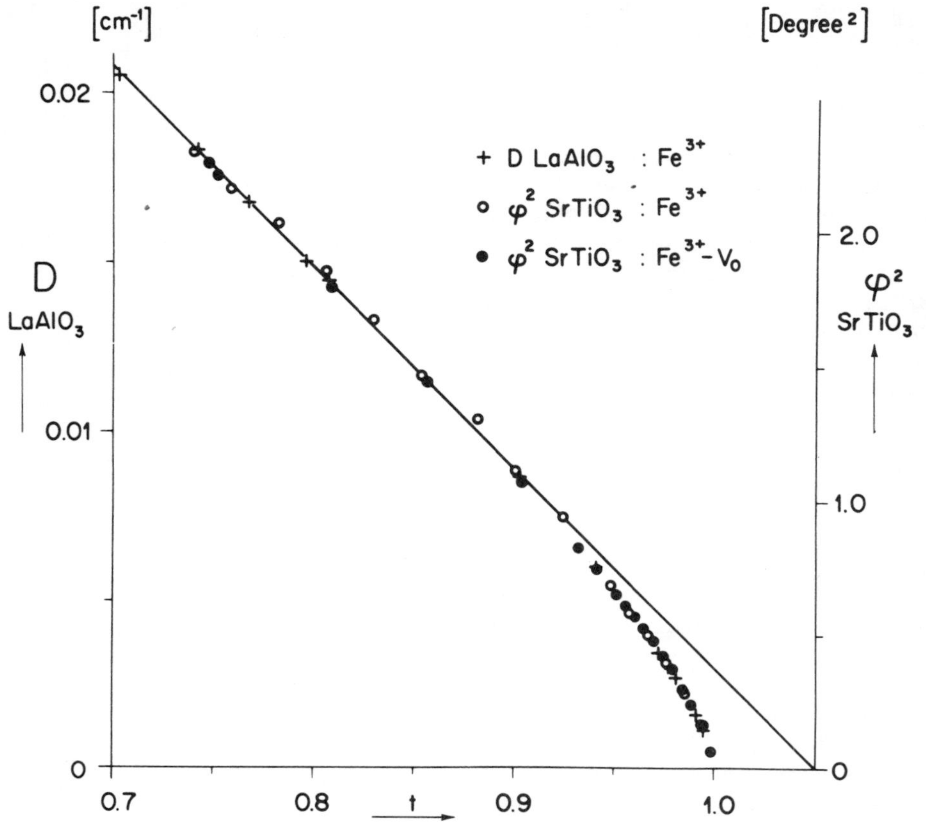

FIG. 3. φ^2 of $SrTiO_3$ and D of $LaAlO_3$ vs t between 0.7 and 1, showing the changeover from Landau to critical behavior.

$SrTiO_3$. The β's observed in this group appear to deviate slightly from the ones in noncubic $CrBr_3$ and isotypes as well as the gas-liquid systems[5]. It is common to all static scaling theories that the critical exponents of two physical variables remain undertermined.[1-4] Apart from β obtained here, a measurement of specific heat possibly yielding α or α' could be of value.

To obtain the limit beyond which the behavior becomes classical, φ^2 of $SrTiO_3$ and D of $LaAlO_3$ were plotted as a function of t. If the systems behaved Landau-like, a straight line should result. From Fig. 3 it is seen that this is approximately the case for $0.7 \leq t < 0.9$ in both systems. Furthermore φ^2 ($SrTiO_3$) and D ($LaAlO_3$) are proportional to each other to an appreciable extent, also outside the critical region, thus their respective rotational parameters are, too. From this a number of conclusions can be drawn.

In scaling, the critical region extends equally above and below T_a, i.e., $t_\pm = 1 \pm \epsilon_c$. With $\epsilon_c \simeq 0.1$ from Fig. 3 it becomes clear why the Landau analysis of Slonczewski and Thomas[12] on several $SrTiO_3$ experiments gave good results except near T_a. Using their calculated specific-heat jump $\Delta c_p = 8.1 \times 10^{-4}$ cal/deg g at T_a and $\rho = 5.13$ g cm^{-3} in the expression of Ginzburg,[13]

$$l = (k/\rho \Delta c_p)^{1/3} \epsilon_c^{-1/6}, \tag{2}$$

which defines a zero-temperature coherence length l, we obtain $l \approx 13$ Å, which is of the order of the distance between equivalent octahedral units (8 Å) as expected. Thus for the structural transitions l is quite short range as evidenced from the large ϵ_c. Further (2) is more quantitative than anticipated and agrees with findings for l in the ordering $CdIn_2S_4$ spinels.[14]

In the megahertz region a strong ultrasound absorption has been observed especially in near-monodomain $SrTiO_3$ peaking at T_a.[15] The absorption clearly occurs in the critical region between t_+ and t_- and thus is due to critical absorption and cannot be explained by classical theories.

The soft optic mode ω_0 associated with these

structural phase transitions is underdamped in $SrTiO_3$ [16] and overdamped in $LaAlO_3$.[17] From our data it follows that static scaling is independent of mode damping. Using the fluctuation-dissipation theorem, one derives that the intensity I' of the diffraction peaks, which are observed below T_a, is proportional to $T\omega_0^2 \propto T\varphi^2$. Thus I'/T should behave like[17] φ^2 or the EPR parameter D. This is observed for the (331) Bragg reflection[18] in $KMnF_3$ undergoing the same transition as $SrTiO_3$ at 184°K, and in $LaAlO_3$ for the (711) reflection as measured by Plakhty and Cochran.[19] In both cases a straight line is followed by a rounding down near T_a. With the more precise data in $LaAlO_3$, an LSF yielded $1/\beta = 2.6$ and 3.5 for the first three and four points, respectively, bracketing our value of 3.0. The classical straight line overshooting T_a as in Fig. 3 was also obtained from Raman[20] and neutron data[17] in $LaAlO_3$ and has been discussed in terms of a possible first-order transition. From the present work it is now assigned in a quantitative way to nonclassical behavior.

It also becomes clear why deviations from Landau behavior are more difficult to discern in displacive ferroelectrics: For perovskite materials the ferroelectric transitions are all first order with $\epsilon_T = (T_0 - T_c)/T_0$ typically of several percent. We can assume that the dipolar forces are rather longer in range than those leading to purely structural transitions, thus we can set $\epsilon_c(\text{ferro}) \lesssim \epsilon_c(LaAlO_3)$ and obtain $\epsilon_c(\text{ferro}) \lesssim \epsilon_T$. This agrees with the results of Vaks,[21] who estimates correlations ξ of the order of unity at T_0. However due to strong soft Cochran-mode-acoustic-mode coupling he concludes that ξ is sufficiently large to induce the first-order ferroelectric transitions. The small rotational-mode-strain coupling in $SrTiO_3$ and $LaAlO_3$ [13,22] allows a near-second-order transformation to take place to such an extent that most of the critical region associated with large fluctuations is observable.

Our work suggests a large variety of high-precision experiments to be undertaken to test the static and dynamic scaling laws near the order-order phase transitions of these compounds as well as in crystals undergoing similar transitions as $KMnF_3$ [18] or $NdAlO_3$ [20] and possibly then allowing systematics in their critical exponents.

We are indebted to Dr. T. Schneider for bringing Vaks's paper to our attention, to Professor M. E. Fisher and Dr. J. F. Scott for a critical reading of and suggestions on the manuscript, and to Dr. P. Wolf for his LSF-APL program.

[1] B. Widom, J. Chem. Phys. 43, 3898 (1965).
[2] L. P. Kadanoff, Physics 2, 263 (1966).
[3] R. B. Griffiths, Phys. Rev. 158, 176 (1967).
[4] M. E. Fischer, Rep. Progr. Phys. 30, 615 (1967).
[5] M. J. Cooper, M. Vicentini-Missoni, and R. I. Joseph, Phys. Rev. Lett. 23, 70 (1969), and references quoted therein.
[6] L. P. Kadanoff, in Critical Phenomena, Proceedings of the International School of Physics "Enrico Fermi," Varenna, Italy, 1970 (to be published).
[7] K. A. Müller, W. Berlinger, and J. C. Slonczewski, Phys. Rev. Lett. 25, 734 (1970).
[8] K. A. Müller, W. Berlinger, and F. Waldner, Phys. Rev. Lett. 21, 814 (1968).
[9] L. Landau and E. Lifshitz, *Statistical Physics* (Addison-Wesley, Reading, Mass., 1969), Chap. 9.
[10] K. A. Müller, W. Berlinger, M. Capizzi, and H. Gränicher, Solid State Commun. 8, 549 (1970).
[11] P. Heller and G. Benedek, Phys. Rev. Lett. 14, 71 (1965).
[12] J. C. Slonczewski and H. Thomas, Phys. Rev. B 1, 3599 (1970).
[13] V. L. Ginzburg, Fiz. Tverd. Tela 2, 2031 (1960) [Sov. Phys. Solid State 2, 1824 (1961)].
[14] W. Czaja, Phys. Kondens. Mater. 10, 299 (1970).
[15] B. Berre, K. Fossheim, and K. A. Müller, Phys. Rev. Lett. 23, 583 (1969); W. Rehwald, Solid State Commun. 8, 607 (1970).
[16] G. Shirane and Y. Yamada, Phys. Rev. 177, 858 (1969).
[17] J. D. Axe, G. Shirane, and K.A. Müller, Phys. Rev. 183, 820 (1969).
[18] V. J. Minkiewicz, Y. Fujii, and Y. Yamada, J. Phys. Soc. Jap. 28, 443 (1970).
[19] V. Plakhty and W. Cochran, Phys. Status Solidi 29, 81 (1968).
[20] J. F. Scott, Phys. Rev. 183, 823 (1969).
[21] V. G. Vaks, Zh. Eksp. Teor. Fiz. 58, 296 (1970) [Sov. Phys. JETP 31, 161 (1970)].
[22] J. Feder and E. Pytte, Phys. Rev. B 1, 4803 (1970).

INFLUENCE OF LATTICE CONTRACTION ON LONG-RANGE ORDER IN CoO NEAR T_N

M. D. Rechtin, S. C. Moss, and B. L. Averbach
Department of Metallurgy and Materials Science, Center for Materials Science and Engineering, Massachusetts Institute of Technology, Cambridge, Massachusetts 02139
(Received 16 April 1970)

Just below the Néel temperature of CoO, the intensity of neutrons scattered by the antiferromagnetic lattice of Co^{2+} ionic spins should vary as $I \propto (T_N-T)^{2\beta}$. We measure $\beta = 0.244 \pm 0.015$, or substantially lower than the Ising prediction of 0.3125 which CoO should assume. By applying a phenomenological correction for the measured tetragonal lattice contraction below T_N, similar in form to Heller's correction for the contraction in MnF_2, we make $T_N=T_N(T)$ which raises β to a value of 0.290 ± 0.025. By thus adjusting for nonrigidity of the lattice we bring β into a range which includes the rigid-lattice Ising prediction.

Below its Néel temperature ($T_N \cong 289°K$) CoO undergoes a phase transition from a paramagnetic to antiferromagnetic state. The change is accompanied by a tetragonal contraction with $c/a = 0.988$ at 93°K [1,2] and a small trigonal elongation of $e_{yz} = e_{zx} = e_{xy} = 5\times 10^{-4}$.[3] Precise dilatometry measurements of the behavior of the tetragonal contraction in our single crystal indicate a continuous change of lattice dimensions on passing through the critical temperature. Neutron diffraction measurements of magnetic peak positions show this same characteristic as do the measurements of sublattice magnetization reported here.

Consequently, the phase transition appears to be of order greater than unity. The Co^{2+} ionic spins order ferromagnetically within any single $\{111\}$ plane while each $\{111\}$ plane is oriented antiferromagnetically with respect to the two adjacent planes. This configuration signifies a predominantly next-nearest-neighbor (nnn) exchange interaction. The 6 nnn ionic moments are oriented antiparallel while of the 12 nearest neighbors (nn), half are parallel and half antiparallel. In a study to be reported elsewhere our measurements of the critical scattering above T_N suggest, via an analysis similar to the Clapp-Moss treatment,[4] that the nn exchange interaction is about one-third to one-half the value of the nnn energy. Although a number of noncollinear spin-axis structures are possible in CoO,[5,6] present experimental evidence points to a collinear structure[7] with the moments tipped out of the {111} planes 7°50′ toward the c axis.[6] This spin-axis orientation was verified by our single-crystal measurements. We also noted that the spin axis does not change orientation with temperature from measurements at 4.2 and 272°K.

Here we shall describe for CoO the decay of antiferromagnetic long-range order in the critical region in terms of the critical exponent, β. The spontaneous sublattice magnetization in the critical region is given by $M(T)/M(0)=D(1-T/T_N)^\beta$. The neutron scattering cross section (or scattered intensity, I) is proportional to the square of the magnetization and thus we may determine 2β from $I=C(T_N-T)^{2\beta}$ much as has been done in other studies of Ising and Heisenberg systems.[8] In order to compare any experimental β value with a Heisenberg or Ising model we must, however, have the system referred to its rigid-lattice state. Essam and Baker[9] have recently developed the effects of lattice compressibility on critical behavior, and current work of Norvell and Als-Nielsen[10] on long-range order in β'-CuZn showed agreement with the compressible bcc lattice calculations, although this result has recently been questioned by Chipman and Walker.[11] We have here measured the decay of spontaneous sublattice magnetization in the critical region for CoO by neutron diffraction and find a β of about $\frac{1}{4}$. By correcting for the lattice compressibility, however, an Ising-like β is obtained and this agrees with the treatment of Salamon[12] who shows that for CoO the behavior should become Ising-like for $(1-T/T_N)\lesssim 0.020$. His specific-heat measurements support this conclusion by yielding nearly Ising values for the C_v indices α and α' of 0.12 and 0.05, respectively.

In our experiment neutrons of $\lambda=1.05$ Å were obtained by transmission from the (111) planes of a mosaic lead single crystal. After monochromatization the beam was collimated by a cadmium slit system with horizontal divergence of 3.2′. The $\frac{1}{2}\lambda$ neutrons were then removed by a Pu^{239} filter, and the neutron beam was diffracted from the CoO single crystal into a BF_3 counter with open collimation. The crystal itself was nearly spherical in shape with a mean diameter of 0.68 cm. The mosaic width of the crystal was 7.1′. Chemical analysis of the specimen showed the Co^{2+} to O^{2-} ratio was 1.001 to 1.000 with a 0.15% Ni^{2+} impurity. The specimen was held in an aluminum clamp attached to a two-circle goniometer, and this assembly was mounted within a cryostat. Temperature was controlled to ±0.05°K with an ice water bath in the liquid nitrogen reservoir and a 5-W heater in the sample chamber. The (111) magnetic peak intensity was measured by a rotating crystal (ω) scan with the peak height being used to characterize the decay of magnetization. Since only the long-range order is varying in the system, the peak breadth remains constant and $M(T)/M(0)$ is still proportional to $I^{1/2}$. However we cannot compare the peak heights of several reflections since instrumental factors invalidate peak height as an absolute measurement.

Nuclear and magnetic multiple-scattering contributions were avoided by first calculating the positions in azimuthal angular setting of the crystal that are free from them. The temperature diffuse scattering (TDS) correction was evaluated on the basis of the Debye-Waller factor for CoO. From measurements of the mean-square vibrational amplitude of the Co^{2+} ion in CoO at room temperature,[13] the Debye-Waller factor e^{-2w} is about 0.991 for the (111) reflection near room temperature. This indicates that much less than 1% of the measured radiation is thrown back into the (111) reflection in the TDS process. We therefore assume a negligible correction for TDS.

Secondary extinction can be an important correction in a highly mosaic single crystal for high-intensity Bragg reflections. In weakly scattering processes, such as critical scattering, there is no extinction. To be free of secondary extinction the measured intensity must be given by the ideally mosaic crystal formula, namely I is proportional to the square of the structure factor. For the given experimental conditions we plotted the integrated intensities for a number of nuclear reflections against the ideal mosaic prediction. From this it was determined that there would be negligible extinction for all but the lowest three temperatures in our long-range-order study. In fact, however, the measured integrated magnetic intensities all fell on the extrapolation of the extinction-free line, indicating the absence of both primary and secondary extinction. The reason for this lies in the presence of the so-called "S" or spin domains in CoO. Upon ordering below T_N, spontaneous magnetization can begin on any one of the four {111} planes with equal probability because the tetragonal contrac-

Table I. Summary of data and corrections on long-range–order measurements in CoO.

T (°K)	⟨I⟩	n	σ(I)	I_{BG}	I_{CS}	I_{BRAGG}	w
278.2	21503	1	146	134	62	21307	3.56
284.4	14623	1	121	134	308	14181	2.73
285.0	13336	6	47	134	328	12874	3.02
285.1	13467	1	116	134	333	13000	2.83
285.8	12149	3	64	134	357	11658	1.75
286.0	11821	11	33	134	367	11320	6.41
287.3	9023	4	47	134	422	8467	2.91
287.5	8617	2	66	134	432	8051	2.43
288.7	4352	19	15	134	550	3715	1.00
288.8	3363	5	25	134	623	2606	0.65
288.9	2940	5	31	134	825	1981	0.38

tion axis is symmetric with respect to these four {111} planes. Thus, there are four distinct domains below T_N with their overall volumes equal.[14] We have also verified this to be so by measurements on the four {333} planes at approximately 100°K. This domain formation results in a distribution of domains much smaller in volume than the perfect mosaic regions and renders all magnetic reflections extinction free.

Besides the "S" domains we can have "T" domains, that is, regions of the crystal with different contraction c axes. Uchida et al.[15] have shown that application of a temperature gradient along one of the ⟨001⟩ directions upon passing through T_N causes most contractions to occur along one axis. For our set of measurements we obtained 96% of one "T" domain. For any case, in the temperature region of our measurements for determining β, the (111) peak splitting due to the three domains was very small so that no correction of intensity was needed and our use of peak height was not invalidated.

To accomplish the removal of the critical scattering contribution below T_N, the (111) position was scanned at a number of temperatures above T_N. The wings of the peak profiles below T_N, free from Bragg intensity, were matched with the diffuse peak from above T_N which had the best fit of intensity and curvature. This diffuse peak was then used as the critical scattering correction. The temperature dependence of the (111) peak intensity below T_N is summarized in Table I. This table gives the temperature of the measurement, T(°K); average intensity for n repetitions, $\langle I\rangle$; the expected Poisson standard deviation, $\sigma(I)$; background intensity, I_{BG}; critical scattering correction, I_{CS}; corrected long-range order intensity, I_{BRAGG}; and the statistical weight w of the observation relative to the measurement at $T = 288.7$°K with unit weight. A weighted least-squares fit (LSF) was carried out to find the parameter β for various values of T_N. The best fit was determined by finding the minimum of the sum of the squares of the standard deviations, $S^2(T_N) = \sum_T w(T)[I_{BRAGG}(T) - C(T_N - T)^{2\beta}]^2$. A typical double-log plot of the Bragg intensity versus $T_N - T$ is shown in the upper line in Fig. 1 for the best T_N. There appears to be no systematic deviation from a straight line throughout the entire temperature range, indicating again that no secondary extinction is present. The lower line in Fig. 1 is the result of correcting for lattice contraction and is discussed below.

This value for β (~0.25) is surprisingly small. On the basis of Salamon's specific-heat measurements we would expect an Ising value near $\frac{5}{16}$. A major correction to our measured β was therefore suspected to be the lattice compressibility, since the tetragonal distortion is extremely large. The effect of pressure on the Néel temperature has been determined by Bloch, Chaisse, and Pauthenet.[16] They found, $dT_N/dP = 0.60$°K kbar^{-1} and for $T_N = 288.9825$°K, $(1/T_N)(dT_N/dP) = (d\ln T_N)/dP = 2.076 \times 10^{-3}$ kbar^{-1}. The depen-

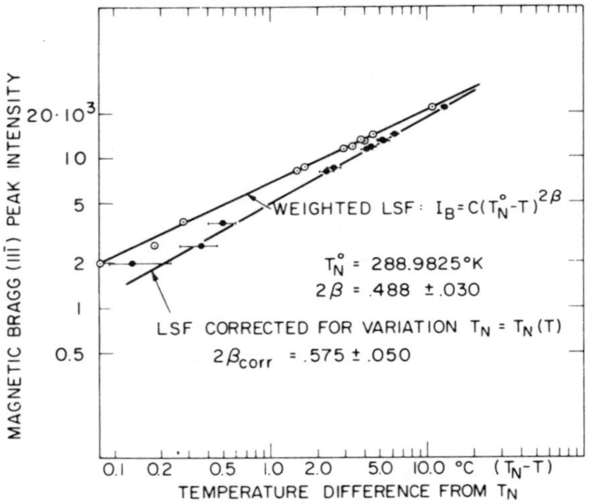

FIG. 1. Double-log plot of magnetic Bragg peak intensity from the (111) planes in CoO versus the temperature difference $(T_N - T)$. The upper limit is a least-squares fit (LSF) to a power law assuming a constant $T_N{}^0$. The fit was made without the first two points near $T_N{}^0$ and the resulting line still includes them indicating that the large critical scattering correction near $T_N{}^0$ was appropriate. The lower line is a similar plot in which the simple variation of $T_N(T)$ was applied and where the error bars are all due to the uncertainty in $T_N(T)$.

dence of the c-axis length on pressure was found from the compressibility, $[1/c(T_N)]dc/dP = d\ln c/dP = -0.605\times 10^{-3}$ kbar^{-1}, thus $(1/T_N)(dT_N/dP)/[1/c(T_N)]dc/dP = d\ln T_N/d\ln c = -3.43$. Expressing the variation of T_N with the c axis, $T_N(T) = T_N{}^0\{1-3.43[c(T)-c(T_N)]/c(T_N)\}$. Knowing $\Delta c/c$ from strain measurements on our crystal (Fig. 2), we may determine T_N as a function of temperature. If we now account for the increase of T_N with decreasing temperature in Fig. 1, we must shift the coordinate $(T_N - T)$ of each point. Upon doing this we note a change of β to 0.290 $\pm$0.025. The shift is made with respect to $T_N{}^0$, the temperature at which the long-range order goes to zero. The value of 288.9825°K obtained in the LSF is within ±0.1°K of the absolute T_N, and this error is included in the error brackets of the corrected data points.

Another approach to correct for the lattice compressibility is taken from Heller[17] in his NMR study of MnF_2 in which we express the intensity in terms of the reduced temperature and $T_N(T)$, $I^{1/(2\beta)}/T_N(T) = A(T)^{1/(2\beta)} - A(T)^{1/(2\beta)}T/T_N(T)$. The β value which linearizes this equation gives the best fit. A plot of the left side of the equation is made against $T/T_N(T)$. Then a

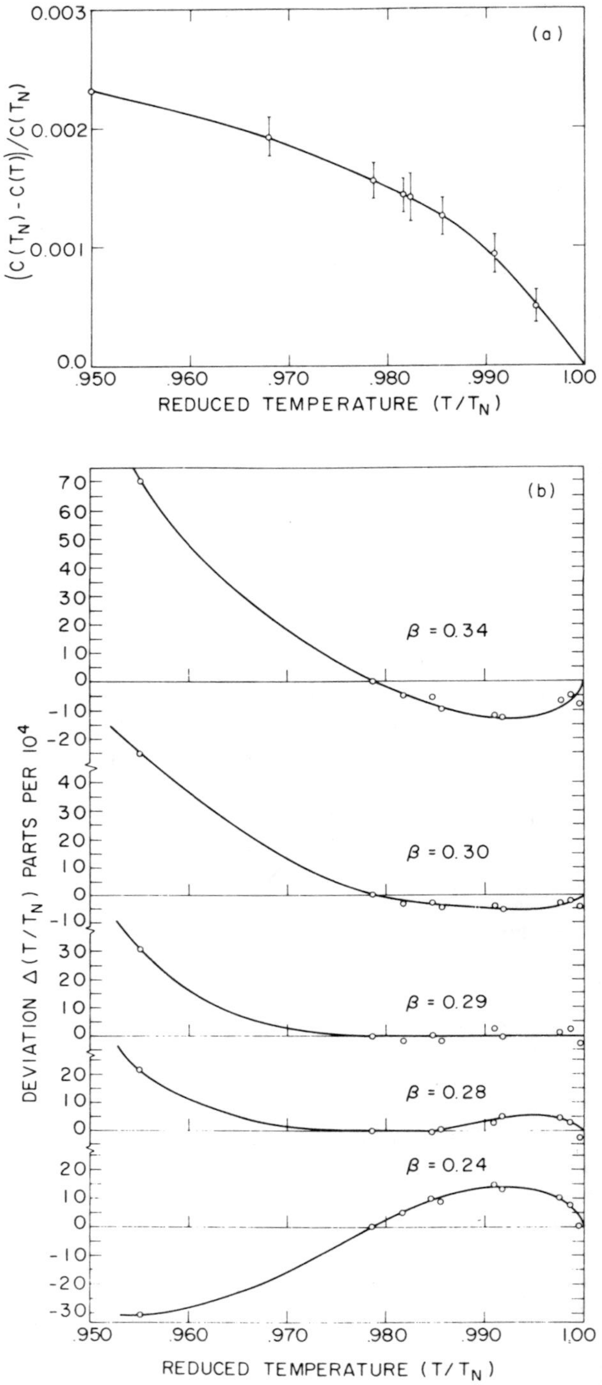

FIG. 2. (a) The variation of the tetragonal c parameter $\Delta c/c$ as a function of reduced temperature T/T_N. (b) The deviation from linearity of the plot of $I^{1/(2\beta)}/T_N(T)$ vs $T/T_N(T)$ for five different β values. This deviation is expressed in terms of separation along the $T/T_N(T)$ coordinate. The appropriate β value leads to no departure.

straight line is plotted through $T/T_N=1.000$ and (somewhat arbitrarily, although it makes little difference) the data point at $T=284.4°K$ for a number of β values. The deviation of the data points from this straight line is given in the lower part of Fig. 2 for five different β values. The best fit is for $\beta \cong 0.29$ which is the same conclusion found by correcting Fig. 1. Within experimental accuracy, set mainly by the tetragonality data, the β thus appears to assume an Ising value in agreement with the experimental specific-heat measurements in the critical region.[12]

We would especially like to thank Professor H. E. Stanley for many helpful discussions and for showing that the simple shift in $T_N(T)$ included in Fig. 1 gives the same result as Fig. 2. We are also indebted to M. B. Salamon for discussions of his work and ours and for his preprint. We would in addition like to thank J. Als-Nielsen, R. Harrison, L. P. Kadanoff, and C. B. Walker for stimulating discussions. This work was supported by the National Science Foundation to whom it is a pleasure to express our appreciation.

---

[1]H. P. Rooksby, Acta Cryst. 1, 226 (1948).

[2]R. C. Lambs and H. P. Rooksby, Nature 165, 442 (1950).

[3]J. Nagamiya, S. Saito, Y. Shimomura, and E. Uchida, J. Phys. Soc. Japan 20, 1285 (1965).

[4]P. C. Clapp and S. C. Moss, Phys. Rev. 142, 418 (1966); 171, 754 (1968).

[5]W. L. Roth, Phys. Rev. 111, 772 (1958).

[6]B. van Laar, Phys. Rev. 138, A584 (1965).

[7]S. Saito, K. Nakahigashi, and Y. Shimomura, J. Phys. Soc. Japan 21, 850 (1966).

[8]J. Als-Nielsen and O. W. Dietrich, Phys. Rev. 153, 717 (1967), on β'-CuZn; L. M. Corliss, A. Delapalme, J. M. Hastings, H. Y. Lau, and R. Nathans, J. Appl. Phys. 40, 1278 (1969), on $RbMnF_3$; J. C. Norvell, W. P. Wolf, L. M. Corlis, J. M. Hastings, and R. Nathans, Phys. Rev. 186, 557 (1969), on DyAlG where $\beta \cong \frac{1}{4}$ due to long-range dipolar forces. This is discussed at greater length by N. E. Frankel and A. E. Spargo, Phys. Letters 31A, 442 (1970).

[9]G. A. Baker, Jr., and J. W. Essam, Phys. Rev. Letters 24, 447 (1970).

[10]J. C. Norvell and J. Als-Nielsen, to be published.

[11]D. R. Chipman and C. B. Walker, Bull. Am. Phys. Soc. 15, 363 (1970).

[12]M. B. Salamon, to be published.

[13]R. Uno, J. Phys. Soc. Japan 18, 1686 (1963).

[14]B. van Laar, Phys. Rev. 141, 538 (1966).

[15]E. Uchida, R. Fukuoka, H. Kondoh, J. Takeda, Y. Nakazumi, and J. Nagamiya, J. Phys. Soc. Japan 19, 2088 (1964).

[16]D. Bloch, F. Chaisse, and R. Pauthenet, J. Appl. Phys. 37, 1401 (1966).

[17]P. Heller, Phys. Rev. 146, 403 (1966).

INELASTIC NEUTRON SCATTERING FROM MnF_2 IN THE CRITICAL REGION

M. P. Schulhof*† and P. Heller*†
Brandeis University, Waltham, Massachusetts 02154

and

R. Nathans
State University of New York at Stony Brook, New York 11790, and
Brookhaven National Laboratory,* Upton, New York 11973

and

A. Linz
Center for Materials Science and Engineering, Massachusetts Institute of Technology,
Cambridge, Massachusetts 02139
(Received 10 April 1970)

Detailed neutron-scattering measurements have yielded the behavior of the scattering function $S(q,\omega)$ for both the transverse and longitudinal fluctuations in the critical region of the uniaxial antiferromagnet MnF_2. The static wavelength-dependent susceptibilities are measured both above and below T_N. The relaxation rates for the longitudinal fluctuations are found to be correctly described by dynamical scaling both above and below T_N. The scaling functions $\Omega_\pm(q/\kappa)$ applying respectively above and below T_N are determined explicitly.

We have made a detailed neutron-scattering study of the static and time-dependent correlation functions in the critical region of the uniaxial antiferromagnet MnF_2. The static susceptibilities, both longitudinal and transverse, and the longitudinal and transverse correlation lengths have been determined for the first time in a magnetic system both above and below the critical temperature $T_N = 67.46°K$. The relaxation rates for the longitudinal and transverse fluctuations have been measured throughout the critical region. It is shown for the first time that the theory of dynamical scaling[1] provides a very good description of the longitudinal relaxation rates both above and below T_N. Our work follows the initial verification of dynamical scaling by Lau et al.[2] for the paramagnetic region of the isotropic antiferromagnet $RbMnF_3$. In the present work on MnF_2, which constitutes the most complete investigation of correlations near a second-order magnetic phase transition yet undertaken, the scaling functions $\Omega_\pm(q/\kappa)$ applying respectively above and below T_N are determined explicitly.

An essential part of the present work is the separation of the transverse and longitudinal fluctuations. To see how this is done, note that the cross section for the magnetic scattering of unpolarized neutrons is given by[3]

$$\frac{d^2\sigma}{d\Omega d\omega}=N\left(\frac{\gamma e^2}{mc^2}\right)\frac{K_f}{K_i}\,|F(K)|^2\Big\{\sum_{\alpha\beta}(\delta_{\alpha\beta}-\hat{K}_\alpha\hat{K}_\beta)S^{\alpha\beta}(q,\omega)\Big\}. \tag{1}$$

Here ω and $\vec{K}=\vec{K}_i-\vec{K}_f$ correspond respectively to the neutron energy and momentum loss, and $\vec{q}=\vec{K}-2\pi\vec{\tau}$, with $2\pi\vec{\tau}$ a magnetic reciprocal lattice vector. With $\vec{\tau}$ along [001], the curly bracketed term in (1) is such that fluctuations in each of the two transverse directions contribute to the scattering, with no contribution from the longitudinal fluctuation. With $\vec{\tau}$ along [100], however, fluctuations in the longitudinal direction and in one of the transverse directions contribute. Thus, by investigating both the [100] and [001] reflections, the transverse and longitudinal scattering functions $S_\perp(q,\omega)$ and $S_\parallel(q,\omega)$, respectively, may be separated.

We take the static cross section to be of the Ornstein-Zernike form,[3] for both the longitudinal and transverse contributions. We assume the frequency dependence of $S_\parallel(q,\omega)$ to be described by a single Lorentzian curve centered at $\omega=0$, while the frequency dependence of $S_\perp(q,\omega)$ is taken to consist of the sum of two Lorentzians displaced symmetrically about $\omega=0$ by an amount $\omega_0(q,T)$ (which may vanish). The cross sections for each of the two reflections studied then take the form

$$\left(\frac{d^2\sigma}{d\Omega d\omega}\right)_{[001]}=B(\omega,T)\frac{K_f}{K_i}\Big\{\frac{A_{[001]}}{(\kappa_\perp^2+q^2)}\frac{1}{2}\left[\frac{\Gamma_\perp}{\Gamma_\perp^2+(\omega-\omega_0)^2}+\frac{\Gamma_\perp}{\Gamma_\perp^2+(\omega+\omega_0)^2}\right]\Big\}, \tag{2}$$

and

$$\left(\frac{d^2\sigma}{d\Omega d\omega}\right)_{[100]}=B(\omega,T)\frac{K_f}{K_i}\Big\{\frac{A_{[100]}}{(\kappa_\parallel^2+q^2)}\left[\frac{\Gamma_\parallel}{\Gamma_\parallel^2+\omega^2}\right]+\frac{\lambda A_{[100]}}{(\kappa_\perp^2+q^2)}\frac{1}{2}\left[\frac{\Gamma_\perp}{\Gamma_\perp^2+(\omega-\omega_0)^2}+\frac{\Gamma_\perp}{\Gamma_\perp^2+(\omega+\omega_0)^2}\right]\Big\}, \tag{3}$$

where the factor $B(\omega,T)=\hbar\omega[1-\exp(-\hbar\omega/k_BT)]^{-1}$ expresses the requirement that the cross section for a neutron energy loss must exceed that for an equal gain by the Boltzman factor. With the exception of λ, which we assume[4] to be unity, the parameters appearing in Eqs. (2) and (3) are determined by means of a least-squares fitting procedure which folds the experimentally determined instrumental resolution function[5] with one of the cross-section expressions written above, and compares the result with the experimental data. Note that parameters $A_{[001]}$, $A_{[100]}$, $\kappa_\perp$, and $\kappa_\parallel$ depend only on the temperature T, while $\Gamma_\perp$, $\Gamma_\parallel$, and ω_0 are to depend on both q and T.

Data were collected for incident neutron energies of 6.6 and 13 meV, at wave vectors from $q=0$ to $q=0.258$ Å$^{-1}\simeq(0.25)2\pi\tau_{[100]}$. This was done at eleven different temperatures in a 16°K region centered on the critical temperature. The cryostat and temperature control techniques, providing 1-mdeg stability at the sample, are described elsewhere.[6] The parameters $A_{[001]}(T)$, $\kappa_\perp(q,T)$, $\Gamma_\perp(q,T)$, and $\omega_0(q,T)$ were first found from a fit to the pure transverse [001] data. Then, a fit to the [100] data permitted the determination of $A_{[100]}(T)$, $\kappa_\parallel(T)$, and $\Gamma_\parallel(q,T)$. All error limits quoted below for these quantities, or for quantities derived from them, correspond to standard deviations. Excellent fits were obtained for all of the observed data.

Using the fluctuation-dissipation theorem,[3] relative values of the static wavelength-dependent susceptibilities may now be obtained from

$$\chi_\parallel(q,T)\propto\frac{(k_BT)^{-1}A_{[100]}}{\kappa_\parallel^2+q^2} \tag{4a}$$

and

$$\chi_\perp(q,T)\propto\frac{(k_BT)^{-1}A_{[001]}}{\kappa_\perp^2+q^2}. \tag{4b}$$

In particular, relative values of the "staggered" susceptibilities are calculated from (4) by setting $q=0$. In Fig. 1 we plot the results for the longitudinal staggered susceptibility $\chi_\parallel(0,T)$ both above and below T_N. This figure includes data obtained in our earlier quasielastic experiment,[6] the constant of proportionality being adjusted for agreement with the present data at the highest temperature. On fitting the combined data of Fig. 1 to $\chi_\parallel(T)=b_+(T-T_N)^{-\gamma}$ for $T>T_N$ and $\chi_\parallel(T)=b_-(T_N-T)^{-\gamma'}$ for $T<T_N$, we obtain $\gamma=1.27\pm0.02$, $\gamma'=1.32\pm0.06$, and $b_+/b_-=4.8\pm0.5$.

The behavior of the longitudinal wavelength-dependent susceptibility for $q\neq0$ may be described

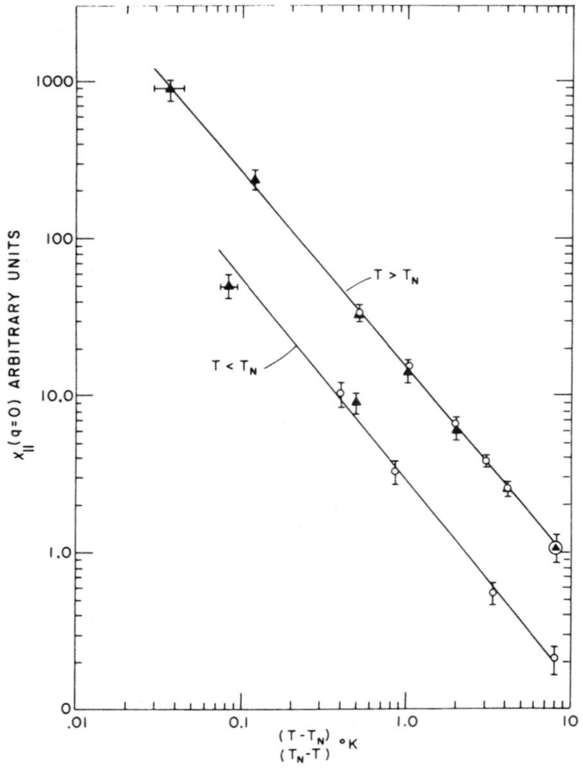

FIG. 1. Temperature dependence of the longitudinal staggered susceptibility above and below T_N. The open circles are obtained from the present inelastic-scattering data, the closed triangles from an earlier energy-unresolved scattering experiment (Ref. 6), normalized to the present data at 76°K.

by Eq. (4a) by specifying $\kappa_{\parallel}(T)$. On fitting our data by $\kappa_{\parallel}(T)=\kappa_{+}(T-T_N)^{\nu}$ for $T>T_N$ and $\kappa_{\parallel}(T)=\kappa_{-}(T_N-T)^{\nu'}$ for $T<T_N$ we find

$$\nu=0.634\pm0.02, \quad \nu'=0.56\pm0.05,$$
$$\kappa_{+}=(0.032\pm0.004)\ \text{Å}^{-1}\ (°\text{K})^{-\nu},$$
$$\kappa_{-}=(0.055\pm0.006)\ \text{Å}^{-1}\ (°\text{K})^{-\nu'}. \tag{5}$$

Our data for the transverse susceptibilities, the transverse relaxation rates, and the spin-wave energy gap, will be presented in a more detailed communication.

Of paramount interest here is the behavior of the longitudinal relaxation rates $\Gamma_{\parallel}(q,t)$. Our results are shown graphically in Fig. 2. Above T_N, the decay rates vary rather slowly with temperature, except for $q=0$ where we find an essentially linear temperature dependence, $T>T_N$:

$$\Gamma_{\parallel}(0,T)=(2.1\pm0.1\ \text{meV})\times[(T-T_N)/T_N]^{0.95\pm0.05}. \tag{6}$$

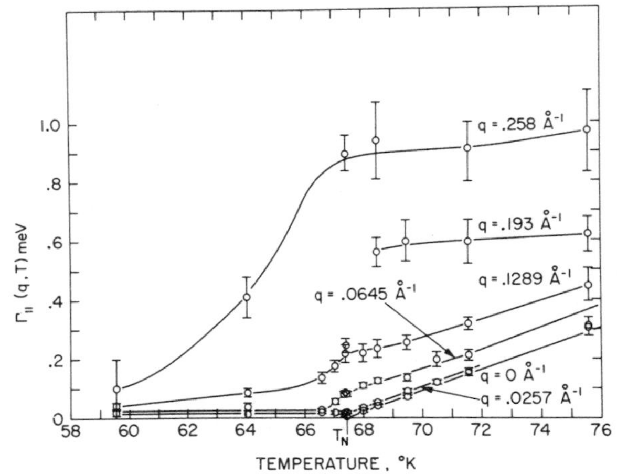

FIG. 2. Relaxation rate of the longitudinal spin fluctuations as a function of temperature and wave vector in the critical region.

The situation is completely different below T_N, where $\Gamma_{\parallel}$ apparently vanishes as $q\rightarrow0$ at all temperatures. Our data are consistent with a diffusive behavior $\Gamma_{\parallel}(q,T)\cong D(T)q^2$. Actually, finite instrumental resolution prevented us from measuring $\Gamma_{\parallel}$ exactly at $q=0$, since a magnetic Bragg peak dominates the scattering there.

For $T=T_N$, our data are well fitted by the expression

$$\Gamma_{\parallel}(q,T_N)=(7.0\pm0.9\ \text{meV})q^{1.6\pm0.2}, \tag{7}$$

where q is expressed in Å^{-1}. Equation (7) correctly describes our data for $0.026\ \text{Å}^{-1}<q<0.2\ \text{Å}^{-1}$. The work of Riedel and Wegner[7] predicts a change in the power-law behavior close to the critical point due to the effects of anisotropy. On account of resolution limitations, it is simply not possible for us to say whether a different power law might apply for smaller q values. Similarly, although Eq. (6) correctly describes our data for $0.5<T-T_N<9°\text{K}$, we simply cannot say whether a different exponent might apply for smaller $T-T_N$ just above the critical point.

Dynamical scaling predicts[1] that the exponent expressing the dependence of $\Gamma_{\parallel}(0,T)$ on $\kappa_{\parallel}(T)$ above T_N should equal the exponent expressing the q dependence of $\Gamma_{\parallel}$ and T_N. Indeed, using (6) and (5) we find, for $T>T_N$,

$$\Gamma_{\parallel}(0,T)=(6.6\pm0.6\ \text{meV})[\kappa_{\parallel}(T)]^{1.49\pm0.07}, \tag{8}$$

where $\kappa_{\parallel}$ is expressed Å^{-1}. The exponents in (7) and (8) do in fact agree to within experimental error. More generally, dynamical scaling predicts that

$$\Gamma_{\parallel}(q,T)=[\kappa_{\parallel}(T)]^{z}\,\Omega_{\pm}[q/\kappa_{\parallel}(T)], \tag{9}$$

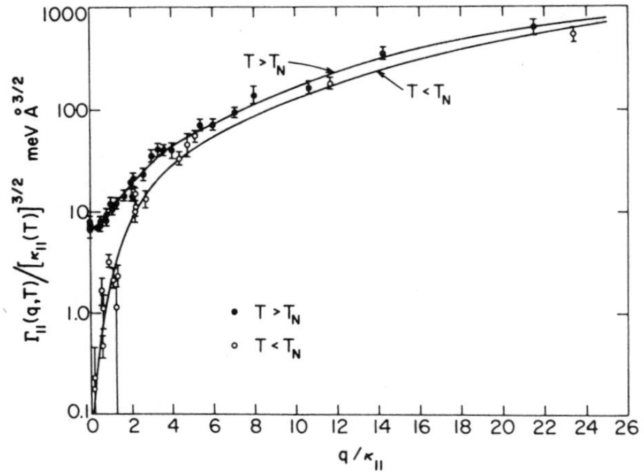

FIG. 3. A replot of the data of Fig. 2 in the form suggested by the theory of dynamical scaling. The ordinate is the scaled longitudinal relaxation rate, the abscissa the scaled wave vector. The two curves are the two branches of the longitudinal dynamical scaling function.

where the ± signs apply respectively above and below T_N. To test Eq. (9), we have replotted the data of Fig. 2 in Fig. 3. Here, the ordinate is $\Gamma_{\parallel}(q, T)/[\kappa_{\parallel}(T)]^z$, and the absicissa is $q/[\kappa_{\parallel}(T)]$. In making this plot, we have taken $z=\frac{3}{2}$, which is consistent with both (7) and (8). The data for $T>T_N$ and $T<T_N$ are represented respectively by the solid and open circles. Note that all of the data of Fig. 2 are present in the two curves of Fig. 3. The two curves represent the two branches of the scaling function; they merge for large q/κ, as must be the case, since this represents the behavior close to T_N. For each branch, the data points are intermixed over the curve: There is no systematic departure from it as a function of T or q. This kind of plot is reminiscent of the original representations of static scaling phenomena.[8]

The two branches of the scaling function have quite different behaviors for small values of q/κ. Thus, above T_N we find

$$\Omega_+(q/\kappa)=[6.9+2.6(q/\kappa)^2+\cdots]\text{ meV Å}^{3/2}, \quad (10a)$$

while below T_N we have

$$\Omega_-(q/\kappa)=[1.8(q/\kappa)^2+\cdots]\text{ meV Å}^{3/2}. \quad (10b)$$

The absence of the constant term in (10b) is a reflection of the diffusive behavior observed below T_N. A theory for this has been advanced by one of us[9] (P.H.) and, independently, by Halperin and Hohenberg.[10] According to this view, the diffusive central peak for the longitudinal fluctuation below T_N is a manifestation of thermal diffusion taking place within the spin system. The diffusion constant $D(T)$ should then equal the ratio of the spin system thermal conductivity to the magnetic specific-heat density, in analogy with the Landau-Placzek theory for fluids.

In conclusion, we have made a complete study of the static and dynamic correlations near the second-order phase transition in MnF_2. We find that dynamical scaling provides a remarkably successful description of our data. The scaling functions applying respectively above and below T_N are found to have quite different behaviors for small values of q/κ.

We gratefully acknowledge many fruitful discussions with Dr. P. Hohenberg, Dr. M. Blume, Professor R. Ferrell, Professor P. Martin, and Professor R. B. Griffiths. Interest and advice were also generously given by Dr. G. Shirane, Dr. L. Corliss, and Dr. J. Hastings.

---

*Work performed under the auspices of the U. S. Atomic Energy Commission.

†Work supported by the Air Force Office of Scientific Research, Grant No. AF68-1480.

[1]B. I. Halperin and P. C. Hohenberg, Phys. Rev. 177, 952 (1969), and references contained therein, including references to the original work of R. Ferrell and coworkers on liquid helium.

[2]H. Y. Lau, L. M. Corliss, A. Delapalme, J. M. Hastings, R. Nathans, and A. Tucciarone, Phys. Rev. Letters 23, 1225 (1969).

[3]W. Marshall and R. Lowde, Rep. Progr. Phys. 21, 705 (1968).

[4]A theoretical discussion of λ, together with an experimental verification of the value $\lambda=1$ will be presented in a more detailed report.

[5]M. J. Cooper and R. Nathans, Acta Cryst. 23, 357 (1967).

[6]M. P. Schulhof, P. Heller, R. Nathans, and A. Linz, Phys. Rev. B 1, 2304 (1970).

[7]E. Riedel and F. Wegner, Phys. Rev. Letters 24, 730 (1970).

[8]See, for example, A. Arrott and J. E. Noakes, Phys. Rev. Letters 19, 786 (1967); M. S. Green, M. Vicentini-Missoni, and J. M. H. Levelt-Sengers, Phys. Rev. Letters 18, 1113 (1967).

[9]P. Heller, to be published.

[10]B. I. Halperin and P. C. Hohenberg, Phys. Rev. 188, 898 (1969).

Scaling of the Thermal Conductivity Near the Gas-Liquid Critical Point*

J. V. Sengers and P. H. Keyes
Institute for Molecular Physics and Department of Physics and Astronomy, University of Maryland, College Park, Maryland 20742
(Received 23 November 1970)

The thermal conductivity of a gas becomes infinite at the critical point. It is shown that the experimental data for the thermal conductivity of CO_2 satisfy scaling-law relations very similar to those previously established for equilibrium properties near the critical point. The analysis also suggests that the values reported in the literature for the critical exponent of the Rayleigh linewidth should be revised downwards.

The thermal conductivity of a gas becomes infinite at the critical point.[1,2] The purpose of the present Letter is (i) to emphasize the importance of background terms in the thermal conductivity and to elucidate a method for estimating their effect; (ii) to demonstrate that the thermal conductivity, after background corrections, satisfies scaling-law relations similar to those previously established for equilibrium properties; (iii) to indicate that background corrections will reduce the values reported in the literature for the critical exponent of the Rayleigh linewidth.

We first recall the scaling laws proposed by Widom, Kadanoff, and others[3-6] for the thermodynamic behavior in the critical region. In this formulation a thermodynamic property $X(\rho, T)$ which becomes zero or infinite at the critical point is represented by

$$A(\rho)X \propto |\epsilon|^{\varphi} f(x) \text{ or } A(\rho)X \propto |\Delta\rho|^{\varphi/\beta} g(x), \quad (1)$$

where $\epsilon = (T-T_c)/T_c$, $\Delta\rho = (\rho-\rho_c)/\rho_c$, and β is the exponent of the coexistence curve. The functions $f(x)$ and $g(x) = x^{\varphi} f(x)$ depend on a single scaling parameter $x = \epsilon/|\Delta\rho|^{1/\beta}$; the exponent φ represents the critical exponent X along the critical isochore. The factor $A(\rho)$ accounts for any asymmetry of X around the critical density ρ_c. In particular,[7] $A(\rho) = \mathrm{sgn}(\Delta\rho)$ for the chemical potential $\Delta\mu = \mu(\rho, T) - \mu(\rho_c, T)$, and $A(\rho) = \rho^2$ for the isothermal compressibility K_T. To the extent that the Ornstein-Zernike relation $\xi^2 \propto kT\rho K_T$ is adequate, the correlation length ξ is expected to scale after multiplying with $\rho^{1/2}$. The validity of the scaling law representation has been demonstrated for several gases.[7]

Several authors have attempted to extend the theory to dynamical properties.[8-11] In particular Kadanoff and Swift,[8] Kawasaki,[10] and Ferrell[11] have predicted that the anomalous thermal conductivity should diverge as

$$\Delta\lambda(\rho_c, T) = \Lambda \epsilon^{-\gamma+\nu}, \quad (2)$$

where γ and ν are the critical exponents of the compressibility and the correlation length, respectively, and Λ is a proportionality constant.

The thermal conductivity of CO_2 in the critical region was measured by one of the authors in collaboration with Michels and Van der Gulik.[1] In interpreting these data we note that the theoretical prediction (2) refers to the anomalous contribution to the thermal conductivity λ. Thus

$$\lambda = \Delta\lambda + \lambda_{id}, \quad (3)$$

where λ_{id} is an ideal thermal conductivity in the absence of any critical anomaly. Empirically, the background λ_{id} can only be estimated by extrapolating data away from the critical point into the critical region.

Our experimental data for CO_2 cover a range of temperatures up to 75°C; at the latter temperature an anomalous contribution to the thermal conductivity is still noticeable. Recently, Le Neindre and coworkers measured the thermal conductivity of CO_2 at temperatures up to 700°C and at densities from zero to twice the critical density.[12] Their data agree with our data in the region of overlap; thus we may confidently use the data of Le Neindre et al. for assessing the background in our thermal conductivity data.

An assessment of the background λ_{id} is greatly simplified by considering a so-called excess thermal conductivity,

$$\tilde{\lambda} = \lambda(\rho, T) - \lambda(0, T), \quad (4)$$

which measures the excess of the actual thermal conductivity $\lambda(\rho, T)$ at density ρ over the dilute-gas value $\lambda(0, T)$ at the same temperature. It turns out that away from the critical point, $\tilde{\lambda}$ is independent of temperature and is a function of density alone. This phenomenon has been noted previously for other gases[13] and is demonstrated for CO_2 in Fig. 1. The values for $\tilde{\lambda}$ were deduced from the experimental data in Ref. 12 and cover a range of temperatures from 200 to 700°C; no

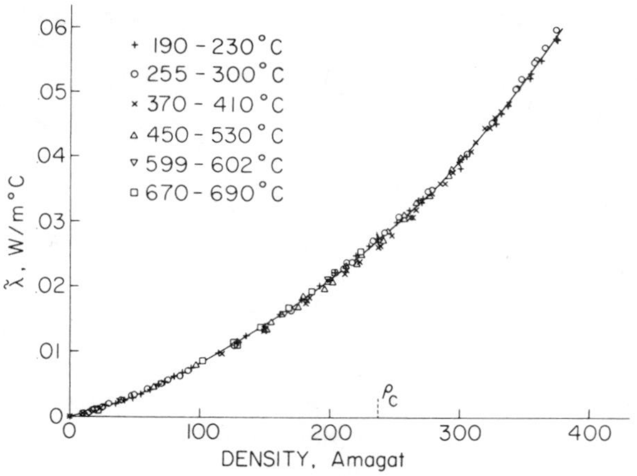

FIG. 1. The excess thermal conductivity $\tilde{\lambda}=\lambda(\rho,T)-\lambda(0,T)$ for CO_2 as a function of density, deduced from the experimental data of Le Neindre *et al.*, Ref. 12. (1 amagat = 0.001 976 4 g cm^{-3}).

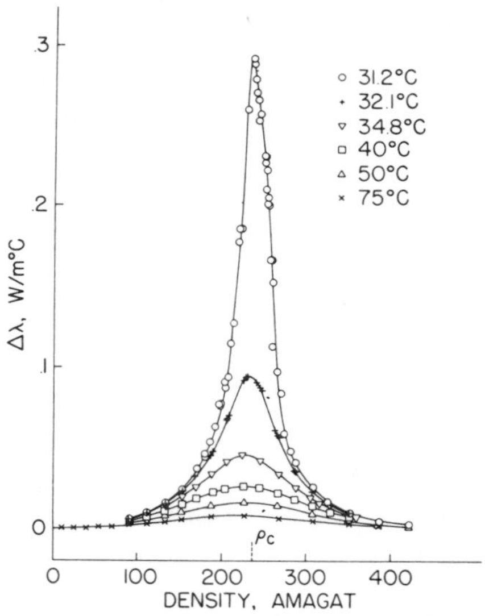

FIG. 2. The anomalous thermal conductivity $\Delta\lambda$ of CO_2.

systematic trend with temperature can be detected and the scatter is of the order of the precision of the experiment.[14] Once $\tilde{\lambda}$ is established as a function of ρ from the high-temperature data, the ideal thermal conductivity can be calculated at any temperature as

$$\lambda_{id}=\tilde{\lambda}(\rho)+\lambda(0,T) \tag{5}$$

with the known experimental dilute-gas values[1] for $\lambda(0,T)$. The empirical values thus deduced for the anomalous thermal conductivity $\Delta\lambda=\lambda-\lambda_{id}$ are shown in Fig. 2.

The values of $\Delta\lambda$ on the critical isochore $\rho=\rho_c$ were fitted to a power law $\Delta\lambda=\Lambda\epsilon^{\varphi}$ with the result

$$\Lambda=0.0030\pm0.0004 \text{ W/m °C},$$

$$\varphi=-0.60\pm0.05, \tag{6}$$

valid in the range $t_c<t\leq 50$°C. The major contribution to the errors quoted in (6) is the estimated uncertainty in the critical temperature $t_c=31.04\pm0.06$°C. The values recently determined for γ and ν are $\gamma=1.26\pm0.05$ [7] and $\nu=0.67\pm0.05$.[15] Thus we note that our experimental data are consistent with the theoretical prediction $\varphi=-\gamma+\nu=-0.59\pm0.10$.

Applications of the dynamical scaling theories to transport properties have thus far been restricted to the behavior along the critical isochore above the critical temperature and along the coexistence curve below the critical temperature. In this Letter we attempt to represent the anomalous thermal conductivity at arbitrary densities and temperatures. In the dynamical scaling theories anomalies in transport properties are related to those in equilibrium properties and to the static correlation length ξ. In the same spirit we have investigated the simplest possibility, namely, whether $\Delta\lambda$ away from the critical isochore is related to $\Delta\lambda$ at the critical isochore by the same scaling-law relations as those for equilibrium properties. That is, in analogy to (1), we consider

$$(\rho/\rho_c)^{1/2}\Delta\lambda=\Lambda|\epsilon|^{\varphi}f(x). \tag{7}$$

A factor $(\rho/\rho_c)^{1/2}$ has been introduced in (7) to account for a small asymmetry in $\Delta\lambda$, which manifests itself in a shift of the peaks in Fig. 2 to lower densities when the temperature is increased. Thus it seems that, in this approximation, the symmetry character of $\Delta\lambda$ is similar to that of the correlation length ξ.

To investigate assumption (7) we calculated the dimensionless quantities $(\rho/\rho_c)^{1/2}\Delta\lambda/\Lambda|\epsilon|^{\varphi}$ and plotted them as a function of x. Since the data cover six decades in the scaling parameter x, we actually prefer to plot the data as a function of x^{β}. The results are presented in Fig. 3 for the parameters $\Lambda=0.003\,05$ W/m °C, $\varphi=-0.60$, as determined from (7), $\beta=0.35$, $\rho_c=236.7$ amagat, and $t_c=31.04$°C. The data in this figure were taken from Table XIV of the original publication[1] after correcting for the background λ_{id}; they cover the density ranges $|\Delta\rho|<40\%$ for $t=31.20$, 32.10, and 34.80°C; $|\Delta\rho|<31\%$ for $t=40.00$°C;

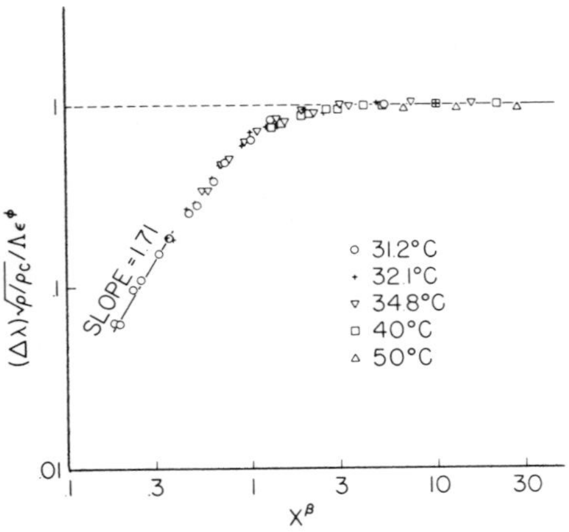

FIG. 3. Reduced thermal conductivities $(\Delta\lambda)(\rho/\rho_c)^{1/2}/\Lambda\epsilon^\varphi$ of CO_2 as a function of x^β.

and $|\Delta\rho| < 6\%$ for $t = 50.00°C$. In the limit $x \to \infty$, $f(x)$ should approach the value for the critical isochore $f(\infty) = 1$. For $x \to 0$, $f(x)$ should approach its asymptotic behavior for the critical isotherm $f(x) \propto (x^\beta)^{-\varphi/\beta}$ with $-\varphi/\beta = 1.71$. Figure 3 demonstrates that the reduced thermal conductivity data can be represented by a single-valued function $f(x)$ of the scaling parameter x. The scatter of the data seems to be within the precision with which the analysis can be carried through. We note that the range of the scaling-law relation, $\epsilon < 0.065$ and $|\Delta\rho| < 0.4$, is similar to the range over which the scaling-law relations have been verified for equilibrium properties.[7]

The analysis presented here is restricted to temperatures above the critical temperature. We do not have enough experimental data below the critical temperature to investigate a similar scaling law relation for $\epsilon < 0$.

We should mention that Murthy and Simon[16,17] have reported thermal conductivity data for CO_2 at the critical isochore which seem to indicate that λ itself diverges as $|\epsilon|^{-0.67 \pm 0.02}$ without any background corrections. We remark, however, that their data do not reproduce at higher temperatures the values for λ established independently by three groups of investigators[1,12,18] and they appear to vary with the magnitude of the temperature gradient, in contrast to our experimental data.[19] Moreover, two runs with the same apparatus yielded values for λ differing by as much as 25%. Thus the accuracy of the data obtained by Murthy and Simon has not been established sufficiently well to contradict the results of our analysis.

The Rayleigh linewidth Γ in light scattered through a fluid is proportional to the thermal diffusivity $\lambda/\rho c_p$. At the critical isochore this linewidth should behave asympototically near the critical point as

$$\Gamma \propto |\epsilon|^{\gamma+\varphi} = |\epsilon|^\nu. \tag{8}$$

Recently, Henry, Swinney, and Cummins[20] reported that the experimental values for the exponent $\gamma + \varphi$ are higher than the predicted value $\gamma + \varphi = \nu$. In particular, they found $\gamma + \varphi = 0.751 \pm 0.004$ for Xe and $\gamma + \varphi = 0.73 \pm 0.02$ for CO_2. The presence of a background term λ_{id}, however, should also affect the behavior of Γ. In fact, a corrected linewidth proportional to $\Delta\lambda/\rho c_p$ will follow a power law with a smaller exponent. (Since c_p diverges more strongly than λ, background effects in c_p are less important than those in λ.)

The effect can be illustrated for the linewidth data on the Rayleigh line in CO_2 determined by Swinney and Cummins.[21] In this case the correction can be estimated by multiplying the measured linewidth Γ by $\Delta\lambda/\lambda$ as deduced from our experiments. The linewidth data, thus corrected, satisfy a power law with exponent $\gamma + \varphi = 0.62 \pm 0.02$ in the range $0.060°C \le t - t_c \le 1.5°C$, to be compared with the uncorrected value $\gamma + \varphi = 0.73 \pm 0.02$. For $t - t_c > 1.5°C$ the corrected linewidth data begin to deviate from this asymptotic behavior. At these temperatures, however, the scattering intensity is low and the light-scattering measurements become less accurate. While the corrected value $\gamma + \varphi$ of 0.62 ± 0.02 is a little low compared with $\nu = 0.67 \pm 0.05$, the analysis does demonstrate that the apparent difference between experiment and theory for the Rayleigh linewidth as mentioned by Henry, Swinney, and Cummins[20] is within the uncertainty caused by neglecting background effects. We suggest that the exponents reported for the Rayleigh linewidth of Xe[20] and of SF_6[22,23] should be revised for the same reason.

The authors are indebted for Mr. J. C. Solberg, Dr. M. J. Cooper, Dr. H. L. Swinney, and Dr. J. H. M. Levelt Sengers for some valuable remarks. They also thank Dr. B. Le Neindre for providing them with a copy of his disertation.

*Research supported by the Office of Naval Research under Contract No. N00014-67-A-0239-0014 and by the Advance Research Project Agency.

[1]A. Michels, J. V. Sengers, and P. S. Van der Gulik, Physica (Utrecht) 28, 1216 (1962).

[2]J. V. Sengers, in *Critical Phenomena, Proceedings of a Conference, Washington, D. C., 1965,* edited by

M. S. Green and J. V. Sengers, National Bureau of Standards Miscellaneous Publication No. 273 (U.S. G.P.O., Washington, D. C., 1966), p. 165.
[3]B. Widom, J. Chem. Phys. 43, 3898 (1965).
[4]L. P. Kadanoff, Physics (Long Is. City, N. Y.) 2, 263 (1966).
[5]R. B. Griffiths, Phys. Rev. 158, 176 (1967).
[6]M. J. Cooper, Phys. Rev. 168, 183 (1968).
[7]M. Vicentini-Missoni, J. M. H. Levelt Sengers, and M. S. Green, J. Res. Nat. Bur. Stand., Sect. A 73, 563 (1969).
[8]L. P. Kadanoff and J. Swift, Phys. Rev. 166, 89 (1968).
[9]B. I. Halperin and P. C. Hohenberg, Phys. Rev. 177, 952 (1969).
[10]K. Kawasaki, Phys. Rev. A 1, 1750 (1970), and to be published.
[11]R. A. Ferrell, Phys. Rev. Lett. 24, 1167 (1970).
[12]B. Le Neindre, P. Bury, R. Tufeu, P. Johannin, and B. Vodar, in *Proceedings of the Ninth Thermal Conductivity Conference*, edited by H. R. Shanks (U.S. Atomic Energy Commission, Division of Technical Information Extension, Oak Ridge, Tenn., 1970), p. 169.
[13]J. V. Sengers, in *Recent Advances in Engineering Science*, edited by A. C. Eringen (Gordon and Breach, New York, 1968), Vol. 3, p. 153.
[14]B. Le Neindre, thesis, University of Paris, 1969 (unpublished).
[15]B. Chu and J. S. Lin, to be published.
[16]M. L. R. Murthy and H. A. Simon, Phys. Rev. A 2, 1458 (1970).
[17]M. L. R. Murthy and H. A. Simon, in *Proceedings of the Fifth Symposium on Thermophysical Properties*, edited by C. F. Bonilla (American Society of Mechanical Engineers, New York, 1970), p. 214.
[18]L. A. Guildner, J. Res. Nat. Bur. Stand., Sect. A 66, 341 (1962).
[19]A. Michels and J. V. Sengers, Physica (Utrecht) 28, 1238 (1962).
[20]D. L. Henry, H. L. Swinney, and H. Z. Cummins, Phys. Rev. Lett. 25, 1170 (1970).
[21]H. L. Swinney and H. Z. Cummins, Phys. Rev. 171, 152 (1968).
[22]G. B. Benedek, in *Polarisation Matière et Rayonnement, Livre de Jubilé en l'Honneur du Professeur A. Kastler*, edited by The French Physical Society (Presses Universitaires de France, Paris, France, 1969), p. 49.
[23]P. Braun, D. Hammer, W. Tsarnuter, and P. Weinzierl, Phys. Lett. 32A, 390 (1970).

PRETRANSITIONAL PHENOMENA IN THE ISOTROPIC PHASE OF A NEMATIC LIQUID CRYSTAL*

T. W. Stinson, III,† and J. D. Litster
Physics Department and Center for Materials Science and Engineering, Massachusetts Institute of Technology, Cambridge, Massachusetts 02139
(Received 7 July 1970)

We have observed a divergence of the magnetic birefringence, and a critical increase and slowing of the fluctuations in order in the isotropic phase of a nematic liquid crystal. Our results are quantitatively described by a mean-field model except for a critical region close to the ordering temperature where the fluctuations are so large that the mean-field approximation fails.

In this Letter we report the results of an experimental study of pretransitional phenomena in the isotropic phase of the nematic liquid crystal *p*-methoxy benzylidene *p*-*n*-butylaniline (MBBA).[1] As the temperature of a liquid crystal is lowered there is a phase change from an isotropic liquid state to a state (also liquid) with long-range orientational order of the molecules. In the ordered phase of a nematic liquid the centers of mass of the molecules remain as randomly distributed as in the isotropic phase, and the molecules align with their long axes parallel. The anisotropy of the molecules in a nematic material is uniaxial; the electric polarizability is usually greater parallel to the long axis of the molecule.[2] The degree of order may therefore be determined by measuring the anisotropy of the dielectric constant, and optical methods are ideal for this purpose. The diamagnetic susceptibility is also usually greater along the axis of the molecule and it is possible to align molecules in the isotropic phase with a magnetic field. The magnetically induced birefringence (Cotton-Mouton effect) then is proportional to the alignment produced. In addition, from the intensity and spectrum of scattered light one may obtain the mean squared amplitude and time dependence of fluctuations in the order.

In the isotropic phase of MBBA we have accurately measured the magnetic birefringence as a function of temperature. We have also measured the intensity and spectrum of light scattered by anisotropic fluctuations in the dielectric constant. We observed a divergence of the Cotton-Mouton coefficient, and a divergence and critical slowing of the fluctuations in order as the phase transition was approached. This behavior is similar to that of materials in the vicinity of a critical point.[3] Although the nematic-isotropic transition is first order[2] (as shown by a latent heat and volume discontinuity), our measurements demonstrate that over a wide temperature range the liquid crystal behaves as if it were going to undergo a second-order phase transition at a critical point.

We provide a theoretical interpretation of our data using a phenomenological model due to Landau[4] and its extension by de Gennes[5] to describe dynamical behavior. We find that this mean-field model adequately describes the behavior over most of the temperature range, but that close to the phase transition there is a critical region where the mean-field approximation fails.

We now discuss our experimental results using the Landau model. We take the ordered nematic liquid crystal as optically uniaxial and so it is necessary to specify only the birefringence and the direction of the optic axis. Except for gradient terms (which we shall see later are negligible) the free energy is independent of the orientation of the optic axis. Therefore for purposes of the Landau model we may specify an order parameter[6] $Q=\frac{3}{2}\langle\cos^2\theta-\frac{1}{3}\rangle$, where θ is the angle between the long axis of a molecule and the local optic axis. For a completely aligned material ($Q=1$) let ϵ_p and ϵ_t be the dielectric constants parallel and transverse to the optic axis. Then the Cartesian dielectric tensor is $\epsilon_{\alpha\beta}=\bar{\epsilon}\delta_{\alpha\beta}$

$+Q(\frac{1}{3}\Delta\epsilon)(3n_\alpha n_\beta-\delta_{\alpha\beta})$, where n_α, n_β are the Cartesian components of a unit vector parallel to the optic axis, $\bar{\epsilon}=\frac{1}{3}(2\epsilon_t+\epsilon_p)$, and $\Delta\epsilon=\epsilon_p-\epsilon_t$. Following Landau, we write the free-energy density in the vicinity of the phase transition as

$$\Phi=\Phi_0+\tfrac{1}{2}AQ^2-\tfrac{1}{3}BQ^3+\tfrac{1}{4}CQ^4-\tfrac{1}{3}\Delta\chi H^2 Q+\cdots. \quad (1)$$

In the model the coefficient of the quadratic term is taken to be $A=a(T-T_c{}^*)$. The anisotropy in the diamagnetic susceptibility is $\Delta\chi=\chi_p-\chi_t$. Minimizing the free energy with respect to Q gives the equation of state,

$$AQ-BQ^2+CQ^3-\tfrac{1}{3}\Delta\chi H^2=0. \quad (2)$$

Identical results may be obtained from the Maier-Saupe mean-field theory of nematic liquid crystals.[6]

The equilibrium value of Q is the root of Eq. (2) corresponding to the lower minimum of the free energy. If the coefficient B were zero the system would have a second-order transition to an ordered phase at $T=T_c{}^*$. The cubic term in Eq. (1) means that in zero field a first-order transition to a state of finite Q will occur at $T_K=T_c{}^*+2B^2/9aC$. The discontinuity in order parameter at T_K is $Q_K=2B/3C$, and the latent heat is $\frac{1}{2}T_K Q_K{}^2$.

We now analyze our birefringence measurements to establish all of the parameters in the free-energy expression, Eq. (1), and then proceed to check the consistency of the model by comparing its predictions with our measurements of the intensity and spectrum of the scattered light. This represents the first quantitative application of the mean-field model to interpret experimental measurements in the isotropic phase of a liquid crystal.

The order produced by a magnetic field in the isotropic phase ($T>T_K$) may be calculated to a first approximation from Eq. (2) as $Q=\Delta\chi H^2/3A$. The magnetic field determines the optic axis, and the magnetically induced birefringence (for $Q\ll 1$) is

$$\Delta n=(\epsilon_{zz})^{1/2}-(\epsilon_{xx})^{1/2}$$
$$=\Delta\epsilon\Delta\chi H^2/6(\bar{\epsilon})^{1/2}a(T-T_c{}^*). \quad (3)$$

In Fig. 1 we present our measured values of $H^2/\Delta n$; the temperature dependence over most of the range is clearly linear in agreement with Eq. (3). [The transition temperature T_K often varies from sample to sample of liquid crystal. This is probably caused by impurities resulting from hydrolysis of the MBBA into its amine and aldehyde constituents. Data are shown for two samples; it is apparent that the behavior is identical except for a shift in temperature scale. Although $T_c{}^*$ shows considerable variation from sample to sample, the other coefficients in Eq. (1) remain constant within the sensitivity of our measurements.] From the slope of the plots (45×10^{12} $G^2/°K$) we may estimate the coefficient a. For MBBA[7] $\bar{\epsilon}=2.605$, Δn at T_K is 0.133, and $\Delta\chi=1.25\times10^{-7}$ erg G^{-2} cm^{-3}. Taking $Q_K=0.40$[6] we estimate $\Delta\epsilon=1.09$ and $a=0.062$ J cm^{-3} $°K^{-1}$. This gives a latent heat of 1.6 J cm^{-3}, which compares favorably with the value 1.5 J cm^{-3} measured for MBBA.[7] Extrapolation of the solid lines in Fig. 1 gives $T_c{}^*$ for each sample; we also measured T_K for each and find $T_K-T_c{}^*=1.0°K$ for both samples. From these results and the relations given earlier (with $Q_K=0.40$) we estimate $B=0.47$ J cm^{-3} and $C=0.79$ J cm^{-3}.

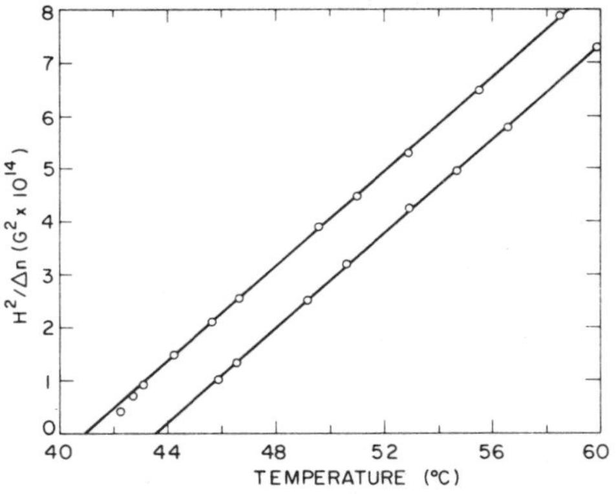

FIG. 1. The reciprocal of the Cotton-Mouton coefficient is shown as a function of temperature for two samples of MBBA. The solid lines are a fit to the mean-field result of Eq. (3).

Our measured temperature dependence of the intensity of light scattered by fluctuations in the order parameter in the isotropic phase is shown in Fig. 2. If $\vec{k}$ is the wave vector of the incident light, then the light is scattered through an angle θ by a Fourier component of the fluctuations in the dielectric constant whose wave vector is $q=2k\sin\frac{1}{2}\theta$. The intensity of scattered light which is polarized parallel to the incident light is proportional to $\langle\delta\epsilon_{zz}{}^2(q)\rangle$, while $\langle\delta_{xz}{}^2(q)\rangle$ gives the scattered intensity polarized perpendicular to the incident polarization.

In the long-wavelength limit for an illuminated volume V the free-energy expression [Eq. (1), with no applied field] leads to the result[8]

$$\langle Q^2(q)\rangle=kT/VA(1+\xi^2q^2)$$
$$=kT/Va(T-T_c{}^*)(1+\xi^2q^2). \quad (4)$$

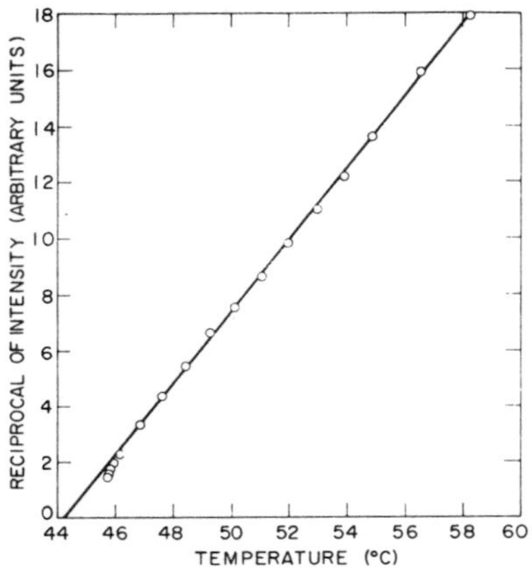

FIG. 2. The reciprocal of the intensity of light scattered by fluctuations in the isotropic phase of MBBA. The solid line is a fit to the mean field result of Eq. (4).

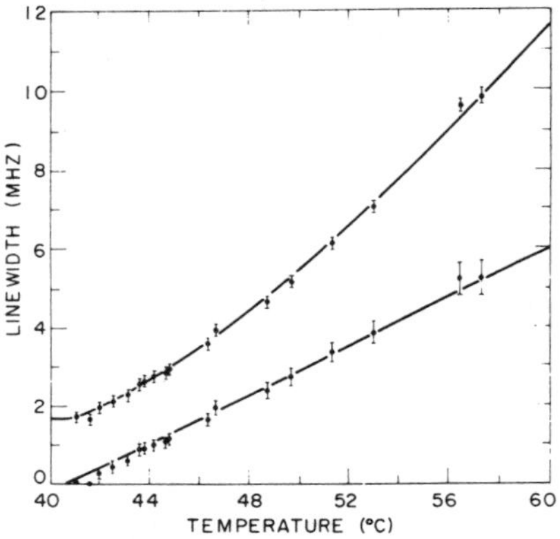

FIG. 3. The half-width of the Lorentzian spectrum of light scattered by fluctuations in the isotropic phase of MBBA. The upper curve is the raw data and includes the instrumental half-width of 1.65 MHz; the solid line is a fit of a $(T-T_c{}^*)^{4/3}$ power law (see Ref. 9). The lower curve shows the results corrected for the temperature dependence of the transport coefficient ν, as described in the text. (The relaxation time for the fluctuations is $\tau = 1/\Gamma$.)

Here ξ is the correlation length for the fluctuations, and in the Landau theory[8] varies as $\xi = \xi_0[(T/T_c)-1]^{-1/2}$. Since fluctuations in the isotropic phase involve all orientations of the optic axis with equal probability, we average Eq. (4) over all angles to obtain

$$\langle\delta\epsilon_{zz}{}^2(q)\rangle = (4/3)\langle\delta\epsilon_{xz}{}^2(q)\rangle = (4/45)\Delta\epsilon^2\langle Q^2(q)\rangle. \tag{5}$$

A reasonable estimate for ξ (we take $\xi_0 = 15$ Å, about the maximum dimension of a molecule) shows the $\xi^2 q^2$ term in Eq. (4) to be negligible for visible light. Our experiments confirm this and justify the omission of gradient terms from Eq. (1). We measured the intensity ratio for polarized and depolarized components of the scattered light to be 1.35 ± 0.06, as predicted by Eq. (5). The temperature dependence of the intensity shown in Fig. 2 is linear over a wide temperature range as predicted by Eq. (4).

In an earlier publication[9] we reported measurements of the relaxation time for fluctuations of the order parameter in the isotropic phase of MBBA. We observed a critical slowing down as the ordering temperature was approached, and de Gennes has extended the Landau theory to explain our results. In de Gennes' theory[5] the relaxation time for long-wavelength fluctuations is $\tau = \nu/A$, where ν is a transport coefficient. We found our measurements of the relaxation time could be fitted by a power law of the form $\tau \sim (T-T_c{}^*)^{-4/3}$ as shown in the upper curve of Fig. 3 ($\tau = 1/\Gamma$). To interpret these data properly it is necessary to know the temperature dependence of ν. We might reasonably expect this transport coefficient to be proportional to the shear viscosity of the isotropic liquid.[5] In this isotropic phase of p-azoxyanisole (a nematic liquid crystal similar to MBBA) the shear viscosity varies as $\exp(W/kT)$ with $W/k \simeq 2800$°K.[10] Assuming this temperature dependence for ν in MBBA, we obtain from our relaxation-time measurements the linear variation of A with temperature shown as the lower curve of Fig. 3. Although the experimental uncertainties are large, the result is consistent with our birefringence and intensity measurements.

A careful examination of Figs. 1 and 2 shows close to T_K a departure from the $(T-T_c{}^*)^{-1}$ dependence that is predicted by the mean-field model for the Cotton-Mouton coefficient and the intensity of the scattered light. This occurs where the precision of our measurements is highest and is definitely a real effect. Analogous observations have been made in many other substances[3] that the mean-field approximation does not correctly describe the detailed behavior in the critical region. The approximation fails because it does not include the effects of the large fluctuations as the phase transition is approached. At the transition temperature T_K, two equal minima in the free energy (corresponding to $Q=0$ and $Q = Q_K$) are separated by a free-energy maximum which we estimate below to be $\sim kT$. This explains why we observed no superheating or supercooling of the phase transition and also means that close to T_K large fluctuations and departures from the mean-field results [e.g., Eq. (3)] are to be expected. However, we may use the mean-field approximation to estimate the temperature at which these departures may be observable.

To discuss these effects in a liquid crystal we use the fact that fluctuations are correlated over a finite range $\xi = \xi_0[(T/T_c^*)-1]^{-1/2}$. The increase in free energy associated with a fluctuation $Q(\vec{r})$ at a point is then approximately given by

$$\delta\Phi = v^*[\tfrac{1}{2}AQ^2 - \tfrac{1}{3}BQ^3 + \tfrac{1}{4}CQ^4 - \tfrac{1}{3}\Delta\chi H^2 Q], \qquad (6)$$

where $v^* \simeq (4\pi/3)\xi^3$ is the correlation volume. We take the probability of a fluctuation to be proportional to $\exp(-\delta\Phi/kT)$ and expand the exponential for cubic and quartic terms of Eq. (6). We obtain thereby the approximate result in zero field[11]:

$$\langle Q^2(q)\rangle = \frac{kT}{VA}\left[1 - \frac{3CkT}{v^*A^2} + \frac{5B^2kT}{v^*A^3} + \cdots\right]. \qquad (7)$$

We find that Eq. (3) should also be multiplied by the correction factor given in brackets in Eq. (7). Since the two free-energy minima (for $Q = Q_K$ and 0) at T_K are separated by a barrier $v^*(B^4/324C^3) \simeq 0.6kT$, we expect the approximations used to derive Eq. (7) will fail very close to T_K. However the result should still serve to indicate where departures from the mean-field results may first be observed. With $\xi_0 = 15$ Å we calculate that the Cotton-Mouton coefficient and the intensity of the scattered light should be 6% larger than the mean-field result when $T = T_K + 1°$K. This is in good agreement with our experimental results.

Our conclusions are as follows: We have observed a divergence in the magnetic birefringence and fluctuations in the order parameter, as well as a critical slowing of these functions, in the isotropic phase of a nematic liquid crystal. These pretransitional phenomena are analogous to the behavior of substances (e.g., ferromagnets) in the vicinity of a critical point. Our measurements may be consistently and quantitatively interpreted using the mean-field approximation, except for a critical region very close to the phase transition. This model provides a more detailed description than the Frenkel concept of heterophase fluctuations[12] and could be used to interpret other observations[13,14] of pretransitional phenomena in liquid crystals. It seems reasonable that the behavior we have observed is typical of nematic liquid crystals, and it would be interesting to learn the results of similar studies of smectic and cholesteric materials.

The model we have used to interpret our results also predicts non-Gaussian fluctuations within a correlation volume. Unfortunately it would be difficult to observe this in the statistics of the scattered radiation. The only sufficiently coherent sources available (visible light) probe many correlation volumes and the central-limit theorem dictates that the scattered light will be Gaussian. We estimate that departures from Gaussian statistics in the fluctuations within a coherence volume v^* will be reduced in the scattered light by a factor v^*/V, where V is the illuminated volume.

We appreciate helpful discussions with P. G. de Gennes, M. J. Freiser, and M. J. Stephen.

---

*Research supported by the Advanced Projects Research Agency under contract No. SD-90.

†National Science Foundation Predoctoral Fellow.

[1]H. Kelker and B. Scheurle, Angew. Chem. 81, 903 (1969).

[2]An excellent review article on liquid crystals is by I. G. Chistyakov, Usp. Fiz. Nauk (USSR) 89, 563 (1966) [Sov. Phys.—Usp. 9, 551 (1967)].

[3]P. Heller, Rep. Prog. Phys. 30, 731 (1967).

[4]L. D. Landau, in *Collected Papers of L. D. Landau*, edited by D. Ter Haar (Gordon and Breach, New York, 1965), p. 193.

[5]P. G. de Gennes, Phys. Lett. A 30, 454 (1969).

[6]W. Maier and A. Saupe, Z. Naturforsch. A 15, 287 (1960).

[7]We are grateful to M. J. Freiser and I. Haller of the IBM liquid-cyrstal group for a private communication of their measurements on MBBA.

[8]L. D. Landau and E. M. Lifshitz, *Statistical Physics*, (Addison-Wesley, Reading, Mass. 1958), paragraph 116.

[9]J. D. Litster and T. W. Stinson, III, J. Appl. Phys. 41, 996 (1970).

[10]R. S. Porter, E. M. Barrall, II, and J. F. Johnson, J. Chem. Phys. 45, 1452 (1966).

[11]A better approximate calculation has been carried out by C.-P. Fan and M. J. Stephen, Phys. Rev. Lett. 25, 500 (1970).

[12]J. Frenkel, J. Chem. Phys. 7, 538 (1939).

[13]V. N. Tsvetkov and S. P. Krozer, zh. Tekh. Fiz. 28, 1444 (1958). [Sov. Phys.—Tech. Phys. 3, 1340 (1958)].

[14]A. Torgalkar, R. S. Porter, E. M. Barrall, II, and J. F. Johnson, J. Chem. Phys. 48, 3897 (1968).

Thermal Diffusivity of CO_2 in the Critical Region*

H. L. SWINNEY AND HERMAN Z. CUMMINS
Department of Physics, The Johns Hopkins University, Baltimore, Maryland 21218
(Received 26 February 1968)

We have measured the Rayleigh linewidth in CO_2 in the critical region using a self-beat spectrometer. The linewidth was measured as a function of both temperature and cell height. The thermal diffusivity χ calculated using the Landau-Placzek equation is in excellent agreement with the values that have been obtained by thermodynamic measurements at three temperatures within the temperature range we investigated ($T-T_c=-1.04$, $+1.06$, and $+3.8$C°). Thus the Landau-Placzek equation is directly verified in the critical region, at least for temperatures not too close to T_c. However, we find that very near T_c [for $\epsilon\equiv(T-T_c)/T_c\lesssim10^{-4}$], the correlation length in CO_2 is of sufficiently long range (~250 Å at $\epsilon=10^{-4}$) to require that the Fixman-modified linewidth equation be used in order to correctly describe the linewidth behavior. The thermal diffusivity was obtained along the critical isochore for the temperature range $0.02\leq(T-T_c)\leq5.3$C° and along both the gas and liquid sides of the coexistence line for $0.02\leq(T_c-T)\leq2.3$C°. The results are (in units of cm²/sec): along the critical isochore, $\chi=(18.1\pm0.5)\times10^{-6}(T-T_c)^{0.73\pm0.02}$; along the gas coexistence line, $\chi=(36.0\pm3.0)\times10^{-6}(T_c-T)^{0.66\pm0.05}$; and along the liquid coexistence line, $\chi=(34.8\pm2.5)\times10^{-6}(T_c-T)^{0.72\pm0.05}$. These exponents are in reasonable quantitative agreement with the prediction of Kadanoff and Swift that $\chi\sim|\epsilon|^{-\nu}(\nu\approx\frac{2}{3})$. Our exponents are also in accord with the thermal-conductivity divergence $\lambda\sim\epsilon^{-1/2}$ predicted by Fixman and by Mountain and Zwanzig, if the isothermal compressibility diverges as $\epsilon^{-5/4}$, as predicted by the Ising model. Thus both theory and experiment indicate a stronger divergence in the thermal conductivity than has heretofore been assumed. Our subcritical exponents are also in agreement with the linewidth measurements by Saxman and Benedek in SF_6; however, above the critical temperature they obtained an exponent of 1.27, in definite disagreement with our result.

I. INTRODUCTION

A. Historical Background

In recent years it has become increasingly apparent that such diverse systems as simple fluids, binary mixtures, ferromagnets, antiferromagnets, and binary metallic alloys exhibit marked similarities in their behavior near critical points.[1] Much of the new information has come from detailed studies of the asymptotic behavior of the thermodynamic properties of these systems as they approach critical points. New experimental techniques and refinements of older ones have revealed the presence of some striking anomalies in these properties which in turn have stimulated the development of the theory of critical phenomena for which the accurate determinations of the asymptotic behavior of the thermodynamic properties serve as crucial tests.[2–4]

It has been recognized for over 50 years that light-scattering measurements can provide information on the critical point in fluids and binary mixtures through the observation of critical opalescence.[5] Although light-scattering measurements were until quite recently limited to observation of the total scattered intensity, it was recognized that the spectrum of the scattered light contains independently useful information.

According to Landau and Placzek, the central (quasi-elastic) component of the Rayleigh-scattered light should be a Lorentzian whose half-width at half-maximum $\Gamma\equiv\Delta\omega_{1/2}$ is given for pure fluids by[6]

$$\Gamma=\chi K^2, \tag{1}$$

where K, the momentum-transfer vector, is given by $|K|=2nK_0(\sin\frac{1}{2}\theta)$, and n, K_0, and θ are the refractive index, the magnitude of the wave vector of the incident light in vacuum, and the scattering angle, respectively. χ, the thermal diffusivity, equals $\lambda/\rho c_p$, where λ, ρ, and c_p are the thermal conductivity, the density, and the specific heat at constant pressure, respectively. The specific heat c_p diverges in the same way as the isothermal compressibility, which is known to be strongly divergent at the critical point, whereas λ is expected to be only weakly divergent. Hence the Rayleigh linewidth should go to zero at the critical point. Thus measurements of the Rayleigh linewidth can be used to study the detailed temperature and density dependence of χ in the critical region.

Until quite recently measurements of the Rayleigh linewidth in the critical region were not possible since the linewidths (typically several kHz) are far too small to be measured with conventional spectroscopic techniques. With the advent of the laser, however, it was recognized that the technique of light-beating spectroscopy could be used to measure the widths of ex-

* This research was supported by the Advanced Research Projects Agency (Project DEFENDER) and was monitored by the U. S. Army Research Office (Durham) under Contract No. DA-31-124-ARO-D-400.

[1] *Critical Phenomena, Proceedings of a Conference, Washington, D. C., 1965*, edited by M. S. Green and J. V. Sengers, Natl. Bur. Std. (U. S.) Misc. Publ. No. **273** (1966).

[2] L. P. Kadanoff, W. Götze, D. Hamblen, R. Hecht, E. A. S. Lewis, V. V. Palciauskas, M. Rayl, and J. Swift, Rev. Mod. Phys. **39**, 395 (1967).

[3] M. E. Fisher, Rept. Progr. Phys. **30**, 615 (1967).

[4] P. Heller, Rept. Progr. Phys. **30**, 731 (1967).

[5] See the review by O. K. Rice, in *Thermodynamics and Physics of Matter*, edited by F. D. Rossini (Princeton University Press, Princeton, N. J., 1955), p. 419.

[6] L. D. Landau and E. M. Lifshitz, *Electrodynamics of Continuous Media* (Addison-Wesley Publishing Co., Reading, Mass., 1960), p. 392; L. Landau and G. Placzek, Physik Z. Sowjetunion **5**, 172 (1934). The linewidth resulting from concentration fluctuations was derived earlier by M. Leontowitsch, Z. Physik **72**, 247 (1931).

TABLE I. Critical parameters.

Physical quantity	Temperature dependence[a]		Value of exponent[b] Theory	Experiment
$c_p \sim \kappa_T$	$T>T_c$	$\epsilon^{-\gamma}$	Classical, 1.00[b]	Green *et al.* data analysis, 1.4[g]
			Ising model, 1.250 ± 0.001[b]	Wilcox and Balzarini, 1.0[h]
	$T<T_c$	$(-\epsilon)^{-\gamma'}$	Classical, 1.00[b]	Green *et al.* data analysis, 1.4[g]
			Ising model, 1.31 ± 0.05[b]	
λ (thermal conductivity)	$T>T_c$	$\epsilon^{-\psi}$	Kadanoff and Swift, $\gamma - \nu$[c]	Sengers, $\lambda \sim c_v$ [i]; thus $\psi = 0.1 \pm 0.1$
			Fixman[d] and Mountain and Zwanzig,[e] 0.5	
	$T<T_c$	$(-\epsilon)^{-\psi'}$	Kadanoff and Swift, $\gamma' -' \nu$[c]	?
$\chi \equiv \lambda / \rho c_p$ (thermal diffusivity)	$T>T_c$	$\epsilon^{\gamma - \psi}$	Kadanoff and Swift, ν[c]	Thermodynamic experiments, 0.78[j]
			Fixman[d] and Mountain and Zwanzig,[e] $\gamma - 0.5$	Present experiment, 0.73 ± 0.02
				Saxman and Benedek, 1.27[k]
	$T<T_c$	$(-\epsilon)^{\gamma' - \psi'}$	Kadanoff and Swift, ν' [c]	Present experiment: gas, 0.66 ± 0.05: liquid, 0.72 ± 0.05
				Saxman and Benedek, 0.64[k]
ξ (correlation length)	$T>T_c$	$\epsilon^{-\nu}$	Classical, 0.50[b]	Ferromagnets and antiferromagnets, 0.65 ± 0.03[b]
			Ising model, 0.643 ± 0.003[b]	
	$T<T_c$	$(-\epsilon)^{-\nu'}$	Classical, 0.50[b]	
			Scaling law result, $\nu' = \nu$[b]	?
			Ising model, 0.675 ± 0.03[f]	

[a] $\epsilon \equiv (T-T_c)/T_c$. The indicated temperature dependence is along the critical isochore for $T>T_c$ and along the coexistence line for $T<T_c$.
[b] Reference 2.
[c] Reference 44.
[d] Reference 41.
[e] Reference 40.
[f] Reference 3.
[g] Reference 19.
[h] Reference 35.
[i] References 27 and 36.
[j] Reference 25.
[k] Reference 34.

tremely narrow spectral lines.[7] A laser heterodyne spectrometer was first used to study the linewidth of light scattered from a polymer solution,[8] and was subsequently employed by Alpert *et al.* to study the binary mixture aniline-cyclohexane in the critical region.[9]

The first measurements of the Rayleigh linewidth of a pure fluid were reported by Ford and Benedek for sulphur hexafluoride,[10] and shortly afterwards Alpert *et al.* reported linewidth measurements for carbon dioxide.[11] These experiments showed that the Rayleigh linewidth along the critical isochore approaches zero approximately linearly with $|T-T_c|$ and also follows the $\sin^2(\frac{1}{2}\theta)$ dependence predicted by the Landau-Placzek equation [Eq. (1)]. Similar behavior was found in the experiments with binary mixtures where the diffusion coefficient D (which plays the role that χ plays for pure fluids) also goes to zero at the critical point.[9,12]

Although these experiments agreed qualitatively with the Landau-Placzek prediction, quantitative verification of Eq. (1) was not possible since direct thermodynamic determinations of the thermal diffusivity χ (or the binary diffusion coefficient D) do not exist in the critical region covered by the light-scattering experiments. However, Lastovka and Benedek[13] measured the Rayleigh linewidth in toluene far from the critical region and found that χ deduced from their linewidths using Eq. (1) agrees within the experimental uncertainty with the value obtained by conventional thermodynamic measurements.

B. Critical Exponents

Many of the thermodynamic properties of systems in the critical region are found to exhibit simple power-law dependences on the reduced differential temperature $\epsilon \equiv (T-T_c)/T_c$. Predictions of the numerical values for the exponents deduced from "classical" theories (e.g., the van der Waals fluid and the Weiss ferromagnet) have been available for many years, and recently numerical approximation techniques have produced predictions for the exponents for the three-dimensional Ising model (and hence for the mathematically identical lattice gas). Some of these predictions along with the experimental values for some of the exponents of interest are shown in Table I. (See the recent review articles for a complete compilation.[2–4])

Above the critical temperature on the critical isochore the specific heat at constant pressure c_p (which diverges in the same way as the isothermal compressibility κ_T) is proportional to $\epsilon^{-\gamma}$, while the thermal conductivity λ is proportional to $\epsilon^{-\psi}$. Therefore χ is proportional to $\epsilon^{\gamma-\psi}$. Similarly, below the critical temperature on the coexistence curve, χ is proportional to $(-\epsilon)^{\gamma'-\psi'}$.

The exponent γ is 1.00 for "classical" models and is

[7] A. T. Forrester, R. A. Gudmundsen, and P. O. Johnson, Phys. Rev. **99**, 1691 (1955); A. T. Forrester, J. Opt. Soc. Am. **51**, 253 (1961).

[8] H. Z. Cummins, N. Knable, and Y. Yeh, Phys. Rev. Letters **12**, 150 (1964).

[9] S. S. Alpert, Y. Yeh, and E. Lipworth, Phys. Rev. Letters **14**, 486 (1965); S. S. Alpert, Ref. 1, p. 157.

[10] N. C. Ford, Jr. and G. B. Benedek, Ref. 1, p. 150; Phys. Rev. Letters **15**, 649 (1965).

[11] S. S. Alpert, D. Balzarini, R. Novick, L. Seigel, and Y. Yeh, in *Physics of Quantum Electronics, Conference Proceedings, 1965*, edited by P. L. Kelly, B. Lax, and P. E. Tannenwald (McGraw-Hill Book Co., New York, 1966), p. 253.

[12] P. Debye, Phys. Rev. Letters **14**, 783 (1965).

[13] J. B. Lastovka and G. B. Benedek, Phys. Rev. Letters **17**, 1039 (1966).

predicted by the Ising model to be 1.25. Direct experimental determination of γ in the critical region of fluids is extremely difficult since the divergent compressibility leads to large gravitationally produced density gradients.[14] The only detailed direct study of the divergence in λ was for CO_2 by Sengers and collaborators; the singularity was thought to be weak, like the singularity in c_v which is proportional to $\epsilon^{0.1\pm0.1}$.[15]

Linewidth determinations of the thermal diffusivity appeared to offer a simple way of choosing between the classical and Ising models since $\gamma-\psi$ would be approximately 0.9 for the former and approximately 1.2 for the latter. By using a beam of very small diameter, linewidth measurements can be performed at essentially a unique density, even when there are large density gradients in the sample. Moreover, unlike thermodynamic determinations of λ which require a temperature gradient in the sample, linewidth measurements can be carried out for an isothermal sample.

While the early experiments demonstrated the usefulness of Rayleigh linewidth measurements, the result $\gamma-\psi=1$ had a large uncertainty for several reasons. First, the mean density was not precisely known. Secondly, the linewidth was not investigated as a function of height in the region near T_c where density gradients are significant. [In plots of the linewidths versus $(T-T_c)$ in the experiments cited, the extrapolated linewidth at $T=T_c$ was nonzero in every case by an amount considerably larger than the instrumental resolution.] And thirdly, the temperature ranges over which the measurements were performed were not sufficiently large to yield accurate exponents.

II. DESCRIPTION OF PRESENT EXPERIMENT

We have measured the Rayleigh linewidth for carbon dioxide along the critical isochore for the temperature range $0.018\leq T-T_c\leq 5.27C°$ and along both the gas and liquid sides of the coexistence line for $-0.019\leq T-T_c\leq -2.31C°$.[16] The linewidth was measured both as a function of the beam height in the sample cell and as a function of temperature. Measurements were made over more than two orders of magnitude in $T-T_c$, both above and below T_c, a sufficiently large range in temperature to yield fairly accurate values for the exponents $\gamma-\psi$ and $\gamma'-\psi'$.

The 77-mm-long sample cell was formed from thick-walled Pyrex tubing with a 6.0×6.0 mm square inside cross section. The all-glass sample cell avoids the problem of contamination sometimes encountered when a critical fluid sample is in contact with valves or O rings, and the square walls avoid the problem of the focusing effect of round tubing. After pumping to high vacuum, the cell was loaded by cryogenic transfer from a cylinder of Matheson research grade CO_2 containing less than 50 ppm impurities and then sealed off. The loading was checked by observing the movement of the meniscus over several degrees below T_c. By employing the predictions of the law of corresponding states[17] the density was then determined to exceed the critical density ρ_c by approximately 0.3%.

The cell was suspended in an oil bath which was matched in refractive index (at the critical temperature and the laser wavelength) to the Pyrex. The temperature of the oil bath was stable within one millidegree for several hours and within several millidegrees per day. The bath was vigorously stirred to minimize temperature gradients; its temperature was measured with a calibrated thermistor and a dc Wheatstone bridge with a relative accuracy of $\pm0.0002C°$. The critical temperature of the carbon dioxide was $31.080\pm0.015°C$.

The problem of density gradients in the critical region was mentioned in Sec. I. Not only is the density a function of height in the sample at a fixed temperature, but also the density at *any* fixed height in the sample varies with temperature. In our spectrometer the diameter of the focused laser beam was 0.2 mm, so that each linewidth measurement corresponded to a density averaged over a height of only 0.2 mm. The sample cell was suspended from a micrometer so that its height in the thermostat could be varied. Thus the linewidth could be measured as a function of height as well as temperature.

We used the self-beat spectrometer technique of Ford and Benedek[10] to measure the linewidth, since the self-beat or homodyne spectrometer is simpler to set up and easier to align than the heterodyne spectrometer used by Cummins *et al.*[8] and by Alpert, Yeh, and collaborators.[9,11] The 75-mW output from a He-Ne laser was attenuated up to 30 dB to avoid heating of the sample. Spectra were obtained at scattering angles (θ) of 90°, 22°, 16°, and 15°, with an acceptance aperture of 0.3° in all cases. The scattered light was focused onto an RCA 7326 photomultiplier and the photocurrent was Fourier-analyzed with a Panoramic spectrum analyzer (Models SB-15a and LP-1a). The output of the spectrum analyzer was electronically squared and the resultant spectra were computer fit to a Lorentzian lineshape.

As a consequence of the critical opalescence, multiple scattering can be a serious problem in the critical region, leading to spurious results. At $T-T_c=0.026C°$, the "photon mean free path" in our sample was 40 mm, many times longer than the 1.5-mm length of the scattering volume. At that temperature and $\theta=90°$ we varied the size of the scattering volume, examined different scattering volumes within the sample cell, and

[14] See discussion by B. Chu and J. A. Duisman, J. Chem. Phys. **46**, 3267 (1967).

[15] J. V. Sengers, Ref. 1, p. 165; Proceedings of the Fourth Technical Meeting of the Society of Engineering Science, 1967 (to be published).

[16] Preliminary results of this experiment were reported by H. L. Swinney and H. Z. Cummins, Bull. Am. Phys. Soc. **12**, 588 (1967); see also a related investigation of the Brillouin components, R. W. Gammon, H. L. Swinney and H. Z. Cummins, Phys. Rev. Letters **19**, 1467 (1967).

[17] E. A. Guggenheim, J. Chem. Phys. **13**, 253 (1945).

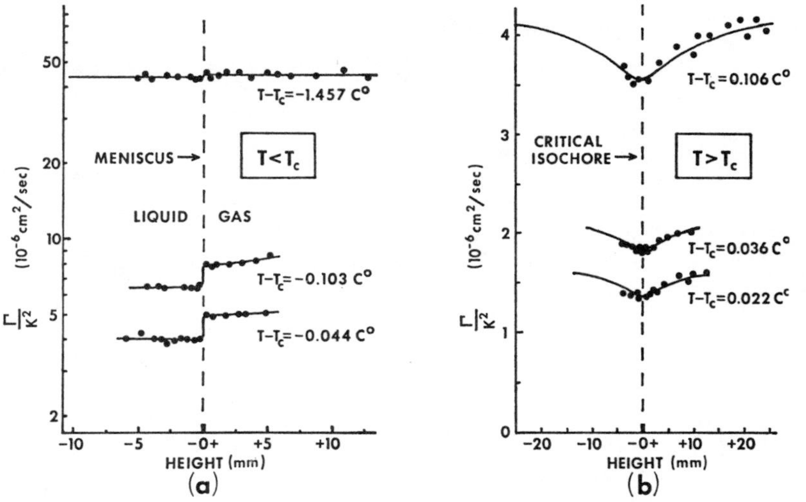

FIG. 1. Γ/K^2 as a function of beam height in the sample cell. The height resolution is equal to the diameter of the focused beam, 0.2 mm. (a) For three temperatures below T_c (height is measured relative to the meniscus height). (b) For three temperatures above T_c (height is measured relative to the height which corresponds to the minimum in Γ/K^2).

measured the depolarization of the scattered light; these tests indicated that multiple scattering effects were not significant in our experiment.

III. EXPERIMENTAL RESULTS

In general, any departure of the spectra from the Lorentzian lineshape was less than our experimental uncertainty. For most of the spectra, the experimental amplitude at any frequency differed from the amplitude of the computer-fit Lorentzian by an amount not greater than 2%, which is approximately equal to the experimental uncertainty. At no temperature or angle were spectra *consistently* obtained with a departure from the Lorentzian shape which was larger than the experimental error; hence the occasional non-Lorentzian spectra were discarded.

The linewidth- (Γ) versus-height data were converted into Γ/K^2-versus-height graphs. [If Eq. (1) is assumed to be valid, $\Gamma/K^2=\chi$.] This required knowledge of the refractive index as a function of height, which was obtained from density-versus-height curves using the Lorentz-Lorenz relation. (Density-versus-height curves were obtained from Straub[18] for a few temperatures and were computed from Fig. 1 of Green, Vicentini-Missoni, and Sengers[19] for other temperatures.)

Figure 1(a) is a graph of Γ/K^2 versus height for three temperatures less than T_c. Figure 1(b) shows Γ/K^2 versus height for three temperatures greater than T_c. Above T_c measurements were not made for heights less than 6 mm below h_c, the height which corresponds to the critical density, because the cell holder blocked the beam. For densities near the critical density, however, $|\rho-\rho_c|$ versus h is symmetrical about h_c,[18,19] and since $\chi_T(\rho-\rho_c)$ has a minimum very near $\rho=\rho_c$, as will be explained later, $\chi_T(\rho-\rho_c)$ is symmetrical about $\rho=\rho_c$ and hence about $h=h_c$. This is indicated in Fig. 1(b) by the extension of the curves to the left beyond the data points. Note that there is a distinct minimum in Γ/K^2 even as far as 0.106C° above T_c. As the critical temperature is approached, the magnitude of the density gradient increases and the minimum in Γ/K^2 as a function of height becomes sharper. At a fixed temperature below T_c the density varies only slightly with height within each phase (e.g., at $T-T_c=-0.049$C°, the density varies only 0.5% for a height variation of 10 mm within each phase above or below the meniscus).[18] Thus Γ/K^2 varies only slightly as a function of height within each subcritical phase. The values of Γ/K^2 on the gas and liquid sides of the coexistence line are given by the limiting value of Γ/K^2 as $h\rightarrow h_{\text{meniscus}}$ within each phase.

IV. FIXMAN'S MODIFICATION

Before attempting to extract the temperature dependence of χ from our Γ/K^2 data, we must consider the validity of Eq. (1) in the critical region.

Fixman has shown that in the immediate neighborhood of the critical temperature the linearized hydrodynamic equations must be modified slightly in order to include the effects of long-range density correlations.[20] Botch has incorporated Fixman's modification in a derivation of the Rayleigh linewidth, and has obtained for the linewidth[21]

$$\Gamma=\chi K^2(1+\xi^2K^2), \tag{2}$$

where ξ is the two-body correlation length. On the critical isochore, $\xi\sim\epsilon^{-\nu}$ as $\epsilon\rightarrow+0$. On the coexistence

[18] J. Straub, Chem. Ingr. Tech. **5/6**, 291 (1967); thesis, Technische Hochschule, München, Germany, 1965 (unpublished).

[19] M. S. Green, M. Vicentini-Missoni, and J. M. H. Levelt Sengers, Phys. Rev. Letters **18**, 1113 (1967). *Note added in proof:* The authors find in a more thorough analysis of the carbon dioxide data alone that $\gamma\approx1.24$ (private communication).

[20] M. Fixman, J. Chem. Phys. **33**, 1357, 1363 (1960); W. D. Botch and M. Fixman, *ibid.* **42**, 199 (1965).

[21] W. D. Botch, Ph.D. dissertation, University of Oregon, 1963, p. 63 (unpublished); see also the discussion and references in Ref. 23.

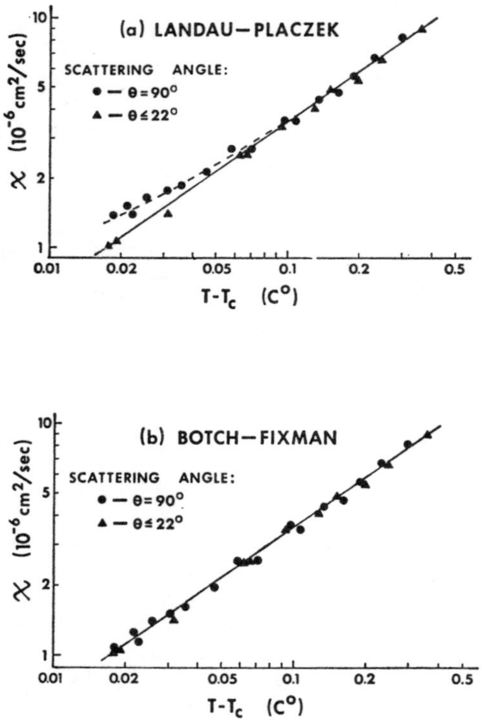

FIG. 2. χ versus $(T-T_c)$ for T above T_c from (a) the Landau-Placzek equation, $\chi=\Gamma/K^2$, and (b) the Botch-Fixman equation, $\chi=(\Gamma/K^2)/(1+\xi^2K^2)$.

line, $\xi\sim(-\epsilon)^{-\nu'}$ as $\epsilon\rightarrow -0$.[2] Thus Eq. (1) is applicable for $K\xi\ll 1$, that is, for sufficiently small $\epsilon^{-\nu}[\sin(\frac{1}{2}\theta)]$ for $T>T_c$, and for sufficiently small $(-\epsilon)^{-\nu'}[\sin(\frac{1}{2}\theta)]$ for $T<T_c$.

In order to explore the possible existence of a Fixman modification term in CO_2, we plotted Γ/K^2 versus $(T-T_c)$ for $\theta=90°$, 22°, and 15° (for $T-T_c\leq 0.4$C°), as shown in Fig. 2(a). Fixman's modification predicts that $\Gamma/K^2=\chi(1+\xi^2K^2)$, so that the Γ/K^2 data will be dependent on the scattering angle. Figure 2(a) exhibits just this effect, since the 90° data curve upward as T approaches T_c.

The temperature range for which we observed a significant Fixman correction was insufficient to deduce ν from our data. For ferromagnets $\nu=\frac{2}{3}$, and for the Ising model $\nu=0.643\pm0.003$.[2,3] With the assumption that $\nu=\frac{2}{3}$ we were able to bring our data into agreement with Fixman's equation by taking $\xi=(0.53\pm0.11)\epsilon^{-2/3}$ Å [$\xi=(150\pm30)$Å at $\epsilon=2\times10^{-4}$]. The result $(\Gamma/K^2)/(1+\xi^2K^2)=\chi$ is plotted in Fig. 2(b). Notice that the correction term becomes unimportant for temperatures over 0.1C° above T_c, where the Landau-Placzek and Fixman equations become essentially equal.

The value of ξ that we have obtained for CO_2 is approximately three times smaller than the value 520 Å obtained by Chu[22] for an isobutyric acid–water mixture at $\epsilon=2\times10^{-4}$; this is qualitatively what one would expect on the basis of angular scattering intensity measurements.[23] However, our ξ is approximately 15 times smaller than the surprisingly large result reported by Yeh for xenon.[24]

V. TEMPERATURE DEPENDENCE OF χ

Following the application of the Fixman modification to the Γ/K^2 data, the resultant values for the thermal diffusivity were plotted as a function of temperature. Values of χ obtained from the limiting values of Γ/K^2 as $h\rightarrow h_{\text{meniscus}}$ [Fig. 1(a)] from either side in the coexistence region (i.e., along the liquid and vapor sides of the coexistence curve) are plotted versus $(T-T_c)$ in Fig. 3. Values obtained from the minima in the Γ/K^2-versus-h curves in the $T>T_c$ region [Fig. 1(b)] are plotted in Fig. 4. The data of Seigel and Wilcox shown in Fig. 4 will be discussed in Sec. VI.

Our data points shown in Figs. 3 and 4 are representative rather than exhaustive; at most temperatures more than one run was made, and on the scale of Figs. 3 and 4 many of those points would superpose.

Note that within the temperature range of our data, there are three thermodynamic data points: At $T-T_c=-1.04$ (on the liquid side of the coexistence line), $T-T_c=1.06$, and $T-T_c=3.76$C°.[25] For those three points the average difference between our values for χ obtained using Eq. (1) and the values for χ measured by classical thermodynamic experiments is 7%. Thus

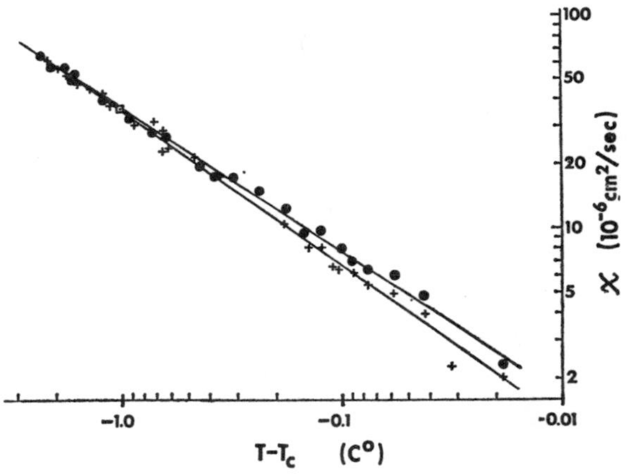

FIG. 3. χ versus $(T-T_c)$ for $T<T_c$. ⊡, thermodynamic data; present experiment: ●, gas; +, liquid. The slope of the lower line, which is for the liquid side of the coexistence curve, is 0.72±0.05; the slope of the upper line, for the gas, is 0.66±0.05. The source of the thermodynamic datum, which is on the liquid side of the coexistence curve, is given in Ref. 25.

[22] B. Chu, Phys. Rev. Letters 18, 200 (1967); J. Chem. Phys. 47, 3816 (1967).

[23] H. Z. Cummins and H. L. Swinney, J. Chem. Phys. 45, 4438 (1966).

[24] Y. Yeh, Phys. Rev. Letters 18, 1043 (1967).

[25] The thermodynamic data used in this paper were compiled by Dr. J. V. Sengers from experiments of the van der Waals Laboratory. The sources of the data for this compilation were given in Refs. 45–51 of R. D. Mountain, Rev. Mod. Phys. 38, 205 (1966).

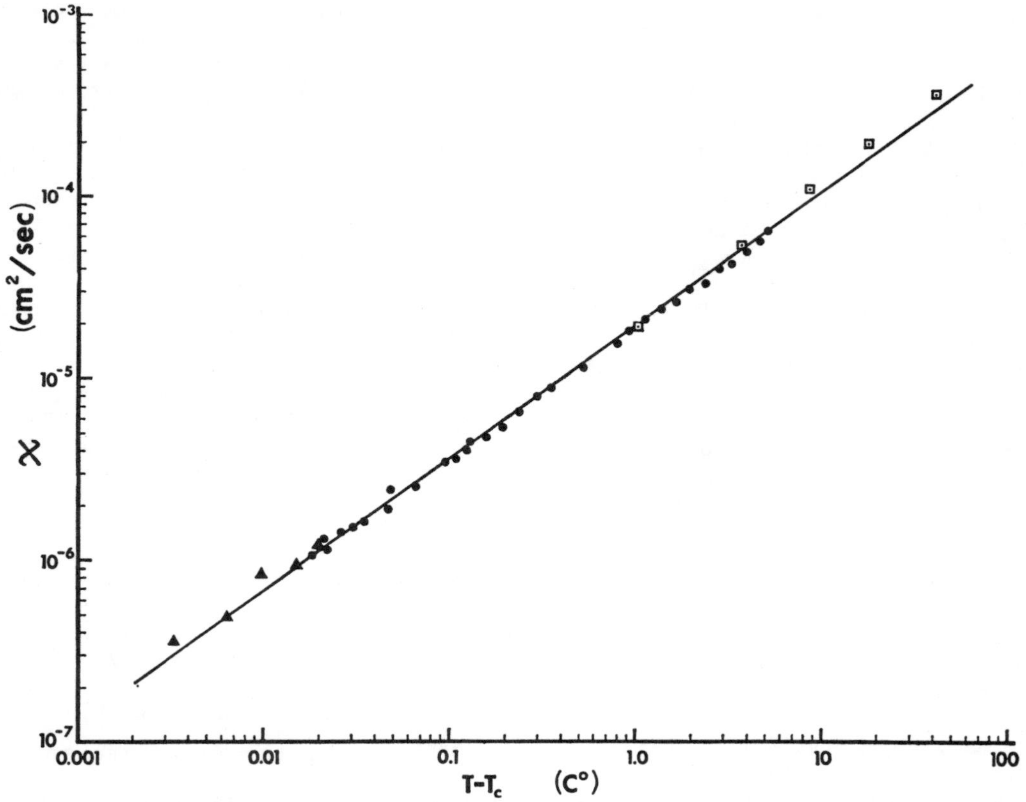

FIG. 4. The thermal diffusivity along the critical isochore versus the difference temperature $T-T_c$. ●, present experiment; ⊡, thermodynamic data; ▲, Seigel and Wilcox. The source of the thermodynamic data is given in Ref. 25; the data of Seigel and Wilcox are from Ref. 28. From a least squares fit of the data from the present experiment to a straight line, the slope is 0.73 ± 0.02. The slope given by combining the three independent sets of data is also 0.73 ± 0.02.

the Landau-Placzek equation [Eq. (1)] appears valid in the critical region, at least for temperatures not too close to T_c.

The exponents for the dependence of χ on $|\epsilon|$ along the critical isochore and along the two sides of the coexistence line are given by the magnitudes of the slopes of the curves in Figs. 3 and 4. Using as the critical temperature the temperature at which the meniscus appeared as the temperature was lowered, a least-squares analysis of our data yields for the thermal diffusivity on the critical isochore, $\chi=(18.1\pm0.5)\times10^{-6}(T-T_c)^{0.73\pm0.02}$ cm²/sec. The results along the liquid and gas sides of the coexistence line are $\chi_L=(34.8\pm2.5)\times10^{-6}(T_c-T)^{0.72\pm0.05}$ cm²/sec and $\chi_G=(36.0\pm3.0)\times10^{-6}(T_c-T)^{0.66\pm0.05}$ cm²/sec. The value for T_c was varied in an attempt to improve the least-squares fit of the data, but no improvement was obtained by using a critical temperature different from that at which the meniscus appeared.

We will now give some justification for our assertion that the minima in our χ-versus-h graphs correspond to χ at densities very near $\rho=\rho_c$. The maximum in the isothermal compressibility as a function of density does not occur at $\rho=\rho_c$ (see discussion by Widom[26]); however, near T_c, the difference between the critical density and the density corresponding to the maximum in κ_T is very small. Also, near T_c the isotherms are so flat that the maximum compressibility and the compressibility at $\rho=\rho_c$ are essentially equal. Our interest is only for the range $T-T_c\lesssim0.2$C°, since only this near T_c do our χ-versus-h data exhibit discernible minima. From the above comments on κ_T and from Sengers's thermodynamic data for λ versus ρ,[27] we conclude that for $T-T_c\lesssim0.2$ the minima in $\lambda/\kappa_T\sim\chi$ occur at densities within 0.4% of the critical density.

VI. COMPARISON WITH OTHER RESULTS

A. Comparison with Other Experiments

The classical thermodynamic measurements of the thermal diffusivity of CO_2 yield, in a least-squares fit,

[26] B. Widom, J. Chem. Phys. **43**, 3898 (1965).

[27] See Ref. 15 and J. V. Sengers, thesis, van der Waals Laboratory, Amsterdam, 1962 (unpublished); A. Michels, J. V. Sengers, and P. S. van der Gulik, Physica **28**, 1201, 1216 (1962); A. Michels and J. V. Sengers, *ibid.* **28**, 1238 (1962); J. V. Sengers, J. Heat Mass Transfer **8**, 1103 (1965).

the exponent $\gamma-\psi=0.78$ for the temperature range $1.06 \leq T-T_c \leq 44C°$.[25] Linewidth determinations of χ for CO_2 reported by Seigel and Wilcox give $\gamma-\psi=0.7\pm0.1$ for $0.0033 \leq T-T_c \leq 0.020C°$.[28] (Using our ξ we compute that for the angles and temperatures they investigated, the Landau-Placzek equation is accurate within a few percent.) Recent linewidth measurements by Osmundson and White have also given $\gamma-\psi=\frac{2}{3}$ for CO_2.[29] Combining our thermal diffusivity data with the thermodynamic data and the Seigel and Wilcox data, we have $\gamma-\psi=0.73\pm0.02$ for $0.0033 \leq T-T_c \leq 44C°$, a range of four orders of magnitude in the difference temperature (see Fig. 4).

Three recent experiments in binary mixtures have also given an exponent of less than one for the dependence of the linewidth upon $T-T_c$. Chu[30] found an exponent of 0.67 for the mixture isobutyric acid and water; in an aniline-cyclohexane mixture Berge and Volochine[31] have obtained an exponent of 0.55 ± 0.05; and Chen and Polonsky[32] obtained an exponent of 0.65 ± 0.05 for the mixture *n*-hexane and nitrobenzene. In measurements of the relaxation time for the concentration fluctuations in the critical mixture nitrobenzene in iso-octane, Gravatt has also obtained an exponent of less than 1 (0.90 ± 0.05) for the dependence of the reciprocal relaxation time upon $T-T_c$.[33] In contrast to these results for four binary mixtures and for carbon dioxide, Saxman and Benedek have found $\gamma-\psi=1.27$ for sulphur hexafluoride.[34] However, for $T<T_c$, Saxman and Benedek obtained $\gamma'-\psi'=0.64$, in agreement with our result for carbon dioxide.

We will now discuss the data for γ and ψ obtained in separate experiments in order to compare our result for $\gamma-\psi$ with the expected temperature dependence of the thermal diffusivity. Green, Vicentini-Missoni, and Sengers have concluded from an analysis of the data from experiments for a variety of gases that $\gamma=\gamma'=1.4$.[19] Heller has analyzed the data from several experiments on carbon dioxide and has concluded that $\gamma=1.35\pm0.15$ and $\gamma'=1.1\pm0.4$.[4,35]

The question of whether or not the thermal conductivity has an anomaly in the critical region was a subject of controversy for thirty years, but the controversy was resolved by some very careful thermodynamic experiments by Sengers, Michels, and van der Gulik.[27] They found that the thermal conductivity does indeed exhibit a pronounced anomaly in the critical region. The nature of the singularity in λ has been investigated by Sengers, who compared λ and c_v data and concluded that, to a first approximation, λ diverges as c_v on the critical isochore.[15,36] That is, $0 \leq \psi \leq 0.2$.

Combining the data for γ and ψ from the experiments cited, we can now compare the resultant $\gamma-\psi$ with our value for $\gamma-\psi$ (the subcritical thermodynamic data are insufficient to permit a comparison with our data). It appears likely that $\gamma=1.3\pm0.1$ and that c_v has a weak singularity (logarithmic to 0.1); thus the assumption that λ behaves as c_v leads to a lower bound of 1.1 for $\gamma-\psi$, far larger than our $\gamma-\psi=0.73$.[37] We are led to the conclusion that the singularity in the thermal conductivity is even greater than that indicated by thermodynamic experiments. For $\gamma=1.3\pm0.1$ our data require that $\psi=0.57\pm0.12$. Possibly $\gamma=\frac{5}{4}$, $\psi=\frac{1}{2}$, and $\gamma-\psi=\frac{3}{4}$. Within three times our experimental error our data also allow $\gamma=\frac{4}{3}$, $\psi=\frac{2}{3}$, and $\gamma-\psi=\frac{2}{3}$.

B. Comparison with Theory

Theoretical investigations of systems near the critical point have very recently been extended to the dynamical properties. Kawasaki, Mountain, Zwanzig, Deutch, and Fixman have applied correlation function techniques to predict the behavior of transport coefficients in the critical region.[38–41] Several investigators, Ferrell *et al.*, Halperin and Hohenberg, and Kadanoff and Swift, have extended scaling-law techniques to dynamic phenomena.[42–44]

Kadanoff and Swift predict that $\lambda \sim \rho c_p \xi^{-1}$, so that $\chi=\lambda/\rho c_p \sim \xi^{-1}$.[44] Thus, measurements of $\chi(\rho,T)$ yield the behavior of ξ^{-1} as a function of density and temperature. Hence our graphs of χ_T versus height, as in Fig. 1, can

---

28 L. Seigel and L. R. Wilcox, Bull. Am. Phys. Soc. **12**, 525 (1967).

29 J. S. Osmundson and J. A. White, Bull. Am. Phys. Soc. **13**, 183 (1968); B. Maccabee and J. A. White, *ibid.* **13**, 182 (1968).

30 B. Chu (private communication); see also Ref. 22.

31 P. Berge and B. Volochine, Phys. Letters **26A**, 267 (1968).

32 S. H. Chen and N. Polonsky (private communication); see also Bull. Am. Phys. Soc. **13**, 183 (1968).

33 C. C. Gravatt, Phys. Rev. Letters **18**, 948 (1967); see also P. Debye, C. C. Gravatt, and M. Ieda, J. Chem. Phys. **46**, 2352 (1967).

34 G. B. Benedek, in *Brandeis University Summer Institute in Theoretical Physics, 1966 Lectures*, edited by M. Chretien, S. Deser, and E. P. Gross (Gordon and Breach Science Publishers, Inc., New York, 1968), Vol. 2; A. C. Saxman and G. B. Benedek (to be published).

35 L. R. Wilcox and D. Balzarini have recently obtained the result $\gamma=1.0$ for xenon and carbon dioxide by using a novel forward-scattering optical interference technique: for xenon, J. Chem. Phys. **48**, 753 (1968); for carbon dioxide, the result is preliminary (private communication). However, see Bull. Am. Phys. Soc. **13**, 579 (1968).

36 The thermodynamic specific-heat data used in that analysis are not sufficiently accurate to make a reliable conclusion concerning the critical exponent. J. V. Sengers (private communication).

37 If $\gamma=1.0$ as found by Wilcox and Balzarini (Ref. 35), and $\psi=0.2$, then $\gamma-\psi=0.8$, in fair agreement with our data.

38 K. Kawasaki, Phys. Rev. **145**, 224 (1966); **148**, 375 (1966); **150**, 291 (1966). In the last of these three references it is shown that the thermal conductivity is not anomalous in the critical region, but Kadanoff and Swift (Ref. 44) have pointed out that Kawasaki's proof fails when sound waves are allowed.

39 J. M. Deutch and R. Zwanzig, J. Chem. Phys. **46**, 1612 (1967).

40 R. D. Mountain and R. Zwanzig, J. Chem. Phys. **48**, 1451 (1968).

41 M. Fixman, J. Chem. Phys. **47**, 2808 (1967). Fixman's estimate for λ_0 for CO_2($\lambda_0 \approx \frac{1}{2}\lambda$ at $T-T_c=3.8C°$) appears high, for our data accurately follow a straight line on log-log plot throughout the temperature range investigated, $0.018 \leq T-T_c \leq 5.3C°$.

42 R. A. Ferrell *et al.*, Phys. Letters **24A**, 493 (1967); Phys. Rev. Letters **18**, 891 (1967).

43 B. I. Halperin and P. C. Hohenberg, Phys. Rev. Letters **19**, 700 (1967).

44 L. P. Kadanoff and J. Swift, Phys. Rev. **165**, 310 (1968); **166**, 89 (1968).

be viewed as isotherms of the reciprocal correlation length. Further, since $\chi\sim\xi^{-1}$, the exponent for the divergence of the thermal diffusivity on the critical isochore, $\gamma-\psi$, equals ν. Our result is then $\nu=0.73\pm0.02$, somewhat higher than either the Ising model $\nu=0.643\pm0.003$ or the experimental result for ferromagnets and antiferromagnets, $\nu=0.65\pm0.03$.[2] There exist no data for ν in pure fluids; however, measurements of either the Rayleigh linewidth or the scattering intensity as a function of angle at various temperatures will yield $\xi(T)$ and hence ν. In our experiment the $K\xi$ term in Eq. (2) was not measurable over a sufficient temperature range to deduce ν from the data, but such measurements are quite feasible.

According to the Kadanoff and Swift prediction, $\chi\sim\xi^{-1}$, the exponent that we have determined along the coexistence line is ν'. There exist practically no experimental data for ν' with which we could compare our results $\nu_G'=0.66\pm0.05$ and $\nu_L'=0.72\pm0.05$. While the scatter in our data dictate the uncertainty ±0.05, it is clear that the data indicate $\nu_L'>\nu_G'$ if ν' is the exponent for the dependence of the total thermal diffusivity on $T-T_c$. This inequality in the gas and liquid exponents can be understood from Fig. 1(a), where very near T_c, $\chi_L<\chi_G$, while further from T_c ($T-T_c\approx-2$C°), $\chi_L=\chi_G$. However, if $\chi=\chi_0+\chi'$, where χ_0 is temperature-independent (and χ_0 may have different values in the gas and liquid), and χ' is the anomalous part of the thermal diffusivity, then it is possible that the scaling-law result $\nu_L'=\nu_G'=\nu$ and our thermal-diffusivity data are both correct. For then a log-log plot of the total thermal diffusivity versus $(T-T_c)$, such as our Fig. 3, could yield $\nu_L'\neq\nu_G'$ very near T_c, even if χ_L' and χ_G' behaved in the same way. Our result $\nu_L'(=\gamma_L'-\psi_L')>\nu_G'(=\gamma_G'-\psi_G')$ is in the same direction as the difference observed by Roach for γ' in the gas and liquid along the coexistence line of He^4: $\gamma_L'=1.22$ and $\gamma_G'=1.07$.[45]

Using time-dependent correlation functions obtained by assuming that the free energy depends quadratically on the density gradient, Fixman calculates that on the critical isochore, $\lambda'\sim\epsilon^{-1/2}$, where $\lambda=\lambda_0+\lambda'$, and λ_0 and λ' are, respectively, the temperature-independent and anomalous parts of the thermal conductivity.[41] For temperatures sufficiently near T_c, $\lambda'\gg\lambda_0$, and then Fixman's prediction becomes $\psi=0.5$. Mountain and Zwanzig have also obtained $\psi=0.5$ in a calculation of the thermal conductivity for a van der Waals gas using the time correlation function method.[40] If we combine the prediction $\psi=0.5$ with the Ising model γ, we have $\gamma-\psi=0.75$, in agreement with our result.[46]

[45] P. R. Roach, Phys. Rev. **170**, 213 (1968).

[46] The result $\psi=0.5$ was obtained by Mountain and Zwanzig for a van der Waals gas, so their result should properly be combined with the classical $\gamma=1.00$, which yields $\gamma-\psi=0.5$. On the other hand, the exponents given by the Ising model, for which there is no result for ψ, are in better agreement with experiment than the classical exponents. Hence we have combined the Ising $\gamma=1.25$ with the prediction $\psi=0.5$.

Swift[47] has predicted that the diffusion coefficient D for a binary mixture should behave in the critical region as $D\sim\xi^{-1}$, which is the same critical behavior that Kadanoff and Swift obtained for the thermal diffusivity of a pure fluid. Since the Rayleigh linewidth in a binary mixture is given by Eq. (2) with χ replaced by D, the temperature dependence of the linewidth in mixtures and pure fluids should be described by the same exponent. The agreement between the exponents obtained for binary mixtures[30–32] and for carbon dioxide support Swift's prediction.

In conclusion, our results are in reasonable agreement with the behavior that Kadanoff and Swift have predicted for the thermal diffusivity, $\chi\sim\xi^{-1}$. Our data are also in accord with the $\lambda\sim\epsilon^{-1/2}$ behavior predicted by Fixman and by Mountain and Zwanzig, if this prediction is combined with the Ising model γ. Thus the predictions of a strong critical point singularity in the thermal conductivity are substantiated by our data.

VII. SUMMARY

The Landau-Placzek equation for the linewidth is accurately obeyed in the critical region for temperatures not too close to T_c, but for $\epsilon\lesssim10^{-4}$, the correlation length in carbon dioxide is sufficiently long-range to require that the Botch-Fixman equation be used in order to describe correctly the linewidth behavior. The magnitude of the Fixman correction that we observe in carbon dioxide is far smaller than seen by Chu in isobutyric acid and water or by Yeh in xenon.

From our data the exponent for the dependence of χ upon ϵ along the critical isochore is 0.73 ± 0.02 for a range of more than two orders of magnitude in ϵ; our data combined with the thermodynamic data and the data of Seigel and Wilcox yields an exponent 0.73 ± 0.02 for four orders of magnitude: $1.1\times10^{-5}\leq\epsilon\leq1.5\times10^{-1}$. Below T_c, our exponent for the gas side of the coexistence line is 0.66 ± 0.05, and for the liquid, 0.72 ± 0.05.

Our exponents both above and below T_c are in reasonable accord with the recent prediction by Kadanoff and Swift that $\chi\sim\xi^{-1}$. Our data are also in agreement with the divergence in the thermal conductivity predicted by Fixman and by Mountain and Zwanzig, if the compressibility diverges with an exponent $\gamma\approx1.25$.

Our results give further verification of the anomalous behavior of the thermal conductivity in the critical region, as observed by Sengers *et al.* In addition, our results indicate a stronger singularity in the thermal conductivity than has been assumed heretofore. The exponent for the divergence is likely in the range $0.5\leq\psi\leq0.7$.

Measurements of the scattered light intensity as a function of ρ and T along with linewidth measurements such as those reported here would yield both γ and $\gamma-\psi$ and thus ψ. Further, the exponent ν could be

[47] J. Swift (to be published).

obtained from measurements of the linewidth or intensity as a function of θ and T very near T_c. Although carbon dioxide was in a sense a fortunate choice for this experiment since it has been studied more thoroughly than any other gas near its critical point, it will be interesting to have these results for a monatomic gas, for then there will be no possibility that the results will be influenced by the internal degrees of freedom.

ACKNOWLEDGMENTS

We are grateful to Donald L. Henry for assistance in the data collection. We wish to thank Dr. J. V. Sengers and Dr. R. D. Mountain for some helpful discussions, and Dr. Sengers for the compilation of thermodynamic data for carbon dioxide used in this paper. We also wish to acknowledge helpful conversations with L. Seigel, D. Balzarini, and Professor L. R. Wilcox.

CRITICAL-REGION SECOND-SOUND DAMPING IN He II

J. A. Tyson*
Department of Physics and The James Franck Institute, The University of Chicago, Chicago, Illinois 60637
(Received 5 August 1968)

Second-sound attenuation very near T_λ is found to be proportional to the square of the frequency, and the damping diverges as $(T_\lambda - T)^{-\sigma}$, $\sigma = 0.34 \pm 0.06$, in agreement with dynamic scaling. The coefficient of the damping is consistent with recent estimates of the coherence length.

Recently Ferrell et al.[1] and others[2,3] have predicted an asymptotic power-law divergence of the second-sound damping coefficient $D_2 \sim (T_\lambda - T)^{-1/3}$ in the critical region of He II on the basis of "dynamic scaling." Extending the static scaling ideas[4] involving a single coherence length $\xi(T) \sim (T_\lambda - T)^{-2/3}$, they describe the dynamics near T_λ in terms of a single characteristic frequency $\omega^* = u_2 \xi^{-1}$, where $u_2(T)$ is the second-sound velocity. For frequencies ω such that $\omega < \omega^*$ and $k\xi \lesssim 1$, and for temperatures within the critical region, the above asymptotic form for the second-sound damping $D_2(T)$ results. In order to test the dynamic scaling hypothesis, we have measured the damping of second sound in the macroscopic critical region $k\xi \ll 1$ between 7×10^{-5} and 3×10^{-3} deg below T_λ and for $\omega < \omega^*$. In addition, we report measurements of the frequency dependence of the corresponding anomalous attenuation very near T_λ.

The second-sound wave is generated in a low-power cw resonant cavity which is excited at various harmonics, and the width of the resonance at each harmonic is measured by quadrature phase-sensitive detection. Absolute measurements of $T_\lambda - T$ accurate to $\pm 10^{-6}$ deg are obtained from the observed $u_2(T)$ at each point. The cavity is immersed in a superleak-purified He^4 bath which is enclosed within another bath for better temperature regulation and stability. Power input density to the cavity is monitored and kept typically less than 2×10^{-5} W/cm^2 in order to avoid nonlinear effects near T_λ. For several power input levels the resonance widths are time-averaged, and the limiting zero-power resonance width at each harmonic is obtained for various temperatures within the critical region. To test for possible systematic effects, two different cavities are used. The cavities are formed by spacing flat lavite or quartz surfaces with fused quartz spacers, and the temperature wave is generated and detected by thin films of silver and carbon. In addition to bulk attenuation, reflection loss and diffraction (beam spreading) loss give rise to enlarged resonance widths. These other effects are also investigated as a function of power, frequency, and temperature, and corrections made to the observed resonance widths—introducing a corresponding error. The second-sound damping coefficient $D_2(T)$ is defined in terms of the bulk attenuation $\alpha(\omega, T)$:

$$\alpha(\omega, T) = (\omega^2/2u_2^3)D_2(T), \tag{1}$$

where classical superfluid hydrodynamics[5] indicates that D_2 is frequency independent. This is also true of the dynamic scaling theories for $k\xi \ll 1$. The observed resonance width $\Delta\omega_n$ (width at half-amplitude) corresponding to mode number n for a cavity of this type ($\omega_n = n\omega_1$) is given by the sum of three terms:

$$\frac{\Delta\omega_n}{\sqrt{2}} = \frac{\pi^2 n^2}{d^2} D_2 + \frac{2\beta}{\pi}\omega_1 + \frac{4d^2}{\pi n d_0^2}\omega_1, \tag{2}$$

where the first term corresponds to the bulk attenuation; the second and third terms correspond to reflection loss and diffraction loss, respectively. $\beta = -\ln R$ where R is the reflection coefficient, and $\omega_1(T)$ is the fundamental resonant frequency ($n = 1$). The dimensions of the cavities (width d_0, spacing d) were chosen to minimize any corrections due to diffraction. The measured reflection coefficient of the lavite cavity is 0.94 and for the quartz cavity 0.97. This agrees well with other measurements[6] in cw cavities, at lower temperatures. Equation (2) shows that this correction to $\Delta\omega_n$ varies with temperature through $\omega_1(T)$. At $n = 20$, this reflection-loss correction to the observed resonance widths varies from 10% at 10^{-4} deg below T_λ to 50% at 3×10^{-3} deg below T_λ. The power dependence of the width (for 10-30 μW/cm^2) is found to be linear and is apparently singular at T_λ; $\partial\Delta\omega_1/\partial P$ varies between 0.3 Hz $(\mu\text{W/cm}^2)^{-1}$ at 10^{-4} deg below T_λ and 4×10^{-2} Hz $(\mu\text{W/cm}^2)^{-1}$ at 10^{-3} deg below T_λ. This is in good agreement with other measurements[6,7] at higher powers and lower temperatures. $\partial\Delta\omega_n/\partial P$ increases approximately linearly with frequency, and approaches zero for $P < 10\ \mu W/cm^2$ within the critical region. The small error in extrapolating $\Delta\omega_n$ to zero power is used in the determination of total error. Equation (2) indicates that data for $\Delta\omega_n$ over a sufficient range of both n and $\omega_1(T)$ give data for $D_2(T)$. As a check the third term may be calculated, and the resulting values agree with the measured diffraction data to better than 10%. However, this diffraction correction is negligible over the temperature region covered in the $D_2(T)$ data.

The classical ω^2 dependence of the bulk attenuation $\alpha(\omega,T)$ in Eq. (1) may be checked near T_λ by observing $\Delta\omega_n$ for various n near T_λ where $\omega_1(T)$ is small. Figure 1 shows the observed zero-power resonance widths in the quartz cavity, for $T=1.6\times10^{-4}$ deg below T_λ and $\omega_1/2\pi=270$ Hz, plotted versus n^2. The n^2 dependence of $\Delta\omega_n$ at large n (where β is wave-number independent) seems to be valid near T_λ, indicating that D_2 is n-independent over this range of frequencies, in agreement with both classical[5] and scaling[1-3] theories. The intercept at $n=1$ of the large-n asymptote gives the reflection loss; here $\beta=3\%$. The slope of the asymptote is proportional to the damping $D_2(T)$. For each such plot of $\Delta\omega_n$ vs n^2, the corrected value of $\Delta\omega_{20}$ is obtained and the damping D_2 computed. For temperatures farther below T_λ, where $\omega_1(T)$ is large, the percent correction to $\Delta\omega_{20}$ due to reflection loss becomes larger, and the error in $D_2(T)$ becomes relatively large for $(T_\lambda-T)\gtrsim5\times10^{-3}$ deg K. Diffraction corrections become important only for $(T_\lambda-T)$ $>10^{-2}$ deg K. Crosstalk plus noise limit observation to $(T_\lambda-T)\gtrsim5\times10^{-5}$ deg K. From a series of graphs like Fig. 1, the damping $D_2(T)$ is obtained as a function of temperature in the critical region.

Figure 2 shows the data for the second-sound damping coefficient $D_2(T)$ for several temperatures within the critical region. Also included in Fig. 2 are data of Hanson and Pellam[8] extending out of the critical region which were obtained using a different method, and two points of Notarys[7] from megahertz cw cavity-resonance data. Thus, both relative and absolute damping may be compared. If our data are fitted by a power-law function, we obtain

$$D_2=D_0(1-T/T_\lambda)^{-\sigma}, \tag{3}$$

with $\sigma=0.34\pm0.06$ and $D_0=1.02\times10^{-4}\ cm^2\ sec^{-1}$. The extrapolated value of $D_2(T)$ to the edge of the critical region overlaps the absolute attenuation measurements of Hanson and Pellam[8] and the resonance data of Notarys.[7] If these additional data are used in the fit with Eq. (3), reduced error results ($\sigma=0.34\pm0.03$), but this procedure is risky, because of possible differential systematic errors. Outside the critical region, $D_2(T)$ should level out and approach the nonsingular Khalatnikov value[8,9] at lower temperatures. This is apparent in Fig. 2 for $(T_\lambda-T)>4\times10^{-2}$ deg K. In addition to the linewidth data, measurements of the signal amplitude were taken at each point. These data, when scaled to input power, are consistent with all the resonance-width results.

The experimentally determined exponent σ $=0.34\pm0.06$ is not an asymptotic value. It is the apparent exponent for the region between 7×10^{-5}

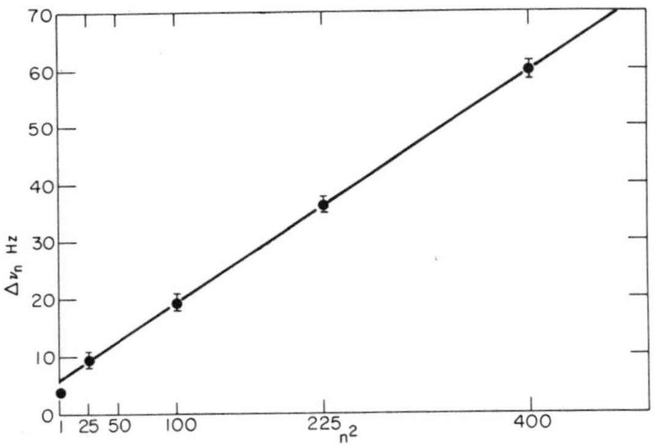

FIG. 1. Plot of the observed resonance width of the nth harmonic at half-amplitude, $\Delta\omega_n/2\pi$, versus the square of the harmonic number n^2 for a single temperature 1.6×10^{-4} deg below T_λ. The ω^2 dependence of $\Delta\omega_n$ is predicted by classical superfluid hydrodynamics and by dynamic scaling for $k\xi\ll1$. The zero-frequency ($n=0$) intercept is proportional to the reflection loss. Slope is proportional to the damping D_2.

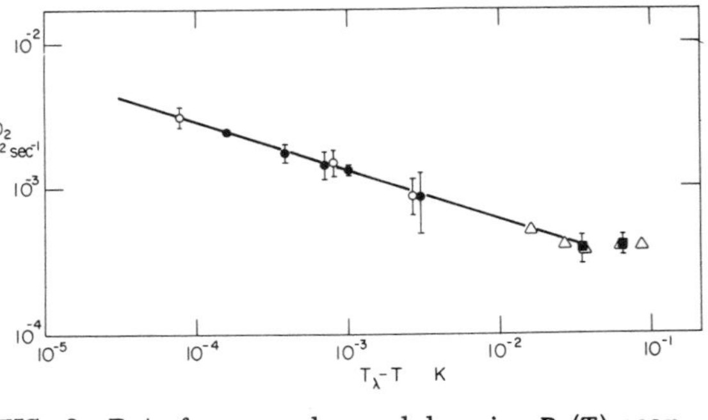

FIG. 2. Data for second-sound damping $D_2(T)$ near T_λ. Our data: solid circles, quartz cavity; open circles, lavite cavity. Data of Hanson and Pellam: open triangles. Notarys' data: solid squares. Outside the critical region, for $(T_\lambda-T)>4\times10^{-2}$ deg, $D_2(T)$ reaches the Khalatnikov minimum. Solid line has slope $-\frac{1}{3}$.

and 4×10^{-2} deg below T_λ. Although Ferrell et al.[1] predict $D_2(T)\sim\xi^{1/2}$, which $\sim(T_\lambda-T)^{-1/3}$ for all T in the critical region, more recent dynamic scaling theories[2,3] indicate $D_2\sim u_2\xi\sim(T_\lambda-T)^{-\sigma}$ where $\sigma=\sigma(T)$ because of the presence of C_p in the second-sound velocity[10]: $u_2\sim(T_\lambda-T)^{\Phi(T)}$ yields the theoretical prediction $D_2\sim(T_\lambda-T)^{-0.28}$ for $5\times10^{-5}<(T_\lambda-T)<5\times10^{-3}$ deg K. In summary, both dynamic-scaling theoretical estimates for the exponent of second-sound damping ($\sigma=0.33, 0.28$) agree to within experimental error with our result $\sigma=0.34\pm0.06$.

The coefficient D_0 in Eq. 3 is also interesting because, according to dynamic scaling, it contains information about the coherence length coefficient ξ_0 in the relation $\xi(T)=\xi_0(1-T/T_\lambda)^{-2/3}$. Ferrell and co-workers' rough calculation[1] of $D_0\approx10^{-5}$ cm^2 sec^{-1} is far from the observed value. However, they calculate D_0 for the limit $\omega>\omega^*$, where the dispersion relation $\omega_k\sim k^{3/2}$ obtains, and also use a value for ξ_0 (0.27 Å) which is $\frac{1}{5}$ the latest estimate[11,12] $\xi_0=1.2$ Å. We repeat this calculation of D_0 for the conditions $k\xi\ll1$, $\omega<\omega^*$ obtaining in this experiment. Using the same frequency scaling idea[1-3] for the limit $\omega<\omega^*$, $\omega_k\sim k+\cdots$, a similarly rough calculation using[1] $\langle k\rangle\approx1/\xi$ and $D_2k^2\approx2\omega^*$, with $\omega^*=u_2\xi^{-1}$ gives $D_0\approx10^{-4}$ cm^2 sec^{-1}, closer to the experimental value.

The author wishes to acknowledge stimulating discussions with P. C. Hohenberg and M. E. Fisher, and the continued interest and assistance of D. H. Douglass, Jr., and G. Johnson. This research was supported in part by the Air Force Office of Scientific Research, The National Science Foundation, and the Advanced Research Projects Agency.

---

*National Research Council–U. S. Air Force Office of Scientific Research Postdoctoral Fellow.

[1]R. A. Ferrell, N. Menyhárd, H. Schmidt, F. Schwabl, and P. Szépfalusy, Phys. Rev. Letters 18, 891 (1967).

[2]B. I. Halperin and P. C. Hohenberg, Phys. Rev. Letters 19, 700 (1967), and to be published.

[3]J. Swift and L. P. Kadanoff, to be published.

[4]B. Widom, J. Chem. Phys. 43, 3892, 3898 (1965); L. P. Kadanoff, Physics 2, 263 (1966); B. D. Josephson, Phys. Letters 21, 608 (1966).

[5]I. M. Khalatnikov, Introduction to the Theory of Superfluidity (W. A. Benjamin, Inc., New York, 1965).

[6]K. N. Zinov'eva, Zh. Eksperim. i Teor. Fiz. 31, 31 (1956) [translation: Soviet Phys.–JETP 4, 36 (1957)].

[7]N. A. Notarys, thesis, California Institute of Technology, 1964 (unpublished).

[8]W. B. Hanson and J. R. Pellam, Phys. Rev. 95, 321 (1954).

[9]I. M. Khalatnikov, Zh. Eksperim. i Teor. Fiz. 20, 243 (1950), and 23, 34 (1952).

[10]J. A. Tyson, thesis, University of Wisconsin, 1967 (unpublished); C. J. Pearce, J. A. Lipa, and M. J. Buckingham, Phys. Rev. Letters 20, 1471 (1968); J. A. Tyson and D. H. Douglass, Jr., in the Proceedings of the Eleventh International Conference on Low Temperature Physics, St. Andrews, Scotland (to be published).

[11]J. A. Tyson, to be published.

[12]M. E. Fisher, private communication.

Fluctuations and Correlations in $SrTiO_3$ for $T \geqslant T_c$

Th. von Waldkirch, K. A. Müller, and W. Berlinger
IBM Zurich Research Laboratory, CH-8803 Rüschlikon, Switzerland

and

H. Thomas
Institut für Theoretische Physik, University of Frankfurt, 6 Frankfurt am Main, Germany
(Received 29 December 1971)

The linewidth broadening of the paramagnetic resonance of the Fe^{3+}-V_O center in $SrTiO_3$ near T_c has been measured accurately and analyzed for $T \to T_c^+$. It is found that the order-parameter susceptibility $\chi(\vec{q}, T)$ is two-dimensionally anisotropic with an anisotropy parameter $\Delta = \frac{1}{60}$. The critical exponent of the correlation length ξ is found to be $\nu = 0.63 \pm 0.07$. The EPR linewidth is cusp shaped for $T \to T_c^+$ and finite at $T = T_c$.

Recently it was found that the local fluctuations near the second-order phase transition in $SrTiO_3$ can be probed by the temperature-dependent linewidth broadening of the paramagnetic-resonance linewidth of the Fe^{3+}-V_O center,[1] i.e., an Fe^{3+} ion with a nearest-neighbor oxygen vacancy.[2] This broadening occurs approximately over the same temperature interval in which Blazey[3] observed simultaneously two electronic band transitions, and over which for $T \geqslant T_c$ Riste *et al.*[4] found a central mode with neutron diffraction. Schwabl[5] has developed a dynamical theory for $T \geqslant T_c$ which yields the central peak and the soft mode. He assumes the following form for the static order-parameter susceptibility for $\hat{\varphi}_{\vec{q}}^{\alpha}$ near $\vec{q}_R = a^{-1}(\pi, \pi, \pi)$:

$$\chi^{\alpha\alpha}(\vec{q}, \epsilon) = \chi_0[\vec{q}^{\,2} - (1-\Delta)q_\alpha^{\,2} + \kappa^2]^{-1+\eta/2}, \qquad (1)$$

where $\hat{\varphi}_{\vec{q}}^{\alpha}$ is the Fourier transform of the rotation angle $\varphi_{\vec{l}}^{\alpha}(t)$ at lattice site $\vec{l}$ of the (BO_6) octahedra around the direction $\alpha = [100]$, $[010]$, $[001]$; χ_0 is proportional to the single-particle susceptibility; and $\kappa(\epsilon) = \kappa_0\epsilon^\nu$ is the inverse of the correlation length $\xi(\epsilon)$. The reduced temperature $(T - T_c)/T_c$ is denoted by ϵ; ν and η are critical exponents.[6] The temperature-independent parameter Δ describes the anisotropy of the dispersion of $\chi^{\alpha\alpha}$ near $\vec{q}_R$. Using Eq. (1) Schwabl has obtained an expression for the linewidth broadening for $T \to T_c^+$ for fast fluctuations,

$$\Delta H(\epsilon) \propto \frac{\kappa(\epsilon)^{-(1-2\eta)}}{\sqrt{\Delta}} \arctan\left(\frac{\pi\sqrt{\Delta}}{\kappa(\epsilon)a}\right), \qquad (2)$$

where $a = 3.9$ Å is the lattice constant.

In the present note we report on high-accuracy measurements of $\Delta H(T)$ which allowed us to determine ν and the anisotropy constant Δ which were so far unknown. The results obtained agree well with Eq. (2) from $T_c + 1°$ to $T_c + 46°$, where $\Delta H \simeq 0$. However, for $T \sim T_c^+$ there occurs a critical slowdown in the fluctuations and the linewidth does not diverge. We show theoretically and experimentally that $\Delta H(T)$ is cusp shaped,[6] behaving as $\Delta H^2(\epsilon) = \Delta H_{max}^2 (1 - C\epsilon^\nu)$ within 0.7° from T_c.

The influence of the rotation angle $\varphi_{\vec{l}}^{\alpha}$ on the resonance field H_r of the Fe^{3+}-V_O complex has been analyzed for all orientations of the center axis and directions of α and H_r.[7] It is found that for $\vec{H}$ in the (001) plane between $[100] + 10°$ and

[100] + 60° the resonance field of the center with axis parallel to [100] (high-field line) is linearly sensitive to the $\varphi^{[001]}$ component. The $\varphi^{[010]}$ component enters only quadratically, and its fluctuation contribution to the linewidth is of the order of 3% of the maximum linear $\varphi^{[001]}$ effect for $\vec{H}$ parallel to [110]. The quadratic $\varphi^{[010]}$ influence has been observed separately by the broadening of the low-field $g \sim 6$ line[7] with $\vec{H}$ parallel to [100] which shows no linear $\varphi^{[001]}$ contribution but is comparable in quadratic dependence. The result confirms the theoretical estimate. The measurements of the high-field resonance linewidth have been carried out with $\vec{H}$ in the (001) plane parallel to [110] − 0.3° [Fig. 1(a)]. The exact [110] direction was avoided because of overlap with centers pointing into the [010] direction.[2,7] Figure 1(b) exhibits the broadening of the low-field $g \sim 6$ resonance for $\vec{H}$ parallel to [110], which for this direction is independent[7] of φ and therefore shows no additional broadening near T_c. The large linewidth difference between these two lines of about 1:15 at T_c also excludes a T_1 relaxation process for the high-field line since even for Fe^{3+} ions in very anisotropic coordination no anisotropic T_1 effects have been reported.

The samples used for the K-band experiments have been described elsewhere.[8] To avoid disturbing influences of different domains for the measurements below T_c, plate-shaped crystals were used which below T_c transform into a monodomain sample.[8] Their domain axis was oriented parallel to the rotation axis of the scanning magnetic field.

For the analysis of the experimental effects we restricted ourselves to temperatures above T_c, since the experimental data below T_c are less reliable because of overlapping lines. Well above T_c the fluctuations are fast and cause the resonance lines to be homogeneously broadened.[9] They are thus Lorentzian as observed, with a width proportional to the fluctuation-correlation time. On approaching T_c from above, the fluctuations are critically slowed down, their correlation time increases and with it the width of the resonance lines. To isolate the influence of the critical rotational fluctuations on the measured linewidth, the background width resulting from different mode fluctuation must be subtracted. This background is by itself temperature dependent and has been interpolated as indicated in Fig. 1(c). The remaining critical linewidth is given by the time integral of the fluctuation-correlation function[9]:

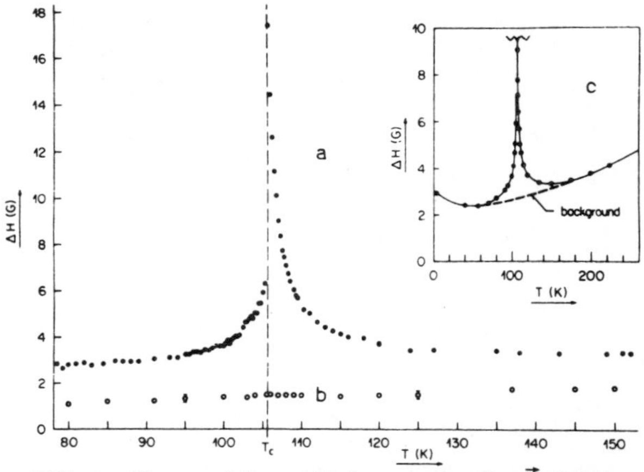

FIG. 1. Observed linewidth broadening for $\vec{H} \parallel [110] - 0.3°$, $\vec{c} \parallel [001]$ of the Fe^{3+}-V_O center at the structural phase transition of $SrTiO_3$ at 19.2 GHz. (a) For the high-field line; (b) for the low-field $g \sim 6$ line. (c) Determination of the background width.

$$\Delta H(\epsilon) \propto \int_0^\infty \langle \varphi^{[001]}(t)\varphi^{[001]}(0)\rangle_\epsilon \, dt = \int_0^{\pi/a} S(q, \epsilon, \omega = 0)\, d^3q, \tag{3}$$

where $S(q, \epsilon, \omega)$ is the temperature-dependent dynamical structure factor of the rotational fluctuations as derived by Schwabl.[5] For zero frequency his expression reduces to

$$S(q, \epsilon, \omega = 0) \propto T[\chi^{\alpha\alpha}(\vec{q}, \epsilon)]^2, \tag{4}$$

where $\chi^{\alpha\alpha}(\vec{q}, \epsilon)$ is given by Eq. (1). Using Eq. (4) in Eq. (3) yields Eq. (2).

Fitting the experimental points with expression (2) in the entire interval from 105.75 to 152°K would require $T_c = 104.5$°K. However, from experiment it is found that T_c is at 105.6°K where the linewidth is maximum and the order parameter vanishes.[10] Thus, for $T \to T_c^+$ the linewidth does not diverge as the fast-fluctuation formula (2) indicates. For the range of T between $T_c + 1°$ and $T_c + 46°$, an excellent fit is obtained (Fig. 2) using the following parameters in (2)[11]:

$$T_c = 105.6°\text{K}, \quad \nu = 0.63 \pm 0.07, \quad \Delta = 0.017 \pm 0.010.$$

η is set equal to zero as it is estimated to be of the order of the experimental error. $\kappa(115°\text{K}) = 0.05$ Å^{-1} was taken from the determination by neutron diffraction.[4] The value of the anisotropy parameter Δ means that sixty octahedral units in a (001) plane are correlated when one in a next (001) plane is correlated to them in the sense of

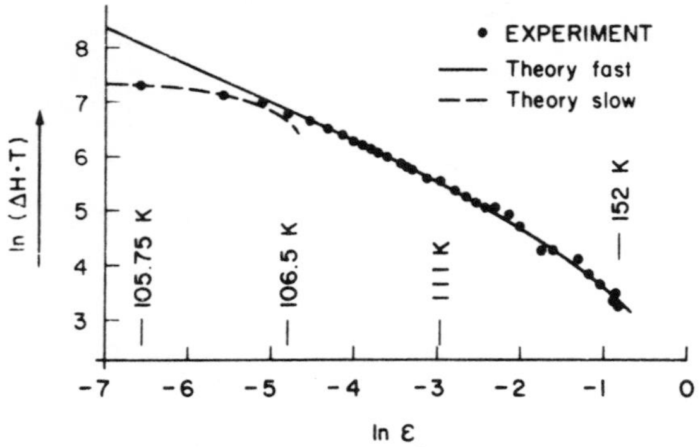

FIG. 2. Comparison between experimental and theoretical linewidths for the fast- and slow-motion regimes. For the latter, the theoretical curve shows $(\langle \Delta H_{\text{crit}}^2 \rangle)^{1/2}$; for convenience, it is adapted to the experimental points determined by linear subtraction of the background width, as in the fast-motion regime.

the $I4/mcm$ structure. This is understood qualitatively by the strong in-plane coupling between the alternately rotating oxygen octahedra through the displacement of the common corners as compared to the weak coupling through the oxygen ions between the planes. The "pancake"-type correlations found here in the cubic phase may, to a certain extent, be compared with the behavior of the planar Heisenberg model as suggested by Stanley[12] although the comparison is limited by the fact that in our case the anisotropy is coupled with the direction of the "spins," i.e., the axis of rotation. This model yields critical exponents $\beta = \frac{1}{3}$, $\gamma = \frac{4}{3}$, and $\nu = 2\beta = \frac{2}{3}$ with $\eta = 0$. The experimentally deduced values for $SrTiO_3$ are $\beta = 0.333 \pm 0.010$ from EPR experiments[10] and $\nu = 0.63 \pm 0.07$ obtained here, as well as $\gamma = 1.29 \pm 0.10$ calculated by Schneider and Stoll[13] using the observed change-over from $\beta \simeq \frac{1}{2}$ outside the critical region, to $\frac{1}{3}$ near T_c.

Schwabl[5] also obtained from his theory an expression for the ultrasound absorption near T_c as carefully measured by Fossheim and Berre[14] for the propagation of ultrasound along a $\langle 100 \rangle$ direction. In view of the anisotropy in $\chi^{\alpha\alpha}(\vec{q}, \epsilon)$, the critical exponents of ultrasound absorption are also anisotropic, as observed by Rehwald.[15]

For temperatures below 106.5°K the fast-fluctuation regime is observed to be no longer valid. The nondivergent width is expected since it is a local variable not depending on the volume of the system considered, and the mechanical coupling between the rotating oxygen octahedra at most allows a maximum local value of $\varphi_{\vec{l}}^{\alpha}$ of the order of $\varphi_{\vec{l}}^{\alpha}(T=0) \sim 2°$. The latter fact prevents $\langle \vec{\varphi}_{\vec{l}}^{2} \rangle$ from diverging for $T \to T_c$, where $\langle \vec{\varphi}_{\vec{l}}^{2} \rangle$ means the ensemble average of the square of the local fluctuations $\varphi_{\vec{l}}^{\alpha}$. On approaching T_c the line broadening changes over to a regime where the fluctuations are critically slowed down and their oscillator strength is more and more concentrated in the central peak[4,5] which contains only frequencies very close to zero. Its width narrows to zero for $T \to T_c^{+}$, and hence the fluctuations are concentrated at frequencies low compared to the characteristic spin precession frequency of the EPR experiment. The latter may be estimated in our experiment to be around 100 MHz.[9] In this temperature range therefore the spins "see" the fluctuations as static. Thus, the instantaneous resonance line is *inhomogeneously* broadened and is essentially characterized by a Gaussian line shape. The change from Lorentzian to more Gaussian-type resonances is found experimentally. The analysis yields 50% Gaussian at $T_c + 1°$. Then,

$$\langle \Delta H_{\text{crit}}^2 \rangle \equiv \langle \Delta H_{\text{expt}}^2 - \Delta H_{\text{background}}^2 \rangle \propto \langle \vec{\varphi}_{\vec{l}}^{2} \rangle = \langle \vec{\varphi}_{\vec{l}}(t) \cdot \vec{\varphi}_{\vec{l}}(t) \rangle = 2 \int_{q=0}^{\pi/a} \int_{\omega=0}^{\infty} S(q, \epsilon, \omega)\, d^3q\, d\omega, \tag{5}$$

where $\langle \vec{\varphi}_{\vec{l}}^{2} \rangle$ is the equal-time correlation function of $\vec{\varphi}_{\vec{l}}$ at site $\vec{l}$. Using Schwabl's expression for $S(q, \epsilon, \omega)$ this yields for the critical contribution ($\kappa \to 0$),

$$\langle \Delta H_{\text{crit}}^2(\epsilon) \rangle \propto \text{const} - [\kappa(\epsilon)/\sqrt{\Delta}] \arctan[\pi\sqrt{\Delta}/\kappa(\epsilon)a]. \tag{6}$$

The second term, which for $\kappa \to 0$ becomes linear in κ, vanishes at $T = T_c$ so that the first term describes the *finite* maximum width at the phase transition. Equation (6) is then rewritten for $T \gtrsim T_c$ as

$$\langle \Delta H_{\text{crit}}^2(\epsilon) \rangle = \Delta H_{\max}^2 (1 - C\epsilon^{\nu}), \tag{7}$$

with $\eta = 0$, and C a proportionality constant.[16] From the experiments we obtain $\Delta H_{\max} = 20.1$ G, $C = 17.7$. In this region the experimental uncertainties of T_c and $\Delta H_{\max}^{\text{expt}}$ do not allow an accurate independent determination of the exponent ν. Therefore $\nu = 0.63$ has been taken from the first analysis. The value of $\Delta H_{\max}$ derived from Eq. (7) lies well within the experimental error. This result shows that for $T \to T_c^{+}$ the linewidth is cusp shaped.[6] The theoretical results according to this regime are included in Fig. 2. The changeover to the fast-fluctuation regime is found near 106.5°K. Strain inhomogeneities in the sample also cause a lowering of ΔH near T_c. However, such effects yield a zero derivative of ΔH with respect to T at T_c, whereas the experiments show a monotonic increasing derivative down to $T_c+0.15°$ as predicted by the cusp-shape behavior. Hence the strain effects are limited to within 0.1°K in our experiment.

In agreement with our quantitative analysis yielding two-dimensional correlations are the critical anisotropic x-ray diffusive scattering data of Denoyer, Comes, and Lambert[17] in $KMnF_3$ and $NaNbO_3$ which undergo an analogous transition to $SrTiO_3$. We also note that the flat dispersion curve of the soft R-corner mode between the R and M points of the Brillouin zone found by Shirane[18] in $KMnF_3$ is indicative of a highly anisotropic correlation in q space. The M_3 point in $SrTiO_3$ lies at 8 meV,[19] about twice as high as in $KMnF_3$ at T_c, indicating a linear anisotropy twice as large in the latter. A recent nuclear magnetic T_1 relaxation study by Bonera, Borsa, and Rigamonti[20] in $NaNbO_3$ can only be interpreted by uncorrelated fluctuations occurring between equivalent and correlated (100) planes near T_c. Nearly two-dimensional correlations for $T \geq T_c$ in systems ordering three-dimensionally at T_c have recently also been found in magnetic systems like K_2NiF_4[21] and in superconducting layer-type compounds.[22] An extended account including the analysis of the linewidth for $T \leq T_c$ as well as its angular dependence in relation to the soft modes[19,23] below T_c will be published elsewhere.

The authors are indebted to F. Schwabl and F. Borsa for communicating to us their unpublished early results, as well as to J. Feder, W. Känzig, J. Lajzerowicz, T. Riste, and H. E. Stanley for pertinent suggestions and comments on this work, and to K. W. Blazey, E. Courtens, and T. Schneider for discussions.

[1]K. A. Müller, in *Structural Phase Transitions and Soft Modes*, edited by E. J. Samuelsen, E. Andersen, and J. Feder (Universitetsforlaget, Oslo, Norway, 1971).

[2]E. S. Kirkpatrick, K. A. Müller, and R. S. Rubins, Phys. Rev. 135, A86 (1964).

[3]K. W. Blazey, Phys. Rev. Lett. 27, 146 (1971).

[4]T. Riste, E. J. Samuelsen, K. Otnes, and J. Feder, Solid State Commun. 9, 1455 (1971).

[5]F. Schwabl, preceding Letter [Phys. Rev. Lett. 28, 500 (1972)].

[6]For a definition of critical exponents, see H. E. Stanley, *Phase Transitions and Critical Phenomena* (Clarendon, Oxford, England, 1971).

[7]Th. von Waldkirch, K. A. Müller, and W. Berlinger, Phys. Rev. B (to be published).

[8]K. A. Müller, W. Berlinger, M. Capizzi, and H. Gränicher, Solid State Commun. 8, 549 (1970).

[9]A. Abragam, *The Principles of Nuclear Magnetism* (Clarendon, Oxford, England, 1961).

[10]K. A. Müller and W. Berlinger, Phys. Rev. Lett. 26, 13 (1971).

[11]The fairly large upper error limit of Δ results mainly from the uncertainty of the background width. Between $T=107$ and 135°K the curve can be approximated by an apparent exponent "ν" $=0.78\pm0.05$ which falls within $\nu=0.76\pm0.08$ reported in Ref. 4.

[12]H. E. Stanley, in *Structural Phase Transitions and Soft Modes*, edited by E. J. Samuelsen, E. Andersen, and J. Feder (Universitetsforlaget, Oslo, Norway, 1971), p. 271.

[13]T. Schneider and E. Stoll, in *Structural Phase Transitions and Soft Modes*, edited by E. J. Samuelsen, E. Andersen, and J. Feder (Universitetsforlaget, Oslo, Norway, 1971), p. 383.

[14]K. Fossheim and B. Berre, in *Structural Phase Transitions and Soft Modes*, edited by E. J. Samuelsen, E. Andersen, and J. Feder (Universitetsforlaget, Oslo, Norway, 1971), p. 255, and Phys. Rev. B (to be published).

[15]W. Rehwald, Solid State Commun. 8, 607 (1970).

[16]A similar approach was made independently by G. Bonera, F. Borsa, and A. Rigamonti (private communication).

[17]F. Denoyer, R. Comes, and M. Lambert, in Proceedings of the Second European Meeting on Ferroelectricity, Dijon, France, 1971 (to be published).

[18]G. Shirane, in *Structural Phase Transitions and Soft Modes*, edited by E. J. Samuelsen, E. Andersen, and J. Feder (Universitetsforlaget, Oslo, Norway, 1971), p. 217.

[19]G. Shirane and Y. Yamada, Phys. Rev. 177, 858 (1969).

[20]G. Bonera, F. Borsa, and A. Rigamonti, in Proceedings of the Second European Meeting on Ferroelectricity, Dijon, France, 1971 (to be published).

[21]R. J. Birgeneau, R. Dingle, M. T. Hutchings, G. Shirane, and S. L. Holt, Phys. Rev. Lett. 26, 718 (1971).

[22]T. H. Geballe, A. Menth, F. J. Di Salvo, and F. R. Gamble, Phys. Rev. Lett. 27, 314 (1971).

[23]P. A. Fleury, J. F. Scott, and J. M. Worlock, Phys. Rev. Lett. 21, 16 (1968).

JOURNAL DE PHYSIQUE *Colloque* C1, *supplément au n*° 2-3, *Tome* 33, *Février-Mars* 1972, *page* C1-135

THE SPECTRUM AND INTENSITY OF LIGHT SCATTERED FROM THE BULK PHASES AND FROM THE LIQUID-VAPOR INTERFACE OF XENON NEAR ITS CRITICAL POINT (*)

J. ZOLLWEG (**), G. HAWKINS, I. W. SMITH, M. GIGLIO (***)
and G. B. BENEDEK

Department of Physics and Center for Materials Science and Engineering
Massachusetts Institute of Technology, Cambridge, Massachusetts 02139

Résumé. — Nous présentons les résultats de la mesure de la longueur de corrélation (ξ), de la compressibilité (κ_T), de la viscosité (η) et de la tension superficielle (σ) du xénon près de son point critique. Ces données ont été utilisées pour vérifier de façon détaillée la validité de plusieurs relations fondamentales reliant la divergence de la longueur de corrélation et la divergence de la compressibilité, pour vérifier aussi la disparition de la tension superficielle et la divergence de la conductivité thermique près du point critique.

Abstract. — We report measurements of the long range correlation length (ξ), the compressibility (κ_T), the viscosity (η), and the surface tension (σ) in xenon near its critical point. Using this data we have been able to examine in detail the validity of several fundamental theoretical relations which connect the divergence of the correlation range with the divergence of the compressibility, the vanishing of the surface tension, and the divergence of the thermal conductivity near the critical point.

The central feature in the description of a system near its critical point is the long range correlation length (ξ) which describes the characteristic distance over which fluctuations in the order parameter are spatially correlated. In the vicinity of the critical point the correlation length becomes very large. Therefore the spatial integral of the correlation function for the order parameter increases extremely rapidly as the critical temperature is approached. The divergence of this integral manifests itself experimentally as the divergence of the equilibrium susceptibilities of the system. Thus the precise nature of the divergence of the equilibrium properties is intimately connected to the magnitude and the temperature dependence of the correlation range. This connection appears explicitly in the static scaling theories of critical phenomena.

On the other hand, non-equilibrium properties of the system are also connected to the divergence of the correlation range. The development of dynamical scaling theories for transport properties near the critical point (expressed in detail in the mode-mode coupling theories of Kadanoff, Swift and Kawasaki) shows that the long range correlation length plays a central role in understanding the transport coefficients.

The purpose of this note is to present our measurements of the correlation length ξ in xenon along the critical isochore and along the coexistence curve and to relate these measurements to the observed behavior of the compressibility, the surface tension, and the thermal diffusivity. In this way we shall be able to examine the validity of theoretical predictions connecting the correlation range to both the static and the dynamic properties of the system.

By observing the angular anisotropy of the intensity of light scattered from xenon, we have measured the magnitude and the temperature dependence of the correlation range ξ along the critical isochore in the temperature range

$$2 \times 10^{-4} < \left[\frac{(T - T_c)}{T_c}\right] < 4 \times 10^{-3} .$$

Our results along the critical isochore are shown in figure 1 and are summarized by the equation

$$\xi = 3.07 \left[\frac{(T - T_c)}{T_c}\right]^{-0.57 \pm 0.03} \text{Å} .$$

Along the vapor side of the coexistence curve our previous measurements [1] of ξ show that the correlation range there obeys the equation

$$\xi = (1.8 \pm 0.2) \left[\frac{(T_c - T)}{T_c}\right]^{-0.57 \pm 0.05} \text{Å} .$$

(*) This research was supported by the Advanced Research Projects Agency under contract DAHC 15 67C 0222 with the Massachusetts Institute of Technology and by the National Science Foundation.

(**) Now at the Department of Chemistry, University of Maine, Orono, Maine.

(***) Now at the University of Milan, Milan, Italy.

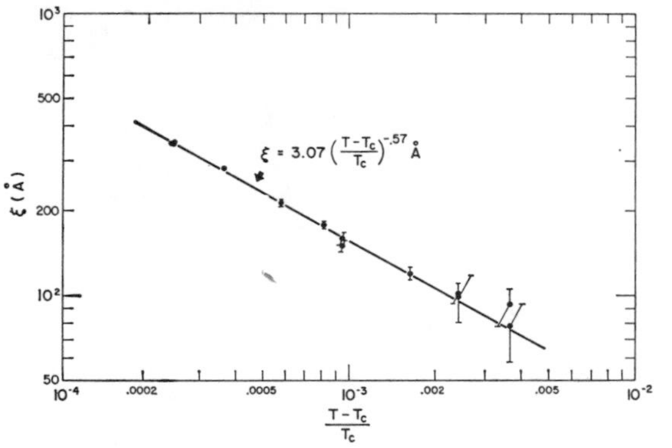

FIG. 1. — The temperature dependence of the long range correlation length along the critical isochore in xenon.

The relationship between the correlation range ξ and the isothermal compressibility (κ_T) is predicted by the Ornstein-Zernike theory to be

$$\frac{\xi^2}{R^2} = \frac{\kappa_T}{\kappa_I}. \tag{1}$$

Here $\kappa_I = 1/(\rho_n\, k_B\, T)$ is the compressibility of an ideal gas, ρ_n is the number density of the fluid, k_B is Boltzman's constant, and R is the direct correlation range. In the Ornstein-Zernike theory the pair correlation function has the form $(e^{-r/\xi}/r)$ from which it follows that R is a constant near the critical point. However, if the pair correlation function has the form $(e^{-r/\xi}/r^{1-\eta})$ as proposed by M. Fisher, then R will diverge weakly near the critical point.

In order to investigate the validity of eq. (1) and to deduce the behavior of R near T_c, we have measured the compressibility κ_T (or equivalently $(\partial\rho/\partial\mu)_T = \rho^2\, \kappa_T$) along both the critical isochore and the vapor side of the coexistence curve. This measurement is made by determining the absolute intensity of light scattered at very small angles in the fluid near its critical point. The results for $(\partial\rho/\partial\mu)_T$ are shown in figure 2. The upper curve corresponds to measurements taken along the critical isochore. Here

$$\left(\frac{\partial\rho}{\partial\mu}\right)_T = 1.43\left[\frac{(T-T_c)}{T_c}\right]^{-1.21} \times 10^{-9}\ \text{g}^2/\text{erg cc}.$$

The lower line corresponds to measurements of $(\partial\rho/\partial\mu)_T$ along the vapor side of the coexistence curve. This data obeys the equation

$$\left(\frac{\partial\rho}{\partial\mu}\right)_T = 0.346\left[\frac{(T_c-T)}{T_c}\right]^{-1.21} \times$$
$$\times\ 10^{-9}\ \text{g}^2/\text{erg cc}.$$

These results may be used to examine the validity of eq. (1). We first consider the limiting behavior as T approaches T_c. If ν and γ are the critical exponents describing respectively the temperature dependence

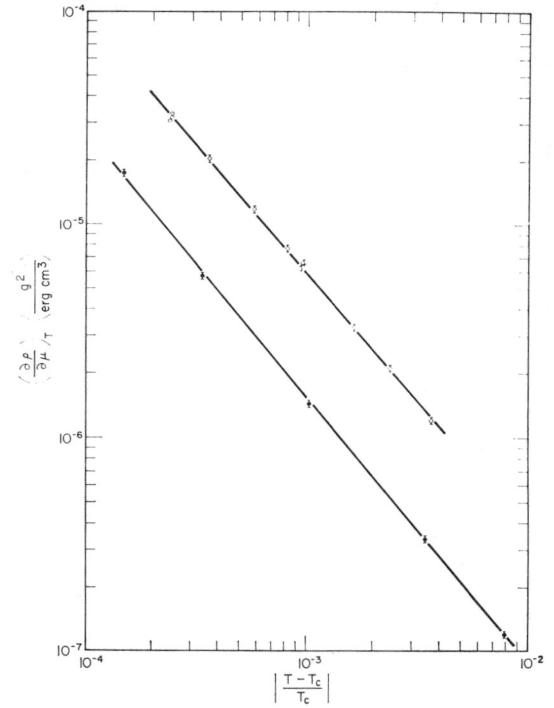

FIG. 2. — $(\partial\rho/\partial\mu)_T$ as a function of temperature along the critical isochore (upper line) and along the vapor side of the coexistence curve (lower line) for xenon.

of ξ and of $(\partial\rho/\partial\mu)_T$, then in the limit $T \to T_c$ we expect from eq. (1) that $\gamma = 2\,\nu$. From our measurements on the critical isochore $\gamma = 1.21 \pm 0.03$ and $2\,\nu = 1.14 + 0.1$. Despite the experimental uncertainties it is possible to estimate a value of Fisher's exponent η. See reference [16].

Secondly, we can use eq. (1) to calculate numerical values of the direct correlation range R at various temperatures in order to determine wheter R is in fact constant. From our data we find that $R = 4.9 \pm 0.1$ Å both above and below T_c. (This value is slightly different from that previously reported [1] because the magnitude of $(\partial\rho/\partial\mu)_T$ has now been obtained more accurately). It is interesting to note that when this value of R is used in analyzing *D.* Cannell's data [2] on the Brillouin components of xenon near T_c, the deduced magnitude and temperature dependence of ξ along the critical isochore are in excellent agreement with the results reported here.

From our data, we have also computed the ratio of $(\partial\rho/\partial\mu)_T$ on the critical isochore to $(\partial\rho/\partial\mu)_T$ on the coexistence curve. We find asymptotically that this ratio is (4.1 ± 0.2). This implies that for equal temperatures above and below T_c, the compressibility on the critical isochore is larger than that in either phase along the coexistence curve by a factor of $4.1(\rho/\rho_c)^2$.

We next examine the implications of our measurements of ξ insofar as the transport properties of xenon are concerned. By measuring the spectral width of monochromatic light scattered quasi-elastically from a fluid, it is possible to deduce the total thermal diffusivity $D = (\Lambda/\rho c_p)$. Here Λ is the thermal conductivity, c_p the specific heat per unit mass, and ρ the mass density. Near the critical point c_p diverges quite strongly, and as a result D approaches zero. This implies that entropy fluctuations in the fluid relax back to equilibrium ever more slowly as the critical point is approached. This increasing relaxation time manifests itself in an extraordinary narrowing of the spectral line width of the quasi-elastically scattered light. Despite this narrowing, the line width (which may be less than 100 Hz) can be measured quite accurately using the techniques of optical mixing spectroscopy [3].

The mode-mode coupling theories of Kadanoff, Swift [4], and Kawasaki [5] predict that in the hydrodynamic limit the so-called « critical contribution » to the thermal diffusivity should vary with temperature in accordance with the formula

$$\left(\frac{\Lambda}{\rho c_p}\right)_{\text{crit}} = \frac{k_B T}{6 \pi \eta \xi} \tag{2}$$

where η is the shear viscosity of the fluid. This result lends itself to a simple physical interpretation. It states that heat is transported by a random walk process in which the elemental units carrying heat have dimension ξ. Eq. (2) is simply the diffusion coefficient for the motion of « particles » of size ξ moving in a medium of viscosity η.

In the case of xenon, the total thermal diffusivity $(\Lambda/\rho c_p)$ has been measured as a function of temperature along the critical isochore by Henry, Swinney and Cummins [6]. In order to compare these measurements with the critical contribution $(k_B T/6 \pi\eta\xi)$ predicted by the mode-mode coupling theories, it is necessary to know the shear viscosity η of xenon. We have measured η (as well as the surface tension σ) along the coexistence curve of xenon by observing the spectrum of light scattered inelastically from thermally excited surface waves on the liquid-vapor interface [7], [8]. Our results show that

$$\frac{\eta}{\rho} = [448 + 12(T_c - T)] \pm 11 \text{ microstokes } [9]$$

from which we may evaluate the shear viscosity at $T = T_c$ and $\rho = \rho_c$. Since the viscosity changes only slightly as the temperature increases along the critical isochore we can, with good accuracy, calculate $k_B T/6 \pi\eta\xi$. In figure 3 we plot the value of $k_B T/6 \pi\eta\xi$ along with the experimental data of Cummins and Swinney for $(\Lambda/\rho c_p)$. In figure 4 we show the same measurements over a narrower range of temperature so as to focus attention on the data closest to the critical point. It is clear from figures 3 and 4 that it is not correct to assume [6] that eq. (2) alone accurately describes the measured thermal diffusivity. In fact there is an additional contribution to the thermal diffusivity that is not included in the mode-mode coupling analysis. This additional term arises from the non-divergent or background part of the thermal conductivity $(\Lambda_{\text{background}})$ and contributes an amount $(\Lambda_{\text{background}}/\rho c_p)$ to the total thermal diffusivity. In order to examine the accuracy of the mode-mode coupling theories one must determine whether $(\Lambda_{\text{background}}/\rho c_p)$ is of such a size as to account for the difference between

$$\left(\frac{\Lambda}{\rho c_p}\right) \quad \text{and} \quad \left(\frac{k_B T}{6 \pi\eta\xi}\right).$$

The background thermal conductivity data of R. Tufeu, B. Le Neindre, and B. Vodar for $\Lambda_{\text{background}}$ in xenon has been recently analyzed by J. V. Sengers [10]. The specific heat c_p in the critical region may be obtained from the thermodynamic relation

$$c_p = c_v + \frac{T(\partial P/\partial T)_v^2(\partial\rho/\partial\mu)_T}{\rho^3}$$

by using our measurements of $(\partial\rho/\partial\mu)_T$, the measure-

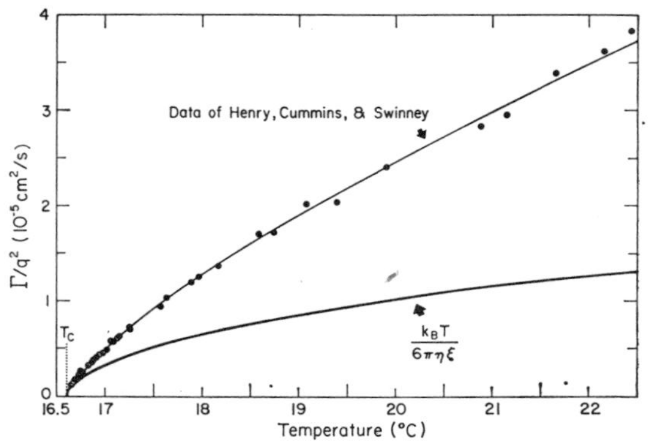

FIG. 3. — The total thermal diffusivity $(\Lambda/\rho c_p)$ (upper curve) compared with the critical part of the thermal diffusivity as predicted by the mode-mode coupling theories (lower curve). The temperature range includes the entire region where experimental data is available : $(T - T_c) \leqslant 6.0$ °K.

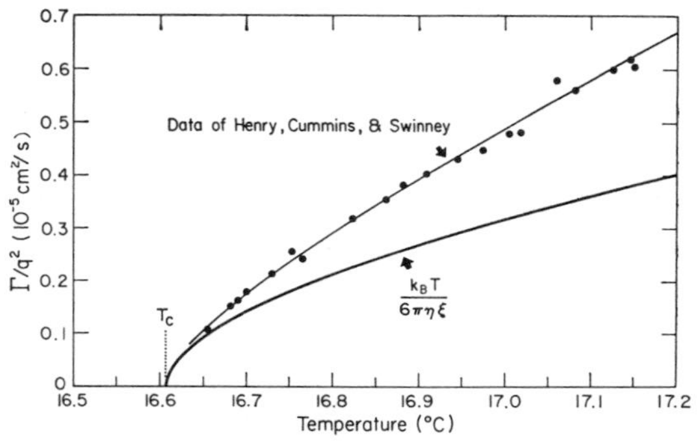

FIG. 4. — $(\Lambda/\rho c_p)$ as compared with $k_B T/6 \pi\eta\xi$ in the temperature range $(T - T_c) \leqslant 0.600$ °K.

ments of Buckingham *et al.* [11] for c_v, and the results of Habgood and Schneider [12] for $(\partial P/\partial T)_v$ on the critical isochore. The contribution of c_v to c_p is small and the primary temperature dependence of c_p is determined by the divergence of $(\partial\rho/\partial\mu)_T$. From these measurements the magnitude and temperature dependence of the background contribution to the thermal diffusivity may be calculated. The results of this calculation are plotted in figure 5 along with the quantity $k_B T/6\pi\eta\xi$. Also shown is the sum of these two quantities, which is seen to agree remarkably well with the data points of Cummins and Swinney for the measured total thermal diffusivity. We therefore conclude that on the critical isochore of xenon the contribution to the total thermal diffusivity predicted by the mode-mode coupling theories is consistent with the observed values of $(\Lambda/\rho c_p)$ when account is taken of the background thermal conductivity.

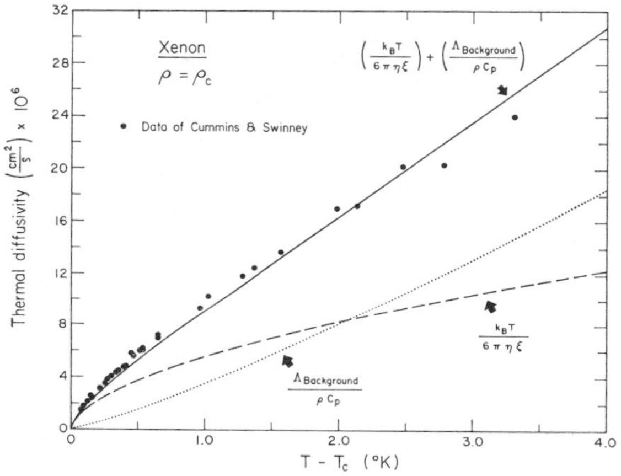

FIG. 5. — Broken line : The Kawasaki critical contribution $(k_B T/6\pi\eta\xi)$ to the total thermal diffusivity $(\Lambda/\rho c_p)$ on the critical isochore of xenon. Dotted line : The non-critical or background contribution $(\Lambda_{\text{background}}/\rho c_p)$ to the total thermal diffusivity. The solid line represents the sum of the critical and the non-critical contributions to $(\Lambda/\rho c_p)$ and the data points correspond to measurements of $(\Lambda/\rho c_p)$ taken by Henry, Cummins, and Swinney.

Finally we shall examine the connection between the divergence of the correlation range ξ and the vanishing of the surface tension σ on the coexistence curve of xenon. We have measured σ by analyzing the spectrum of light scattered inelastically from thermally excited surface waves on the liquid vapor interface [9]. Our data for the temperature dependence of σ is shown in figure 6 as open circles. The closed circles correspond to the data of Smith, Gardner, and Parker [13], corrected by using the more accurate density formulae of Garside, Molgaard, and Smith [14]. These results may be summarized by the equation

$$\sigma = (62.9 \pm 1.8)\left(1 - \frac{T}{T_c}\right)^{1.302\pm0.006} \text{ dynes/cm}$$

for $0.070\ °K < (T_c - T) < 5\ °K$.

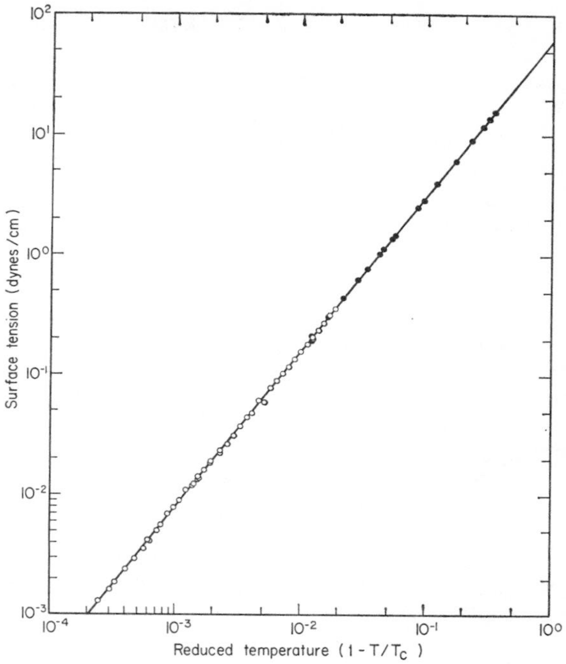

FIG. 6. — Surface tension σ for xenon as a function of the reduced temperature $(1 - T/T_c)$. The open circles represent our measurements. The closed circles correspond to the corrected data of Smith, Gardner and Parker.

The connection between our measurements of the surface tension and the behavior of the correlation range is expressed in the Fisk-Widom formula [15] which relates the interface thickness (L) to the surface tension σ by :

$$L = \frac{1}{c}\frac{1}{\beta^2}\frac{\sigma(\partial\rho/\partial\mu)_T}{(\rho_l - \rho_v)^2}. \qquad (3)$$

Here ρ_l and ρ_v are the densities of the liquid and vapor phases respectively, β is the critical exponent describing the temperature dependence of $(\rho_l - \rho_v)$, and the parameter c is a constant predicted to be of the order of unity. Furthermore, Fisk and Widom demonstrate that in the limit $T \to T_c$, L should be equal to the correlation range ξ. Using literature values and our own measurements of each of the terms in eq. (3), we find that

$$c = 0.93 \pm 0.1 \quad \text{at} \quad \left(1 - \frac{T}{T_c}\right) = 1.477 \times 10^{-4},$$

a temperature at which the value of ξ is well known.

If we keep c constant at this value and deduce the temperature dependence of L from the behavior of $(\partial\rho/\partial\mu)_T$, σ, and $(\rho_l - \rho_v)$ then we find that

$$L = 1.06\left(1 - \frac{T}{T_c}\right)^{-0.63} \text{Å}.$$

This value of the interface thickness is in good numerical agreement with our direct measurements of the correlation range ξ along the vapor side of the coexistence curve for $0.025\,^{o}K < (T_c - T) < 0.125\,^{o}K$. We must conclude that the Fisk-Widom relation is in good agreement with the experimental results in the temperature range $(T_c - T) < 0.125\,^{o}K$. Farther from the critical point, the interfacial thickness L is likely to be a combination of the correlation range along the liquid side of the coexistence curve and that along the vapor side. Consideration of this effect in fact improves the agreement between the values of ξ deduced from the surface tension data and the values measured directly by observing the angular dependence of the intensity of scattered light.

References

[1] GIGLIO (M.) and BENEDEK (G. B.), *Phys. Rev. Letters*, 1969, **23**, 1145.

[2] CANNELL (D.) and BENEDEK (G. B.), *Phys. Rev. Letters*, 1970, **25**, 1157.

[3] BENEDEK (G. B.), « Optical Mixing Spectroscopy with Applications to Problems in Physics, Chemistry, Biology, and Engineering », pp. 49-84, in *Polarization*, *Matter* and *Radiation*, Jubilee Volume published by Presses Universitaires, Paris, 1969.

[4] KADANOFF (L. P.) and SWIFT (J.), *Phys. Rev.*, 1968, **166**, 89.

[5] KAWASAKI (K.), *Phys. Letters*, 1969, **30A**, 325, *Phys. Rev.*, 1970, **A 1**, 1750.

[6] HENRY (D. H.), SWINNEY (H. L.) and CUMMINS (H. Z.), *Phys. Rev. Letters*, 1970, **25**, 1170 and Henry (D. H.), Thesis (Ph. D.), Johns Hopkins University.

[7] KATYL (R. H.) and INGARD (U.), *Phys. Rev. Letters*, 1968, **20**, 248.

[8] BOUCHIAT (M. A.) and MEUNIER (J.), *Phys. Rev. Letters*, 1969, **23**, 752.

[9] ZOLLWEG (J.), HAWKINS (G.) and BENEDEK (G. B.), *Phys. Rev. Letters*, 1971, **27**, 1182.

[10] SENGERS (J. V.), private communication.

[11] EDWARDS (C.) LIPA (J. A.) and BUCKINGHAM (M. J.), *Phys. Rev. Letters*, 1968, **20**, 496.

[12] HABGOOD (H. W.) and SCHNEIDER (W. G.), *Can. J. Chem.*, 1954, **32**, 164.

[13] SMITH (D. L.), GARDNER (P. R.) and PARKER (E. H. C.), *J. Chem. Phys.*, 1967, **47**, 1148.

[14] GARSIDE (D. H.), MØLGAARD (H. V.) and SMITH (G. L.), *J. Phys. B* (*Proc. Phys. Soc.*), 1968, **1**.

[15] FISK (S.) and WIDOM (B.), *J. of Chem. Phys.*, 1969, **50**, 3219.

[16] SMITH (I. W.), GIGLIO (M.) and BENEDEK (G. B.), *Phys. Rev. Letters*, 1971, **27**, 1556.

III
ABSTRACTS OF THEORETICAL ARTICLES

Expansion of a Critical Exponent in Inverse Powers of Spin Dimensionality

Ryuzo ABE

Department of Pure and Applied Sciences
University of Tokyo, Komaba, Meguro-Ku, Tokyo

A systematic expansion method is developed for studying a critical exponent. The expansion parameter in the present theory is $1/n$ where n is the spin dimensionality. Since the limit $n \to \infty$ corresponds to the spherical model, the present method means that the deviation of critical exponent from its spherical model value is expanded in powers of $1/n$. The susceptibility exponent γ is discussed and the terms exact up to $1/n$ are derived for the d-dimensional simple hypercubical system with nearest-neighbor interaction (d is assumed to take a continuous value). The result is proved to be consistent with recent Wilson's ϵ expansion.

PHYSICAL REVIEW VOLUME 122, NUMBER 5 JUNE 1, 1961

One-Dimensional Order-Disorder Model Which Approaches a Second-Order Phase Transition

GEORGE A. BAKER, JR.
Los Alamos Scientific Laboratory, University of California, Los Alamos, New Mexico
(Received January 30, 1961)

The calculation of the partition function for a simple one-dimensional order-disorder model is reduced to the solution of a certain functional equation. This equation is solved rigorously and it is shown that in the limit of indefinitely long-range interactions the model exhibits a finite discontinuity in the specific heat.

Reprinted from THE PHYSICAL REVIEW, Vol. 124, No. 3, 768–774, November 1, 1961
Printed in U. S. A.

Application of the Padé Approximant Method to the Investigation of Some Magnetic Properties of the Ising Model

GEORGE A. BAKER, JR.
Los Alamos Scientific Laboratory, University of California, Los Alamos, New Mexico
(Received June 19, 1961)

On the basis of the Padé approximant method we deduce from the exact series expansions for the Ising model that the reduced magnetic susceptibility behaves at the critical point as $\chi_{fcc} \approx [0.09923/(0.101767-w)]^{5/4}$, $\chi_{bcc} \approx [0.152773/(0.1561789-w)]^{5/4}$, $\chi_{sc} \approx [0.22138/(0.218156-w]^{5/4}$, $\chi_t \approx [0.2432/(2-\sqrt{3}-w)]^{7/4}$, $\chi_{sq} \approx [0.35724/(\sqrt{2}-1-w)]^{7/4}$, and $\chi_h \approx [0.4506/(1/\sqrt{3}-w)]^{7/4}$, where $w = \tanh(J/kT)$ and the last figure quoted is somewhat uncertain. The spontaneous magnetization is found to behave as $(I_0/I_\infty)_{fcc} \approx [12.5(0.664658-z^2)]^{0.3}$, $(I_0/I_\infty)_{bcc} \approx [10.4(0.5326607-z^2)]^{0.3}$, $(I_0/I_\infty)_{sc} \approx [10.9(0.411940-z^2)]^{0.3}$, where $z = \exp(-2J/kT)$ and again the last place quoted is somewhat uncertain. The numbers 5/4 and 7/4 have an error of at most 10^{-3}, and 0.3 of at most 10^{-2}. The lattices referred to are fcc, face-centered cubic; bcc, body-centered cubic; sc, simple cubic; t, triangular; sq, simple quadratic; and h, honeycomb.

Reprinted from ANNALS OF PHYSICS
All Rights Reserved by Academic Press, New York and London

Vol. 70, No. 1, March 1972
Printed in Belgium

Partition Function of the Eight-Vertex Lattice Model

RODNEY J. BAXTER

Research School of Physical Sciences, The Australian National University, Canberra, A.C.T. 2600, Australia

Received May 20, 1971

The partition function of the zero-field "Eight-Vertex" model on a square M by N lattice is calculated exactly in the limit of M, N large. This model includes the dimer, ice and zero-field Ising, F and KDP models as special cases. In general the free energy has a branch point singularity at a phase transition, with an irrational exponent.

Reprinted from THE PHYSICAL REVIEW, Vol. 86, No. 6, 821–835, June 15, 1952
Printed in U. S. A.

The Spherical Model of a Ferromagnet

T. H. BERLIN
Department of Physics, The Johns Hopkins University, Baltimore, Maryland

AND

M. KAC
Department of Mathematics, Cornell University, Ithaca, New York

(Received February 28, 1952)

A mathematical model, the spherical model, of a ferromagnet is described. The model is a generalization of the Ising model; and one-, two-, and three-dimensional lattices of infinite extent can be extensively discussed. A three-dimensional lattice shows ferromagnetic behavior and provides a statistical model of the Weiss phenomenological theory. The limiting free energy appears in a form which contains two of the essential features of the exactly known Ising model results in one and two dimensions. This suggests the probable form of the limiting free energy for the three-dimensional Ising model. A simplified model, the Gaussian model, is briefly discussed because this model also contains some of the significant features of the Ising model. However, the Gaussian model, unlike the spherical model, is not defined for all temperatures.

J. Phys. C: Solid St. Phys., 1971, Vol. 4. Printed in Great Britain

Lattice–lattice scaling and the generalized law of corresponding states

D. D. BETTS A. J. GUTTMANN and G. S. JOYCE
Wheatstone Physics Laboratory, King's College, London
MS. received 11*th February* 1971, *in revised form* 8*th April* 1971

Abstract. Lattice-dependent scaled temperature and field variables are shown to produce a generalization of the law of corresponding states. This generalization equates the appropriately scaled free energy of a system on different lattices in the critical region. The theory is tested by making use of the results of series analysis—in particular the critical amplitudes—for the Ising, XY and Heisenberg models, as well as the exact results known for the spherical model. The theory is found to be consistent with all the available data. The scaled field variable appears to be model independent, depending only on the underlying lattice, while the scaled temperature variable is found to be model dependent. It is shown that lattice–lattice scaling is a weaker form of scaling than that which includes the homogeneity hypothesis. It is therefore possible for lattice–lattice scaling to hold for those systems for which the homogeneity argument does not apply. The inclusion of homogeneity into the theory is shown to give rise to the critical invariants recently introduced by Watson.

PHYSICAL REVIEW A   VOLUME 4, NUMBER 3   SEPTEMBER 1971

Ising Model for the λ Transition and Phase Separation in He^3-He^4 Mixtures

M. Blume and V. J. Emery
Brookhaven National Laboratory, Upton, New York 11973
and
Robert B. Griffiths
Physics Department, Carnegie-Mellon University, Pittsburgh, Pennsylvania 15213
(Received 23 March 1971)

A spin-1 Ising model, which simulates the thermodynamic behavior of He^3-He^4 mixtures along the λ line and near the critical mixing point, is introduced and solved in the mean-field approximation. For reasonable values of the parameters of the model the phase diagram is qualitatively similar to that observed experimentally and the phase separation appears as a consequence of the superfluid ordering. Changing the parameters produces many different types of phase diagram, including as features λ lines, critical points, tricritical points, and triple points. Certain thermodynamic features which differ from the He^3-He^4 experiments may be artifacts of the mean-field theory.

Order-disorder statistics.
II. A two-dimensional model

By C. Domb, *The Clarendon Laboratory, University of Oxford*

(*Communicated by D. R. Hartree, F.R.S.—Received* 1 *February* 1949)

The present paper is concerned with the detailed calculation of the partition functions for a two-dimensional quadratic lattice from the matrix derived previously. A brief discussion is given of the general methods of calculation available. The method here employed is the expansion of the eigenvector and eigenvalue as power series about zero temperature. This is readily applicable to finite matrices of quite high order, and, when a suitable notation has been introduced, to the infinite case. It is found more convenient to deal with the general problem of an unsymmetric net with different interactions in the two directions; for a ferromagnetic in the absence of a magnetic field a symmetry relation observed empirically enables terms to be derived successively, and it is assumed that this solution is equivalent to Onsager's (1944). The method is applied in the presence of a magnetic field, and several terms of a generalized series are deduced. Several terms are hence obtained of a series for the spontaneous magnetization. An inspection of the generalized series leads one to conjecture that the specific heat curve becomes continuous in the presence of a magnetic field. A rearrangement of the terms of the generalized series enables several terms of the high-temperature expansion to be deduced. Finally, the results are applied to the theory of binary solid solutions; the solubility curve for the two substances is formally closely related to the spontaneous magnetization. The separation into two phases is established, and the corresponding specific heat singularities are analyzed.

Reprinted from
Proceedings of the Physical Society, Vol. 86, Part 5, No. 553, pp. 1147–1151
1965
Printed in Great Britain by J. W. Arrowsmith Ltd., Bristol 3

On the critical behaviour of ferromagnets

C. Domb
D. L. Hunter

Abstract. Using high-temperature series expansions, an attempt is made to characterize the behaviour of the Ising model of a ferromagnet in a non-zero magnetic field. The functional form derived can be used to pass to the low-temperature side of the Curie point, and hence to determine the low-temperature critical behaviour. As a result, it is suggested that there are two basic parameters, in terms of which critical indices for all thermodynamic properties can be expressed.

On the susceptibility of a ferromagnetic above the Curie point

By C. Domb, *Wheatstone Laboratory, King's College, London*

and M. F. Sykes, *Clarendon Laboratory, University of Oxford*

(*Communicated by J. T. Randall, F.R.S.—Received* 8 *December* 1956)

Series expansions for the susceptibility of a ferromagnetic at high temperatures are examined in detail to estimate the curvature of the reciprocal of the susceptibility immediately above the Curie point. For an Ising model with spin $\frac{1}{2}$ a substantial number of terms of the series are available, and the series is well behaved (i.e. the coefficients vary smoothly). The asymptotic behaviour of the coefficients can be conjectured with fair confidence, and a closed formula can be derived based on this conjecture. For an Ising model with spin greater than $\frac{1}{2}$ fewer terms of the series are available, but the series is still well behaved, and an extrapolation can again be undertaken fairly confidently. For the Heisenberg model with spin $\frac{1}{2}$ the behaviour of the coefficients is considerably more erratic and the predictions are more tentative.

It is concluded generally that the curvature of the reciprocal susceptibility depends primarily on the lattice structure and little on the type of interaction; that there is little variation between the different types of two-dimensional, and different types of three-dimensional lattice, although there is a marked difference between two- and three-dimensional lattices. The experimental curve of Weiss & Forrer for nickel is examined, and it is found that the data can be fitted quite well by the extrapolation formula for a three-dimensional lattice. It is thus possible to account for the experimental results assuming only short-range interatomic forces.

Reprinted from the Journal of Mathematical Physics, Vol. 2, No. 1, 63–67, January–February, 1961
Printed in U. S. A.

Use of Series Expansions for the Ising Model Susceptibility and Excluded Volume Problem

C. Domb and M. F. Sykes
King's College, University of London, London, England

(Received September 16, 1960)

The present paper discusses the problem of making the most effective use of the coefficients of series expansions for the Ising model and excluded volume problem in estimating critical behavior. It is shown that after initial irregularities the coefficients appear to settle down to a smooth asymptotic behavior. Alternative methods of analysis are considered for the provision of a steady series of approximations to the critical point. Numerical conclusions are drawn for particular lattices for which additional terms have recently become available.

Commun. math. Phys. 12, 91—107 (1969)

Existence of a Phase-Transition in a One-Dimensional Ising Ferromagnet

FREEMAN J. DYSON
Institute for Advanced Study, Princeton, New Jersey

Received October 28, 1968

Abstract. Existence of a phase-transition is proved for an infinite linear chain of spins $\mu_j = \pm 1$, with an interaction energy

$$H = - \Sigma J(i - j)\, \mu_i \mu_j ,$$

where $J(n)$ is positive and monotone decreasing, and the sums $\Sigma J(n)$ and $\Sigma (\log \log n)\, [n^3 J(n)]^{-1}$ both converge. In particular, as conjectured by KAC and THOMPSON, a transition exists for $J(n) = n^{-\alpha}$ when $1 < \alpha < 2$. A possible extension of these results to Heisenberg ferromagnets is discussed.

Commun. math. Phys. 21, 269—283 (1971)

An Ising Ferromagnet with Discontinuous Long-Range Order

FREEMAN J. DYSON
Institute for Advanced Study, Princeton, New Jersey

Received February 16, 1971

Abstract. An infinite one-dimensional Ising ferromagnet M with long-range interactions is constructed and proved to have the following properties. (1) M has an order-disorder phase transition at a finite temperature. (2) Any Ising ferromagnet of the same structure as M, but with interactions tending to zero with distance more rapidly than those of M, cannot have a phase-transition. (3) The long-range-order parameter (thermal average of the spin-spin correlation at infinite distance) jumps discontinuously from zero in the disordered phase to a finite value in the ordered phase. All three properties have been conjectured by Anderson and Thouless to hold for a particular Ising ferromagnet which is relevant to the theory of the Kondo effect. Although M is not identical to Anderson's model, the results proved for M support the validity of the physical arguments of Anderson and Thouless.

THE JOURNAL OF CHEMICAL PHYSICS VOLUME 38, NUMBER 4 15 FEBRUARY 1963

Padé Approximant Studies of the Lattice Gas and Ising Ferromagnet below the Critical Point

JOHN W. ESSAM AND MICHAEL E. FISHER
Wheatstone Physics Laboratory, King's College, London, W.C.2, England
(Received 11 October 1962)

The Padé approximant procedure is used to study the low-temperature series for the Ising problem. The critical behavior of the spontaneous magnetization of an Ising ferromagnet and of the liquid-vapor coexistence curve of the corresponding lattice gas is found to be $I_0(T) \sim |\rho_{liq} - \rho_{gas}| \sim |T_c - T|^{\beta}$ where (in three dimensions) $0.303 \leq \beta \leq 0.318$. It is conjectured that $\beta = 5/16 = 0.31250$. The low-temperature initial susceptibility and the compressibility of the lattice gas at condensation are found to diverge as $\chi_0(T) \sim \kappa_0(T) \sim |T_c - T|^{-\gamma'}$ where in two dimensions $\gamma' = 1.75 \pm 0.01$ while in three dimensions $\gamma' = 1.25$ ($+0.07$, -0.02). Consideration of a heuristic model partition function suggests the identity $\alpha' + 2\beta + \gamma' = 2$ where α' is the index of divergence of the specific heat below T_c. The amplitudes of the singularities are derived and extrapolation formulas are presented for the plane triangular, square, and honeycomb lattices and the three-dimensional face-centered, body-centered, and simple cubic lattices. The results are compared briefly with experimental evidence.

VOLUME 22, NUMBER 25 PHYSICAL REVIEW LETTERS 23 JUNE 1969

SCALING FORM OF THE SPIN-SPIN CORRELATION FUNCTION OF THE THREE-DIMENSIONAL ISING FERROMAGNET ABOVE THE CURIE TEMPERATURE

M. Ferer, M. A. Moore, and Michael Wortis,
Department of Physics, University of Illinois, Urbana, Illinois 61801
(Received 28 April 1969)

The spin-spin correlation function for the $S = \frac{1}{2}$ Ising ferromagnet is evaluated in zero magnetic field and for $T \geq T_c$ by the method of series expansions. Evidence is presented showing that the scaling of correlations is valid in the (weak) limit $r \to \infty$, κr fixed, but is not valid in the (strong) limit $\kappa a \to 0$, $r \gg a$, with κr arbitrary.

Reprinted from ANNALS OF PHYSICS
Vol. 47, No. 3, May 1968
Printed in Belgium

Fluctuations and Lambda Phase Transition in Liquid Helium

R. A. FERRELL, N. MENYHÀRD, H. SCHMIDT, F. SCHWABL, AND P. SZÉPFALUSY

Department of Physics, University of Virginia, Charlottesville, Virginia 22904

A general dynamical scaling theory of phase transitions is established by exploiting the absence of a characteristic length in an extended system at its phase transition. This similarity property imposes strong constraints on the frequency and wave-number dependence of the fluctuation spectrum and leads to unambiguous predictions concerning the critical properties. The theory is worked out in detail for the lambda transition of liquid Helium, as a prototype. The fluctuations resulting from density waves at low temperature and from both first and second sound at higher temperatures, are closely examined. The connection between the divergent fluctuations in the second sound modes (as the temperature T approaches the lambda temperature T_λ) and the critical variation of the damping coefficients is established. The critical temperature dependences of the thermal conductivity of He I and the damping of first and second sound in He II are predicted to be essentially $(T - T_\lambda)^{-1/3}$, $(T_\lambda - T)^{-1}$, and $(T_\lambda - T)^{-1/3}$. (There occur also logarithmic factors.) Recent experimental measurements of the first and second of these give quantitative verification of the theory. In addition to the temperature dependence, the dynamical scaling theory does not employ any adjustable parameter and consequently also predicts the absolute magnitude of these quantities. Although this is a less conclusive check on the theory, the agreement is also satisfactory.

Physics Vol. 3, No. 5, pp. 255-283, 1967. Physics Publishing Co. Printed in Great Britain.

THE THEORY OF CONDENSATION AND THE CRITICAL POINT

MICHAEL E. FISHER

Baker Laboratory, Cornell University, Ithaca New York 14850

(*Received* 27 *March* 1967)

Abstract

The droplet or cluster theory of condensation is reviewed critically and extended. It is shown to imply that the condensation point is marked by a singularity of the thermodynamic potential as conjectured by Mayer. The singularity turns out to be an essential singularity at which all derivatives of the thermodynamic variables remain finite. The theory also yields an understanding of the uniqueness of the critical point (in contrast to an extended critical region or Derby-hat type of behaviour) and leads to relations between the various critical point singularities.

A one-dimensional model is described with a Hamiltonian containing short-range many-body potentials. The exact solution of the model is sketched and shown to exhibit condensation and critical phenomena for suitable (fixed) potentials. The analysis confirms the conclusions of the cluster theory and thereby lends support to the validity of its underlying assumptions.

PHYSICAL REVIEW VOLUME 156, NUMBER 2 10 APRIL 1967

Theory of Critical-Point Scattering and Correlations. I. The Ising Model

MICHAEL E. FISHER
Baker Laboratory, Cornell University, Ithaca, New York

AND

ROBERT J. BURFORD
Wheatstone Physics Laboratory, King's College, London, England

(Received 19 October 1966)

The theory of the correlations and critical scattering of two- and three-dimensional nearest-neighbor Ising models is discussed critically. A distinction is drawn between $\kappa(T)$, the true inverse range of exponential decay of the correlations, and $\kappa_1(T)$, the effective range determined from the low-angle scattering intensity. Ten to eleven terms of appropriate high-temperature series exapnsions for κ and κ_1 are determined for the square and simple cubic lattices, and shorter series are given for the triangular, fcc, and bcc lattices. For the former lattices, the complete correlation expansions are obtained to the same order. It is shown that κ and κ_1 vary as $(T-T_0)^\nu$ when $T\to T_c$, with $\nu=1$ for dimensionality $d=2$, but $\nu=0.6430\pm0.0025\simeq9/14$ for $d=3$. The asymptotic decay of correlation *at* $T=T_c$ is found to be $1/r^{d-2+\eta}$, where η is related to the exponent γ of the divergence of the susceptibility by $(2-\eta)\nu=\gamma$, Numerical values are $\eta=\frac{1}{4}$ for $d=2$ and $\eta=0.056\pm0.008\simeq1/18$ for $d=3$. The relative scattering intensity $\hat{\chi}$ as a function of wave number $\mathbf{k}$ is given to high accuracy for all $T\geqslant T_c$ by

$$\hat{\chi}(\mathbf{k},T)\simeq\left(\frac{a}{r_1}\right)^{2-\eta}\frac{[(\kappa_1a)^2+\phi^2a^2K^2(\mathbf{k})]^{\eta/2}}{[(\kappa_1a)^2+\psi a^2K^2(\mathbf{k})]},$$

where (i) a is the lattice spacing, (ii) $a^2K^2=2d[1-q^{-1}\sum\exp(i\mathbf{k}\cdot\mathbf{r})]\simeq(ka)^2$, the sum being over the q nearest-neighbor lattice sites, (iii) $r_1(T)$ is a slowly-varying decreasing function near T_c, (iv) $\psi=1+\frac{1}{2}\eta\phi^2$, and (v) $\phi(T)$ is slowly varying with a magnitude at T_c of 0.03 for $d=2$ and of 0.06 to 0.09 for $d=3$. Explicit formulas are given for κ_1, r_1, and ϕ as functions of T. The correlations and the scattering are isotropic near T_c. The critical scattering isotherm is curved for low k according to $\hat{\chi}^{-1}\sim k^{2-\eta}$ and it *intersects* the isotherms for $T>T_c$. Correspondingly, $\hat{\chi}(\mathbf{k},T)$ exhibits a maximum for fixed $\mathbf{k}$, at a temperature *above* T_c; for $d=2$ the maxima are very well marked, but for $d=3$ they are smaller and occur closer to T_c. The theory is compared favorably with recent neutron-scattering experiments on pure beta-brass.

PHYSICAL REVIEW B VOLUME 5, NUMBER 7 1 APRIL 1972

Theory of Critical-Point Scattering and Correlations. II. Heisenberg Models

Douglas S. Ritchie and Michael E. Fisher
Baker Laboratory, Cornell University, Ithaca, New York 14850

(Received 4 September 1971)

The spin-spin correlation functions and the critical-scattering intensity for Heisenberg models of general spin, $S=\frac{1}{2}$ to ∞, on the sc, bcc, and fcc lattices are studied on the basis of high-temperature series expansions along the lines developed in Paper I [M. E. Fisher and R. J. Burford, Phys. Rev. 156, 583 (1967)]. Subject to increased uncertainties for low spin, it is concluded that the exponents $\gamma=1.375^{+0.02}_{-0.01}$, $2\nu=1.405^{+0.02}_{-0.01}$, and $\eta=0.043\pm0.014$ describe all lattices and all spin. Explicit formulas are presented for the susceptibility/zero-angle scattering $\chi_0(T)$, for the inverse correlation length $\kappa_1(T)$, for the effective interaction range $r_1(T)$, and using the Fisher-Burford approximant, for the total scattering $\hat{\chi}(\vec{k},T)$. The shape parameter ϕ_c attains the "universal" value $\phi_c\simeq0.11$ for large spin but shows signs of spin dependence (and lattice dependence) for low spin. At fixed $\vec{k}$ the scattering is predicted to display a maximum *above* T_c determined by $\kappa_1(T_{max})/k\simeq0.10$ (for $S\gtrsim2$) to 0.15. A detailed study is made of the structure dependence of the critical-point correlations $\langle S_0^zS_r^z\rangle_c$ for various models. This leads to the revised, universal estimate $\phi_c\simeq0.15$ for all three cubic lattice, spin-$\frac{1}{2}$ Ising models. The results are compared d briefly with various experiments which support $\eta\gtrsim0.05$.

Reprinted without change of pagination from the
Proceedings of the Royal Society, A, *volume* 275, pp. 257–270, 1963

An application of Padé approximants to Heisenberg ferromagnetism and antiferromagnetism

By J. Gammel, W. Marshall and L. Morgan
Atomic Energy Research Establishment, Harwell, Berkshire, England

(*Communicated by R. E. Peierls, F.R.S.—Received* 7 *March* 1963)

The method of successive Padé approximants is applied to the high-temperature series for the susceptibility of Heisenberg systems with nearest-neighbour interactions only. It is concluded that the susceptibility of a ferromagnet diverges near the critical temperature with a law $(T-T_c)^{-r}$, where r is either exactly $\frac{4}{3}$ or indistinguishable from $\frac{4}{3}$ by the method used with the power series at present available. The susceptibility of an antiferromagnet is also discussed. The specific heat near the critical point is considered but the results are inconclusive.

PHYSICAL REVIEW B VOLUME 1, NUMBER 3 1 FEBRUARY 1970

Low-Temperature Critical Exponents from High-Temperature Series: The Ising Model

David S. Gaunt and George A. Baker, Jr.
Applied Mathematics Department, Brookhaven National Laboratory, Upton, New York 11973
(Received 1 July 1969)

First, a new method proposed by Baker and Rushbrooke is used to study the simple ferromagnetic Ising model *at* and *below* the Curie temperature. Of course, the properties of the Ising model are already well known, so that the main aim here is to assess the potential and reliability of the new method, since it has wide applicability to other models which have not been otherwise studied. Between 8 and 16 coefficients of exact *high-temperature* expansions for fixed values of the magnetization are derived for various two- and three-dimensional lattices. A Padé-approximant analysis of these expansions at the critical isotherm and magnetic phase boundary enables us to estimate the critical exponents β, γ', and δ, and plot the spontaneous magnetization. The results are in good agreement with previous calculations. Secondly, an analysis of the exact series expansions provides no support for the conjecture that the phase boundary is a line of essential singularities. However, the same expansions strongly suggest the existence of a "spinodal" curve, whose properties are in reasonable agreement with the predictions of various heuristic arguments (based essentially upon analyticity at the phase boundary and one-phase homogeneity in the critical region). Finally, structure and a mild extension of the proven analyticity of the free energy are used to show the $\Delta \leq \Delta', \gamma \leq \gamma'$.

PHYSICAL REVIEW B VOLUME 2, NUMBER 3 1 AUGUST 1970

Magnetic Phase Boundary of the Spin-½ Heisenberg Ferromagnetic Model

George A. Baker, Jr.
Applied Mathematics Department, Brookhaven National Laboratory, Upton, Long Island, New York 11973

and

J. Eve
Computing Laboratory, University of Newcastle upon Tyne, Newcastle upon Tyne, England

and

G. S. Rushbrooke
School of Physics, University of Newcastle upon Tyne, Newcastle upon Tyne, England
(Received 16 December 1969)

We compute in this paper the exact coefficients, in a magnetic field, of the first eight powers of $x = J/\kappa T$ for the spin-$\frac{1}{2}$ Heisenberg ferromagnetic model for five lattices. We use this result to obtain the magnetic equation of state in the form $\tanh(H/\kappa T) = Mg(x, M^2)$; again we obtain the exact coefficients for the first eight powers of x. From this equation, we are able to extrapolate and determine the magnetic phase boundary with reasonable accuracy for $M < 0.8$. We have investigated the critical exponents ι, β, and δ. We are unable to determine ι; we find an effective β over a range of M of $\beta = 0.35 \pm 0.05$, and we estimate $\delta = 5.0 \pm 0.2$.

JOURNAL OF MATHEMATICAL PHYSICS VOLUME 4, NUMBER 2 FEBRUARY 1963

Time-Dependent Statistics of the Ising Model

Roy J. Glauber
Lyman Laboratory of Physics, Harvard University, Cambridge, Massachusetts

The individual spins of the Ising model are assumed to interact with an external agency (e.g., a heat reservoir) which causes them to change their states randomly with time. Coupling between the spins is introduced through the assumption that the transition probabilities for any one spin depend on the values of the neighboring spins. This dependence is determined, in part, by the detailed balancing condition obeyed by the equilibrium state of the model. The Markoff process which describes the spin functions is analyzed in detail for the case of a closed N-member chain. The expectation values of the individual spins and of the products of pairs of spins, each of the pair evaluated at a different time, are found explicitly. The influence of a uniform, time-varying magnetic field upon the model is discussed, and the frequency-dependent magnetic susceptibility is found in the weak-field limit. Some fluctuation–dissipation theorems are derived which relate the susceptibility to the Fourier transform of the time-dependent correlation function of the magnetization at equilibrium.

Extended Thermodynamic Scaling from a Generalized Parametric Form

M. S. Green
Temple University, Philadelphia, Pennsylvania 19122

and

M. J. Cooper and J. M. H. Levelt Sengers
Institute for Basic Standards, National Bureau of Standards, Washington, D. C. 20234
(Received 15 January 1971)

A generalization of the parametric representation for thermodynamic scaling is proposed, introducing a new critical exponent ϵ. Expansions about the critical point are deduced for the fluids, and to lowest order the asymptotic power-law forms are recovered. An exponent $1-\alpha'$ is obtained for the diameter of the coexistence curve. Experimental data are shown to support the predicted forms.

Reprinted from THE JOURNAL OF CHEMICAL PHYSICS, Vol. 43, No. 6, 1958–1968, 15 September 1965
Printed in U. S. A.

Ferromagnets and Simple Fluids near the Critical Point: Some Thermodynamic Inequalities

ROBERT B. GRIFFITHS
Physics Department, Carnegie Institute of Technology, Pittsburgh, Pennsylvania
(Received 24 May 1965)

It is often assumed that thermodynamic variables have a simple power-law behavior near the liquid–vapor critical point of a simple fluid; for example, that the difference in density of saturated liquid and vapor decreases as $(T_c-T)^{\beta}$ as the temperature T approaches its critical value T_c. Thermodynamic arguments are used to derive several inequalities relating exponents which describe the behavior of the isothermal compressibility, specific heat at constant volume, and various other quantities near the critical point. Many of the inequalities apply equally to the analogous problem of a ferromagnet near its Curie point. They are based on the usual "stability" or "convexity" conditions on the thermodynamic potentials together with other plausible, but less general, hypotheses.

Reprinted from THE PHYSICAL REVIEW, Vol. 158, No. 1, 176–187, 5 June 1967
Printed in U. S. A.

Thermodynamic Functions for Fluids and Ferromagnets near the Critical Point

ROBERT B. GRIFFITHS
Physics Department, Carnegie Institute of Technology, Pittsburgh, Pennsylvania
(Received 6 January 1967)

Modifications to the classical analysis of equilibrium thermodynamic properties near the liquid-vapor critical point are proposed in order to allow for infinite singularities in C_v, nonclassical behavior of the coexistence curve, etc. A requirement that all thermodynamic functions for the homogeneous fluid be analytic is retained and turns out to be necessary in order to justify Maxwell's prescription for modifying the Van der Waals equation of state. A corresponding analysis is presented for ferromagnets near the Curie point. Widom's proposed "homogeneous" equation of state is discussed with special attention to requirements of thermodynamic stability. Several examples of such homogeneous functions are constructed, including cases where the critical indices agree (very nearly, at least) with current estimates for two- and three-dimensional Ising models.

VOLUME 24, NUMBER 13 PHYSICAL REVIEW LETTERS 30 MARCH 1970

THERMODYNAMICS NEAR THE TWO-FLUID CRITICAL MIXING POINT IN He^3-He^4

Robert B. Griffiths
Department of Physics, Carnegie-Mellon University, Pittsburgh, Pennsylvania 15213
(Received 20 February 1970)

The two-fluid critical mixing point in He^3-He^4 differs from ordinary critical points in that it occurs at the intersection of three lines of critical points, in a suitable variable space. A free-energy function is proposed which removes certain discrepancies between classical (Landau) theory and experimental thermodynamic measurements. Certain solid-state transitions (e.g., the metamagnetic-antiferromagnetic transition in $FeCl_2$) are thermodynamic analogs of critical mixing in He^3-He^4.

VOLUME 24, NUMBER 26 PHYSICAL REVIEW LETTERS 29 JUNE 1970

DEPENDENCE OF CRITICAL INDICES ON A PARAMETER

Robert B. Griffiths
Physics Department, Carnegie-Mellon University, Pittsburgh, Pennsylvania 15213
(Received 4 May 1970)

A number of lattice models of phase transitions and critical points can be related to one another by variation of linear parameters appearing in the Hamiltonian. It is suggested that the critical indices as functions of these parameters remain constant except possibly at points where the nature of the associated first-order phase transition is itself altered by varying a parameter. Several models exhibiting critical phenomena are consistent with this proposal.

PHYSICAL REVIEW A VOLUME 2, NUMBER 3 SEPTEMBER 1970

Critical Points in Multicomponent Systems

Robert B. Griffiths
Physics Department, Carnegie-Mellon University, Pittsburgh, Pennsylvania 15213
and
John C. Wheeler
Chemistry Department, University of California at San Diego, La Jolla, California 92038
(Received 27 March 1970)

The thermodynamics of critical points in multicomponent systems, more generally systems with more than two independent variables (including binary fluid mixtures, the helium λ transition, order-disorder transitions in alloys, and antiferromagnetism) are discussed from a unified geometrical point of view, in analogy with one-component (liquid-vapor and simple-ferromagnetic) systems. It is shown that, from a few simple postulates, the qualitative behavior near the critical point of quantities such as compressibilities, susceptibilities, and heat capacities, with different choices of the variables held fixed, can be easily predicted. A number of seemingly exceptional cases (such as critical azeotropy), which arise when critical or coexistence surfaces bear an "accidental" geometrical relationship with the thermodynamic coordinate axes, are explained in terms of the same postulates. The predicted results are compared with several theoretical models and experimental data for a variety of systems.

The Principle of Corresponding States

E. A. GUGGENHEIM
Imperial College, London

(Received May 3, 1945)

The principle of corresponding states in modern form has been applied to the following properties: the critical state, the virial coefficient, the Boyle point, the densities of coexistent phases, the vapor pressure of the liquid, the entropies of evaporation and of fusion, the coefficient of thermal expansion of the liquid, the triple-point temperature and pressure, the heat capacity of the liquid, and the surface tension of the liquid. It has been shown that argon, krypton, xenon, and with less accuracy neon, follow the principle with respect to all these properties. It has further been shown that nitrogen, oxygen, carbon monoxide, and methane follow the principle with fair accuracy as vapors and as liquids, but not as solids. The relations between surface tension, temperature, and densities have been analyzed empirically. For the "ideal" substances under consideration Katayama's modification of Eötvos' relation holds good, but McLeod's relation does not; in the relation $\gamma \propto (\rho_l - \rho_c)^s$, the exponent s is not 4 but much more nearly $3\frac{2}{3}$.

PHYSICAL REVIEW VOLUME 166, NUMBER 1 5 FEBRUARY 1968

Condensation of the Ideal Bose Gas as a Cooperative Transition

J. D. GUNTON AND M. J. BUCKINGHAM
Department of Physics, University of Western Australia, Nedlands, Western Australia

(Received 25 August 1967)

The thermodynamic properties of the noninteracting Bose gas in the neighborhood of its transition are examined in detail. The order parameter is a complex extensive variable, but the thermodynamic properties depend only on its amplitude under simple boundary conditions. As the dimensionality or the single-particle energy spectrum is varied, the critical singularity displays a variety of forms. The equation of state has a simple structure, different from the homogeneous form often discussed for critical systems but asymptotically reducing to the latter *except* when logarithmic singularities are involved. The correlation function in the critical region is a homogeneous function of the distance and a correlation length. Only for a quadratic energy spectrum is the Ornstein–Zernike theory result valid at the critical temperature. A precise correspondence is noted between the asymptotic properties of the ideal Bose gas transition and those of the spherical model of ferromagnetism.

PHYSICAL REVIEW VOLUME 177, NUMBER 2 10 JANUARY 1969

Scaling Laws for Dynamic Critical Phenomena

B. I. HALPERIN AND P. C. HOHENBERG
Bell Telephone Laboratories, Murray Hill, New Jersey 07974

(Received 13 August 1968)

The usual static scaling laws are generalized to nonequilibrium phenomena by making assumptions on the behavior of time-dependent correlation functions near the critical point of second-order phase transitions. At any temperature different from T_c, the correlation functions are assumed to reflect the hydrodynamic behavior of the system, for sufficiently long wavelengths and low frequencies. As the critical temperature is approached, however, the range of spatial correlations in the system diverges, and the domain of applicability of hydrodynamics is reduced to a vanishingly small region of wavelengths and frequencies. The dynamic-scaling assumptions lead to predictions for the behavior of the hydrodynamic parameters near T_c, as well as for the form of the correlation functions for macroscopic distances and times, outside the hydrodynamic range. In particular, singularities are predicted to occur in the temperature dependence of transport coefficients, and anomalies are expected in the frequency spectrum of certain operators, which are observable by inelastic scattering of neutrons or light. A distinction is made between the restricted dynamic-scaling hypothesis, which refers to the order parameter only, and extended dynamic scaling, which applies to other operators and involves stronger assumptions. Applications are discussed to antiferromagnets, ferromagnets, the gas-liquid critical point, and the λ transition in superfluid helium. Specific experiments are suggested to test the scaling assumptions, and existing experimental evidence is briefly reviewed. Finally, a comparison is made with other theories of dynamical behavior near critical points.

PHYSICAL REVIEW B VOLUME 6, NUMBER 9 1 NOVEMBER 1972

Systematic Application of Generalized Homogeneous Functions to Static Scaling, Dynamic Scaling, and Universality

Alex Hankey and H. Eugene Stanley
Physics Department, Massachusetts Institute of Technology, Cambridge, Massachusetts 02139
(Received 12 August 1971; revised manuscript received 24 May 1972)

A function $f(x_1, x_2, \ldots, x_n)$ is a generalized homogeneous function (GHF) if we can find numbers $a_1, a_2, \ldots, a_n$ such that for all values of the positive number λ, $f(\lambda^{a_1}x_1, \lambda^{a_2}x_2, \ldots, \lambda^{a_n}x_n) = \lambda^{a_f} f(x_1, x_2, \ldots, x_n)$. We organize the properties of GHFs in four theorems. These are used to systematically examine the consequences of various scaling hypotheses. An advantage of this approach is that the same formalism may be used to treat thermodynamic functions, static correlation functions, dynamic correlation functions, and "universality." The simple case of thermodynamic scaling (two independent variables) is first generalized to static and dynamic correlation functions (three and four variables), and then to scaling with a parameter (for which the critical subspace becomes higher dimensional). In this last case, where a second GHF hypothesis is made, the necessity of crossover lines is demonstrated. The assumption of homogeneity is clearly separated from any extra assumptions that may also be called scaling (or "strong scaling"), but are independent of and different from that of homogeneity. One practical insight gained from the present approach is that *all experimentally measured exponents are expressible as the ratio of two scaling powers*, a_f (which refers to the function) and a_j (which refers to the path of approach to the critical point). A second practical advantage is that, since a GHF can be scaled with respect to any of its arguments, one can immediately write a variety of scaling functions for each type of scaling hypothesis. The GHF approach thereby permits data to be plotted in a variety of convenient fashions, and is found to facilitate computation of the relevant scaling functions (in particular, the GHF approach led directly to the recent calculation of the Heisenberg model scaling function by Milošević and Stanley).

VOLUME 29, NUMBER 5 PHYSICAL REVIEW LETTERS 31 JULY 1972

Geometric Predictions of Scaling at Tricritical Points

Alex Hankey, H. Eugene Stanley, and T. S. Chang
Physics Department, Massachusetts Institute of Technology, Cambridge, Massachusetts 02139
(Received 24 May 1972)

The shapes of the three critical lines meeting at a tricritical point are discussed in the light of the homogeneity hypothesis. We give a complete listing of possible shapes consistent with scaling. Certain possible geometries are thereby shown to be *inconsistent* with scaling. Application of these ideas to the crossover regions is shown to considerably simplify the "metric problem" posed by Riedel.

PHYSICAL REVIEW VOLUME 158, NUMBER 2 10 JUNE 1967

Existence of Long-Range Order in One and Two Dimensions

P. C. Hohenberg
Bell Telephone Laboratories, Murray Hill, New Jersey
(Received 24 October 1966)

It is pointed out that a rigorous inequality first proved by Bogoliubov may be used to rule out the existence of quasi-averages (or long-range order) in Bose and Fermi systems for one and two dimensions and $T \neq 0$.

PHYSICAL REVIEW A VOLUME 6, NUMBER 1 JULY 1972

Gravity Effects near the Gas-Liquid Critical Point

P. C. Hohenberg and M. Barmatz

Bell Laboratories, Murray Hill, New Jersey 07974

(Received 16 December 1971)

The "linear-model" parametric equation of state of Schofield, Litster, and Ho is used to analyze the effect of gravity near the gas-liquid critical point of a fluid. Detailed results are presented on the density distribution as a function of height, the constant-volume specific heat, and the low-frequency sound velocity, for arbitrary points in the (ρ, T) plane near the critical point. The influence of gravity on the determination of critical exponents is also considered. It is concluded that even for the thinnest practical samples, gravity corrections may have a significant effect on the exponents. The present theory permits gravity corrections to be made in a self-consistent way, with high accuracy.

Beitrag zur Theorie des Ferromagnetismus

Von **Ernst Ising** in Hamburg.

(Eingegangen am 9. Dezember 1924.)

Es wird im wesentlichen das thermische Verhalten eines linearen, aus Elementarmagneten bestehenden Körpers untersucht, wobei im Gegensatz zur W e i s s schen Theorie des Ferromagnetismus kein molekulares Feld, sondern nur eine (nicht magnetische) Wechselwirkung benachbarter Elementarmagnete angenommen wird. Es wird gezeigt, daß ein solches Modell noch keine ferromagnetischen Eigenschaften besitzt und diese Aussage auch auf das dreidimensionale Modell ausgedehnt.

PHYSICAL REVIEW VOLUME 176, NUMBER 2 10 DECEMBER 1968

High-Temperature Critical Indices for the Classical Anisotropic Heisenberg Model

DAVID JASNOW AND MICHAEL WORTIS

Department of Physics, University of Illinois, Urbana, Illinois 61801

(Received 15 July 1968)

High-temperature series expansions for the spin-spin correlation function of the classical anisotropic Heinsenberg model are calculated for various lattices and anisotropies through order T^{-8} (close-packed lattices) and T^{-9} (loose-packed lattices). These series are combined and then extrapolated to give the high-temperature critical indices γ (susceptibility), ν (correlation range), and α (specific heat) as functions of anisotropy. Our results are consistent with the hypothesis that the critical indices change only when there is a change in the symmetry of the system, e.g., in interpolating between the Ising and isotropic Heisenberg models, indices remain Ising-like until the system becomes isotropic, at which point they appear to change discontinuously. Previous results for the limiting cases are confirmed and extended.

J. PHYS. C (SOLID ST. PHYS.), 1969, SER. 2, VOL. 2. PRINTED IN GREAT BRITAIN

Equation of state near the critical point

B. D. JOSEPHSON
Solid State Theory Section, Cavendish Laboratory, University of Cambridge
MS. received 9*th January* 1969, *in revised form 2nd April* 1969

Abstract. Two functions of state ξ and η are derived, such that any equation of state for the critical region consistent with the scaling hypothesis can be expressed in the form $\eta = \phi(\xi)$. The representation has the useful property that the function ϕ has no unphysical singularities (such as the branch points associated with the critical isotherm or elsewhere in the usual representations of scaling-law equations of state).

VOLUME 22, NUMBER 12 PHYSICAL REVIEW LETTERS 24 MARCH 1969

PARAMETRIC REPRESENTATION OF THE EQUATION OF STATE NEAR A CRITICAL POINT

P. Schofield
Department of Physics and Center for Materials Science and Engineering, Massachusetts Institute of Technology, Cambridge, Massachusetts
(Received 2 January 1969)

A parametric representation of the usual thermodynamic variables in the neighborhood of a critical point is proposed in terms of new variables r and θ. The representation is chosen so that the "scaling-law" behavior is entirely contained in the r dependence, the θ dependence being free of critical singularities. Preliminary calculations show that the θ dependence may be chosen to have a rather simple form both in a ferromagnet and in fluids.

Reprinted from THE PHYSICAL REVIEW, Vol. 146, No. 1, 349–358, 3 June 1966
Printed in U. S. A.

Spherical Model with Long-Range Ferromagnetic Interactions

G. S. JOYCE
Wheatstone Physics Laboratory, King's College, London, England
(Received 17 January 1966)

The behavior of the spherical model of a ferromagnet with an interaction energy between the magnetic spins which varies with distance as $1/r^{d+\sigma}$ (where d is the dimensionality of the lattice and $\sigma>0$) is analyzed. It is shown that the model exhibits a ferromagnetic transition in one and two dimensions, providing $0<\sigma<d$. (The usual spherical model with nearest-neighbor interactions does not have a transition in one and two dimensions.) The critical-point behavior is investigated. It is found that the singularities in the specific heat and susceptibility are dependent on σ and d, but the behavior of the magnetization is independent of σ and d. In three dimensions the susceptibility diverges as $(T-T_c)^{-\gamma}$, where $\gamma=1$ for $0<\sigma<\frac{3}{2}$, $\gamma=\sigma/(3-\sigma)$ for $\frac{3}{2}<\sigma<2$ and $\gamma=2$ for $\sigma>2$. The asymptotic form of the spin-spin correlation function $\Gamma(\mathbf{r})$ is studied in the neighborhood of the critical temperature T_c. At $T=T_c$, $\Gamma(\mathbf{r})$ decays for large r as $1/r^{d-\sigma}$. Several two-dimensional models with long-range interactions falling off as $1/r^2$ in certain directions only are also investigated.

JOURNAL OF MATHEMATICAL PHYSICS VOLUME 4, NUMBER 2 FEBRUARY 1963

On the van der Waals Theory of the Vapor–Liquid Equilibrium. I. Discussion of a One-Dimensional Model

M. KAC, G. E. UHLENBECK, AND P. C. HEMMER
The Rockefeller Institute, New York, New York
(Received 17 September 1962)

For a one-dimensional fluid model where the pair interaction potential between the molecules consists of a hard core and an exponential attraction, Kac has shown that the partition function can be determined exactly in the thermodynamic limit. In Sec. II this calculation is reviewed and further discussed. In Sec. III, we show that in the so-called van der Waals limit when the range of the attractive force goes to infinity while its strength becomes proportionally weaker, a phase transition appears which is described exactly by the van der Waals equation plus the Maxwell equal-area rule. In Sec. IV the approach to the van der Waals limit is discussed by an appropriate perturbation method applied to the basic integral equation. The perturbation parameter is the ratio of the size of the hard core to the range of the attractive force. It is seen that the phase transition persists in any order of the perturbation. The two-phase equilibrium is characterized by the fact that in this range of density, the maximum eigenvalue of the integral equation is doubly degenerate and that the corresponding two eigenfunctions do not overlap. In Sec. V we comment on the relevance of our results for the three-dimensional problem.

Physics Vol. 2, No. 6, pp. 263-272, 1966. Physics Publishing Co. Printed in Great Britain.

SCALING LAWS FOR ISING MODELS NEAR T_c

LEO P. KADANOFF
Department of Physics, University of Illinois
Urbana, Illinois

(Received 3 February 1966)

A model for describing the behavior of Ising models very near T_c is introduced. The description is based upon dividing the Ising model into cells which are microscopically large but much smaller than the coherence length and then using the total magnetization within each cell as a collective variable. The resulting calculation serves as a partial justification for Widom's conjecture about the homogeneity of the free energy and at the same time gives his result $s\nu' = \gamma' + 2\beta$.

VOLUME 23, NUMBER 25 PHYSICAL REVIEW LETTERS 22 DECEMBER 1969

OPERATOR ALGEBRA AND THE DETERMINATION OF CRITICAL INDICES

Leo P. Kadanoff
Department of Physics, Brown University, Providence, Rhode Island 02912
(Received 8 September 1969)

The "reduction hypothesis" proposes that a product of nearby fluctuating local variables can be replaced by a linear combination of individual local variables. The linear combinations thereby produced are a kind of algebra of the reduction of products. A particular algebra is proposed for the two-dimensional Ising model. It is shown that a knowledge of which coefficients in the algebra are nonvanishing is sufficient to determine all critical indices.

Reprinted from ANNALS OF PHYSICS, Volume 24, October 1963

ANNALS OF PHYSICS: **24,** 419–469 (1963)

Hydrodynamic Equations and Correlation Functions

LEO P. KADANOFF

Physics Department, University of Illinois, Urbana, Illinois

AND

PAUL C. MARTIN

Lyman Laboratory of Physics, Harvard University, Cambridge, Massachusetts

The response of a system to an external disturbance can always be expressed in terms of time dependent correlation functions of the undisturbed system. More particularly the linear response of a system disturbed slightly from equilibrium is characterized by the expectation value in the equilibrium ensemble, of a product of two space- and time-dependent operators. When a disturbance leads to a very slow variation in space and time of all physical quantities, the response may alternatively be described by the linearized hydrodynamic equations. The purpose of this paper is to exhibit the complicated structure the correlation functions must have in order that these descriptions coincide. From the hydrodynamic equations the slowly varying part of the expectation values of correlations of densities of conserved quantities is inferred. Two illustrative examples are considered: spin diffusion and transport in an ordinary one-component fluid.

Since the descriptions are equivalent, all transport processes which occur in the nonequilibrium system must be exhibited in the equilibrium correlation functions. Thus, when the hydrodynamic equations predict the existence of a diffusion process, the correlation functions will include a part which satisfies a diffusion equation. Similarly when sound waves occur in the nonequilibrium system, they will also be contained in the correlation functions.

The description in terms of correlation functions leads naturally to expressions for the transport coefficients like those discussed by Kubo. The analysis also leads to a number of sum rules relating the dissipative linear coefficients to thermodynamic derivatives. It elucidates the peculiarly singular limiting behavior these correlations must have.

PHYSICAL REVIEW VOLUME 166, NUMBER 1 5 FEBRUARY 1968

Transport Coefficients near the Liquid-Gas Critical Point

LEO P. KADANOFF AND JACK SWIFT

Department of Physics, University of Illinois, Urbana, Illinois

(Received 23 August 1967; revised manuscript received 23 October 1967)

A perturbation theory for the determination of transport coefficients near the critical point is presented. This perturbation theory is based upon processes in which one transport mode decays into several low-wave-number modes. Scaling-law concepts are used to calculate the order of magnitude of the matrix elements and frequency denominators which appear in this theory. This permits the estimation of the order of magnitude of the transport coefficients near the critical point. In particular, this approach indicates that the thermal conductivity should diverge roughly as $(T-T_c)^{-2/3}$ on the critical isochore and coexistence curve, while the viscosity η should be either weakly divergent or strongly cusped at the critical point. On the other hand, the bulk viscosity ζ should diverge roughly as $(T-T_c)^{-2}$ for low frequencies, and as $(T-T_c)^{-2/3}$ for higher frequencies on the critical isochore near the critical point. Specific predictions are made for these quantities in terms of critical indices, and the connection between these relations and the scaling of frequencies is discussed.

Reprinted from ANNALS OF PHYSICS
Vol. 61, No. 1, November 1970
Printed in Belgium

Kinetic Equations and Time Correlation Functions of Critical Fluctuations

KYOZI KAWASAKI

Department of Physics and,
Department of Chemistry and Chemical Engineering,

University of Illinois, Urbana, Illinois

and

Bell Telephone Laboratories, Inc., Murray Hill, New Jersey

Received November 20, 1969

Near the critical point the characteristic time of motion associated with certain degrees of freedom experiences an enormous slowing-down. This circumstance gives rise to a possibility of constructing a kinetic equation to describe such slowed-down motions of the system, which corresponds to the Boltzmann equation of dilute gases which describes slow variations of the single-particle distribution function. We accomplish the derivation of kinetic equations with the aid of a generalized Langevin equation due to Mori. The theory is illustrated by deriving kinetic equations obeyed by critical fluctuations in isotropic and planar Heisenberg ferromagnets, an isotropic Heisenberg antiferromagnet, the liquid helium near the λ point, and a binary critical mixture. The kinetic equations conform to the dynamical scaling whenever it holds, and are valid in the hydrodynamic as well as in the critical regimes. The kinetic equations are then used to obtain selfconsistent closed equations to determine time correlation functions of critical fluctuations. In particular, in the case of a binary critical mixture, the selfconsistent equation can be used to obtain the diffusion constant and the decay rate of concentration fluctuation in the critical regime, which are expressed in terms of shear viscosity, and the results are in good numerical agreement with those of the recent light scattering experiments. The theory also leads to the modification of the Fixman correction $[1 + (q/\kappa)^2]$ to $[1 + (3/5)(q/\kappa)^2]$ when $q \ll \kappa$ where q and κ are the wavenumber and inverse correlation range of the concentration fluctuation, respectively.

AUGUST 1, 1941 PHYSICAL REVIEW VOLUME 60

Statistics of the Two-Dimensional Ferromagnet. Part I

H. A. KRAMERS, *University of Leiden, Leiden, Holland*

AND

G. H. WANNIER, *University of Texas, Austin, Texas*[1]

(Received April 7, 1941)

In an effort to make statistical methods available for the treatment of cooperational phenomena, the Ising model of ferromagnetism is treated by rigorous Boltzmann statistics. A method is developed which yields the partition function as the largest eigenvalue of some finite matrix, as long as the manifold is only one dimensionally infinite. The method is carried out fully for the linear chain of spins which has no ferromagnetic properties. Then a sequence of finite matrices is found whose largest eigenvalue approaches the partition function of the two-dimensional square net as the matrix order gets large. It is shown that these matrices possess a symmetry property which permits location of the Curie temperature if it exists and is unique. It lies at

$$J/kT_c = 0.8814$$

if we denote by J the coupling energy between neighboring spins. The symmetry relation also excludes certain forms of singularities at T_c, as, e.g., a jump in the specific heat. However, the information thus gathered by rigorous analytic methods remains incomplete.

AUGUST 1, 1941 PHYSICAL REVIEW VOLUME 60

Statistics of the Two-Dimensional Ferromagnet. Part II

H. A. KRAMERS, *University of Leiden, Leiden, Holland*

AND

G. H. WANNIER, *University of Texas, Austin, Texas*

(Received June 12, 1941)

The study of the two-dimensional Ising model is continued. Its specific heat at the Curie point is investigated. The quantity in question is computed for six successive finite matrix problems and the conclusion is drawn that the specific heat is infinite at the Curie point. A new closed form approximation of the partition function λ is then developed by using the matrix method in its variational form. The two power series for λ at extreme temperatures are used as a test for this and various other approximations, and it is found that the new result is a considerable improvement over the existing solutions. Finally it is pointed out that these closed form solutions support our conclusion as to the place and nature of the Curie point transition.

PHYSICAL REVIEW VOLUME 87, NUMBER 3 AUGUST 1, 1952

Statistical Theory of Equations of State and Phase Transitions. II. Lattice Gas and Ising Model

T. D. LEE AND C. N. YANG
Institute for Advanced Study, Princeton, New Jersey

(Received March 31, 1952)

The problems of an Ising model in a magnetic field and a lattice gas are proved mathematically equivalent. From this equivalence an example of a two-dimensional lattice gas is given for which the phase transition regions in the $p-v$ diagram is exactly calculated.

A theorem is proved which states that under a class of general conditions the roots of the grand partition function always lie on a circle. Consequences of this theorem and its relation with practical approximation methods are discussed. All the known exact results about the two-dimensional square Ising lattice are summarized, and some new results are quoted.

VOLUME 18, NUMBER 24 PHYSICAL REVIEW LETTERS 12 JUNE 1967

EXACT SOLUTION OF THE *F* MODEL OF AN ANTIFERROELECTRIC

Elliott H. Lieb
Physics Department, Northeastern University, Boston, Massachusetts

(Received 8 May 1967)

The Rys F model of an antiferroelectric is solved by the transfer-matrix method. The result is different in many respects from the analogous Ising antiferromagnet, i.e., an infinite-order phase transition and a natural boundary in the complex T plane. It can also be solved when an external electric field is included. Above the transition temperature the behavior is normal while below T_c there is no polarization unless the electric field is sufficiently large.

VOLUME 19, NUMBER 3 PHYSICAL REVIEW LETTERS 17 JULY 1967

EXACT SOLUTION OF THE TWO-DIMENSIONAL SLATER KDP MODEL OF A FERROELECTRIC

Elliott H. Lieb
Physics Department, Northeastern University, Boston, Massachusetts
(Received 31 May 1967)

The Slater KDP model is solved for all temperatures and with an electric field. Above T_c the specific heat behaves like $(T-T_c)^{-1/2}$ and the polarizability like $(T-T_c)^{-1}$. There is a first-order phase transition at T_c (latent heat). Below T_c the free energy is simply $-|\mathcal{E}|d$ ($\mathcal{E}$ = electric field, d = dipole moment).

VOLUME 29, NUMBER 14 PHYSICAL REVIEW LETTERS 2 OCTOBER 1972

Some Results Concerning the Crossover Behavior of Quasi–Two-Dimensional and Quasi–One-Dimensional Systems

Luke L. Liu and H. Eugene Stanley
Physics Department, Massachusetts Institute of Technology, Cambridge, Massachusetts 02139
(Received 27 April 1972)

A magnetic system with intraplanar and interplanar interaction strengths J and RJ is is treated. Rigorous relations are established concerning the first few derivatives with respect to R of the susceptibility $\chi(R)$. Considering $\chi(R) = b_0 + b_1R + b_2R^2 + \cdots$, we find b_1 and the order of magnitude of b_2. Hence we can predict when the system "crosses over" from d-dimensional to $\bar{d}$-dimensional behavior (e.g., for quasi–two-dimensional systems, $d=2$, $\bar{d}=3$, while for quasi–one-dimensional systems, $d=1$, $\bar{d}=3$). These results also support scaling with respect to the anisotropy parameter R.

VOLUME 21, NUMBER 8 PHYSICAL REVIEW LETTERS 19 AUGUST 1968

RANDOM IMPURITIES AS THE CAUSE OF SMOOTH SPECIFIC HEATS NEAR THE CRITICAL TEMPERATURE

Barry M. McCoy
Institute for Theoretical Physics, State University of New York, Stony Brook, New York 11790

and

Tai Tsun Wu
Gordon McKay Laboratory, Harvard University, Cambridge, Massachusetts 02138
(Received 1 July 1968)

We present a modification of the two-dimensional Ising model which incorporates random impurities. The specific heat of this model is infinitely differentiable even at the critical temperature where it possesses an essential singularity. We find this specific heat to be in perfect quantitative agreement with the smooth peak recently observed by van der Hoeven, Teaney, and Moruzzi for $T \gtrsim T_c$ in the specific heat of EuS.

SOVIET PHYSICS JETP VOLUME 28, NUMBER 5 MAY, 1969

A DIAGRAM TECHNIQUE NEAR THE CURIE POINT AND THE SECOND ORDER PHASE TRANSITION IN A BOSE LIQUID

A. A. MIGDAL

Submitted April 19, 1968

Zh. Eksp. Teor. Fiz. 55, 1964–1979 (November, 1968)

The behavior of a Bose liquid as the phase-transition curve is approached from the side of the normal phase is considered by means of finite-temperature Green's functions. A renormalization method for diagrams is proposed allowing to separate in all quantities their dependence on the distance ξ from the phase-transition curve. It is proved that the zero-frequency Green's functions which determine the static properties of the system and the singularities of the thermodynamic quantities are universal homogeneous functions of the momenta and of ξ. Thus, the momentum distribution of the particles (the one-particle Green's function) has the form $n(p) = \xi^{-\alpha} g(p/\xi^{\beta})$. The exponents of ξ in the thermodynamic correlation functions (the critical indices) can be expressed in terms of α and β. The function $g(x)$ and the numbers α and β are simply related to the renormalized phonon emission vertex, for which the equation does not contain any free parameters and can, in principle, be solved numerically. Various asymptotic behaviors of the Green's function are investigated. It is shown how these results can be generalized to the case of an arbitrary system near its phase-transition curve, using the unitarity and analyticity properties in exactly the same manner as for relativistic particles of mass $M = \xi^{\beta}$.

PHYSICAL REVIEW B VOLUME 6, NUMBER 3 1 AUGUST 1972

Equation of State near the Critical Point. I. Calculation of the Scaling Function for $S=\frac{1}{2}$ and $S=\infty$ Heisenberg Models Using High-Temperature Series Expansions

Sava Milošević and H. Eugene Stanley

Physics Department, Massachusetts Institute of Technology, Cambridge, Massachusetts 02139

(Received 3 January 1972)

In recent years there have been many measurements of the scaling-law equation of state for different materials, and the "scaling function" so obtained has generally been fit by an empirical equation involving the selection of several adjustable parameters. We propose a method for calculating, directly from high-temperature series expansions, the function $h(x)$ that determines the scaling-law equation of state $H = M^{\delta} h(x)$. Previously, $h(x)$ has been calculated only for the $S=\frac{1}{2}$ Ising model, but the method is not generalizable to the case of the Heisenberg model because it relies upon the use of low-temperature expansions as well, and these are not known for the Heisenberg model. First we calculate $h(x)$ for the Ising model (bcc, fcc, and simple cubic lattices) in order to assess the utility and credibility of our method. Our Ising model $h(x)$ agrees well with the previous calculation that used both high- and low-temperature expansions. Next we calculate $h(x)$ in its entire region of definition for the $S=\frac{1}{2}$ Heisenberg model (fcc and bcc lattices) and the $S=\infty$ Heisenberg models (fcc lattice), where S denotes the spin quantum number. The accuracy of our resulting expressions is limited by the finite number of known terms in the corresponding high-temperature series expansions, but it is generally of the order of a few percent. In Paper II the scaling functions calculated here are compared with experiment and with the predictions of the universality hypothesis.

PHYSICAL REVIEW B VOLUME 6, NUMBER 3 1 AUGUST 1972

Equation of State near the Critical Point. II. Comparison with Experiment and Possible Universality with Respect to Lattice Structure and Spin Quantum Number

Sava Milošević and H. Eugene Stanley
Physics Department, Massachusetts Institute of Technology, Cambridge, Massachusetts 02139
(Received 3 January 1972)

The scaling functions for the $S=\frac{1}{2}$ Ising, $S=\frac{1}{2}$ Heisenberg, and $S=\infty$ Heisenberg models are compared with experimental measurements of $h(x)$ on the insulating ferromagnets $CrBr_3$ ($S=\frac{3}{2}$) and EuO ($S=\frac{7}{2}$), with the metallic ferromagnet Ni and with the conducting alloy Pd_3Fe (in both *disordered* and *ordered* states). The data agree much better with our Heisenberg model $h(x)$ than with the Ising model $h(x)$. Comparison with experimental data is made by using plots of scaled magnetization $M_H \equiv M/H^{1/\delta}$ vs scaled temperature $\epsilon_H \equiv \epsilon/H^{1/\beta\delta}$, which has the virtue that all points fall upon a *single* curve and log-log plots need not be resorted to. We proceed to consider the extent to which the equation of state depends upon parameters appearing in the Hamiltonian (the "universality" question). We find that our calculations indicate that $h(x)$, if properly normalized, does *not* depend on lattice structure and is very likely independent of spin quantum number S. Moreover, the fact that our calculated $h(x)$ agrees with data on $CrBr_3$ (for which there exists considerable *lattice anisotropy*) and on EuO (for which there exists nonnegligible *next-nearest-neighbor interactions*) suggests the further conjecture that $h(x)$ might be independent of these features of the Hamiltonian. By contrast, the change in $h(x)$ on going from Ising coupling ($D=1$) to Heisenberg coupling ($D=3$) was quite substantial, and indeed would seem to account for the fact that earlier workers did not obtain agreement between the Ising model $h(x)$ and experimental data on magnetic systems. We conclude with the working hypothesis that the scaling function depends principally on spin dimensionality D and on lattice dimensionality d.

VOLUME 22, NUMBER 18 PHYSICAL REVIEW LETTERS 5 MAY 1969

SPIN-SPIN CORRELATION FUNCTION OF THE THREE-DIMENSIONAL ISING FERROMAGNET ABOVE THE CURIE TEMPERATURE

M. A. Moore, David Jasnow, and Michael Wortis
Department of Physics, University of Illinois, Urbana, Illinois 61801
(Received 20 March 1969)

High-temperature series expansions for the Ising model's spin-spin correlation function are found to order $(J/kT)^{12}$ on the simple cubic, bcc, and fcc latices. Analysis of moment series indicates that the scaling form of the correlation function is correct, at least in the limit κr fixed, $\kappa \to 0$. Values are obtained for the correlation indices ν and η—the latter by two methods, thereby providing a direct check on the scaling law.

PHYSICAL REVIEW VOLUME 65, NUMBERS 3 AND 4 FEBRUARY 1 AND 15, 1944

Crystal Statistics. I. A Two-Dimensional Model with an Order-Disorder Transition

LARS ONSAGER
Sterling Chemistry Laboratory, Yale University, New Haven, Connecticut

(Received October 4, 1943)

The partition function of a two-dimensional "ferromagnetic" with scalar "spins" (Ising model) is computed rigorously for the case of vanishing field. The eigenwert problem involved in the corresponding computation for a long strip crystal of finite width (n atoms), joined straight to itself around a cylinder, is solved by direct product decomposition; in the special case $n=\infty$ an integral replaces a sum. The choice of different interaction energies ($\pm J, \pm J'$) in the (0 1) and (1 0) directions does not complicate the problem. The two-way infinite crystal has an order-disorder transition at a temperature $T=T_c$ given by the condition

$$\sinh(2J/kT_c)\sinh(2J'/kT_c)=1.$$

The energy is a continuous function of T; but the specific heat becomes infinite as $-\log|T-T_c|$. For strips of finite width, the maximum of the specific heat increases linearly with $\log n$. The order-converting dual transformation invented by Kramers and Wannier effects a simple automorphism of the basis of the quaternion algebra which is natural to the problem in hand. In addition to the thermodynamic properties of the massive crystal, the free energy of a (0 1) boundary between areas of opposite order is computed; on this basis the mean ordered length of a strip crystal is

$$(\exp(2J/kT)\tanh(2J'/kT))^n.$$

PHYSICAL REVIEW VOLUME 76, NUMBER 8 OCTOBER 15, 1949

Crystal Statistics. II. Partition Function Evaluated by Spinor Analysis

BRURIA KAUFMAN*
Columbia University, New York City, New York

(Received May 11, 1949)

The partition function for a two-dimensional binary lattice is evaluated in terms of the eigenvalues of the 2^n-dimensional matrix **V** characteristic for the lattice. Use is made of the properties of the 2^n-dimensional "spin"-representation of the group of rotations in $2n$-dimensions. In consequence of these properties, it is shown that the eigenvalues of **V** are known as soon as one knows the angles of the $2n$-dimensional rotation represented by **V**.

Together with the eigenvalues of **V**, the matrix $\mathbf{\Psi}$ which diagonalizes **V** is obtained as a spin-representation of a known rotation. The determination of $\mathbf{\Psi}$ is needed for the calculation of the degree of order.

The approximation, in which all the eigenvalues of **V** but the largest are neglected, is discussed, and it is shown that the exact partition function does not differ much from the approximate result.

PHYSICAL REVIEW VOLUME 76, NUMBER 8 OCTOBER 15, 1949

Crystal Statistics. III. Short-Range Order in a Binary Ising Lattice

BRURIA KAUFMAN
The Institute for Advanced Study, Princeton, New Jersey

AND

LARS ONSAGER
Yale University, New Haven, Connecticut

(Received May 11, 1949)

The degree of order in a binary lattice is described in terms of a family of "correlation" functions. The correlation function for two given lattice sites states what is the probability that the spins of the two sites are the same; this probability is, of course, a function of temperature, as well as of the distance and orientation of the atoms in the pair. It is shown that each correlation function is given by the trace of a corresponding 2^n-dimensional matrix. To evaluate this trace, we make use of the apparatus of spinor analysis, which was employed in a previous paper to evaluate the partition function for the lattice. The trace is found in terms of certain functions of temperature, Σ_a, and these are then calculated with the aid of an elliptic substitution.

Correlations for the five shortest distances (without restriction as to the orientation of the pair within the plane) are plotted as functions of temperature. In addition, the correlation for sites lying within the same row is given to any distance. For the critical temperature this correlation is plotted as a function of distance. It is shown that the correlation tends to zero as the distance increases, that is to say: there is no long-range order at the critical temperature.

SOVIET PHYSICS JETP VOLUME 23, NUMBER 2 AUGUST, 1966

BEHAVIOR OF ORDERED SYSTEMS NEAR THE TRANSITION POINT

A. Z. PATASHINSKIĬ and V. L. POKROVSKIĬ

Semiconductor Physics Institute, Siberian Division, Academy of Sciences, U.S.S.R.

Submitted to JETP editor August 2, 1965

J. Exptl. Theoret. Phys. (U.S.S.R.) **50**, 439–447 (February, 1966)

The form of the variation of the free energy with temperature and external field near the transition point is derived by estimating the correlation functions of a partially ordered system. The behavior of various thermodynamic quantities in the transition region is related to the binary correlation function. In particular, the magnetic susceptibility in a strong magnetic field is determined; for a plane Ising model it behaves as $H^{-14/15}$. The results of physical and mathematical experiments are discussed.

PHYSICAL REVIEW B VOLUME 5, NUMBER 7 1 APRIL 1972

Partial Test of the Universality Hypothesis: The Case of Different Coupling Strengths in Different Lattice Directions

Gerald Paul and H. Eugene Stanley

Physics Department, Massachusetts Institute of Technology, Cambridge, Massachusetts 02139

(Received 19 July 1971; revised manuscript received 2 November 1971)

High-temperature series expansions are used to examine the dependence of critical-point exponents upon *lattice anisotropy* (different interaction strengths in different directions in the lattice). The two-spin correlation function $C_2(\vec{r})$ is calculated to tenth order in $(1/k_BT)$ for the Ising Hamiltonian

$$\mathcal{H}_{I\,\mathrm{anis}} = -J_{xy}\sum_{\langle ij\rangle}^{xy} S_i^z S_j^z - J_z \sum_{\langle ij\rangle}^{z} S_i^z S_j^z$$

for a wide range of anisotropy parameters $R \equiv J_z/J_{xy}$ and for both the sc and fcc lattices; here the first summation is over all pairs of nearest-neighbor sites whose relative displacement vector $\vec{r}_{ij}$ has no z component, while the second summation is over all other pairs of nearest-neighbor sites. Hence for $R=0$, both the sc and fcc lattices reduce to the two-dimensional square lattice, while in the limit $R \to \infty$, the sc becomes a one-dimensional linear chain and the fcc becomes two noninteracting three-dimensional bcc lattices. The series for $C_2(\vec{r})$ are then used to obtain series of corresponding lengths for the specific heat, susceptibility, and second moment. Analysis of these series yields results consistent with the universality hypothesis of critical-point exponents. Specifically, it is found that when lattice anisotropy is introduced, the critical-point exponents studied (the susceptibility exponent γ and the correlation length exponent ν) do not appear to change from their values for an isotropic lattice. The problem of *next-nearest-neighbor interactions* is treated using similar methods in Paper II of this series (and briefly discussed in this paper).

PHYSICAL REVIEW B VOLUME 5, NUMBER 9 1 MAY 1972

Partial Test of the Universality Hypothesis: The Case of Next-Nearest-Neighbor Interactions

Gerald Paul and H. Eugene Stanley
Physics Department, Massachusetts Institute of Technology, Cambridge, Massachusetts 02139

(Received 19 July 1971; revised manuscript received 9 December 1971)

High-temperature series expansions are used to examine the dependence of critical-point exponents upon the presence of second-neighbor interactions. We consider the Hamiltonian

$$\mathcal{H}_{\rm nnn} = -J_1 \sum_{\langle ij\rangle}^{\rm nn} \vec{S}_i^{(D)} \cdot \vec{S}_j^{(D)} - J_2 \sum_{\langle ij\rangle}^{\rm nnn} \vec{S}_i^{(D)} \cdot \vec{S}_j^{(D)},$$

where the first and second sums are over pairs of nearest-neighbor (nn) and next-nearest-neighbor (nnn) sites, and where the spins $\vec{S}^{(D)}$ are D-dimensional unit vectors. The two-spin correlation function, $C_2(\vec{r})$, is calculated to tenth, ninth, and eighth order in $1/k_BT$ for the Ising ($D=1$), classical-planar ($D=2$), and classical-Heisenberg ($D=3$) models, respectively, for various values of the parameter $R' \equiv J_2/J_1$ and for various cubic lattices (fcc, bcc, and sample cubic). These represent the first series expansions of the spin correlation function for nnn interactions. From $C_2(\vec{r})$ we obtain series for the specific heat, susceptibility, and second moment. Analysis of these series and detailed comparisons with the exactly soluble spherical model ($D=\infty$) lead us to conclude that the exponents γ (susceptibility) and ν (correlation length) may be independent of R'; this suggestion is consistent with the universality hypothesis.

SOVIET PHYSICS JETP VOLUME 28, NUMBER 3 MARCH 1969

MICROSCOPIC DESCRIPTION OF CRITICAL PHENOMENA

A. M. POLYAKOV

Submitted April 4, 1968

Zh. Eksp. Teor. Fiz. **55**, 1026–1038 (September 1968)

We have been able to give a basis for and to generalize the earlier proposed phenomenological scaling hypotheses by considering phase transitions using the methods of quantum field theory. We show that the critical behavior of most quantities of physical interest such as the specific heat, correlation radius, magnetic susceptibility, etc., is determined by two unknown parameters which are connected with an infinite number of Feynman diagrams. The parameters are independent of the properties of the binary interaction potential but may change when non-binary interactions are included. These parameters are different for systems of differing symmetry (Bose gas, Ising model, Heisenberg ferromagnet). The behavior of the correlations near the transition point is described not only by the above-mentioned parameters but also by unknown functions. We have calculated the asymptotic behavior of these functions at distances $r \gg r_c$ (where r_c is the correlation radius). Depending on the symmetry of the correlated quantities two kinds of asymptotic behavior may occur: (4.4) or (4.12) in three-dimensional systems and (4.4′) or (4.13) in two-dimensional systems. In the two-dimensional case this result agrees with calculations for the Ising model. We give in the Appendix a simple diagram technique for classical systems such as the Ising model, a lattice of planar dipoles, or a Boltzmann gas.

Z. Physik 225, 195—215 (1969)

Scaling Approach to Anisotropic Magnetic Systems Statics

EBERHARD RIEDEL and FRANZ WEGNER
Institut Max von Laue — Paul Langevin, Garching bei München, Germany

Received March 12, 1969

Scaling laws are stated for anisotropic magnetic systems, where the anisotropy parameters are either scaled or held fixed. Combining the two ways of scaling, the critical behavior of thermodynamic quantities in anisotropic systems is determined. Particular attention is drawn to the temperature range where the anisotropy becomes important, and to the dependence there of the different quantities on the anisotropy parameters. In a transverse magnetic field the phase transition of an anisotropic magnet takes place along a λ-line. Assuming the singular part of the free enthalpy to depend on the distance from the λ-line, anomalous corrections to the transverse susceptibility and magnetization are calculated. For an experimental verification of many of the results, experiments including a variation of the anisotropy parameters or a finite transverse field are necessary.

VOLUME 28, NUMBER 11 PHYSICAL REVIEW LETTERS 13 MARCH 1972

Scaling Approach to Tricritical Phase Transitions

Eberhard K. Riedel
Department of Physics, Duke University, Durham, North Carolina 27706
(Received 26 January 1972)

Scaling laws for multicomponent systems near the tricritical point are derived by relating the "scales" of the competing second-order and tricritical phase transitions. The approach leads to a consistent description of the thermodynamics of the superfluid and phase-separation transitions in He^3-He^4 mixtures.

Reprinted from THE JOURNAL OF CHEMICAL PHYSICS, Vol. 39, No. 3, 842 1 August 1963
Printed in U. S. A.

On the Thermodynamics of the Critical Region for the Ising Problem

G. S. RUSHBROOKE*
Chemistry Department, University of Oregon, Eugene, Oregon
(Received 1 April 1963)

THE recent paper by Essam and Fisher[1] which, *inter alia*, reports numerical work, by Padé approximants, on the nature of the specific heat and susceptibility singularities for the Ising problem, on the low-temperature side of the Curie temperature, allows us to test the thermodynamic formula

$$C_H - C_M = \frac{T(\partial M/\partial T)^2_H}{(\partial M/\partial H)_T} \tag{1}$$

in the case $H=0$ and T less than, but approximately equal to, T_c. The equation itself, of course, is not in doubt: it is simply the more familiar formula for $C_P - C_V$ when the work term pdV is replaced by the work term $-HdM$, where H is the magnetic field and M the magnetization of the system. The only question is whether (1) yields useful information.

At $H=0$, $(\partial M/\partial T)_H$ is the slope of the magnetic phase boundary, $M(T)$, and $(\partial M/\partial H)_T$ is the zero-field susceptibility, χ. Likewise C_H is then the zero-field specific heat. Using the notation of Essam and Fisher, we may suppose that for T less than but approximately equal to T_c,

$$M(T) \text{ behaves like } (T_c - T)^{\beta},$$

and

$$\chi(T) \text{ behaves like } (T_c - T)^{-\gamma'}.$$

The right-hand side of (1) then behaves like $(T_c - T)^{-\epsilon}$ where $\epsilon = 2 - 2\beta - \gamma'$. If $\epsilon \leq 0$, Eq. (1) tells us nothing about C_H other than that C_H and C_M have equal singularities. On the other hand, should we find $\epsilon > 0$ then, since C_M is necessarily positive (being essentially a mean-square energy fluctuation) we can conclude that C_H has a singularity at T_c at least as strong as $(T_c - T)^{-\epsilon}$. In the notation of Essam and Fisher, C_H behaves like $(T_c - T)^{-\alpha'}$, and a logarithmic singularity may be regarded as corresponding to $\alpha' = 0$. We thus infer

$$\alpha' + 2\beta + \gamma' \geq 2 \tag{2}$$

as a thermodynamic necessity: in contrast to the equation

$$\alpha' + 2\beta + \gamma' = 2,$$

which Essam and Fisher derive heuristically from a simple model.

For two-dimensional lattices we know rigorously[2] that $\beta = 1/8$. The work of Essam and Fisher gives $\gamma' = 7/4$. In this case $\epsilon = 0$, and the known logarithmic singularity[3] in C_H is not in conflict with Eq. (1); indeed, Eq. (1) and the known logarithmic singularity in C_H allow us to infer $\gamma' \geq 7/4$.

For three-dimensional lattices the situation is more interesting. Originally Baker[4] proposed the value 0.3 for the index β, but Essam and Fisher make a strong case for $\beta = 5/16$. Essam and Fisher also favor $\gamma' = 5/4$, the same index as on the high-temperature side of the susceptibility singularity (for which see Baker[4]). Adopting these values we find $\epsilon = 1/8$, i.e., $\alpha' \geq 1/8$. On the other hand, Baker[5] presents fairly strong evidence for $\alpha' = 0$ (a logarithmic singularity in C_H). There would seem to be a definite conflict here with thermodynamic requirements.

There are two possibilities: either the specific heat singularity is not logarithmic, or $\epsilon = 0$ [in which case we can infer nothing about the specific heat singularity from Eq. (1)]. Keeping $\beta = 5/16$, $\epsilon = 0$ demands $\gamma' = 11/8$, as distinct from the exponent 10/8 on the high-temperature side of the susceptibility singularity. Although this lack of symmetry is esthetically displeasing, for the body-centered cubic lattice a value of γ' as high as 11/8 does not seem to be ruled out by Essam and Fisher's numerical data (Table IV of their paper). For the simple cubic and face-centered lattices, however, so high a value for γ' seems unlikely. The discrepancy will probably not be finally resolved until more adequate estimates of α' are practicable.

I am indebted to Dr. Baker for discussion and to Dr. Fisher for correspondence on this matter.

* On leave of absence from King's College, Newcastle upon Tyne, England, and indebted to the National Science Foundation for financial support.

[1] J. W. Essam and M. E. Fisher, J. Chem. Phys. **38**, 802 (1963).
[2] C. N. Yang, Phys. Rev. **85**, 808 (1952).
[3] L. Onsager, Phys. Rev. **65**, 117 (1944).
[4] G. A. Baker, Phys. Rev. **124**, 768 (1961).
[5] G. A. Baker (preprint, 1962); see also A. J. Wakefield, Proc. Cambridge Phil. Soc. **47**, 799 (1951).

Reprinted from
MOLECULAR PHYSICS, Vol. 1, No. 3, p. 257, July 1958

On the Curie points and high temperature susceptibilities of Heisenberg model ferromagnetics

by G. S. RUSHBROOKE and P. J. WOOD
Physics Department, King's College, Newcastle upon Tyne,
University of Durham

(*Received* 14 *May* 1958)

The first six coefficients in the expansion of the susceptibility χ, and its inverse, χ^{-1}, in ascending powers of the reciprocal temperature, have been determined for the Heisenberg model of a ferromagnetic, for any spin value, S, and any lattice. The first five coefficients appropriate to the magnetic specific heat, C, have also been found. For the body-centred and face-centred cubic lattices, the χ and C coefficients are tabulated for half-integral S from $\frac{1}{2}$ to 3.

From these coefficients estimates have been made of the reduced Curie temperatures, $\theta_c = \boldsymbol{k}T_c/J$. It is found that for the simple, body-centred and face-centred cubic lattices the formula

$$\theta_c = \tfrac{5}{96}(z-1)(11X-1)$$

reproduces the estimated Curie temperatures fairly accurately. Here $X = S(S+1)$ and z is the lattice coordination-number.

It is found that, suitably scaled, the theoretical curves for inverse susceptibility against temperature above the Curie point are rather insensitive to the spin value and to the precise lattice structure. The ratio of their initial to their final gradients is approximately 0·3. A comparison is made with the experimental values of χ^{-1} for both iron and nickel. If iron is represented by the Heisenberg model with $S=1$, then the observed Curie temperature corresponds to a J value of $1{\cdot}19 \times 10^{-2}$ ev.

Brief consideration is given to the use of the tabulated coefficients for antiferromagnetic problems.

Reprinted from THE PHYSICAL REVIEW, Vol. 176, No. 2, 718–722, 10 December 1968
Printed in U. S. A.

Spherical Model as the Limit of Infinite Spin Dimensionality

H. E. STANLEY
Lincoln Laboratory, Massachusetts Institute of Technology, Lexington, Massachusetts
and
Physics Department, University of California, Berkeley, California

(Received 20 April 1968)

The Berlin-Kac spherical model (or "spherical approximation to the Ising model")

$$\mathcal{H}^{\mathrm{SM}} = -J \sum_{\langle ij \rangle} \mu_i \mu_j, \quad \text{with} \quad \sum_{j=1}^{N} u_j^2 = N,$$

is found to be equivalent to the $\nu \to \infty$ limit of the Hamiltonian

$$\mathcal{H}^{(\nu)} = -J \sum_{\langle ij \rangle} \mathbf{S}_i^{(\nu)} \cdot \mathbf{S}_j^{(\nu)},$$

where $\mathbf{S}_j^{(\nu)}$ are isotropically interacting ν-dimensional classical spins.

VOLUME 17, NUMBER 17 PHYSICAL REVIEW LETTERS 24 OCTOBER 1966

POSSIBILITY OF A PHASE TRANSITION FOR THE TWO-DIMENSIONAL HEISENBERG MODEL

H. E. Stanley
Lyman Laboratory of Physics, Harvard University, Cambridge, Massachusetts, and
Lincoln Laboratory, Massachusetts Institute of Technology, Lexington, Massachusetts

and

T. A. Kaplan
Lincoln Laboratory, Massachusetts Institute of Technology, Lexington, Massachusetts
(Received 16 August 1966)

We point out that the existence of a phase transition—as indicated by extrapolation from high-temperature expansions—is as well-founded for two-dimensional lattices with nearest-neighbor ferromagnetic Heisenberg interactions as for three-dimensional lattices, and that the "well-known result" that there exists no phase transition in two dimensions is not a valid conclusion from the standard spin-wave argument.

VOLUME 17, NUMBER 22 PHYSICAL REVIEW LETTERS 28 NOVEMBER 1966

ABSENCE OF FERROMAGNETISM OR ANTIFERROMAGNETISM IN ONE- OR TWO-DIMENSIONAL ISOTROPIC HEISENBERG MODELS

N. D. Mermin and H. Wagner
Laboratory of Atomic and Solid State Physics, Cornell University, Ithaca, New York
(Received 17 October 1966)

It is rigorously proved that at any nonzero temperature, a one- or two-dimensional isotropic spin-S Heisenberg model with finite-range exchange interaction can be neither ferromagnetic nor antiferromagnetic. The method of proof is capable of excluding a variety of types of ordering in one and two dimensions.

PHYSICAL REVIEW VOLUME 184, NUMBER 1 5 AUGUST 1969

Some Critical Properties of Ornstein-Zernike Systems

G. Stell
Department of Mechanics, State University of New York, Stony Brook, New York 11790
(Received 24 February 1969)

A study is made of some critical properties of systems that satisfy the Ornstein-Zernike (OZ) condition that the direct correlation function $c(\vec{r})$ behaves like $-(kT)^{-1}$ times the pair potential $V(\vec{r})$ for $\vec{r}$ such that $V(\vec{r}) \ll kT$, even at the critical point. It is pointed out that a number of models of interest that satisfy this condition exist. The relationship (and great difference) between the implication of this condition and the results of the van der Waals-Bragg-Williams-Weiss approach is clarified, and it is noted that in systems satisfying the OZ condition both Widom's homogeneity condition and Kadanoff's scaling hypothesis can be violated, although a self-similarity condition is in general satisfied. The importance of the subtle interplay between small $|\vec{r}|$ and large $|\vec{r}|$ correlations, which is lost in both the mean field and scaling picture, is discussed.

JOURNAL OF THE PHYSICAL SOCIETY OF JAPAN, Vol. 24, No. 1, JANUARY, 1968

Dynamics of the Ising Model near the Critical Point. I

Masuo SUZUKI and Ryogo KUBO

Department of Physics, Faculty of Science, University of Tokyo, Tokyo

(Received September 19, 1967)

A simple dynamical model of interacting Ising spins is discussed. Each spin is assumed to flip spontaneously with a transition probability which depends on the temperature and the configuration of surrounding spins, but its functional form is assumed to be the simplest. The frequency-wave number dependent susceptibility $\chi(\boldsymbol{q}, \omega)$ is given exactly in the one-dimensional case. In two-and three-dimensional cases the model is treated in the molecular field and the generalized approximations.

SOVIET PHYSICS JETP VOLUME 22, NUMBER 3 MARCH, 1966

ON PHASE TRANSITIONS OF SECOND ORDER

V. G. VAKS and A. I. LARKIN

Submitted to JETP editor April 21, 1965

J. Exptl. Theoret. Phys. (U.S.S.R.) **49**, 975-989 (September, 1965)

It is shown that, subject to certain relations between the parameters, the singularities in the thermodynamic quantities at a second-order transition point in a binary alloy, at which there is a change in the crystal symmetry, are of the same type as in the Ising model, for which a number of results are known from computer calculations. Arguments are presented to show that subsequent terms do not alter these results. A phase transition in a Bose gas is found to be equivalent to a transition in a lattice of plane dipoles.

PHYSICAL REVIEW B VOLUME 5, NUMBER 11 1 JUNE 1972

Corrections to Scaling Laws

Franz J. Wegner

Department of Physics, Brown University, Providence, Rhode Island 02912

(Received 14 January 1972)

The effects of higher-order contributions to the linearized renormalization group equations in critical phenomena are discussed. This analysis leads to three quite different results: (i) An exact scaling law for redefined fields is obtained. These redefined fields are normally analytic functions of the physical fields. Corrections to the standard power laws are derived from this scaling law. (ii) The theory explains why logarithmic terms can exist in the free energy. (iii) The case in which the energy scales like the dimensionality is analyzed to show that quite anomalous results may be obtained in this special situation.

THE JOURNAL OF CHEMICAL PHYSICS VOLUME 43, NUMBER 11 1 DECEMBER 1965

Surface Tension and Molecular Correlations near the Critical Point

B. Widom
Department of Chemistry, Cornell University, Ithaca, New York
(Received 21 June 1965)

The van der Waals, Cahn–Hilliard theory of interfacial tension is reformulated for a fluid in the neighborhood of its critical point. The reformulated theory becomes equivalent to the Ornstein–Zernike, Debye theory of molecular correlations when the interface thickness is identified with the correlation length. When account is taken of the nonclassical behavior of the compressibility and coexistence curve, the theory is found to be in good agreement with independently known facts in three-dimensional systems, yet slightly but unambiguously wrong in two dimensions. When one of the hypotheses of the original theory is replaced by an alternative hypothesis, the resulting theory is found to be correct in both two and three dimensions.

THE JOURNAL OF CHEMICAL PHYSICS VOLUME 43, NUMBER 11 1 DECEMBER 1965

Equation of State in the Neighborhood of the Critical Point

B. Widom
Department of Chemistry, Cornell University, Ithaca, New York
(Received 15 July 1965)

A specific form is proposed for the equation of state of a fluid near its critical point. A function $\Phi(x, y)$ is introduced, with x a measure of the temperature and y of the density. Fluids obeying an equation of state of van der Waals type ("classical" fluids) are characterized by Φ being a constant. It is suggested that in a real fluid $\Phi(x, y)$ is a homogeneous function of x and y, with a positive degree of homogeneity (Sec. 2). This leads to a nonclassical compressibility, the behavior of which is determined by the degree of homogeneity of Φ (Sec. 3). A previously derived relation connecting the degree of the critical isotherm, the degree of the coexistence curve, and the compressibility index, again follows, this time without the restrictive assumption of effective isochore linearity (Sec. 4). The locus in the temperature–density plane of the points of inflection in the pressure–density isotherms, as determined experimentally by Habgood and Schneider, is accounted for (Sec. 5). It is shown that if a certain combination of the compressibility and coexistence curve indices is an integer, then the constant-volume specific heat on the critical isochore has a logarithmic singularity at the critical temperature with, in general, a superimposed finite discontinuity (Sec. 6).

PHYSICAL REVIEW B VOLUME 4, NUMBER 9 1 NOVEMBER 1971

Renormalization Group and Critical Phenomena. I. Renormalization Group and the Kadanoff Scaling Picture

Kenneth G. Wilson
Laboratory of Nuclear Studies, Cornell University, Ithaca, New York 14850
(Received 2 June 1971)

The Kadanoff theory of scaling near the critical point for an Ising ferromagnet is cast in differential form. The resulting differential equations are an example of the differential equations of the renormalization group. It is shown that the Widom-Kadanoff scaling laws arise naturally from these differential equations if the coefficients in the equations are analytic at the critical point. A generalization of the Kadanoff scaling picture involving an "irrelevant" variable is considered; in this case the scaling laws result from the renormalization-group equations only if the solution of the equations goes asymptotically to a fixed point.

Renormalization Group and Critical Phenomena. II. Phase-Space Cell Analysis of Critical Behavior

Kenneth G. Wilson
Laboratory of Nuclear Studies, Cornell University, Ithaca, New York 14850
(Received 2 June 1971)

A generalization of the Ising model is solved, qualitatively, for its critical behavior. In the generalization the spin $s_{\vec{n}}$ at a lattice site $\vec{n}$ can take on any value from $-\infty$ to ∞. The interaction contains a quartic term in order not to be pure Gaussian. The interaction is investigated by making a change of variable $s_{\vec{n}} = \sum_m \psi_m(\mathrm{n}) s_m'$, where the functions $\psi_m(\vec{n})$ are localized wave-packet functions. There are a set of orthogonal wave-packet functions for each order-of-magnitude range of the momentum $\vec{k}$. An effective interaction is defined by integrating out the wave-packet variables with momentum of order 1, leaving unintegrated the variables with momentum <0.5. Then the variables with momentum between 0.25 and 0.5 are integrated, etc. The integrals are computed qualitatively. The result is to give a recursion formula for a sequence of effective Landau-Ginsberg-type interactions. Solution of the recursion formula gives the following exponents: $\eta = 0$, $\gamma = 1.22$, $\nu = 0.61$ for three dimensions. In five dimensions or higher one gets $\eta = 0$, $\gamma = 1$, and $\nu = \frac{1}{2}$, as in the Gaussian model (at least for a small quartic term). Small corrections neglected in the analysis may make changes (probably small) in the exponents for three dimensions.

Critical Exponents in 3.99 Dimensions

Kenneth G. Wilson and Michael E. Fisher
Laboratory of Nuclear Studies and Baker Laboratory, Cornell University, Ithaca, New York 14850
(Received 11 October 1971)

Critical exponents are calculated for dimension $d = 4 - \epsilon$ with ϵ small, using renormalization-group techniques. To order ϵ the exponent γ is $1 + \frac{1}{6}\epsilon$ for an Ising-like model and $1 + \frac{1}{5}\epsilon$ for an XY model.